Principles and Applications of Quantum Chemistry

Principles and Applications of Quantum Chemistry

V.P. Gupta

Department of Physics, University of Lucknow,
Lucknow, Uttar Pradesh, India

ELSEVIER

Amsterdam • Boston • Heidelberg • London • New York • Oxford
Paris • San Diego • San Francisco • Singapore • Sydney • Tokyo

Academic Press is an imprint of Elsevier

Academic Press is an imprint of Elsevier
125 London Wall, London EC2Y 5AS, UK
525 B Street, Suite 1800, San Diego, CA 92101-4495, USA
225 Wyman Street, Waltham, MA 02451, USA
The Boulevard, Langford Lane, Kidlington, Oxford OX5 1GB, UK

ISBN: 978-0-12-803478-1

British Library Cataloguing in Publication Data
A catalogue record for this book is available from the British Library

Library of Congress Cataloging-in-Publication Data
A catalog record for this book is available from the Library of Congress

For information on all Academic Press publications
visit our website at http://store.elsevier.com/

In loving memory of my beloved mother
Smt. Ram Kali
Whose life has always been a source of inspiration to me

Contents

List of Figures

List of Tables

Biography

Professor V.P. Gupta

Department of Physics, University of Lucknow, Lucknow, Uttar Pradesh, India

Born on December 30, 1942; PhD (Moscow, USSR) 1967.

Presently, Principal Investigator, Department of Science and Technology (DST) Project, Govt. of India; Formerly, Professor and Chairman, Department of Physics, University of Jammu, Jammu Tawi, India; Visiting Professor of Chemistry, Université de Provence, Marseille, France; Professor and Chairman, Department of Physics, University of Calabar, Nigeria; Professor Emeritus (Emeritus Fellow), University Grants Commission (UGC), India; Emeritus Scientist, Council of Scientific and Industrial Research (CSIR), India; Emeritus Scientist, All India Council of Technical Education (AICTE), India; Visiting Scientist/Fellow, University of Helsinki, Helsinki, Finland and International Center for Theoretical Physics, Trieste, Italy.

Over the past four decades successfully executed several major and minor Scientific Research Projects granted by National Funding Agencies such as DST, UGC, CSIR, AICTE, and Indian Space Research Organization (ISRO), Bangalore, India.

Experience of Teaching and Research—45 years.

Research Publications—99; Books Published—2 (including translation from Russian to English).

Major areas of research interest: Quantum chemistry, molecular spectroscopy and molecular structure, matrix isolation infrared studies, astrochemistry, and laser spectroscopy.

Preface

The main purpose of this book is to share knowledge about the upcoming theories of quantum chemistry and the quantum chemical tools, which have emerged as a part of computational chemistry, and their applications to the wide and varied areas of chemistry. The primary concern of chemistry has always been the interpretation of the structures of molecules and the chemical reactions they undergo. This is also the concern of this book as it attempts to explain the principles of quantum chemistry and their application to study the molecular structure and molecular properties, thermodynamics, reaction mechanisms, reactivity indices, molecular spectroscopy, the intramolecular and intermolecular forces, etc., using ab initio, semiempirical, and density functional theory (DFT) methods. All these topics have become an integral part of the chemistry curriculum in universities. Practicing chemists, material scientists, biochemists, and other professionals have also shown immense interest in the use of quantum chemical tools for understanding the problems related to their research work. A great interest in quantum chemistry has also been generated in chemists, material scientists, biochemists, and other professionals, who wish to use quantum chemical tools to understand the problems related to their work. The present book, which is mostly based on my lectures to the graduate and postgraduate students in several universities in India and abroad over along period of time, has been written with twin objectives in mind: firstly, to serve as a text book on quantum chemistry for postgraduate students in India and the senior undergraduate and postgraduate students in foreign universities and secondly, to serve a utilitarian purpose for all others who are only interested in the tools of quantum chemistry.

The book covers most recent advances in the field in its various chapters and includes, besides others, chapters on most current topics such as: DFT and time-dependent DFT (TDDFT), quantum chemical treatments of vibrational and electronic spectra and CIS theory, characterization of chemical reactions, molecular electrostatic potential, and quantum theory of atoms in molecules. Every attempt has been made to make the treatment of the subject simple and clear. The introductory chapter on "Basic Principles of Quantum Chemistry" catalyzes the process of recapitulation of topics covered in undergraduate courses in quantum mechanics and provides a background knowledge of some of the basic concepts and mathematical tools of quantum mechanics and matrix mechanics that are used in subsequent chapters. Derivations, where needed, are given with enough details for better assimilation of content to enable the users to have a fuller understanding of the physical and mathematical aspects of quantum chemistry and molecular electronic structure. A large number of examples to support different applications have been given in the book as illustrations to make the subject matter more understandable and also to serve as a practical guide to all those interested in using the quantum chemical tools in their research. Bibliography at the end of each chapter aims at opening the door for those who intend to pursue quantum chemistry more deeply by referring to some of the texts that are discussed therein.

My sincere thanks are due to Professor H. Bodot, Université de Provence, Marseilles, France, Professors Juhani Murto and M. Räsännen, Helsinki University, Helsinki, and Professor S. Califano, University of Florence, Florence, Italy with whom I had the opportunity to work and to have very fruitful interactions. Professor Krishan Lal, Ex-Director, National Physical Laboratory, New Delhi, Professor G. Govil, Tata Institute of Fundamental Research, Mumbai, India, and Professor Anindya Dutta, Indian Institute of Technology, Mumbai, India deserve a special mention for their constant help and support at every stage of this work. Though it would be a difficult task for me to individually acknowledge the efforts of those who have contributed toward shaping this book, I would express my appreciation for my students who interacted with me with their searching queries and colleagues who read and commented on various parts of the manuscript. Also, in a very special and personal way, I acknowledge my wife, Madhu, for rendering her years of loyal support and being a perennial source of inspiration and encouragement for this project. I also appreciate my children and grandchildren, Manjari, Vikas, Ashish, Nidhi, Pulkit, Divayum, and Shubhang for their constant motivation and emotional support during this project. Finally I would like to record my appreciation for the assistance of Ms. Neha Singh Chauhan who, so carefully and painstakingly, typed the entire manuscript.

Acknowledgment

This work has been catalyzed and supported by the Science and Engineering Research Board, Department of Science and Technology, Government of India, under its Utilization of Scientific Expertise of Retired Scientists Scheme. A due acknowledgment is also made of the infrastructural facilities and administrative support provided by the Department of Physics, University of Lucknow, Lucknow, where the work was carried out.

1

Basic Principles of Quantum Chemistry

CHAPTER OUTLINE

Principles and Applications of Quantum Chemistry. http://dx.doi.org/10.1016/B978-0-12-803478-1.00001-7

1.1 Introduction

Origin of quantum mechanics took place towards the end of the nineteenth century at a time when most of the fundamental physics laws had been worked out. The motions of mechanical objects, both terrestrial and celestial, were successfully discussed in terms of Newton's equations and the wave nature of light as suggested by interference and diffraction experiments was put on a firmer footing by Maxwell's equations which explained connection between the optical and electrical phenomena. The inadequacies of classical mechanics in explaining large volume of experimental data related to the behavior of very small particles like electrons and nuclei of atoms and molecules lead to the origin of quantum mechanics. The first milestone in this direction was laid by Planck by explaining the distribution of thermal radiation emitted by heated solids in terms of discrete quanta of electromagnetic radiation, later named as photon. He proposed that a photon can have energy in multiples of frequency, $E = nh\nu$. He calculated h to be 6.626×10^{-27} Js for reproducing the experimental data. The quantum idea was later on used by Einstein to explain some of the experimental observations on photoelectric effect. Photoelectric effect shows that light can exhibit particle-like behavior in addition to the wave-like behavior it shows in diffraction experiments. Einstein also showed that not only light but atomic vibrations too are quantized and used this concept to explain the variation of specific heat of solids with temperature. This demonstrated that the energy quantization concept was important even for a system of atoms in a crystal. Bohr introduced the concept of quantization of angular momentum and energy and used to explain the origin of discrete lines seen in the spectrum of hydrogen atom for which only a continuous spectrum could be predicted by the electromagnetic theory. Further studies showed that quantum mechanics departs from classical mechanics primarily in the realm of atomic and subatomic length scales and is able to provide mathematical description of much of the particle-like and wave-like behavior and interactions of energy and matter. The early versions of quantum mechanics were significantly reformulated by Heisenberg, Born, Jordan, and Schrödinger. Much of the nineteenth-century physics can now be treated as classical limit of quantum mechanics which has since resulted in the creation of other disciplines such as quantum chemistry, quantum electronics, quantum optics, and quantum informatics, etc.

1.2 Particle–Wave Duality

Since light can behave both as a wave (it can be diffracted and it has a wavelength) and as a particle (it contains packets of energy $h\nu$), de Broglie reasoned in 1924 that matter also can exhibit this wave–particle duality. He further reasoned that matter would obey the same equation for wavelength as light namely, $\lambda = h/p$, where p is the linear momentum, as shown by Einstein. This relationship easily follows from the consideration that $E = h\nu$ for a photon and $\lambda\nu = c$ for an electromagnetic wave. If we use Einstein's relativity result $E = mc^2$, we find that $\lambda = h/mc$ which is equivalent to $\lambda = h/p$. Here, m refers to the relativistic mass, not the rest mass, since the rest mass of a photon is zero. For a particle moving with a velocity v, the momentum $p = mv$ and hence $\lambda = \frac{h}{mv}$. If a wave is associated with a particle (the de Broglie wave), the phase velocity of the wave v_p shall be $v_p = \nu\lambda = c^2/v$, where v is the particle velocity. Since $v < c$, v_p shall be greater than c which means that, the de Broglie wave associated with the particle moves faster than the particle itself. This result is absurd. It was subsequently shown that instead of a single wave, a particle need to be associated with a wave packet. A wave packet consists of a group of waves with phases and amplitudes so chosen that they undergo constructive interference only over a small region of space where the particle can be located; outside this region they undergo destructive interference so that the amplitude tends to reduce to zero rapidly.

Simplest type of a wave is a plane monochromatic wave

$$\Psi(\mathbf{r}, t) = e^{[i(\mathbf{k}\cdot\mathbf{r}-\omega t)]} \tag{1.2.1}$$

Using relations $E = \hbar\omega$ and $p = \hbar\mathbf{k}$, for the energy and linear momentum, this equation may also be written as,

$$\Psi(\mathbf{r}, t) = e^{\left[\frac{i}{\hbar}(\mathbf{p}\cdot\mathbf{r}-Et)\right]} \tag{1.2.2}$$

where ω and $\mathbf{k}$ are the angular frequency and wave vector of the plane wave, respectively.

A wave packet is constructed by superposition of waves by Fourier relation. Thus, for one spatial dimension the wave packet is

$$\Psi(x, t) = \frac{1}{\sqrt{2\pi}} \int_{-\infty}^{\infty} A(\mathrm{k}) e^{[i(\mathrm{k}x-\omega t)]} dk \tag{1.2.3}$$

or in terms of energy and linear momentum,

$$\Psi(x, t) = \frac{1}{\sqrt{2\pi\mathrm{h}}} \int_{-\infty}^{\infty} A(p) e^{\left[\frac{i}{\hbar}(px-Et)\right]} dp \tag{1.2.4}$$

The behavior of such a wave group in time is determined by the way in which the angular frequency ω $(= 2\pi\nu)$ depends upon the wave number $\mathbf{k}\left(= \frac{2\pi}{\lambda}\right)$, i.e., by the law of dispersion.

Such a wave packet moves with its own velocity v_g, called the group velocity which is equal to the particle velocity. The association of a wave packet with a particle provided an explanation of the Heisenberg's uncertainty principle. In 1927, Davisson and Germer observed diffraction patterns by bombarding metals with electrons, confirming de Broglie's proposition.

1.3 Matrix Mechanics and Wave Mechanics

At the end of 1925, Werner Heisenberg, in collaboration with Born and Jordan, proposed theory that replaces physical quantities like coordinates of particles, momenta, and energies by matrices, and was therefore called matrix mechanics. Rules for manipulating these matrices then lead to predictions that could be compared to experiments.

In conformity with the dual nature of matter proposed by de Broglie and subsequently confirmed through the experiments of Davisson and Germer, Erwin Schrödinger proposed another theory based on the Hamilton–Jacobi formulation of classical mechanics. In this formulation the behavior of particles is described by a wave equation. A modification in this equation led to what's now called the Schrödinger equation. It too agreed well with experiments. A system, for example, an atom or molecule, is described by a so-called wavefunction in Schrödinger's theory, and the theory is called wave mechanics. Wave mechanics was much more readily accepted than matrix mechanics, in part because the wavefunction could be visualized, and the theory was based on well-established classical mechanics.

Matrix mechanics and wave mechanics predict exactly the same results for experiments. This suggests that they are really different forms of a more general theory. In 1930, Paul Dirac gave a more general formulation of quantum mechanics; the one that is still used today. Matrix and wave mechanics can be derived from this formulation and are then called the Heisenberg and the Schrödinger picture, respectively. Other pictures can also be derived. Which picture one actually uses in practice depends on which one is the most convenient to work with. In quantum chemistry the Schrödinger picture is generally the easier.

1.4 Relativistic Quantum Mechanics

While the non-relativistic quantum mechanics (non-RQM) refers to the mathematical formulation of quantum mechanics in the context of Galilean relativity and quantizes the equations of classical mechanics by replacing dynamical variables by operators, the relativistic quantum mechanics (RQM) is the development of quantum mechanics incorporating the concepts of the special theory of relativity. The relativistic formulation has been more successful than the original quantum mechanics in some contexts, like the prediction of antimatter, electron spin, spin magnetic moments of elementary $-1/2$ fermions, fine structure, and quantum dynamics of charged particles in electromagnetic fields. Relativistic effects in chemistry can be considered to be perturbations, or small corrections, to the nonrelativistic theory of chemistry, which is developed from the solutions of the Schrödinger equation. These corrections affect the electrons differently depending on the electron speed relative to the speed of light. Relativistic effects are more prominent in heavy elements because only in these elements do electrons attain relativistic speeds. Quantum chemistry uses the RQM to explain elemental properties and structure, especially for the heavy metals of the periodic table. A prominent example of such an explanation is the fact that the color of gold (it is not silvery like almost all other metals) is explained via such relativistic effects.

1.5 Schrödinger Wave Equation

The time-dependent Schrödinger equation for particle wave for one spatial dimension is of the form

$$-\frac{\hbar^2}{2m}\frac{\partial^2\Psi(x,t)}{\partial x^2}+V(x)\Psi(x,t)=i\hbar\frac{\partial\Psi(x,t)}{\partial t} \tag{1.5.1}$$

where $V(x)$ represents the potential field in which the particle moves.

This equation may be derived from Eq. (1.2.3) by appropriate differentiation. Differentiation of Eq. (1.2.3) with respect to t gives

$$\frac{\partial\Psi}{\partial t}=\frac{1}{\sqrt{2\pi\hbar}}\int_{-\infty}^{\infty}A(p)\left(-\frac{i}{\hbar}E\right)e^{\left[\frac{i}{\hbar}(px-Et)\right]}dp$$

or

$$i\hbar\frac{\partial\Psi}{\partial t}=\frac{1}{\sqrt{2\pi\hbar}}\int_{-\infty}^{\infty}E\,A(p)e^{\left[\frac{i}{\hbar}(px-Et)\right]}dp \tag{1.5.2}$$

and second differentiation with respect to x gives

$$-\hbar^2\frac{\partial^2\Psi}{\partial x^2}=\frac{1}{\sqrt{2\pi\hbar}}\int_{-\infty}^{\infty}p^2A(p)e^{\left[\frac{i}{\hbar}(px-Et)\right]}dp \tag{1.5.3}$$

If we consider that the total energy (E) of the particle, which in classical expression is the Hamiltonian (H), is given as:

$$H=E=\frac{p^2}{2m}+V(x) \tag{1.5.4}$$

then, on using Eqs (1.5.2) and (1.5.3) with Eq. (1.5.4), we get Schrödinger wave Eq. (1.5.1).

For a free particle $V(x)=0$ and hence Eq. (1.5.1) reduces to

$$\frac{-\hbar^2}{2m}\frac{\partial^2\Psi(x,t)}{\partial x^2}=i\hbar\frac{\partial\Psi(x,t)}{\partial t} \tag{1.5.5}$$

Here, $\Psi(x, t)$ is called wavefunction of the particle wave which, though replaces the amplitude of a mechanical wave, is a complex quantity for which a physical interpretation was provided by Born.

Differentiation of Eq. (1.2.2) with respect to position and time gives,

$$-i\hbar\frac{\partial\Psi}{\partial x} = p\Psi \quad (1.5.6a)$$

$$i\hbar\frac{\partial\Psi}{\partial t} = E\Psi \quad (1.5.6b)$$

and

$$-\hbar^2\frac{\partial^2\Psi}{\partial x^2} = p^2\Psi \quad (1.5.6c)$$

From these equations it may be seen that the operator $-i\hbar\frac{\partial}{\partial x}$ represents p_x—the x-component of linear momentum and the operator $i\hbar\frac{\partial}{\partial t}$ represents the energy E.

1.5.1 Time-Independent Schrödinger Wave Equation

When the Hamiltonian is independent of time the general solution (Ψ) of the Schrödinger Eq. (1.5.1) can be expressed as a product of function of spatial position and time. Thus

$$\Psi(x,t) = \Phi(x)\mathrm{f}(t) \quad (1.5.7)$$

Substitution of this equation into Eq. (1.5.1) leads to the time-independent Schrödinger equation for one-dimension

$$-\frac{\hbar^2}{2m}\frac{d^2\Phi(x)}{dx^2} + \mathrm{V}(x)\Phi(x) = E\Phi(x) \quad (1.5.8)$$

and

$$\mathrm{f}(t) = Ce^{\frac{-iEt}{\hbar}} \quad (1.5.9)$$

where C is a constant.

The total wavefunction is therefore

$$\Psi(x,t) = \Phi(x)\cdot e^{\frac{-iEt}{\hbar}} \quad (1.5.10)$$

Equation (1.5.8) can also be written as

$$H\Phi = E\Phi \quad (1.5.11)$$

where the Hamiltonian

$$H = \frac{p^2}{2m} + \mathrm{V} = -\frac{\hbar^2}{2m}\frac{\partial^2}{\partial x^2} + \mathrm{V}(x) \quad (1.5.12)$$

Equation (1.5.11) is an eigenvalue equation and E is the energy eigenvalue. A state with a well-defined energy therefore has a wavefunction of the form of Eq. (1.5.10).

Equation (1.5.1) leads to an equation of continuity in quantum mechanics.

$$\frac{\partial}{\partial t}(\Psi^*\Psi) + \frac{\hbar}{2im}\nabla\cdot(\Psi^*\nabla\Psi - \Psi\nabla\Psi^*) = 0 \quad (1.5.13)$$

This equation is similar to the equation of continuity in electrodynamics.

$$\frac{\partial\rho}{\partial t} + \nabla\cdot\mathrm{J} = 0 \quad (1.5.14)$$

where $\rho = \Psi^*\Psi$ is the charge density and J is the current density. Comparing Eqs (1.5.3) and (1.5.4) we get in our case

$$\rho = \Psi^*\Psi = \Phi^*\Phi = |\Phi|^2 \tag{1.5.15}$$

which is called the position probability density.

$$\mathrm{J} = \frac{\hbar}{2mi}(\Psi^*\nabla\Psi - \Psi\nabla\Psi^*) \tag{1.5.16}$$

by analogy to Eq. (1.5.14) is called the probability current density.

From the above, it follows that if

$$\mathrm{div}\,\mathrm{J} = \nabla \cdot \mathrm{J} = 0,$$

the probability density ρ will be a constant in time. Such states are called stationary states and are independent of time.

$\Psi^*\Psi$ defines the probability of finding a particle in unit volume element. Since the probability of finding the particle somewhere in the region must be unity,

$$\int_{-\infty}^{\infty} \Psi^*(\mathbf{r}, t)\Psi(\mathbf{r}, t)d\tau = 1 \tag{1.5.17}$$

where $d\tau$ is the three-dimensional volume element $dx\,dy\,dz$. The function Ψ is now said to be normalized and the above equation is said to be the normalization condition.

1.5.2 Schrödinger Equation in Three-Dimensions

In a three-dimensional space the wave packet can be written as

$$\Psi(\mathbf{r}, t) = \frac{1}{\sqrt{(2\pi)^3}} \int_{-\infty}^{\infty} A(\mathbf{k})\exp^{[i(\mathbf{k}\cdot\mathbf{r}-\omega t)]}dk \tag{1.5.18}$$

and the time-dependent Schrödinger Eq. (1.5.1) is replaced by

$$-\frac{\hbar^2}{2m}\nabla^2\Psi(\mathbf{r}, t) + \mathrm{V}(\mathbf{r})\Psi(\mathbf{r}, t) = i\hbar\frac{\partial\Psi(\mathbf{r}, t)}{\partial t} \tag{1.5.19}$$

where,

$$\nabla^2 = \frac{\partial^2}{\partial x^2} + \frac{\partial^2}{\partial y^2} + \frac{\partial^2}{\partial z^2} \tag{1.5.20}$$

is known as Laplacian operator. The time-independent Schrödinger equation is written as

$$-\frac{\hbar^2}{2m}\nabla^2\Psi(\mathbf{r}) + \mathrm{V}(\mathbf{r})\Psi(\mathbf{r}) = E\Psi(\mathbf{r}) \tag{1.5.21}$$

If the potential of the physical system to be examined is spherically symmetric then, instead of Cartesian coordinates, the Schrödinger equation in spherical polar coordinates can be used to advantage. For a three-dimensional problem, the Laplacian in spherical polar coordinates is used to express the Schrödinger equation in the condensed form

$$-\frac{\hbar^2}{2m}\nabla^2\Psi + \mathrm{V}(r, \theta, \phi)\Psi(r, \theta, \phi) = E\Psi(r, \theta, \phi) \tag{1.5.22}$$

Expanded, it takes the form

$$-\frac{\hbar^2}{2m}\frac{1}{r^2\sin\theta}\left[\sin\theta\frac{\partial}{\partial r}\left(r^2\frac{\partial}{\partial r}\right)+\frac{\partial}{\partial\theta}\left(\sin\theta\frac{\partial}{\partial\theta}\right)+\frac{1}{\sin\theta}\frac{\partial^2}{\partial\phi^2}\right]\Psi(r,\theta,\phi)+\mathrm{V}(\mathbf{r})\Psi(r,\theta,\phi)=E\Psi(r,\theta,\phi)$$

or

$$-\frac{\hbar^2}{2m}\left[\frac{1}{r^2}\frac{\partial}{\partial r}\left(r^2\frac{\partial}{\partial r}\right)+\frac{1}{r^2\sin\theta}\frac{\partial}{\partial\theta}\left(\sin\theta\frac{\partial}{\partial\theta}\right)+\frac{1}{r^2\sin^2\theta}\frac{\partial^2}{\partial\phi^2}\right]\Psi(r,\theta,\phi)+\mathrm{V}(\mathbf{r})\Psi(r,\theta,\phi)=E\Psi(r,\theta,\phi) \quad (1.5.23)$$

This is the form best suited for the study of the hydrogen atom.

1.6 Operators—General Properties, Eigenvalues, and Expectation Values

Each measurable parameter in a physical system is represented by a quantum mechanical operator. Such operators arise because in quantum mechanics we are describing nature with waves (the wavefunction) rather than with discrete particles whose motion and dynamics can be described with the deterministic equations of Newtonian physics. Quantities such as coordinates and components of velocity, momentum and angular momentum of particles, and the functions of these quantities—in fact variables in terms of which classical mechanics is built up are described by linear operators.

An operator operates upon a function and may transform it into another function. Thus, for example,

$$\hat{Q}\,\mathrm{f}(x)=\mathrm{g}(x)$$

If the effect of operating some function f(x) with an operator $\hat{Q}$ is simply to multiply it by a certain constant c, i.e.,

$$\hat{Q}\,\mathrm{f}(x)=c\,\mathrm{f}(x) \quad (1.6.1)$$

we can then say that f(x) is an eigenfunction of $\hat{Q}$ with eigenvalue c. Equation (1.6.1) is called an eigenvalue equation.

Thus, Schrödinger Eq. (1.5.8) is an eigenvalue equation. Here, $H=-\frac{\hbar^2}{2m}\frac{\partial^2}{\partial x^2}+\mathrm{V}(x)$ is the operator, called the Hamilton operator for the system, the values of energy E are the eigenvalues and the wavefunctions $\Phi(x)$ are the eigenfunctions of the operator.

An operator $\hat{Q}$ is said to be linear if and only if it has the following two properties:

$$\hat{Q}\,[\mathrm{f}(x)+\mathrm{g}(x)]=\hat{Q}\,\mathrm{f}(x)+\hat{Q}\,\mathrm{g}(x) \quad (1.6.2)$$

and

$$\hat{Q}\,[c\mathrm{f}(x)]=c\hat{Q}\,\mathrm{f}(x) \quad (1.6.3)$$

where f and g are two arbitrary functions and c is an arbitrary constant.

Linear operators are, in general, complex quantities since one can multiply them by complex numbers and get other quantities of the same nature. Hence, they must correspond to complex dynamical variables, i.e., to complex functions of coordinates, velocity, etc.

In quantum mechanics, the operators operate on functions called wavefunctions which are characteristic of the state of the system being considered. Certain conditions are imposed on the wavefunctions to make them conform to our ideas about the properties of physically realizable states. It is not easy to describe explicitly all the requirements that the wavefunction must meet but it is essential that it is well behaved.

Well-behaved wavefunctions are those which by themselves or their derivatives are single valued, continuous, and finite. It must also be square integrable, which means that $\int |\Psi(\mathbf{r},t)|^2 dx\,dy\,dz$ is finite for all physically meaningful values of x, y, z like $-\infty < x < \infty$, $-\infty < y < \infty$, $-\infty < z < \infty$. They must be normalizable. This implies that the wavefunction approaches zero as the distance tends to infinity. Thus, e^{-x^2}, $\sin x$, $\cos x$, etc., are well behaved but $\tan x$, $\cot x$, e^x, e^{-x}, etc., are not.

1.6.1 Some Operators in Quantum Mechanics

In classical mechanics, most dynamical variables are expressed in terms of position and momentum coordinates like x, y, z, p_x, p_y, or p_z, etc. In quantum mechanics, the operators can be constructed by writing the classical expressions and replacing the position coordinates and linear momenta by the corresponding operators.

Operators corresponding to any function of position, such as x, or potential V(x) are simply the functions themselves.

As shown above, linear momentum is represented in the operator form as

$$\widehat{p}_x = -i\hbar\frac{\partial}{\partial x} \quad \text{(similarly for the } y \text{ and } z \text{ components).}$$

The time-dependent Hamiltonian is represented as $\widehat{H}(t) = i\hbar\frac{\partial}{\partial t}$, while in the time-independent form, if may be written as

$$\widehat{H} = \frac{p_x^2}{2m} + \frac{p_y^2}{2m} + \frac{p_z^2}{2m} + \mathrm{V}(\mathbf{r}) = \frac{-\hbar^2}{2m}\left(\frac{\partial^2}{\partial x^2} + \frac{\partial^2}{\partial y^2} + \frac{\partial^2}{\partial z^2}\right) + \mathrm{V}(\mathbf{r}) = -\frac{\hbar^2}{2m}\nabla^2 + \mathrm{V}(\mathbf{r}) \tag{1.6.4}$$

For a classical particle with linear momentum $\mathbf{p}$ and position vector $\mathbf{r}$, the orbital angular momentum $\mathbf{L}$ is:

$$\widehat{L} = \mathbf{r} \times \widehat{p} = -i\hbar\mathbf{r} \times \nabla \tag{1.6.5}$$

or in Cartesian components,

$$\begin{aligned}
\widehat{L}_x &= y\widehat{p}_z - z\widehat{p}_y = -i\hbar\left(y\frac{\partial}{\partial z} - z\frac{\partial}{\partial y}\right), \\
\widehat{L}_y &= z\widehat{p}_x - x\widehat{p}_z = -i\hbar\left(z\frac{\partial}{\partial x} - x\frac{\partial}{\partial z}\right), \\
\widehat{L}_z &= x\widehat{p}_y - y\widehat{p}_x = -i\hbar\left(x\frac{\partial}{\partial y} - y\frac{\partial}{\partial x}\right).
\end{aligned} \tag{1.6.6}$$

Without going into mathematical rigors it may be mentioned that in terms of spherical polar coordinates

$$\widehat{L}_x = i\hbar\left(\sin\phi\frac{\partial}{\partial\theta} + \cot\theta\cos\phi\frac{\partial}{\partial\phi}\right)$$

$$\widehat{L}_y = -i\hbar\left(\cos\phi\frac{\partial}{\partial\theta} - \cot\theta\sin\phi\frac{\partial}{\partial\phi}\right)$$

$$\widehat{L}_z = -i\hbar\frac{\partial}{\partial\phi}$$

and

$$\widehat{L}^2 = \widehat{L}_x^2 + \widehat{L}_y^2 + \widehat{L}_z^2 \quad (1.6.7)$$

$$\widehat{L}^2 = -\hbar^2\left\{\frac{1}{\sin\theta}\frac{\partial}{\partial\theta}\left(\sin\theta\frac{\partial}{\partial\theta}\right) + \frac{1}{\sin^2\theta}\frac{\partial^2}{\partial\phi^2}\right\} \quad (1.6.8)$$

1.6.2 Properties of Operators

1. Two operators $\widehat{A}$ and $\widehat{B}$ are said to be equal if $\widehat{A}f = \widehat{B}f$ for all functions f.
2. Product of two operators is defined by the equation $\widehat{A}\widehat{B}f(x) = \widehat{A}[\widehat{B}f(x)]$.
3. Operators obey the associative law of multiplication $\widehat{A}(\widehat{B}\widehat{C}) = (\widehat{A}\widehat{B})\widehat{C}$.
4. Unlike ordinary algebra, the operators do not obey the commutative law of multiplication. While, $ab = ba$ in ordinary algebra, $\widehat{A}\widehat{B}$ and $\widehat{B}\widehat{A}$ are not necessarily equal operators.

   Commutator $[\widehat{A}, \widehat{B}]$ of operators $\widehat{A}$ and $\widehat{B}$ is defined as

   $$[\widehat{A}, \widehat{B}] = \widehat{A}\widehat{B} - \widehat{B}\widehat{A}$$

   If $\widehat{A}\widehat{B} = \widehat{B}\widehat{A}$, then $[\widehat{A}, \widehat{B}] = 0$, the operators $\widehat{A}$ and $\widehat{B}$ are said to commute. If $[\widehat{A}, \widehat{B}] \neq 0$, they do not commute. Thus, for example, if $\widehat{A} = x$ and $\widehat{B} = \frac{\partial}{\partial x}$, then

   $$[\widehat{A}, \widehat{B}]f(x) = \widehat{A}\widehat{B}f(x) - \widehat{B}\widehat{A}f(x) = x\frac{\partial f}{\partial x} - \frac{\partial}{\partial x}(xf) = -f \neq 0$$

   So $\widehat{A}$ and $\widehat{B}$ do not commute.

   It is found that the physical quantities represented by commuting operators alone can be measured simultaneously with certainty. In all other cases, there is an uncertainty in simultaneous measurements. The extent of uncertainty in a measurement is given by the Heisenberg's uncertainty principle.

5. Eigenfunctions of commuting linear operators are simultaneous eigenfunctions or in the case of degeneracy can be constructed by superposition principle to be simultaneous.

An eigenfunction is said to be a simultaneous eigenfunction of two linear operators $\widehat{A}$ and $\widehat{B}$ for eigenvalues α and β, respectively, if it satisfies the condition

$$\widehat{A}\Psi = \alpha\Psi$$

$$\widehat{B}\Psi = \beta\Psi$$

Conversely, two operators shall commute if they have simultaneous eigenfunctions. Operators having simultaneous eigenfunctions are said to be compatible.

6. Some other properties of commutators are:

$$[\widehat{A}, \widehat{B}] = -[\widehat{B}, \widehat{A}]$$

$$[k\widehat{A}, \widehat{B}] = [\widehat{A}, k\widehat{B}] = k[\widehat{A}, \widehat{B}]$$

$$[\widehat{A}, \widehat{B} + \widehat{C}] = [\widehat{A}, \widehat{B}] + [\widehat{A}, \widehat{C}] \text{ and } [\widehat{A} + \widehat{B}, \widehat{C}] = [\widehat{A}, \widehat{C}] + [\widehat{B}, \widehat{C}] \tag{1.6.9}$$

$$[\widehat{A}, \widehat{B}\widehat{C}] = [\widehat{A}, \widehat{B}]\widehat{C} + \widehat{B}[\widehat{A}, \widehat{C}] \text{ and } [\widehat{A}\widehat{B}, \widehat{C}] = [\widehat{A}, \widehat{C}]\widehat{B} + \widehat{A}[\widehat{B}, \widehat{C}]$$

where k is a constant and the operators are assumed to be linear.

1.6.2.1 Commutation Properties of Linear and Angular Momentum Operators

The x, y, and z components of the linear momentum operators commute with each other but not with the corresponding position coordinates.

$$[\widehat{p}_x, \widehat{p}_y] = [\widehat{p}_y, \widehat{p}_z] = [\widehat{p}_z, \widehat{p}_x] = 0$$

$$[\widehat{p}_x, x] = [\widehat{p}_y, y] = [\widehat{p}_z, z] = -i\hbar \tag{1.6.10}$$

The physical significance of these relations follows from Heisenberg's uncertainty principle according to which if ΔA and ΔB are the uncertainty of measurement of two dynamical variables represented by operators $\widehat{A}$ and $\widehat{B}$ then

$$\Delta A \Delta B \geq \frac{1}{2}\left|\left\langle [\widehat{A}, \widehat{B}] \right\rangle\right|, \tag{1.6.11}$$

If $[\widehat{A}, \widehat{B}] \neq 0$, then $\Delta A \Delta B \neq 0$.

This shows that while there is no uncertainty in the measurement of the components of momentum, the position and the corresponding component of the momentum of a particle cannot be measured with certainty.

The commutation relations for the angular momentum operators are:

$$[\hat{L}_x, \hat{L}_y] = i\hbar\,\hat{L}_z,$$

$$[\hat{L}_z, \hat{L}_x] = i\hbar\,\hat{L}_y,$$

$$[\hat{L}_y, \hat{L}_z] = i\hbar\,\hat{L}_x,$$

and

$$[L^2, L_x] = [L^2, L_y] = [L^2, L_z] = 0 \tag{1.6.12}$$

Thus, the x, y, and z components of the angular momentum operators do not commute but the square of the angular momentum operator L^2 commutes with L_x, L_y, and L_z. The length of the angular momentum vector can therefore be measured with perfect precision along with its z-component.

1.7 Postulates of Quantum Mechanics

Heisenberg made some postulates in the development of quantum mechanics. These are:

Postulate 1:
The state of a quantum mechanical system is represented by a mathematical function Ψ to be called a state function or a wavefunction. For a particle moving in a conservative field of force the wavefunction determines everything that can be known about the system. The wavefunction must be a single-valued function of position and time as that is sufficient to guarantee an unambiguous value of probability of finding the particle at a particular position and time.

Postulate 2:
With every physical observable q there is associated an operator $\hat{Q}$. It turns out that the operators occurring in quantum mechanics are linear operators.

The only possible values which a single measurement of an observable associated with an operator can yield are the eigenvalues q of the equation

$$\hat{Q}\Psi = q\Psi \tag{1.7.1}$$

Here, q are just numbers called eigenvalues of operator $\hat{Q}$ corresponding to eigenfunction Ψ. In fact, this equation can be satisfied for a number of (often an infinite number) different wavefunctions called eigenfunctions and eigenvalues. Two possible cases that may exist are:

1. There is only one independent eigenfunction for one eigenvalue. In this case the physical state is said to be nondegenerate.

2. There are a number of eigenfunctions for a given eigenvalue. In this case, the state is said to be degenerate.

Postulate 3:
Any operator $\hat{Q}$ associated with a physically measurable property q will be Hermitian.

An operator is said to be Hermitian if for any two functions Ψ and Φ it satisfies the condition

$$\int \Psi^* \hat{Q}\Phi d\tau = \int \Phi\left(\hat{Q}\Psi\right)^* d\tau \tag{1.7.2}$$

The asterisk as superscript indicates the complex conjugate of the quantity immediately to its left. Hermitian operators are linear operators and have real eigenvalues.

Postulate 4:
The set of eigenfunctions of an operator Q form a complete set of linearly independent functions. Also, if an eigenvalue q of the operator $\hat{Q}$ is degenerate, any linear combination of the linearly independent eigenfunctions is also an eigenfunction. Thus,

$$\hat{Q}\left(\sum_n C_n\Psi_n\right) = q\left(\sum_n C_n\Psi_n\right) \tag{1.7.3}$$

This is known as the superposition principle.

If $\Psi_1, \Psi_2, \ldots \Psi_n$ are a complete set of linearly independent eigenfunctions of the operator Q representing a physical system, then these functions are said to be normal if

$$\int_{-\infty}^{\infty} \Psi_i^*\Psi_i d\tau = 1$$

and orthogonal, if

$$\int_{-\infty}^{\infty} \Psi_i^*\Psi_j d\tau = 0$$

These two equations combined together may be written as

$$\int_{-\infty}^{\infty} \Psi_i^*\Psi_j d\tau = \delta_{ij}, \tag{1.7.4}$$

where $\delta_{ij} = 1,$ if $i = j$ and

$= 0,$ if $i \neq j$

Equation (1.7.4) represents the orthonormality conditions of the eigenfunctions.

Postulate 5:
This postulate establishes relationship between the quantum mechanical and classical measurements and states that the calculated "expectation value" of a measurable parameter in quantum mechanics can be related with its average value from a large number of physical measurements.

According to this postulate, if a system at a given time t is characterized by normalized wavefunction Ψ of the observable $\widehat{Q}$ then $\int \Psi^* \widehat{Q}\Psi d\tau$ gives the average of a large number of measurements performed on this system under the same initial physical conditions.

$\int \Psi^* \widehat{Q}\Psi d\tau$ is known as the expectation value of $\widehat{Q}$ in the state Ψ at time t.

$$\langle q \rangle = \int \Psi^* Q \Psi d\tau = Lt_{n\to\infty} \frac{q_1 + q_2 + \ldots + q_n}{n} \tag{1.7.5}$$

The meaning of the expectation value in terms of the classical measurements can be understood in the following manner:

It is a practice to repeat uncertain experiments several times and to determine the average of several measurements on a chosen parameter.

If a certain value x_k for the parameter x occurs n_k times such that $\sum_{k=1}^{m} n_k = n$ then the average value of x may be written as

$$\overline{x} = \frac{\sum_{k=1}^{m} n_k x_k}{\sum_{k=1}^{m} n_k} = \sum_{k=1}^{m} \frac{1}{n} n_k x_k \tag{1.7.6}$$

If Q_k is the probability that a measurement yields a value x_k, then

$$Q_k = \frac{n_k}{n} \tag{1.7.7}$$

From Eqs (1.7.6) and (1.7.7), we get

$$\overline{x} = \sum_{k=1}^{m} Q_k x_k \tag{1.7.8}$$

In quantum mechanics, the average value $\overline{x}$ is called the expectation value and is written as $\langle x \rangle$.

If the distribution of measurements is continuous then

$$\overline{x} = \int_{-\infty}^{\infty} Q(x) x \, dx = \int_{-\infty}^{\infty} x \, dW \tag{1.7.9}$$

where $dW = Q(x)dx$ is the probability that x has a value between x and $x + dx$.

Since the total probability for x to have some value is 1, $Q(x)$ must satisfy the condition

$$\int_{-\infty}^{\infty} Q(x) dx = 1 \tag{1.7.10}$$

$Q(x)$ is called the probability density of x because it gives the probability per unit interval.

$$Q(x) = \frac{dW}{dx} \tag{1.7.11}$$

In forming the average of x according to Eq. (1.7.9) those values of x that are more probable contribute more heavily to the integral. The value of $\overline{x}$ obtained from this equation is therefore called the weighted average of x.

The above procedure can be used to get expectation value of a function of one or several parameters.

$$\langle f(x,y,z)\rangle = \iiint Q(x,y,z)f(x,y,z)dx\,dy\,dz \tag{1.7.12}$$

If $Q(x, y, z) = \Psi^*(\mathbf{r})\Psi(\mathbf{r})d\tau$ then

$$\langle f\rangle = \int \Psi^*(\mathbf{r})f(x,y,z)\Psi(\mathbf{r})d\tau \tag{1.7.13}$$

To relate a quantum mechanical calculation to something one observes in the laboratory, the "expectation value" of the measurable parameter is calculated. For the position x, the expectation value is defined as

$$\langle x\rangle = \int_{-\infty}^{\infty} \Psi^*(x,t)x\Psi(x,t)dx \tag{1.7.14}$$

This integral can be interpreted as the average value of x that we would expect to obtain from a large number of measurements.

1.8 Hydrogen Atom

The hydrogen atom consists of a proton of charge $+e$ and an electron of charge $-e$. These two particles attract each other according to Coulomb law of electrostatic attraction. Instead of treating just the hydrogen atom, we shall consider a more general problem of a hydrogen-like atom, i.e., a system consisting of one electron and a nucleus of charge $+Ze$. For $Z=1$, we have the hydrogen atom, for $Z=2$, the He^+ ion, for $Z=3$, the Li^{++} ion, and so on. The importance of a hydrogen-like atom in quantum chemistry arises from the fact that an exact solution of Schrödinger equation is possible for a one-electron system. In atoms having more than one electron, interelectronic repulsions make the solution becomes very difficult. If, as an approximation, we neglect the electron repulsions, the electrons can be treated independently and the wavefunction for a many-electron atom can be approximated by a product of hydrogen-like one-electron wavefunctions. A one-electron wavefunction is called an orbital. An orbital for an electron in an atom is called atomic orbital. These atomic orbitals can be used to construct approximate wavefunctions for many-electron atoms and molecules.

1.8.1 Solution of Schrödinger Equation for Hydrogen-Like Atoms

The hydrogen problem is best solved in spherical polar coordinates (Figure 1.1). Let $+Ze$ be the charge on the nucleus of mass m_N of a hydrogen-like atom and $-e$ the charge on the electron of mass m_e. Also, let (x, y, z) be the coordinates of the electron relative to the nucleus and r the distance between two particles. According to the Coulomb's law, the

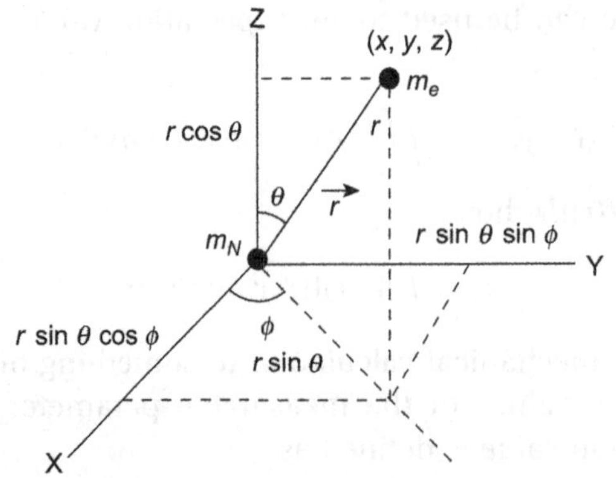

FIGURE 1.1 Hydrogen-like atom in spherical polar coordinates.

attractive force between the two charged particles will be operative in the direction of the line joining the two particles and shall have the magnitude.

$$F = -\frac{Ze^2}{r^2} \tag{1.8.1}$$

The potential energy resulting from this force is

$$\mathrm{V}(r) = -\int_{\infty}^{r} F dr = \int_{\infty}^{r} \frac{Ze^2}{r^2} dr$$

or

$$\mathrm{V}(r) = -\frac{Ze^2}{r} \tag{1.8.2}$$

In view of the fact that the problem is a central force field two-body problem, it can be reduced to a one-body problem by introducing the reduced mass μ, where

$$\mu = \frac{m_e m_N}{m_e + m_N}$$

On replacing m by μ and using Eq. (1.8.2), the Schrödinger Eq. (1.5.23) in spherical polar coordinates can therefore be written as

$$\frac{1}{r^2}\frac{\partial}{\partial r}\left(r^2\frac{\partial \Psi}{\partial r}\right) + \frac{1}{r^2 \sin\theta}\frac{\partial}{\partial \theta}\left(\sin\theta \frac{\partial \Psi}{\partial \theta}\right) + \frac{1}{r^2 \sin\theta}\frac{\partial^2 \Psi}{\partial^2 \phi} + \frac{2\mu}{\hbar^2}\left(E + \frac{Ze^2}{r}\right)\Psi = 0 \tag{1.8.3}$$

where Ψ is a function of r, θ, and ϕ.

This equation can be separated into the radial and angular parts by writing

$$\Psi(r, \theta, \phi) = \mathrm{R}(r) \cdot \mathrm{Y}(\theta, \phi) \tag{1.8.4}$$

Substituting Eq. (1.8.4) into Eq. (1.8.3), we get

$$\frac{1}{\mathrm{R}}\frac{\partial}{\partial r}\left(r^2\frac{\partial \mathrm{R}}{\partial r}\right) + \frac{1}{\mathrm{Y}\sin\theta}\frac{\partial}{\partial \theta}\left(\sin\theta \frac{\partial \mathrm{Y}}{\partial \theta}\right) + \frac{1}{\mathrm{Y}\sin^2\theta}\frac{\partial^2 \mathrm{Y}}{\partial \phi^2} + \frac{2\mu r^2}{\hbar^2}\left(E + \frac{Ze^2}{r}\right) = 0 \tag{1.8.5}$$

or

$$\frac{1}{R}\frac{\partial}{\partial r}\left(r^2\frac{\partial R}{\partial r}\right)+\frac{2\mu r^2}{\hbar^2}\left(E+\frac{Ze^2}{r}\right)=-\left[\frac{1}{Y\sin\theta}\frac{\partial}{\partial\theta}\left(\sin\theta\frac{\partial Y}{\partial\theta}\right)+\frac{1}{Y\sin^2\theta}\frac{\partial^2 Y}{\partial\phi^2}\right]=A \tag{1.8.6}$$

Since the two sides are functions of independent variable they must be equal to a constant, say A. Thus we get two equations one depending on the radial coordinate (r) and the other on spherical coordinates (θ, ϕ)

$$\frac{\partial}{\partial r}\left(r^2\frac{\partial R}{\partial r}\right)+\frac{2\mu r^2}{\hbar^2}\left(E+\frac{Ze^2}{r}\right)R-AR=0 \tag{1.8.7}$$

$$\frac{1}{\sin\theta}\frac{\partial}{\partial\theta}\left(\sin\theta\frac{\partial Y}{\partial\theta}\right)+\frac{1}{\sin^2\theta}\frac{\partial^2 Y}{\partial\phi^2}+AY=0 \tag{1.8.8}$$

Equation (1.8.8) can also be separated into a polar part (θ) and an azimuthal part (ϕ) by writing

$$Y(\theta,\phi)=\Theta(\theta)\Phi(\phi) \tag{1.8.9}$$

Substituting Eq. (1.8.9) in Eq. (1.8.8) and multiplying by $\frac{\sin^2\theta}{\Theta\Phi}$,

$$\frac{\sin\theta}{\Theta}\frac{\partial}{\partial\theta}\left(\sin\theta\frac{\partial\Theta}{\partial\theta}\right)+\frac{1}{\Phi}\frac{\partial^2\Phi}{\partial\phi^2}+A\sin^2\theta=0 \tag{1.8.10}$$

or

$$\frac{\sin\theta}{\Theta}\frac{\partial}{\partial\theta}\left(\sin\theta\frac{\partial\Theta}{\partial\theta}\right)+A\sin^2\theta=-\frac{1}{\Phi}\frac{\partial^2\Phi}{\partial\phi^2}=m^2 \tag{1.8.11}$$

Since the two sides of Eq. (1.8.7) are functions of different variables, they must be equal to some constant, say m^2

We therefore get,

$$\frac{1}{\sin\theta}\frac{\partial}{\partial\theta}\left(\sin\theta\frac{\partial\Theta}{\partial\theta}\right)+\left(A-\frac{m^2}{\sin^2\theta}\right)\Theta=0 \tag{1.8.12}$$

and

$$\frac{\partial^2\Phi}{\partial\phi^2}=-m^2\Phi \tag{1.8.13}$$

Thus the Schrödinger Eq. (1.8.3) breaks into three equations—the radial Eq. (1.8.7), the angular Eq. (1.8.12), and the azimuthal Eq. (1.8.13).

1.8.1.1 Solution of the ϕ Equations

The normalized solution to the azimuthal Eq. (1.8.13) is simple and may be written as

$$\Phi_m(\phi)=\frac{1}{\sqrt{2\pi}}e^{\pm im\phi} \tag{1.8.14}$$

The boundary conditions $(\phi)=(\phi+2\pi)$ requires that m can have only integer values

$$m=0,\pm1,\pm2\ldots\pm l \tag{1.8.15}$$

Table 1.1 $\Phi_m(\phi)$ functions

$\Phi_0(\phi)=\frac{1}{\sqrt{2\pi}}$	or	$\Phi_0(\phi)=\frac{1}{\sqrt{2\pi}}$
$\Phi_1(\phi)=\frac{1}{\sqrt{2\pi}}e^{i\phi}$	or	$\Phi_{1\cos}(\phi)=\frac{1}{\sqrt{\pi}}\cos\phi$
$\Phi_{-1}(\phi)=\frac{1}{\sqrt{2\pi}}e^{-i\phi}$	or	$\Phi_{1\sin}(\phi)=\frac{1}{\sqrt{\pi}}\sin\phi$
$\Phi_2(\phi)=\frac{1}{\sqrt{2\pi}}e^{i2\phi}$	or	$\Phi_{2\cos}(\phi)=\frac{1}{\sqrt{\pi}}\cos 2\phi$
$\Phi_{-2}(\phi)=\frac{1}{\sqrt{2\pi}}e^{(-i2\phi)}$	or	$\Phi_{2\sin}(\phi)=\frac{1}{\sqrt{\pi}}\sin 2\phi$

m is known as the magnetic quantum number. It refers to the orientation of the orbital angular momentum **L** toward some external coordinate system and has no effect on the energy of an isolated one-electron atom. If this coordinate system is referred to an external magnetic field, it should be obvious that differing orientations of the orbital angular momentum will give rise to different energies. In the absence of external effects, it is impossible to assign any preferred orientation to orbitals of a given m value.

The $\Phi_m(\phi)$ functions can be given both in the complex and real forms, either set being satisfactory. Some of these are given in Table 1.1.

1.8.1.2 Solution of the θ Equations

The formal solutions to Eq. (1.8.12) can be found in terms of the associated Legendre polynomial. The expression for the solutions of Θ function after normalization is

$$\Theta_{lm}(\theta)=\left\{\frac{(2l+1)(l-|m|)!}{2(l+|m|)!}\right\}^{1/2}P_l^{|m|}(\cos\theta) \tag{1.8.16}$$

$P_l^{|m|}(\cos\theta)$ is associated Legendre Polynomial of degree l and order $|m|$ and has the form of an infinite series expansion. These polynomials are the solutions of a certain type of second-order differential equations known as Legendre's equation.

It can be shown that for the series expansion (Eq. (1.8.16)) to terminate the constant A in Eq. (1.8.10) must be equal to $l(l+1)$ where l must be an integer with values $l=0$, 1, 2,...$(n-1)$. l is called the orbital quantum number of an electron as it moves about the nucleus. The angular momentum is equal to $\hbar\sqrt{l(l+1)}$. Owing to this relationship l is also referred to as angular momentum quantum number. In one-electron systems, in the absence of magnetic fields, it offers no contribution to the overall energy of the system. However, if electric charges are moving in such a manner that there is a net angular momentum associated with them, then they shall produce a magnetic moment. If an external magnetic field is present or else if the nucleus has a magnetic moment then the interaction of the orbital magnetic moment with the magnetic field will affect the energy. This gives rise to splitting of energies of orbitals of different l values in polyelectron

Table 1.2 $\Theta_{lm}(\theta)$ functions

***l* = 0, *s* orbitals**
$\Theta_{00}(\theta) = \frac{\sqrt{2}}{2}$
***l* = 1, *p* orbitals**
$\Theta_{10}(\theta) = \frac{\sqrt{6}}{2}\cos\theta$
$\Theta_{1\neq 1}(\theta) = \frac{\sqrt{3}}{2}\sin\theta$
***l* = 2, *d* orbitals**
$\Theta_{20}(\theta) = \frac{\sqrt{10}}{4}(3\cos^2\theta - 1)$
$\Theta_{2\neq 1}(\theta) = \frac{\sqrt{15}}{2}\sin\theta\cos\theta$
$\Theta_{2\neq 2}(\theta) = \frac{\sqrt{15}}{4}\sin^2\theta$

atoms. Some of the $\Theta_{lm}(\theta)$ functions corresponding to $l = 0$ (*s* orbitals), $l = 1$ (*p* orbitals) and $l = 2$ (*d* orbitals) frequently used in quantum chemistry of organic systems are given in Table 1.2.

1.8.1.3 Solution of the Radial Equation

An insight into the behavior of the radial part of the Schrödinger Eq. (1.8.7) may be obtained by looking at its limiting behavior for large values of r ($r \to \infty$). In this limit the equation reduces to

$$\frac{d^2\mathrm{R}}{dr^2} = -\frac{2\mu E}{\hbar^2}\mathrm{R} \tag{1.8.17}$$

The most general solution to this equation is an exponential function:

$$\mathrm{R}(r) = \exp\left(\pm\sqrt{-\frac{2\mu E}{\hbar^2}}r\right) \tag{1.8.18}$$

Since an acceptable wavefunction must be finite everywhere consequently, we are allowed to use only the negative exponent. We thus see that at large values of r our wavefunction decays exponentially.

For small values of r the radial function can be assumed to have some form such as

$$\mathrm{R}(r) = \exp\left(-\sqrt{-\frac{2\mu E}{\hbar^2}}r\right)\mathrm{F}(r) \tag{1.8.19}$$

F(r) being simply some arbitrary function of r. If F(r) is evaluated it takes the form of what is known as the associated Laguerre polynomial which again is an infinite series expansion. To terminate the series an integer n must be introduced. The final form of the radial function after normalization is given in Eq. (1.8.20).

$$R_{nl}(r) = -\left(\frac{2Z}{na_0}\right)^{3/2}\left(\frac{(n-l-1)!}{2n\{(n+l)!\}^3}\right)^{1/2}\rho^l e^{-\rho/2}L_{n+l}^{2l+1}(\rho) \tag{1.8.20}$$

$$a_0 = \frac{h^2}{4\pi^2\mu e^2} \tag{1.8.21}$$

$$\rho = \frac{2Z}{na_0}r \tag{1.8.22}$$

a_0 is known as Bohr's radius having value 0.52918 Å.

L_{n+l}^{2l+1} is the associated Laguerre polynomial of degree $\{(n+l)-(2l+1)\}$ and order $(2l+1)$.

For a satisfactory solution of the radial Eq. (1.8.20), the degree of the Laguerre Polynomial must be a positive integer. So, $(n+l)-(2l+1)=(n-l-1)\geq 0$ or $n \geq l+1$. Since l may have values 0, 1, 2,...; n may have values 1, 2, 3,... ∞.

The energy for a hydrogenic system is

$$E = -\frac{2\pi^2\mu e^4 Z^2}{n^2h^2},$$

or

$$E = -\frac{e^2Z^2}{2n^2a_0}, \quad n = 1,2,3,\ldots\infty \tag{1.8.23}$$

The quantum number n is called the principal quantum number or the total quantum number. For one-electron atoms, in the absence of any perturbing external influences, it is the only quantum number upon which the energy depends. Since the quantum number n is associated only with the radial part of the wavefunction, it appears that this must be the determining factor in the energy of the hydrogen atom. This is logical since the potential energy is a function of only the distance of the electron from the nucleus, the r coordinate. It turns out that the most probable electron distributions for the differing values of n occur at r values exactly equal to the radii of the Bohr orbits of corresponding energies.

Some of the hydrogen-like radial wavefunctions for $n=1$ (K shell), $n=2$ (L shell), and $n=3$ (M shell) are given in Table 1.3.

1.8.2 The Charge-Cloud Interpretation of Ψ

Earlier we have seen the probability interpretation of the wavefunction, according to which $\Psi^2 d\tau$ is the probability that it is found in the volume $d\tau$. However, an alternative and more pictorial (though less strictly accurate) interpretation of Ψ has been provided in terms of the charge cloud. If we are dealing with a moving electron then we suppose that this electron is spread out in the form of a cloud, called the charge cloud, whose

Table 1.3 Hydrogen-like radial wavefunctions $R_{nl}(r)$

$n = 1$, K shell	
$l = 0$, 1s	$R_{10}(r) = (Z/a_0)^{3/2} \cdot 2e^{-\frac{\rho}{2}}$
$n = 2$, L shell	
$l = 0$, 2s	$R_{20}(r) = \frac{(Z/a_0)^{3/2}}{2\sqrt{2}}(2 - \rho)e^{-\frac{\rho}{2}}$
$l = 1$, 2p	$R_{21}(r) = \frac{(Z/a_0)^{3/2}}{2\sqrt{6}} \cdot \rho e^{-\frac{\rho}{2}}$
$n = 3$, M shell	
$l = 0$, 3s	$R_{30}(r) = \frac{(Z/a_0)^{3/2}}{9\sqrt{3}}(6 - 6\rho + \rho^2)e^{-\frac{\rho}{2}}$
$l = 1$, 3p	$R_{31}(r) = \frac{(Z/a_0)^{3/2}}{9\sqrt{6}}(4 - \rho)\rho e^{-\frac{\rho}{2}}$
$l = 2$, 3d	$R_{32}(r) = \frac{(Z/a_0)^{3/2}}{9\sqrt{30}}\rho^2 e^{-\frac{\rho}{2}}$
where, $\rho = \frac{2Z}{na_0}r$	

density at any point being proportional to Ψ^2. Larger the value of Ψ^2, denser the charge cloud, and hence greater the negative charge that shall be found. Hence, the essential difference between the two interpretations is that instead of speaking of the probability density or the chance of finding the electron in any given region we now speak of the actual particle density. This charge-cloud picture, though very useful, is not strictly correct because if a single electron is a particle, it cannot possibly be distributed over regions of the size of atom or molecule.

As Coulson (1961) explained, a link between the two viewpoints may be found by a simple experiment: Perform a large number of measurements to exactly determine the position of an electron and each time represents this position by a minute dot in a three-dimensional space. If the dots are so small that we cannot distinguish them from each other then the general effect of this diagram will be exactly the same as that of a cloud. In this case, the densest parts of the cloud will be those where there are more dots and obviously these are the places where there is greater likelihood or probability of discovering the electron. Hence the density of the charge cloud is a direct measure of the probability function. Although the concept of charge cloud lacks validity it is very useful. The wavefunctions of an electron are not strictly confined but stretched to infinite distance from the nucleus. Obviously, there is a finite chance of finding the electron even at large distances. Using the charge-cloud picture we may say that there is a certain contour for each Ψ such that most of the charge, say 90–95%, lies within this contour.

The concept of charge cloud presents a mental picture for the charge distribution in space and the shapes of contours largely determine the stereochemical disposition of the atoms in a polyatomic system. This concept shall be used to understand the ground state of the hydrogen atom.

1.8.3 Normal State of the Hydrogen Atom

The total wavefunction $\Psi(r, \theta, \phi)$ for the hydrogen atom in its ground state $n = 1$, $l = 0$, and $m = 0$ can be obtained by using Tables 1.1–1.3 and Eqs (1.8.14), (1.8.16), and (1.8.20). Since,

$$\Psi_{nlm}(r, \theta, \phi) = \mathrm{R}_{nl}(r)\Theta_{ln}(\theta)\Phi_m(\phi) \tag{1.8.24}$$

we get in this case,

$$\Psi_{100} = \mathrm{R}_{10}\Theta_{00}\Phi_0 = \frac{1}{\sqrt{2\pi}} \cdot \frac{1}{\sqrt{2}} \cdot \frac{2}{\sqrt{a_0^3}} e^{-\rho/2} \tag{1.8.25}$$

or

$$\Psi = \sqrt{\left(\frac{1}{\pi a_0^3}\right)} e^{-r/a_0} \tag{1.8.26}$$

From this it follows that

$$\rho = \Psi^*\Psi = \frac{1}{\pi a_0^3} e^{-2r/a_0} \tag{1.8.27}$$

The values of Ψ and Ψ^2 are graphically represented in various forms in Figure 1.2. Since the atom is spherically symmetrical, ρ is a function of the radial distance r only. Instead of plotting the density $\rho(r)$, we plot the radial density $4\pi r^2\rho(r)$. Since $4\pi r^2 dr$ is the volume element lying between the two spheres r and $r + dr$, it follows that $4\pi r^2\rho(r)dr$ is the total probability that the electron is at a distance between r and $r + dr$ of the origin. Figure 1.3 shows the graph of this radial density. It is interesting to see that the maximum radial density is at $r = a_0$, which is in agreement with the classical Bohr's theory; the difference being that instead of an orbit we are now talking in terms of the locus of points of maximum probability density.

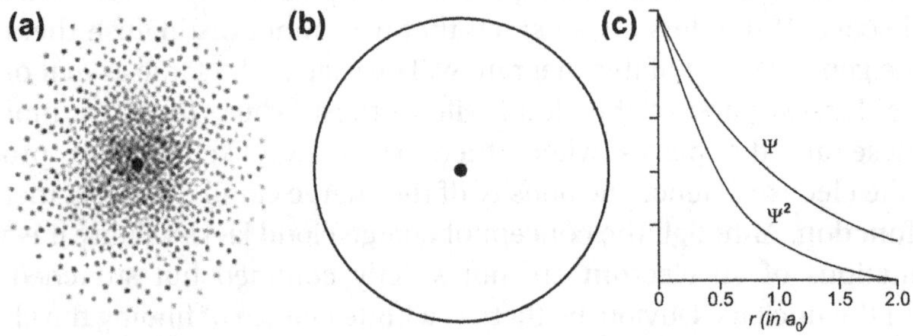

FIGURE 1.2 (a) Charge cloud, (b) boundary surface, and (c) variation of probability function $|\Psi|^2$ in space for hydrogen atom.

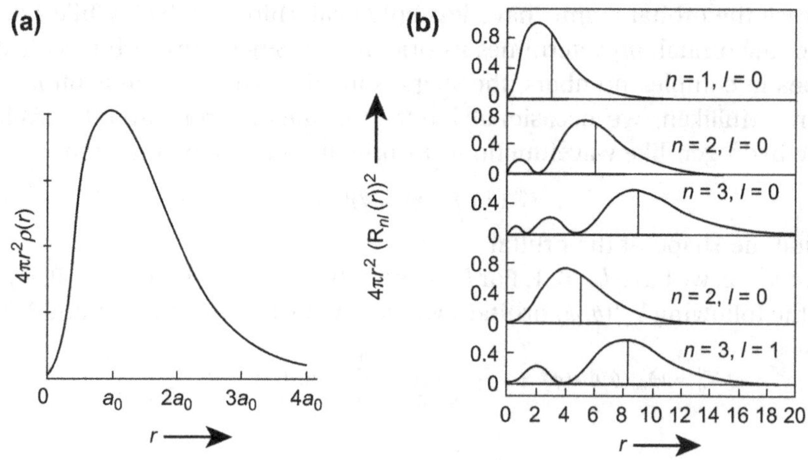

FIGURE 1.3 Radial distribution function or density for the ground state of (a) hydrogen and (b) hydrogen-like atoms.

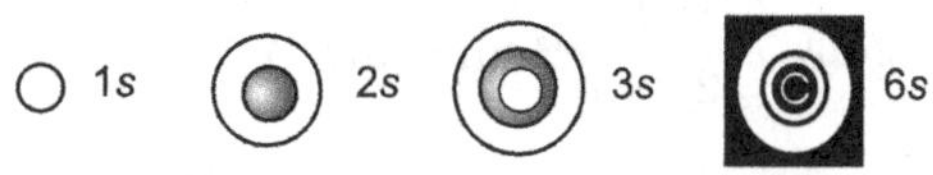

FIGURE 1.4 Shapes of atomic orbitals.

Figure 1.3(b) shows the radial distribution for different states of hydrogen-like atoms. For higher values of the quantum number n, the curve touches the radial axis $(n - l - 1)$ times between $\rho = 0$ and $\rho = \infty$. These are the nodal points. Thus, if the normal state (1*s*) of hydrogen is considered as a concentric spherical ball around the nucleus, the 2*s* state is a ball and an outer shell, the 3*s* state is a ball and two concentric shells, and a 6*s* state is a ball and 5 concentric shells (Figure 1.4).

1.9 Atomic Orbitals

The wavefunction Ψ_{nlm} describes the motion of the electron. It can be used to calculate the probability of finding an electron of an atom in any specific region around the atom's nucleus. For this reason, we can call it an atomic orbital. Following spectroscopy practices, the symbols *s, p, d, f, g,*... are used to represent states characterized by the values of the orbital quantum number *l*. Thus by *s, p, d, f,*... orbitals we mean an orbital with $l = 0, 1, 2, 3, \ldots$. The coordinate systems chosen for atomic orbitals are usually the spherical coordinates (r, θ, ϕ) in atoms and Cartesian (x, y, z) in polyatomic molecules.

As in the case of hydrogen atom, simple pictures showing the orbital shapes or contours containing 90–95% of the electron probability density are used to represent atomic orbitals. Sometimes the $\Psi(r, \theta, \phi)$ instead of the $|\Psi(r, \theta, \phi)|^2$ functions, are plotted to show its phases. The $\Psi(r, \theta, \phi)$ and $|\Psi(r, \theta, \phi)|^2$ graphs are very similar except that in the latter case the orbital graphs have less spherical, thinner lobes. While l determines the shape of the orbital, m_l determines its orientation. Since some orbitals are described by equations in complex numbers, the shape sometimes also depends on m_l.

Following Mulliken, we occasionally refer to, one-electron orbital wavefunctions, such as the hydrogen-like wavefunctions, as orbital and use the function

$$Y_l^{m_l}(\theta,\phi) = \Theta_{l,m}(\theta)\Phi_m(\phi) \tag{1.9.1}$$

to determine the shape of the orbital.

Thus, for $n = 2$, we have $l = 0, 1$. For $l = 0$, $m_l = 0$ and for $l = 1$, $m_l = +1, 0, -1$. Thus we may have the following $Y_l^m(\theta,\phi)$ orbitals, whose values from Tables 1.1 and 1.2 are:

$$\begin{aligned}
Y_0^0 &= \Theta_{00}(\theta)\Phi_0(\phi) = \frac{\sqrt{2}}{2}\cdot\frac{1}{\sqrt{2\pi}} = \frac{1}{2\sqrt{\pi}} \text{ and } |Y_0^0|^2 = \left(\frac{1}{2\sqrt{\pi}}\right)^2\\
Y_1^0 &= \Theta_{10}(\theta)\Phi_0(\phi) = \frac{1}{2}\sqrt{\frac{3}{\pi}}\cos\theta \text{ and } |Y_1^0|^2 = \left(\frac{1}{2}\sqrt{\frac{3}{\pi}}\cos\theta\right)^2\\
Y_1^1 &= \Theta_{11}(\theta)\Phi_1(\phi) = \frac{1}{2}\sqrt{\frac{3}{2\pi}}e^{i\phi}\sin\theta \text{ and } |Y_1^1|^2 = \left(\frac{1}{2}\sqrt{\frac{3}{2\pi}}\sin\theta\right)^2\\
Y_1^{-1} &= \Theta_{1-1}(\theta)\Phi_{-1}(\phi) = \frac{1}{2}\sqrt{\frac{3}{2\pi}}e^{-i\phi}\sin\theta \text{ and } |Y_1^{-1}|^2 = \left(\frac{1}{2}\sqrt{\frac{3}{2\pi}}\sin\theta\right)^2
\end{aligned} \tag{1.9.2}$$

Since Y_0^0 is independent of θ and ϕ, it follows that the probability density distribution for the s-orbital is spherically symmetrical.

The functions $Y_1^{0,\pm 1}$ representing p orbitals have the form of two spheres with centers on the z-axis and tangent at the origin. They resemble the shape of "dumbbell." The three p orbitals are at right angles to each other. They show a marked directional character which is exhibited by suffix p_x, p_y, p_z. If we exclude the part of the electron cloud which is very near the origin, and not effective in bond formation, then in one-half of the dumbbell the wave function is positive and in the other half it is negative (Figure 1.5).

For $l = 2$, the magnetic quantum number m_l can take five values 0, ±1, ±2. We thus get five d orbitals (Figure 1.5). Four of these five orbitals look similar, each with four pear-shaped lobes, each lobe tangent to two others, and the centers of all four lying in one plane between a pair of axes. Three of these planes are xy-, xz-, and yz-planes, and the fourth has center on the x and y axes. The fifth d orbital consists of the three regions of high probability density: a torus with two pear-shaped regions placed symmetrically on its z-axis.

Since the shapes of the atomic orbitals in one-electron atom are related to three-dimensional spherical harmonics $Y_l^m(\theta,\phi)$ given in Tables 1.1 and 1.2 they are not unique and any linear combination is also valid. Thus, for example, it is possible to generate sets where all the d orbitals are of the same shape just like the p_x, p_y, and p_z orbitals which are of the same shape.

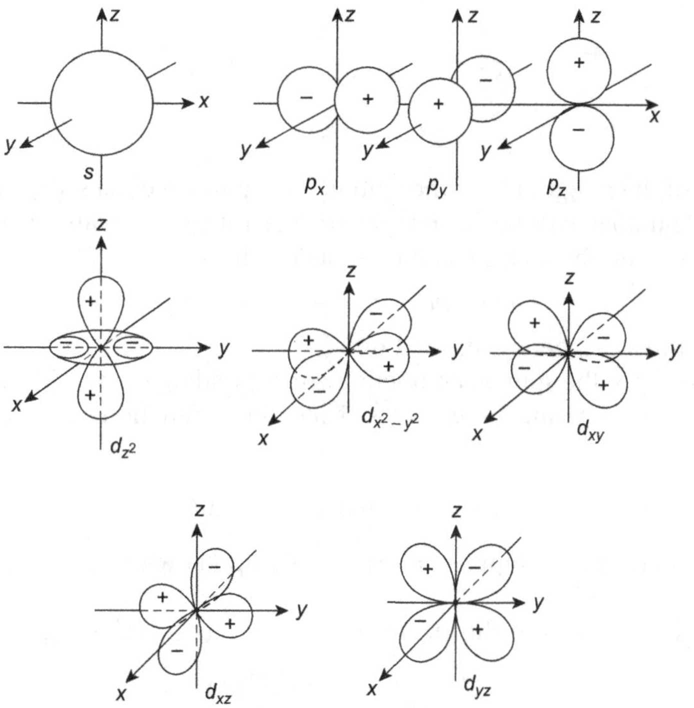

FIGURE 1.5 Types of atomic orbitals - approximate boundary surfaces.

1.10 Electron Spin

While the quantum members *n*, *l*, and *m* arise out of the mathematical derivation of the hydrogen problem, the spin quantum member, m_s, was introduced for empirical reasons. In order to explain the spectral properties of alkali metals and certain properties of other atoms when placed in a magnetic field, it was necessary to assume that the electron itself had an intrinsic angular momentum, giving rise to magnetic moment. The magnitude of this spin angular momentum had to be $\hbar\sqrt{s(s+1)}$, where *s* is an integer. In addition to the orbital angular momentum **L**, which arises from motion through space, elementary particles such as electrons, protons, μ mesons, etc., possess the spin angular momentum **s**. The spin *s* has no classical analog and is a relativistic effect. In nonrelativistic quantum mechanics, the existence of spin is introduced as an additional postulate. Like the orbital angular momentum operators $\widehat{L}_x$, $\widehat{L}_y$, $\widehat{L}_z$, and $\widehat{L}^2$, we have spin angular momentum operators $\widehat{s}_x$, $\widehat{s}_y$, $\widehat{s}_z$, and $\widehat{s}^2$. Also, like angular momentum operators they obey the same commutation relations:

$$\left[\widehat{s}_x, \widehat{s}_y\right] = i\hbar\widehat{s}_z, \quad \left[\widehat{s}_y, \widehat{s}_z\right] = i\hbar\widehat{s}_z, \quad \text{and} \quad \left[\widehat{s}_z, \widehat{s}_x\right] = i\hbar\widehat{s}_y \tag{1.10.1}$$

and

$$[\hat{s}^2, \hat{s}_x] = [\hat{s}^2, \hat{s}_y] = [\hat{s}^2, \hat{s}_z] = 0 \qquad (1.10.2)$$

$$\hat{s}^2 = \hat{s}_x^2 + \hat{s}_y^2 + \hat{s}_z^2 \qquad (1.10.3)$$

The square of the magnitude of the spin $|s|^2$ of a particle equals $s(s+1)\hbar^2$, where the spin quantum number can be an integer or half integer. For an electron $s = \frac{1}{2}$. The component of s along the z-axis has the possible values

$$m_s\hbar, \quad m = -s, -s+1, \ldots, s-1, s \qquad (1.10.4)$$

For an electron $s = \frac{1}{2}$, hence $m_s = -\frac{1}{2}$ or $+\frac{1}{2}$

Let us denote by α, the spin eigenfunction corresponding to $m_s = \frac{1}{2}$ and by β, the spin eigenfunction corresponding to $m_s = -\frac{1}{2}$. Then these functions satisfy the following equations:

$$\hat{s}_z\alpha = \frac{1}{2}\hbar\alpha \text{ and } \hat{s}_z\beta = -\frac{1}{2}\hbar\beta \qquad (1.10.5)$$

$m_s = \frac{1}{2}$ is sometimes said to represent a state spin up (↑) while $m_s = -\frac{1}{2}$ spin down (↓) (Figure 1.6).

Also, since $[\hat{s}^2, \hat{s}_z] = 0$, and the eigenvalue of $\hat{s}^2$ is $s(s+1)\hbar^2$, we get,

$$\hat{s}^2\alpha = \frac{3}{4}\hbar^2\alpha, \quad \hat{s}^2\beta = \frac{3}{4}\hbar^2\beta \qquad (1.10.6)$$

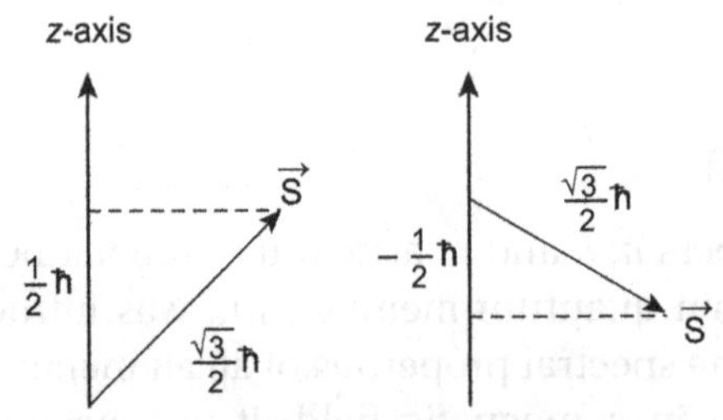

FIGURE 1.6 Orientations of electron spin vector with respect to the z-axis.

1.10.1 Spin Orbitals

In a complete quantum mechanical description of the motion of an electron, both its position and spin coordinate must be considered. Thus, its wavefunction must be a function of x, y, z or r, θ, ϕ, and s:

$$\Psi = \Psi(r, \theta, \phi, s) \qquad (1.10.7)$$

Taking cognizance of the independence of spatial and orbital motions, we can set up wavefunctions as products of the spatial $\Psi_{nlm}(r, \theta, \phi)$ and spin functions $u_{m_s}(s)$,

$$\Phi(r, \theta, \phi, s) = \Psi_{nlm}(r, \theta, \phi)u_{m_s}(s) \qquad (1.10.8)$$

Orbitals like $\Phi(r, \theta, \phi, s)$ are known as spin orbitals.

Since the quantum number m_s takes only two values $\frac{1}{2}$ and $-\frac{1}{2}$, we may have two spin functions $u_{\frac{1}{2}}(s)$ and $u_{-\frac{1}{2}}(s)$. Pauli defines $u_{\frac{1}{2}}(s)$ as $\alpha(s)$ and $u_{-\frac{1}{2}}(s)$ as $\beta(s)$.

Thus, for the ground state of hydrogen, the spin orbitals are

$$\Phi_{1s} = \Psi_{1s}(r)\alpha(s) \text{ or } \Phi'_{1s} = \Psi_{1s}(r)\beta(s) \tag{1.10.9}$$

If we consider the excited state of a two electron the atom, for example, helium atom, in which one electron is in 1*s* state and the other in 2*s* state, we can write

$$\Phi_{1s} = \Psi_{1s}(r)\alpha(s) \text{ or } \Phi'_{1s} = \Psi_{1s}(r)\beta(s)$$

and

$$\Phi_{2s} = \Psi_{2s}(r)\alpha(s) \text{ or } \Phi'_{2s} = \Psi_{2s}(r)\beta(s) \tag{1.10.10}$$

From these spin orbitals, we can obtain four possible wavefunctions for this particular configuration of the helium atom.

$$\Phi_1 = \Phi_{1s}\Phi_{2s} = \Psi_{1s}(r_1)\alpha(s_1)\Psi_{2s}(r_2)\alpha(s_2) \tag{1.10.11a}$$

$$\Phi_2 = \Phi_{1s}\Phi'_{2s} = \Psi_{1s}(r_1)\alpha(s_1)\Psi_{2s}(r_2)\beta(s_2) \tag{1.10.11b}$$

$$\Phi_3 = \Phi'_{1s}\Phi_{2s} = \Psi_{1s}(r_1)\beta(s_1)\Psi_{2s}(r_2)\alpha(s_2) \tag{1.10.11c}$$

$$\Phi_4 = \Phi'_{1s}\Phi'_{2s} = \Psi_{1s}(r_1)\beta(s_1)\Psi_{2s}(r_2)\beta(s_2) \tag{1.10.11d}$$

Since the two electrons are indistinguishable, a similar set of equations can be written by considering electron 2 in 1*s* orbital and electron 1 in 2*s* orbital.

1.11 Linear Vector Space and Matrix Representation

The operators representing dynamical variables and wavefunctions representing the state of a system can be represented as vectors in some linear space.

An *n*-dimensional vector space can be constructed by using an orthonormal set of functions $\Psi_1, \Psi_2, \ldots \Psi_n$, which may be treated as vectors. By analogy with vector algebra, the length of a base vector Ψ_n is given as

$$\text{Length} = \sqrt{\int \Psi_n^* \Psi_n d\tau}$$

If the length of a vector is unity, it will be called a unit vector. Thus, for a unit base vector

$$\int \Psi_n^* \Psi_n d\tau = 1$$

The set of functions satisfying this condition form the normal set of base vectors. Given two vectors Ψ_i and Ψ_j, the scalar product is defined as $\int \Psi_i^* \Psi_j d\tau$. This may be a complex number, denoted in standard linear algebra notation as $(\Psi_i, \Psi_j) = (\Psi_j, \Psi_i)^*$.

If the scalar product of these vectors is zero, they are called orthogonal or perpendicular. Thus, a set of functions Ψ_1, Ψ_2,...Ψ_n is said to constitute an orthogonal set of base vectors, if

$$\int \Psi_i^* \Psi_j d\tau = \delta_{ij}, \tag{1.11.1}$$

where $\delta_{ij} = 1$ for $i = j$ and $\delta_{ij} = 0$ for $i \neq j$.

Any vector Ψ may be represented in the vector space as

$$\Psi = c_1\Psi_1 + c_2\Psi_2 + \ldots c_n\Psi_n = \sum_{i=1}^{n} c_i\Psi_i \tag{1.11.2}$$

where c_1, c_2,...c_n are the components of vector Ψ along the base vectors. The wavefunction Ψ can then be represented as a column matrix

$$C = \begin{pmatrix} c_1 \\ c_2 \\ \vdots \\ c_n \end{pmatrix} \tag{1.11.3}$$

This is called matrix representation of wavefunction. The components of function Ψ in the orthonormal vector space can be obtained by multiplication of Eq. (1.11.2) by Ψ_j^*, followed by integration

$$c_j = \int \Psi_j^* \Psi d\tau. \tag{1.11.4}$$

The orthonormality condition in matrix representation is

$$\tilde{C}_a^* C_b = C_a^\dagger C_b = \delta_{ab}$$

$$\begin{aligned} \delta_{ab} &= 1 \text{ for } a = b \\ &= 0 \text{ for } a \neq b \end{aligned} \tag{1.11.5}$$

where C_a and C_b are the matrix representation of wavefunction Ψ_a and Ψ_b. An operator in matrix representation in the vector space of base vectors Ψ_1, Ψ_2,...Ψ_n is given as

$$A = \begin{pmatrix} A_{11} & A_{12} & \cdots A_{1n} \\ A_{21} & A_{22} & \cdots A_{2n} \\ \vdots & \vdots & \vdots \\ A_{n1} & A_{n2} & \cdots A_{nn} \end{pmatrix}$$

where,

$$A_{ij} = \int \Psi_i^* A \Psi_j d\tau, \tag{1.11.6}$$

The matrix representation of an operator does not change the properties of an operator.

In operator form, the Schrödinger equation is

$$\widehat{H}\Psi = E\Psi$$

If Ψ is represented by the column matrix of elements c_i and $\widehat{H}$ by a square matrix of elements H_{ij} such that

$$H_{ij} = \int \Psi_i^* \widehat{H}\Psi_j d\tau$$

then the Schrödinger equation may be written as

$$\begin{pmatrix} H_{11} & H_{12} & \cdots H_{1n} \\ H_{21} & H_{22} & \cdots H_{2n} \\ \vdots & \vdots & \vdots \\ H_{n1} & H_{n2} & \cdots H_{nn} \end{pmatrix} \begin{pmatrix} c_1 \\ c_2 \\ \vdots \\ c_n \end{pmatrix} = E \begin{pmatrix} c_1 \\ c_2 \\ \vdots \\ c_n \end{pmatrix}$$

or,

$$HC = EC \tag{1.11.7}$$

The time-dependent Schrödinger equation is written as

$$HC = i\hbar \frac{\partial}{\partial t} C \tag{1.11.8}$$

1.11.1 Dirac's Ket and Bra Notations

Every dynamical state is represented by a certain type of vector which we call, following Dirac, ket vector represented as $|\ \rangle$. In order to distinguish ket's from each other we insert a label, for example, $|A\rangle$.

Thus, we define a ket as

$$|\Psi\rangle = \begin{pmatrix} c_1 \\ c_2 \\ \vdots \\ c_n \end{pmatrix} \tag{1.11.9}$$

Ket's form a linear vector space. Any linear combination of several kets is also a ket vector.

$$c_1|A\rangle + c_2|B\rangle = |D\rangle$$

There is a one to one correspondence between the states of a dynamical system and the directions of the ket vectors. Thus $|A\rangle$ and $c|A\rangle$, where c is a complex number, correspond to the same state.

In linear algebra, every vector space can be associated with a dual vector space such that one can obtain a scalar product of the two vectors, one from each space. The vectors of space dual to that of ket vectors is called bra vectors or bras and is denoted by $\langle\ |$. Thus

$$\langle\Psi| = \left(c_1^* c_2^* \ldots c_n^*\right) \tag{1.11.10}$$

The scalar product of $|A\rangle$ and $\langle B|$ is denoted as $\langle B|A\rangle$. The scalar product is a complex number. Any state of a dynamical system at a particular time may be specified by the direction of a bra vector, just as well as by the direction of a ket vector,

$$|A\rangle + |B\rangle \rightarrow \langle A| + \langle B|$$

and

$$c|A\rangle \rightarrow c^*\langle A|$$

Further

$$\langle A|B\rangle = \langle B|A\rangle \tag{1.11.11}$$

Sometimes, in superficial treatments of Dirac notation, the symbol $\langle \Psi_a|\Psi_b\rangle$ is alternatively defined as

$$\langle \Psi_a|\Psi_b\rangle = \int \Psi_a^*(x)\Psi_b(x)dx \tag{1.11.12}$$

Often only the subscript of the vector is used to denote a bra or ket. Thus Eq. (1.11.12) may also be written as

$$\langle a|b\rangle = \int \Psi_a^*(x)\Psi_b(x)dx$$

The overlap integral between two functions Ψ_a and Ψ_b therefore, be written as

$$\int \Psi_a^*\Psi_b d\tau = \langle a|b\rangle \text{ or } \langle \Psi_a|\Psi_b\rangle$$

Operation of an operator on a ket produces another ket vector. Thus

$$A|\Psi\rangle = |\Psi'\rangle$$

In the case of operation on a ket the operator is always placed on the left of the k. Likewise, an operation on a bra from the right with operator A produces another bra.

$$\langle \Phi|A = \langle \Phi'|$$

The Schrödinger equation $\widehat{H}\Psi_n = E_n\Psi_n$ in this notation reads as

$$\widehat{H}|\Psi_n\rangle = E_n|\Psi_n\rangle$$

or

$$(\widehat{H} - E_n)|\Psi_n\rangle = 0 \tag{1.11.13}$$

and the matrix elements H_{nm} of $\widehat{H}$ are

$$H_{nm} = \int \Psi_n^*\widehat{H}\Psi_m d\tau = (\Psi_n, \widehat{H}\Psi_m) = \langle \Psi_n|\widehat{H}|\Psi_m\rangle \tag{1.11.14}$$

or in short notation, it is also written as

$$H_{nm} = \langle n|\widehat{H}|m\rangle \tag{1.11.15}$$

Similarly, the expectation value of an operator A in the eigenstate Ψ_n may be written as

$$\langle A\rangle = \int \Psi_n^* A \Psi_n d\tau = \langle n|A|n\rangle \tag{1.11.16}$$

1.12 Atomic Units

In order to simplify the representation of equations in quantum mechanics and to reduce the use of powers of 10, a set of units called atomic units (au or a. u.) are used. Two different kinds of atomic units: the Hartree atomic units and Rydberg atomic units which differ in the choice of the unit of mass and charge are in common use but the Hartree atomic units, with which we shall be presently concerned, are especially useful for quantum mechanical calculations as the numerical values of four fundamental physical constants: electronic mass (m_e), electronic charge (e), reduced Planck's constant $\left(\hbar = \frac{h}{2\pi}\right)$, and the Coulomb's constant $\left(K_e = \frac{1}{4\pi\varepsilon_0}\right)$ are taken as unity by definition.

In Hartree unity by definition atomic units, the unit of length is a_0 which is equal to the radius of first Bohr orbit of hydrogen atom, and is usually called the Bohr and the unit of energy is Hartree (E_H) which is twice the energy of the ground state of the hydrogen atom. It is also conventional to rewrite the basic equations of quantum chemistry in dimensionless form. Thus, for example, a reduced length r' is given by $\frac{r}{a_0}$ and the reduced energy E' as $\frac{E}{E_H}$, respectively. Since on switching to the atomic units, $\hbar$, m_e, and e and $\frac{1}{4\pi\varepsilon_0}$ each have a numerical value of 1, the Schrödinger equation for the hydrogen-like atom

$$\left[-\frac{\hbar^2}{2m}\nabla^2 - \frac{1}{4\pi\varepsilon_0}\frac{Ze^2}{r}\right]\Psi(\mathbf{r}) = E\Psi(\mathbf{r}) \tag{1.12.1}$$

can be rewritten in a simplified form as

$$\left[-\frac{1}{2}\nabla'^2 - \frac{Z}{r'}\right]\Psi(\mathbf{r}') = E'\Psi(\mathbf{r}') \tag{1.12.2}$$

The atomic units of some common physical quantities and their equivalents in cgs and SI units are given in Table 1.4.

1.13 Approximate Methods of Solution of Schrödinger Equation

Schrödinger equation can be solved completely only for some very simple atomic and molecular systems like the hydrogen and hydrogen-like atoms, harmonic oscillator, rigid rotator, etc. For the vast majority of chemical applications, the Schrödinger's equation must be solved by approximate methods. The two primary approximation techniques that are used for the purpose are the perturbation theory and the variational method.

Table 1.4 Atomic units and equivalents in cgs and SI units

Quantity	Unit	cgs Equivalent	SI	Name
Charge	$e = 1$	4.803×10^{-10} esu	1.60217×10^{-19} C	Electron charge
Angular Momentum	$\hbar = 1$	1.05×10^{-27} erg s	1.05457×10^{-34} Js	"h-bar"
Mass	$m_e = 1$	9.11×10^{-28} g	9.10938×10^{-31} kg	Electron mass
Length	$a_0 = \frac{\hbar^2}{m_e e^2} = 1$	5.29×10^{-9} cm	5.2918×10^{-11} m	Bohr or "atomic unit"
Velocity	$\frac{e^2}{\hbar} = 1$	2.188×10^{8} cm/s	2.1877×10^{6} m/s	Velocity in first Bohr orbit
Energy	$\frac{m_e e^4}{\hbar^2} = 1$	4.36×10^{-11} ergs or 627.509 kcal/mol or 27.211 eV	4.3597×10^{-18} J	Hartree (=2 Rydbergs)
Electric field	$\frac{e}{a_0^2} = 1$	5.142×10^{9} V/cm	5.142×10^{11} V/m	Internal field of H atom
Electric constant^{-1}	$K_e = \frac{1}{4\pi\varepsilon_0} = 1$	–	8.9875×10^{9} kg m^3/s^2 c^2	Coulomb force constant

1.13.1 Perturbation Theory

The perturbation theory is used in those cases where the wave equation differs from the true one only in the omission of a few terms whose effect on the system is small. Thus, for example, to name a few, the hydrogen atom problem in the presence of electric or magnetic field, the helium atom problem, the anharmonic oscillator problem, and the Møller–Plesset corrections in the molecular orbital theory can been easily treated by this method.

In perturbation theory, the Hamiltonian is split into two parts—the one for which we know how to solve the equation and the other for which we do not know how to solve it. This theory is applicable only in those cases where the perturbation effects are very small.

We assume that the Hamiltonian $\widehat{H}$ can be expanded in terms of some parameter λ, yielding

$$\widehat{H} = \widehat{H}^{0} + \lambda\widehat{H}^{(1)} + \lambda^2\widehat{H}^{(2)} + \ldots, \tag{1.13.1}$$

where λ can be so chosen that the equation to which the above equation reduces when $\lambda \to 0$ can be solved directly.

As $\lambda \to 0$, the equation reduces to $\widehat{H} = \widehat{H}^{0}$ and hence the Schrödinger equation becomes

$$\widehat{H}^{0}\Psi^{(0)} = E^{(0)}\Psi^{(0)} \tag{1.13.2}$$

This equation is said to be the wave equation for the unperturbed system, while the terms $\lambda\widehat{H}^{(1)} + \lambda^2\widehat{H}^{(2)}$ are called the perturbations.

The problem is solved in two steps. In the first step, the eigenfunctions and eigenvalues of the unperturbed Hamiltonian $\widehat{H}^{(0)}$ are obtained:

$$\widehat{H}^{(0)}\Psi_n^{(0)} = E_n^{(0)}\Psi_n^{(0)} \tag{1.13.3}$$

In the second step, these eigenfunctions and eigenvalues are corrected to account for the effect of the perturbation. These corrections in perturbation theory are obtained as an infinite series of terms which become smaller and smaller for the well-behaved systems:

$$E_n = E_n^{(0)} + \lambda E_n^{(1)} + \lambda^2 E_n^{(2)} + \ldots \tag{1.13.4}$$

$$\Psi_n = \Psi_n^{(0)} + \lambda \Psi_n^{(1)} + \lambda^2 \Psi_n^{(2)} + \ldots \tag{1.13.5}$$

Quite frequently, the corrections are only taken through first or second order (i.e., superscripts (1) or (2)). According to perturbation theory, the first-order correction to the energy is

$$E_n^{(1)} = \int \Psi_n^{(0)*} \hat{H}^{(1)} \Psi_n^{(0)} d\tau \tag{1.13.6}$$

and the second-order correction is

$$E_n^{(2)} = \int \Psi_n^{(0)*} \hat{H}^{(1)} \Psi_n^{(1)} d\tau \tag{1.13.7}$$

Here, $\Psi_n^{(0)}$ is the normalized zeroth-order wavefunction and $\Psi_n^{(1)}$ is the first-order correction to the wavefunction.

In order to calculate the second-order correction in energy we need to know $\Psi_n^{(1)}$. Which can be written in terms of the zeroth-order wavefunction as:

$$\Psi_n^{(1)} = \sum_{i \neq n} \Psi_i^{(0)} \frac{\int \Psi_n^{(0)*} \hat{H}^{(1)} \Psi_i^{(0)} d\tau}{E_n^{(0)} - E_i^{(0)}} \tag{1.13.8}$$

Substituting this in the expression for $E_n^{(2)}$ (Eq. (1.13.7)), we obtain

$$E_n^{(2)} = \sum_{i \neq n} \frac{\left| \int \Psi_n^{(0)*} \hat{H}^{(1)} \Psi_i^{(0)} d\tau \right|^2}{E_n^{(0)} - E_i^{(0)}} \tag{1.13.9}$$

As, an example, we can consider the case of an anharmonic oscillator with Hamiltonian $\hat{H}$ given by the equation

$$\hat{H} = -\frac{\hbar^2}{2\mu} \frac{d^2}{dx^2} + \frac{1}{2} kx^2 + ax^3 + bx^4 \tag{1.13.10}$$

Comparison with Eq. (1.13.1) shows that the unperturbed Hamiltonian $\hat{H}^{(0)}$ is the same as that for a harmonic oscillator

$$\hat{H}^{(0)} = -\frac{\hbar^2}{2\mu} \frac{d^2}{dx^2} + \frac{1}{2} kx^2 \tag{1.13.11}$$

and

$$\hat{H}^{(1)} = ax^3 + bx^4 \tag{1.13.12}$$

is the perturbation. If the constants a and b are small, we expect the eigenfunctions and eigenvalues of the anharmonic oscillator to be closely related to those of the harmonic oscillator.

The exact solution of the harmonic oscillator problem gives for the ground state energy and wavefunction the expressions:

$$E_0^{(0)} = \frac{1}{2}\hbar\omega, \tag{1.13.13}$$

$$\Psi_0^{(0)}(x) = \left(\frac{\alpha}{\pi}\right)^{1/4} e^{-\alpha x^2/2}. \tag{1.13.14}$$

where $\alpha = \sqrt{\frac{k\mu}{\hbar^2}}$ and ω is the angular frequency.

The first-order correction to the ground state energy shall therefore be

$$E_0^{(1)} = \left(\frac{\alpha}{\pi}\right)^{1/2} \int_{-\alpha}^{\infty} (ax^3 + bx^4)e^{-\alpha x^2} dx \tag{1.13.15}$$

The first term in the integral in this equation vanishes because the integrand is odd. Hence, there is no contribution of ax^3 term to the energy of the harmonic oscillator. The second term containing bx^4, however, has a value $\frac{3b}{4\alpha^2}$ and so makes a contribution towards the ground state energy of the oscillator. The total corrected ground state energy of the harmonic oscillator, that is, the energy of the anharmonic oscillator, shall be

$$E = E_0^{(0)} + E_0^{(1)} = \frac{1}{2}\hbar\omega + \frac{3b\hbar^2}{4k\mu} \tag{1.13.16}$$

1.13.2 Variation Method

Variation method is a more robust method than the perturbation method. In contrast to the perturbation method, it is useful even in those cases where it is difficult to determine a good unperturbed Hamiltonian. It is based on the following theorem:

If Ψ is any well-behaved wavefunction such that $\int \Psi^*\Psi d\tau = 1$, and if the lowest eigenvalue of the operator $\hat{H}$ is E_0, then,

$$\int \Psi^* \hat{H} \Psi d\tau \geq E_0 \tag{1.13.17}$$

The function Ψ may be chosen arbitrarily but the more wisely it is chosen, the more closely E will approach the actual energy E_0. If we choose for our function Ψ, called the variation function, the true function Ψ_0 of the lowest state, E would be equal to E_0.

$$E = \int \Psi_0^* \hat{H} \Psi_0 d\tau = E_0$$

The variation method needs a trial wavefunction which may consist of some adjustable parameters called "variational parameters." These parameters are then adjusted until the energy of the trial wavefunction is minimized. The resulting trial wavefunction and its corresponding energy are then the upper bound approximations of the exact wavefunction and energy.

In practice, a trial wavefunction Ψ is expanded as a linear combination of a set of exact functions Φ_i, which may preferably form an orthonormal set.

$$\Psi = \sum_{i=1}^{N} c_i \Phi_i \tag{1.13.18}$$

where c_i represents the set of variational parameters.

If the wave equation for the system under consideration is $\widehat{H}\Psi = E\Psi$, we get

$$E = \frac{\int \Psi^* \widehat{H} \Psi d\tau}{\int \Psi^* \Psi d\tau} = \frac{\sum_{ij} c_i^* c_j \int \Phi_i^* \widehat{H} \Phi_j d\tau}{\sum_{ij} c_i^* c_j \int \Phi_i^* \Phi_j d\tau} \tag{1.13.19}$$

The equation can be simplified by using notation

$$H_{ij} = \int \Phi_i^* \widehat{H} \Phi_j d\tau \tag{1.13.20}$$

$$S_{ij} = \int \Phi_i^* \Phi_j d\tau \tag{1.13.21}$$

to yield

$$E = \frac{\sum_{ij} c_i^* c_j H_{ij}}{\sum_{ij} c_i^* c_j S_{ij}} \tag{1.13.22}$$

In order to find values of the variation parameters c_1, c_2,...c_N that minimize the energy we differentiate Eq. (1.13.22) with respect to the variational parameters and impose the condition

$$\frac{\partial E}{\partial c_k} = 0 \quad \text{for } k = 1, 2, \dots N$$

This leads to set of Eq. (1.13.23) that will provide nontrivial solution, if the determinant constructed from them equals 0.

$$\begin{pmatrix} H_{11} - ES_{11} & H_{12} - ES_{12} & \dots & H_{1N} - ES_{1N} \\ H_{21} - ES_{21} & H_{22} - ES_{22} & \dots & H_{2N} - ES_{2N} \\ \vdots & \vdots & \vdots & \vdots \\ H_{N1} - ES_{N1} & H_{N2} - ES_{N2} & \dots & H_{NN} - ES_{NN} \end{pmatrix} \begin{pmatrix} c_1 \\ c_2 \\ \vdots \\ c_N \end{pmatrix} = 0 \tag{1.13.23}$$

If an orthonormal set of functions ϕ_i is used such that $\int \Phi_i^* \Phi_j d\tau = \delta_{ij}$, then $\delta_{ij} = 1$ for $i = j$ and 0 if $i \neq j$. In this case, the secular determinant reduces to

$$\begin{vmatrix} H_{11} - E & H_{12} & \cdots & H_{1N} \\ H_{21} & H_{22} - E & \cdots & H_{2N} \\ \vdots & \vdots & \vdots & \vdots \\ H_{N1} & H_{N2} & \cdots & H_{NN} - E \end{vmatrix} = 0 \tag{1.13.24}$$

The secular determinant for N basis functions gives an N-order polynomial in E which is solved for different roots, each of which approximates a different eigenvalue. Several applications of the variation theory shall be discussed in subsequent chapters.

1.14 Molecular Symmetry

Symmetry plays an important role in the structure of molecules. Some systems are highly symmetrical, for example a sphere which is more symmetrical than a cube, and some have no symmetry. An action which transforms a body into its identical self without bringing about any deformation is called a symmetry operation. If a molecule has symmetry we can say something about its wavefunction and make some general conclusions about molecular properties without calculations. Thus, for example, we can decide about the presence of a permanent dipole moment or chirality in the molecule and know about the degeneracy of molecular states and the selection rules for transitions in polyatomic molecules. The mathematical manipulation of symmetry operations is a part of group theory for which detailed treatments are given in several excellent books. In particular see books by Cotton (1963) and Bishop (1993) for general treatment, Jaffe and Orchin (2003) for applications in electronic spectroscopy, and Colthup (1990) in vibrational spectroscopy (details given in bibliography). We shall here give only a very brief outline of some of its general principles.

1.14.1 Symmetry Elements

Symmetry element is a point, line, or plane with respect to which a symmetry operation is carried out. There are five kinds of symmetry elements and so also the symmetry operations. These are:

1. Identity (E). It is a trivial symmetry element, which corresponds to doing nothing. It is introduced for the purposes of mathematical group theory and is possessed by all molecules.
2. p-field rotation axis of symmetry designated as C_p. It corresponds to rotation through an angle $\frac{360°}{p}$. Taking the example of water (H_2O), a rotation by 180° about an axis dividing the HOH angle transforms the molecule to itself. So, we say it has a twofold axis C_2. Similarly, benzene (C_6H_6) has one sixfold axis C_6 perpendicular to the molecular plane and six twofold axes C_2 in the molecular plane. The axis having the largest value of p is called the principal axis.
3. Plane of symmetry—usually designated as σ with subscripts v, h, or d depending on whether the plane is a vertical, horizontal, or diagonal plane of symmetry. This corresponds to reflection in a mirror plane. If the plane contains the principal axis, it is called vertical, σ_v. If the plane is perpendicular to the principal axis, it is called horizontal, σ_h. The diagonal plane, σ_d, is a vertical plane that bisects the angle

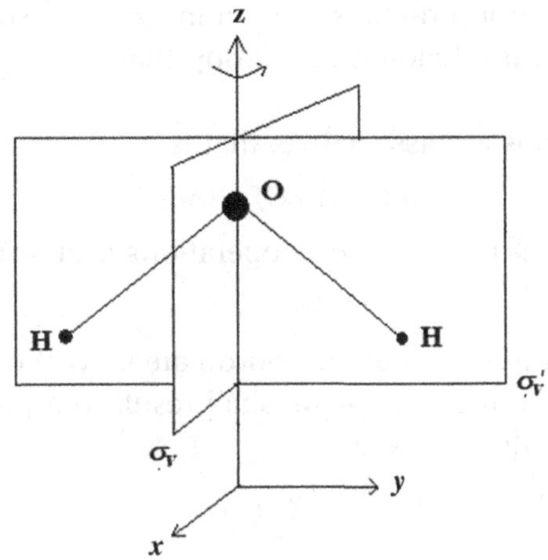

FIGURE 1.7 Symmetry elements for water molecule—$C_2(z)$ rotational axis and σ_v and σ_v' planes of symmetry.

between two C_2 axes. Thus, H_2O has two vertical planes of symmetry designated as σ_v and σ_v' (Figure 1.7), and CH_3Cl has three.

4. Center of symmetry—designated as *i*. A molecule has a center of symmetry, *i*, if by reflection at the center, the molecule transforms into a configuration indistinguishable from the original one. This operation transforms all atoms with coordinates (x, y, z) to identical atoms with coordinate $(-x, -y, -z)$. Typical examples are benzene, carbon dioxide.
5. *p*-fold rotation–reflection axis of symmetry designated as S_p. This is a combination of two successive operations—a rotation through $\frac{360°}{p}$ followed by reflection at a plane perpendicular to the axis of rotation. Neither operation alone need to be a symmetry operation. Thus, CH_4 molecule has three S_4 axes, and borontrifluoride has one S_3 axis.

1.14.2 Symmetry Point Groups

A set of operations together with the identity operation (E), which leaves the molecule unchanged forms a group. According to group theory, only certain combinations of symmetry elements which follow certain rules are possible. Such a restricted combination of symmetry elements that leaves at least one point unchanged is called a point group. All systems belonging to the same point group have their symmetry described by the same set of symmetry elements; thus, H_2O, F_2CO, CH_2Cl_2 all belong to the same point group C_{2v}. Some of the requirements for the symmetry elements to form a point group are:

1. One of the operations in the group is identity E, for which $EA = AE = A$, where A represents an element in the group.

2. For every element in a group there is also an inverse A^{-1} such that $A^{-1}A = I$, where I is a unit matrix. Thus, if A belongs to a group then, $A^{-1} = B$ will also belong to the same group.
3. The symmetry operations are associative, that is

$$ABC = A(BC) = (AB)C$$

4. Successive application of two symmetry operations is also a symmetry operation.

$$AB = C$$

For example, a twofold rotation (C_2) of the coordinate (x, y, z) of a point about the z-axis followed by a reflection about the xz plane shall result in a point that can also be obtained by reflection about the yz-plane.

Thus, $(X, Y, Z) \xrightarrow{C_2(Z)} (-X, -Y, Z) \xrightarrow{\sigma(xz)} (-X, Y, Z)$
Also, $(X, Y, Z) \xrightarrow{\sigma(yz)} (-X, Y, Z)$
Hence, $\sigma(xz)C_2(Z) = \sigma(yz)$

In general however, $AB \neq BA$ as may be seen from the following symmetry transformations:

$$(x, y, z) \xrightarrow{C_4(z)} (y, -x, z) \xrightarrow{C_2(x)} (-y, -x, -z)$$

and

$$(x, y, z) \xrightarrow{C_2(x)} (x, -y, -z) \xrightarrow{C_4(z)} (y, x, -z)$$

So, in this case, $C_2(x)C_4(z) \neq C_4(z)C_2(x)$

The effect of successive symmetry operations is sometimes also represented by a multiplication table. Thus, in the case of C_{2v} symmetry, there are four symmetry elements E, $C_2(z)$, $\sigma_v(xz)$, and $\sigma'_v(yz)$. It can be seen from the multiplication Table 1.5 that these elements fulfill the requirements for a group.

Table 1.5 can easily be verified for water Figure 1.7. It can be seen that the product of any two symmetry transformations leads to another member of the group.

Table 1.5 Multiplication table for C_{2v} symmetry operations[a]

C_{2v}	E	$C_2(z)$	$\sigma_v(xz)$	$\sigma'_v(yz)$
E	E	C_2	σ_v	σ'_v
$C_2(z)$	C_2	E	σ'_v	σ_v
$\sigma_v(xz)$	σ_v	σ'_v	E	C_2
$\sigma'_v(yz)$	σ'_v	σ_v	C_2	E

[a]The first operation at the top row to be followed by the second operation at the left column.

1.14.3 Classification of Point Groups

We can now list the possible symmetry point groups of molecules:

1. **Groups with no C_p axis.** In this case the molecule has no symmetry element except E. The molecules are then said to belong to group C_1. But, if has a plane of symmetry it belongs to C_s group.
2. **Groups with a single C_p axis.** Molecule with a single C_p axis belongs to group C_p ($p = 2, 3, 4...$). Thus, H_2O_2 and $CH_2Cl–CH_2Cl$ (gauche) belong to C_2 group.

   If in addition to the identity E, and p-fold axis of rotation the molecule also has p-vertical planes of symmetry each containing the rotation axis, it is said to belong to C_{pv} group. Thus, ammonia belongs to C_{3v} group as it has symmetry elements E, C_3 and three σ_v planes. Some other molecules belonging to this group are $CHCl_3$, $CCl_3–CBr_3$, etc.

   If besides E and the C_p axis it has a horizontal plane of symmetry, the molecule belongs to C_{ph} group. Thus, molecules like H_2O, Cl $H_2C–CH_2Cl$ (trans) and butadiene belong to C_{2h}, while H_3BO_3 and $B(OH)_3$ belong to C_{3h} group.

   If a molecule has only a p-fold rotation–reflection axis of symmetry, it is said to belong to S_p group. In molecules having a center of symmetry, it is also called C_i group, for example, *trans*-dichlorodibromoethane.

   All linear molecules which do not have a plane of symmetry perpendicular to molecular axis belong to point group $C_{\infty v}$ as they possess an infinite-fold axis and an infinite number of planes through it. Examples of this group are OCS, OCN^-, HCN, etc. However, linear molecules having center of symmetry belong to $D_{\infty h}$ group, for example, dicyanodiacetylene.
3. **Groups with p-fold principal axis C_p and pC_2 axes perpendicular to the C_p axis** constitute the D_p, D_{ph}, and D_{nd} group. For D_p group, the symmetry elements are C_p axis and nC_2 axes perpendicular to the C_p axis. D_1 is of course equivalent to C_2 and the molecules of this symmetry are classified as C_2.

   If in addition to D_p operations, the molecule also has a horizontal symmetry plane σ_h, it belongs to D_{nh} group. Benzene (C_6H_6) has the elements E, C_6, $3C_2$, $3C_2'$, and σ_h and so belongs to D_{6h} group.

   If in addition to D_p operations, the molecule possesses p diagonal mirror planes σ_d (planes) passing through the p-fold axis and bisecting the angles between two consecutive C_2 axes, it belongs to the D_{pd} group. Thus, allene $H_2C{=}C{=}CH_2$, where one CH_2 group is rotated by 90° with respect to the other, belongs to the D_{2d} group as it has the symmetry elements E, $3C_2$, S_4, $2\sigma_d$.
4. **Groups with more than one C_p axis, $p > 2$.** Many important molecules such as methane, CCl_4, SF_6, etc., have more than one principal axes. Thus, methane has four threefold axes ($4C_3$) and three mutually perpendicular twofold axes ($2C_2$). Such molecules form cubic groups such as tetrahedral group (T, T_d, T_h) and octahedral group (O, O_h). They possess rotational symmetry of tetrahedron or octahedron.

Molecules belong to the T group if they have four threefold and three mutually perpendicular twofold axes. If in addition, they have a center of symmetry, they are said to belong to T_h group. Molecules having symmetry elements of the T group and also two mutually perpendicular planes of symmetry through each twofold axis, or a total of six planes of symmetry, are said to belong to T_d group. A typical example is CH_4 having symmetry elements $3C_2$, $4C_3$, $3S_4$ (or six undesignated planes).

Molecules belonging to the O group have three mutually perpendicular C_4 axis and four C_3 axis, while those belonging to the O_h group have in addition a center of symmetry and nine planes of symmetry. Typical examples of this symmetry group are SF_6, $(COCl_6)^{-4}$, etc.

1.14.4 Representation of Point Groups and Character Tables

A representation of a group may be defined as a set of matrices each corresponding to a single symmetry operation in the group. These matrices obey the group multiplication table like Table 1.5. While the matrices may be one-, two-, or three-dimensional but more simply they are expressed as a set of numbers which are the traces (or sum of diagonal elements) of these representative matrices. In general, there can be a large number of such representations but they may not all be mutually independent. The simplest set of mutually independent representations of a point group is called an irreducible representation. The individual entries in a representation are called character and so the table of irreducible representation is called a character table. As an example, we consider the character table of the C_{2v} symmetry group, shown in Table 1.6.

The first row in the table shows the symmetry operations E, C_2, σ_v and σ'_v of the group and the first column shows the irreducible representations A_1, A_2, B_1, and B_2 of the group. List of characters for the individual irreducible representations: $+1$ and -1 are used to indicate the symmetric and antisymmetric transformations. The sixth column indicates the group order $h = 4$, and the simple functions of the coordinates x, y, z which belong to a certain irreducible representation. These coordinates help in understanding the species to which a normal vibrational, translational, or rotational mode or

Table 1.6 Character table for the C_{2v} point group

C_{2v}	E	$C_2(z)$	$\sigma_v(xz)$	$\sigma'_v(yz)$	$h = 4$
A_1	1	1	1	1	z, T_z
A_2	1	1	−1	−1	R_z
B_1	1	−1	1	−1	x, T_x, R_y
B_2	1	−1	−1	1	y, T_y, R_x

wavefunction may belong. The irreducible representations of a symmetry group have been given the name symmetry type or species which is commonly used in molecular spectroscopy. Some standard notations are used in the character tables to represent irreducible representations or symmetry species and the type of symmetry operation. These are:

A—for symmetric with respect to the principal axis of symmetry
B—for antisymmetric with respect to the principal axis of symmetry
E—for doubly degenerate vibrations or wavefunctions. These are represented by (2 × 2) matrix
F—for triply degenerate vibrations or wavefunctions.

In order to differentiate among the various representations of the same type, some subscripts or superscripts are used. These are:

Subscripts 1 and 2—to represent symmetric or antisymmetric with respect to the plane of symmetry.

Superscripts prime (′) and double prime (″)—to represent symmetric or antisymmetric with respect to a plane of symmetry.

Subscripts *g* and *u*—to represent symmetric or antisymmetric with respect to center of symmetry.

In case of linear molecules having symmetry group $C_{\infty v}$ or $D_{\infty h}$, the notations used are:

Σ^+—to represent symmetric with respect to plane of symmetry through the molecular axis
Σ^-—to represent antisymmetric with respect to plane of symmetry through the molecular axis
π, Δ, ϕ—to represent degenerate vibrations or wavefunctions of order (degree 2, 3, or 4), in increasing order.

In addition, subscripts g and u are used to show symmetry with respect to center of inversion.

Let us now try to understand the concepts of representations and characters given in Table 1.6 for the C_{2v} group by the examples of normal vibrational modes and the electronic wavefunctions of water molecule.

1.14.4.1 Symmetry of Normal Vibrations of Water Molecule

Water is a three-atom molecule belonging to the C_{2v} point group having three vibrational, three rotational, and three translational degrees of freedom. For the sake of simplicity let us confine only to the vibrational modes and consider the effect of symmetry operations E, $C_2(z)$, $\sigma_v(xz)$, and $\sigma'_v(yz)$, taking yz as the molecular plane. This is given in Figure 1.8.

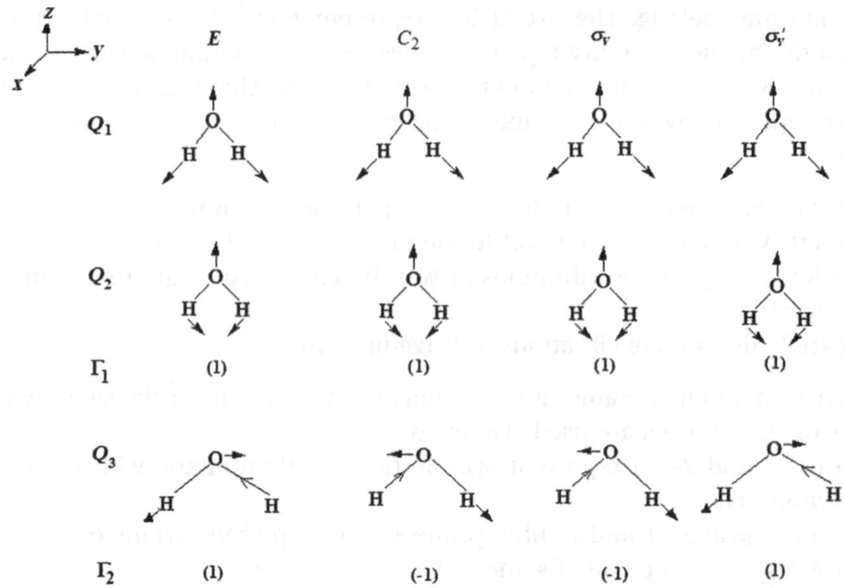

FIGURE 1.8 Symmetry operations on the Q_1, Q_2, and Q_3 vibrational modes of water. Arrows show displacement vectors for the three atoms.

If follows from the figure that there is no effect of symmetry operations on the forms of Q_1 and Q_2 whereas in case of Q_3, the symmetry operations C_2 and σ_v reverse the displacement vectors. This may be mathematically written as

$$Q_i \xrightarrow{\text{Symmetry operation}} (\pm 1)Q_i.$$

In this case, ($\pm$1) may be called transformation number which indicates the effect of symmetry operation. Thus, for the case of nondegenerate vibrations, the normal coordinate is either symmetrical (+1) or antisymmetrical ($-$1) with respect to each vibrational mode. The normal modes Q_1 and Q_2 may be given the same representation Γ_1 and the mode Q_3 another representation Γ_2.

	E	C_2	σ_v	σ'_v
$\Gamma_1(A_1)$	(1)	(1)	(1)	(1)
$\Gamma_2(B_2)$	(1)	($-$1)	($-$1)	(1)

A comparison with Table 1.6 shows that Γ_1 and Γ_2 belong to species A_1 and B_2, respectively. Thus, water molecule has three vibrational modes belonging to species $2A_1 + B_2$.

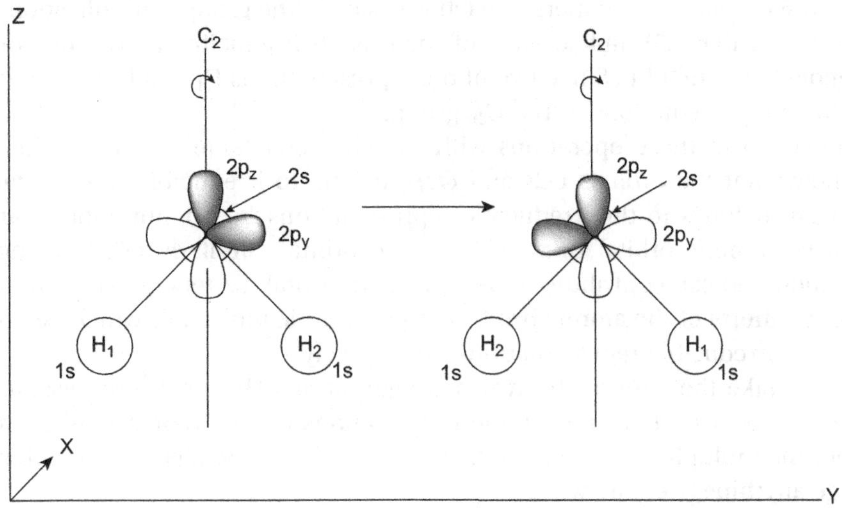

FIGURE 1.9 Symmetry transformation of 2*s* and 2*p* orbitals of oxygen and 1*s* orbital of hydrogen under $C_2(z)$ operation ($2p_x$ not shown).

1.14.4.2 Symmetry of Electronic Orbitals of Water Molecule

Let us now consider the affect of various operations of the C_{2v} group on the 2*s*, $2p_x$, $2p_y$, and $2p_z$ orbitals of the oxygen atom and the 1*s* orbitals of the hydrogen atoms of the water molecule. The effect of $C_2(z)$ operation on these orbitals is shown in Figure 1.9.

The effect of the symmetry operations $E, C_2(z), \sigma_v(xz)$, and $\sigma'_v(yz)$ of the C_{2v} group is as follows:

$$E(2s)\rightarrow 2s, \quad C_2(2s)\rightarrow 2s, \quad \sigma_v(2s)\rightarrow 2s, \quad \sigma'_v(2s)\rightarrow 2s$$

$$E(2p_x)\rightarrow 2p_x, \quad C_2(2p_x)\rightarrow -2p_x, \quad \sigma_v(2p_x) = 2p_x, \quad \sigma'_v(2p_x) = -2p_x$$

$$E(2p_y)\rightarrow 2p_y, \quad C_2(2p_y)\rightarrow -2p_y, \quad \sigma_v(2p_y)\rightarrow -2p_y, \quad \sigma'_v(2p_y)\rightarrow 2p_y$$

$$E(2p_z)\rightarrow 2p_z, \quad C_2(2p_z)\rightarrow 2p_z, \quad \sigma_v(2p_z)\rightarrow 2p_z, \quad \sigma'_v(2p_z)\rightarrow 2p_z$$

The transformation numbers for these operations can be represented in the following tabular form:

	E	**$C_2(z)$**	**$\sigma_v(xz)$**	**$\sigma'_v(yz)$**	
O2s	1	1	1	1	Γ_1
O2p_x	1	−1	1	−1	Γ_2
O2p_y	1	−1	−1	1	Γ_3
O2p_z	1	1	1	1	Γ_4

It can be seen that the numbers in each row satisfy the group multiplication table of the C_{2v} group (Table 1.5) and so each of them is an irreducible representation of the group denoted by symbol Γ. Since two of the representations Γ_1 and Γ_4 are identical, we get only three representations of the C_{2v} group.

A comparison of these operations with the character table (Table 1.6) for the C_{2v} groups shows that the orbitals $O2s$ and $O2p_z$ belong to irreducible representation A_1, whereas $O2p_x$ belongs to the irreducible representations B_1. So, any molecular orbital built with the atomic orbital $O2p_x$ will be a b_1 orbital. Similarly $O2p_y$ belongs to B_2 representation and can contribute to b_2 molecular orbital. Letters a_1, a_2, b_1, b_2 are used to denote symmetry of the atomic orbitals in polyatomic molecules which belong to A_1, A_2, B_1, and B_2 irreducible representations, respectively.

If now we take the case of the two hydrogen atoms H_1 and H_2 represented by $1s$ orbitals, it can be seen that the symmetry operations can be represented by matrices which obey the multiplication table of the C_{2v} group. Since the identity operation E does not change anything, we may write

$$E\begin{pmatrix} H_1 \\ H_2 \end{pmatrix} = \begin{pmatrix} 1 & 0 \\ 0 & 1 \end{pmatrix}\begin{pmatrix} H_1 \\ H_2 \end{pmatrix}$$

The $C_2(z)$ rotation interchanges the positions of H_1 and H_2, so

$$C_2\begin{pmatrix} H_1 \\ H_2 \end{pmatrix} = \begin{pmatrix} 0 & 1 \\ 1 & 0 \end{pmatrix}\begin{pmatrix} H_1 \\ H_2 \end{pmatrix}$$

and similarly,

$$\sigma_v\begin{pmatrix} H_1 \\ H_2 \end{pmatrix} = \begin{pmatrix} 0 & 1 \\ 1 & 0 \end{pmatrix}\begin{pmatrix} H_1 \\ H_2 \end{pmatrix}$$

and

$$\sigma'_v\begin{pmatrix} H_1 \\ H_2 \end{pmatrix} = \begin{pmatrix} 1 & 0 \\ 0 & 1 \end{pmatrix}$$

It can be seen that the (2×2) matrices representing the operations E, C_2, σ_v, and σ'_v satisfy the multiplication rule of the C_{2v} group and so constitute a representation.

1.14.5 Symmetry Properties of Eigenfunctions of Hamiltonian

The connection between the eigenfunctions of the Hamiltonian and the representations of the group can be expressed mathematically. It is based on the fact that the Hamiltonian for a molecule is invariant with respect to three types of symmetry operations:

1. Symmetry arising out of permutation of electrons. Due to indistinguishability of electrons their permutation changes identical terms in the Hamiltonian, leaving the Hamiltonian unchanged.

2. Spin symmetry arising out of the fact that the Hamiltonian does not contain spin-dependent terms.
3. Spatial symmetry. A symmetry operation such as rotation about an axis of symmetry or reflection about a plane of symmetry interchanges identical nuclei leaving the Hamiltonian invariant.

Thus, for example, if R is a symmetry operation (say, σ_v) in H_2O which interchanges the hydrogen atom, then there will be no change in the potential field acting on the electrons. The Hamiltonian H of H_2O will therefore remain invariant under such symmetry operation. So, if

$$H\Psi_k = E_k\Psi_k \tag{1.14.1}$$

Then,

$$\mathrm{R}H\Psi_k = \mathrm{R}E_k\Psi_k \tag{1.14.2}$$

Since R leaves H unchanged and E_k is a constant

$$H(\mathrm{R}\Psi_k) = E_k(\mathrm{R}\Psi_k) \tag{1.14.3}$$

So, the function $\mathrm{R}\Psi_k$ is also an eigenfunction of H with the same eigenvalue E_k. It, therefore, follows that if E_k is a nondegenerate eigenvalue with eigenfunctions Ψ_k then $\mathrm{R}\Psi_k$ must either be equal to Ψ_k or just differ from it by a change of sign, i.e., $\mathrm{R}\Psi_k = \Psi_k$ or $-\Psi_k$. If E_k is an m-fold degenerate eigenvalue with orthonormal set of eigenfunctions $\Psi_k^{(i)}(i = 1, 2, \ldots m)$ then $\mathrm{R}\Psi_k$ will also belong to the same set or shall be some linear combination of the members of set.

$$\mathrm{R}\Psi_k^{(i)} = \sum_{i=1}^{m} C_{ki}\Psi_k^{(i)} \tag{1.14.4}$$

where $\Psi_k^{(1)}, \Psi_k^{(2)}, \ldots \Psi_k^{(m)}$ are members of the degenerate set Ψ_k. Same conclusions can be derived by using the properties of commuting operators (Section 1.6.2).

The eigenfunctions are therefore restricted by the symmetry of the molecule. It can be proved that they have the symmetry properties of the irreducible representations of the group.

This property of the eigenfunctions has been used to construct molecular orbitals and predict the allowed and forbidden transitions. Group theory is a powerful mathematical tool and has been used for many important applications in chemistry and spectroscopy.

Further Reading

[1] C.A. Coulson, Valence, second ed., Oxford University Press, 1961.

[2] L. Pauling, E.B. Wilson, Introduction to Quantum Mechanics with Applications to Chemistry, McGraw Hill, 1935.

[3] L.I. Schiff, Quantum Mechanics, McGraw Hill Co., 1968.

[4] D.A. McQuarrie, Quantum Chemistry, University Science Books, Mill Valey, CA, 1983.

[5] E. Merzbacher, Quantum Mechanics, second ed., Wiley, New York, 1970.

[6] I.N. Levine, Quantum Chemistry, fourth ed., Prentice Hall, Englewood Cliffs, NJ, 1991.

[7] A. Szabo, N.S. Ostlund, Modern Quantum Chemistry: Introduction to Advanced Electronic Structure Theory, McGraw-Hill, New York, 1989.

[8] F.A. Cotton, Chemical Applications of Group Theory, Wiley (Inter science), New York, 1963.

[9] H.H. Jaffe, M. Orchin, Symmetry in Chemistry, Dover, 2003. David M. Bishop, Group Theory, Dover, 1993.

[10] A. Vincent, Molecular Symmetry and Group Theory: A Programmed Introduction to Chemical Applications, second ed., 2000.

[11] N.B. Colthup, L.H. Daly, S.E. Wiberley, Introduction to Infrared and Raman Spectroscopy, third ed., Academic Press, 1990.

2

Many-Electron Atoms and Self-consistent Fields

CHAPTER OUTLINE

In Chapter 1 we discussed the development and solution of the time-independent Schrödinger equation for a one-electron atom. We shall now extend the treatment to a many-electron system.

2.1 Wavefunction of Many-Electron Atoms

In a many-electron system, we need to include the electron repulsion in the potential energy term of the wave-equation. Thus, the potential of a many-electron atom having nuclear charge $+Ze$, in atomic unit representation is

$$V = -\sum_i \frac{Z}{r_i} + \frac{1}{2}\sum_{i \neq j} \frac{1}{r_{ij}} \tag{2.1.1}$$

where, r_i the distance of the ith electron from the nucleus and r_{ij} is the interelectronic distance. A factor of $\frac{1}{2}$ in the second term is included to avoid each r_{ij} term being counted twice in the summation. The Hamiltonian, H, for this system is

$$H = -\frac{1}{2}\sum_i \nabla_i^2 - \sum_i \frac{Z}{r_i} + \frac{1}{2}\sum_{i \neq j} \frac{1}{r_{ij}} \tag{2.1.2}$$

Principles and Applications of Quantum Chemistry. http://dx.doi.org/10.1016/B978-0-12-803478-1.00002-9

The first term in the equation is the kinetic energy operator for the electron, the second one-electron term corresponds to electron–nucleus attraction, and the third term is sum of two electron repulsions.

The Schrödinger equation for an N-electron atom may then be written as

$$H\Psi(1,2,\ldots N) = E\Psi(1,2,\ldots N) \tag{2.1.3}$$

with H given by Eq. (2.1.2). The solution to this equation is not straightforward due to the presence of electron repulsion terms as, unlike hydrogen atom, it is not possible to separate out the variables. Only if we assume that the electrons are independent of each other we can approximate the Hamiltonian, H, as the sum of one-electron Hamiltonians,

$$\hat{H} = \sum_i \hat{H}_i = -\frac{1}{2}\sum_i \nabla_i^2 - \sum_i V_i \tag{2.1.4}$$

In order to guess a solution of Eq. (2.1.3), let us ignore the electron repulsion term. The equation may then be written as a sum of one-electron Hamiltonians, H_1, H_2,...H_N

$$(H_1 + H_2\ldots H_N)\Psi(1,2,\ldots N) = E\Psi(1,2,\ldots N) \tag{2.1.5}$$

We can now solve the equation by the standard "separation of variables" technique used earlier for the hydrogen atom problem.

Assume that,

$$\Psi(1,2,\ldots N) = \Phi_1(1)\Phi_2(2)\ldots\Phi_N(N) \tag{2.1.6}$$

where $\Phi_1(1)$, $\Phi_2(2)$,...$\Phi_N(N)$ are the one-electron orbitals and $\Phi_i(j)$ shows that j^{th} electron is in the ith orbital. Substituting Eq. (2.1.6) into (2.1.5), we get

$$\frac{1}{\Phi_1(1)}H_1\Phi_1(1) + \frac{1}{\Phi_2(2)}H_2\Phi_2(2) + \ldots + \frac{1}{\Phi_N(N)}H_N\Phi_N(N) = E \tag{2.1.7}$$

Each term on the left-hand side is independently variable and so each of them must individually be equal to a constant.

$$\begin{aligned} H_1\Phi_1(1) &= E_1\Phi_1(1) \\ H_2\Phi_2(2) &= E_2\Phi_2(2) \\ &\vdots \\ H_N\Phi_N(N) &= E_N\Phi_N(N) \end{aligned}$$

Thus, $E = E_1 + E_2 + \ldots E_N$ and,

$$H\Psi(1,2,\ldots N) = (E_1 + E_2 + \ldots E_N)\Psi(1,2,\ldots N) \tag{2.1.8}$$

Here, $\Psi(1,2,\ldots N)$ is defined by Eq. (2.1.6).

We thus recover an orbital picture. A many-electron wavefunction can be given as a product of orbitals as in Eq. (2.1.6), also known as Hartree product. However, since we cannot ignore the electron repulsion terms the many-electron Hamiltonian operator cannot be written simply as a sum of one-electron operators; due to the term $\frac{1}{r_{ij}}$ which depends on the instantaneous relative coordinates of the two electrons i and j, the total wavefunction cannot be written as a simple product of orbitals. While constructing many-electron wavefunction we need to include electron spin and thus use spin orbitals.

Also, we need to take into account the symmetry property of the electronic wavefunction under interchange of electron coordinates, also known as *Pauli's exclusion principle.*

Since electrons are essentially indistinguishable particles, no physical property of the system can be affected if we simply renumber or rename the electrons. If we consider the many-electron density function, $\rho(1, 2, \ldots N)$,

$$\rho(1,2,\ldots N) = |\Psi(1,2,\ldots N)|^2$$

it shall not be affected by the interchange of two electrons.

The indistinguishability of the electrons also leads to symmetry properties of the wavefunction. Suppose P_{ij} is a permutation operator which interchanges all the coordinates (including spin coordinates) of electrons i and j, then

$$P_{ij}\Psi(1,2,\ldots i,j\ldots N) = \Psi(1,2,\ldots j,i\ldots N) \tag{2.1.9}$$

$$P_{ij}^2\Psi(1,2,\ldots i,j\ldots N) = P_{ij}\Psi(1,2,\ldots j,i\ldots N) = \Psi(1,2,\ldots i,j\ldots N) \tag{2.1.10}$$

or $P_{ij} = \pm 1$

Hence

$$\Psi(1,2,\ldots i,j\ldots N) = \pm\Psi(1,2,\ldots j,i\ldots N) \tag{2.1.11}$$

Thus, on interchanging two electrons, the wavefunction, Ψ, changes only by a factor of $+1$ or -1. In the former case, the wavefunction Ψ is said to be symmetric with respect to interchange and in the latter case, it is antisymmetric. These are the only two possibilities compatible with the invariance of Ψ^2. The antisymmetric property is appropriate for electrons, since it leads to the Pauli's exclusion principle which states that no two electrons can be assigned to identical spin orbitals.

Since a single product function like Eq. (2.1.6) does not satisfy the antisymmetry principle, it cannot be a suitable approximate form to use. A combination of two functions can however, satisfy this criterion. As an example, consider a two-electron atom, say the excited helium atom, in which the two lowest energy orbitals are, say, $1s$ and $2s$. Since the electrons are identical any one of them can be either in $1s$ or $2s$ orbital. If there is only one electron in each orbital, then on ignoring electron spin, the two-electron wavefunction is

$$\Psi(1,2) = \Phi_{1s}(1)\Phi_{2s}(2)$$

If we apply the two-electron permutation operator, then

$$P_{12}\Psi(1,2) = \Phi_{1s}(2)\Phi_{2s}(1)$$

Clearly, $P_{12}\Psi(1,2)$ is not the negative of $\Psi(1,2)$. However, a combination of Hartree products may be constructed which has antisymmetry. Consider now a wavefunction like

$$\Psi(1,2) = \Phi_{1s}(1)\Phi_{2s}(2) - \Phi_{1s}(2)\Phi_{2s}(1) \tag{2.1.12}$$

Operation by the operator P_{12} shall, therefore, give

$$P_{12}\Psi(1,2) = \Phi_{1s}(2)\Phi_{2s}(1) - \Phi_{1s}(1)\Phi_{2s}(2) = -[\Phi_{1s}(1)\Phi_{2s}(2) - \Phi_{1s}(2)\Phi_{2s}(1)] = -\Psi(1,2) \quad (2.1.13)$$

The wavefunction, therefore, changes sign. The correct form of wavefunction for the two-electron system under consideration will therefore be given by Eq. (2.1.12).

If we now include electron spin and consider that the electron in $1s$ orbital has spin up (α) and in $2s$ orbital spin down (β), we may write the spin orbital for the first electron as $\Phi_{1s}(1)\alpha(1)$ and for the second electron as $\Phi_{2s}(2)\beta(2)$. In this case, the Hartree product for a singlet state is

$$\Psi(1,2) = \Phi_{1s}(1)\alpha(1)\Phi_{2s}(2)\beta(2) \quad (2.1.14a)$$

and

$$\Psi(2,1) = \Phi_{1s}(2)\alpha(2)\Phi_{2s}(1)\beta(1) \quad (2.1.14b)$$

We may then construct an antisymmetric function as

$$\Psi = \Phi_{1s}(1)\alpha(1)\Phi_{2s}(2)\beta(2) - \Phi_{1s}(2)\alpha(2)\Phi_{2s}(1)\beta(1) \quad (2.1.15a)$$

or in the normalized form

$$\Psi = \frac{1}{\sqrt{2}}[\Phi_{1s}(1)\alpha(1)\Phi_{2s}(2)\beta(2) - \Phi_{1s}(2)\alpha(2)\Phi_{2s}(1)\beta(1)] \quad (2.1.15b)$$

This will be the correct form of antisymmetric wavefunction inclusive of electron spin. The wavefunction can also written in matrix form as

$$\Psi = \frac{1}{\sqrt{2}}\begin{vmatrix} \Phi_{1s}(1)\alpha(1) & \Phi_{2s}(1)\beta(1) \\ \Phi_{1s}(2)\alpha(2) & \Phi_{2s}(2)\beta(2) \end{vmatrix} \quad (2.1.16)$$

2.2 Slater Determinants for Wavefunctions

The form of wavefunction given by Eq. (2.1.16) can be generalized to describe wavefunctions with any number of electrons. Thus, the general form for a $2N$-electron system with two electrons per spatial orbital can be written as a determinant of $2N$ spin orbitals.

$$\Psi(1,2,\ldots N) = \frac{1}{\sqrt{2N!}}\begin{vmatrix} \Phi_1(1)\alpha(1) & \Phi_1(1)\beta(1) & \Phi_2(1)\alpha(1) & \ldots & \Phi_N(1)\beta(1) \\ \Phi_1(2)\alpha(2) & \Phi_1(2)\beta(2) & \ldots & \ldots & \ldots \\ \vdots & \vdots & \vdots & \vdots & \vdots \\ \Phi_1(2N)\alpha(2N) & \ldots & \ldots & \ldots & \Phi_N(2N)\beta(2N) \end{vmatrix} \quad (2.2.1)$$

with the normalization constant appropriately adjusted. In this determinant, the various spin orbitals appear in the rows and the various electrons in the columns. Such determinants of spin orbitals are called Slater determinants [1]. A single Slater determinant is the simplest orbital wavefunction which satisfies the antisymmetry principle. In these determinants one electron is allocated to each row or column.

Several abbreviations are in common use in the literature for writing Slater determinants. One example is the replacement of α and β spin functions by unbarred and barred spatial functions. For example, $\Phi_j(j)\alpha(j)$ is represented by $\Phi_j(j)$ and $\Phi_j(j)\beta(j)$ by $\overline{\Phi}_j(j)$. Equation (2.2.1) may then be written as

$$\Psi(1,2,\ldots N) = \frac{1}{\sqrt{2N!}}\begin{vmatrix} \Phi_1(1) & \overline{\Phi}_1(1) & \Phi_2(1) & \ldots & \overline{\Phi}_N(1) \\ \Phi_1(2) & \overline{\Phi}_1(2) & \ldots & \ldots & \ldots \\ \vdots & \vdots & \vdots & \vdots & \vdots \\ \Phi_1(2N) & \overline{\Phi}_1(2N) & \ldots & \ldots & \overline{\Phi}_N(2N) \end{vmatrix} \tag{2.2.2}$$

or in a notational form as

$$\Psi(1,2,\ldots N) = \left|\Phi_1\overline{\Phi}_1\ldots\Phi_N\overline{\Phi}_N\right| \tag{2.2.3}$$

Several important conclusions for the orbital wavefunctions follow from the properties of a determinant. The antisymmetry property of a wavefunction follows from the fact that the interchange of two rows changes sign of determinant. Similarly, the Pauli's exclusion principle follows from the fact that a determinant with two equivalent rows or columns vanish. So, a nonzero function cannot be constructed if two electrons are assigned to the same spin orbital. Further, a determinant may be subjected to orthogonal transformation. This property allows transformation of molecular orbitals delocalized over an entire molecule into orbitals localized in regions associated with a classical bond [2].

2.3 Central Field Approximation

The methods that are used to describe many-electron atoms are based on the central field approximation. In this approximation, the interaction of all electrons is replaced by an averaged charge distribution acting on one-electron. These charge distributions are nearly spherically symmetric. The procedure used in quantum chemistry is to assume an initial charge distribution and then to solve the Schrödinger equation for one electron moving in the averaged field of all the other electrons. The functional solution for each electron is then used to recompute the average electron distribution. This procedure is repeated until a final distribution is obtained that agrees to some desired degree of accuracy with the preceding approximation. In other words, the solution to the problem is obtained iteratively.

As an illustration of this method, let us consider a two-electron atom problem, label these electrons as 1 and 2, and ignore electron spin. If electron 1 occupies orbital Φ_1, it shall have charge density $\rho_1 = |\Phi_1(1)|^2$ and a charge of $-e\rho_1$. Electron 2 occupies orbital Φ_2.

The potential seen by electron 2 in the field of electron 1 expressed in atomic units (where $e = 1$) shall be

$$V = -\frac{Z}{r_2} + \int \frac{\rho_1(1)}{r_{12}}\,d\tau_1 = -\frac{Z}{r_2} + \int \frac{\left|\Phi_1^2(1)\right|^2}{r_{12}}\,d\tau_1 \tag{2.3.1}$$

where r_2 is the distance between electron 2 and the atomic nucleus and r_{12} is the interelectron distance. We can then write the one-electron eigenvalue equation for electron 2 as,

$$\left(-\frac{1}{2}\nabla_2^2 + V\right)\Phi_2(2) = E\Phi_2(2)$$

or,

$$H_2^{eff}\Phi_2(2) = E\Phi_2(2) \tag{2.3.2}$$

where H_2^{eff} is the effective Hamiltonian for electron 2.

Obviously we need to know Φ_1 in order to calculate Φ_2 and so some kind of iterative calculation shall be necessary in order to calculate both Φ_1 and Φ_2.

2.4 Self-consistent Field (SCF) Approximation—Hartree Theory

We can now extend central field approximation to a many-electron system and develop the Hartree theory which is a powerful method to calculate the energy and wavefunction using iterative process. If we ignore Pauli's exclusion principle, the wavefunction can be written as a product of spatial orbitals:

$$\Psi = \Phi_1(1)\Phi_2(2)\ldots\Phi_N(N)$$

where

$$\Phi_i = Y_{lm}(\theta_i, \phi_i)u(s_i)$$

$u(s_i)$ is the spin function of the ith electron. The effective Hamiltonian associated with each electron is now the total Hamiltonian averaged over the other electrons. Each electron will have a kinetic energy, Coulomb attraction to the nucleus and repulsion by other electrons. The average repulsion by each electron (say, electron 1) due to charge distribution associated with another electron (say, electron j) is given by the integral of the product of $\frac{1}{r_{1j}}$ and the charge density of electron j. The charge density on electron j in atomic units is

$$\rho_j = |\Phi_j(j)|^2$$

The average repulsion induced in electron 1 by electron j is then

$$V_{1j} = \int |\Phi_j(j)|^2 \frac{1}{r_{1j}} d\tau_j = J_j(1) \tag{2.4.1}$$

where r_{1j} is the distance between the electrons 1 and j and J_j is called a Coulomb operator or Coulomb integral. The integration in Eq. (2.4.1) is taken over the coordinates of electron j.

The overall effective one-electron Hamiltonian for the electron 1 is then

$$\hat{H}_1^{eff} = -\frac{1}{2}\nabla_1^2 - \frac{Z}{r_1} + \sum_{j\neq 1}\int \frac{|\Phi_j(j)|^2}{r_{j1}} d\tau_j = -\frac{1}{2}\nabla_1^2 - \frac{Z}{r_1} + \sum_{j\neq 1} J_j(1) \tag{2.4.2}$$

where $\Phi_j(j)$ is the spatial function for the jth electron.

The variationally best wavefunction for electron 1 can be obtained by solving the one-electron Schrödinger equation.

$$\hat{H}_1^{eff}(r_1)\Phi_1(r_1) = \varepsilon_1\Phi_1(r_1) \tag{2.4.3}$$

To solve this equation we need wavefunction, Φ_j, of all the electrons except electron 1. However, to get those functions we need to solve equations similar to Eq. (2.4.3). For this, an iteration process is adopted. In the first step, a guess for all the wavefunctions, Φ_j, is made and these are used in Schrödinger equation like Eq. (2.4.3) for each orbital. These equations are then solved and the results are used to redefine new one-electron Hamiltonians and to solve equations equivalent to Eq. (2.4.3) once again. This iteration process is continued until the orbitals and energies that we get from one step of iteration to the next are the same, within some tolerance. This procedure is called the *Self-consistent field (SCF) approximation*. This theory which uses spatial orbitals is known as the Hartree theory.

The shortcoming of the Hartree method is that it does not use electron spin explicitly and hence neglects electron correlation terms. The wavefunction, being a simple product of one-electron wavefunctions, is not properly antisymmetric and so does not obey Pauli principle. In addition, the one-electron orbitals are not mutually orthogonal. Fock modified the Hartree theory to include electron spin and used spin orbitals instead of the spatial orbitals. The Hartree method [3] and the Hartree–Fock method [4] are the basic methods which give the most sophisticated theoretical results for many-electron atoms and molecules.

2.4.1 Hartree–Fock Method

Hartree–Fock (HF) method uses spin orbitals instead of the spatial orbitals in SCF theory proposed by Hartree. In order to gain an insight of the Hartree–Fock method we shall first consider a two-electron system in which the wavefunction is expressed as an antisymmetrized product of spin orbitals, Φ_1 and Φ_2:

$$\Psi = \frac{1}{\sqrt{2}}[\Phi_1(1)\Phi_2(2) - \Phi_2(1)\Phi_1(2)] = \frac{1}{\sqrt{2}}\begin{vmatrix} \Phi_1(1) & \Phi_2(1) \\ \Phi_1(2) & \Phi_2(2) \end{vmatrix} \tag{2.4.4}$$

Here, we also presume that the spin orbitals are exact solution of the Schrödinger equation. This is however, not true in most of the cases.

The Schrödinger equation for the two-electron atom is

$$\hat{H}\Psi = E\Psi \tag{2.4.5a}$$

$$E = \left\langle \Psi | \hat{H} | \Psi \right\rangle \tag{2.4.5b}$$

where

$$\hat{H} = -\frac{1}{2}\nabla_1^2 - \frac{1}{2}\nabla_2^2 - \frac{Z}{r_1} - \frac{Z}{r_2} + \frac{1}{r_{12}} \tag{2.4.6}$$

The first four terms depend upon the coordinates of only one electron but the last term depends upon the coordinates of two electrons. In order to calculate E, we need to calculate expectation values of all the operators.

$$E = \left\langle \hat{H} \right\rangle = -\frac{1}{2}\langle \nabla_1^2 \rangle - \frac{1}{2}\langle \nabla_2^2 \rangle - z\left\langle \frac{1}{r_1} \right\rangle - z\left\langle \frac{1}{r_2} \right\rangle + \left\langle \frac{1}{r_{12}} \right\rangle \tag{2.4.7}$$

Using Eq. (2.4.4), we get

$$\langle \nabla_1^2 \rangle = \langle \Psi | \nabla_1^2 | \Psi \rangle = \frac{1}{2}\langle \Phi_1(1)\Phi_2(2) - \Phi_2(1)\Phi_1(2) | \nabla_1^2 | \Phi_1(1)\Phi_2(2) - \Phi_2(1)\Phi_1(2) \rangle \tag{2.4.8}$$

On expanding Eq. (2.4.8) and remembering that ∇_1^2 and ∇_2^2 operate on functions of only electron 1 and 2, respectively, and after taking into account the condition of orthonormality of functions, namely

$$\langle \Phi_1(1)|\Phi_1(1)\rangle = \langle \Phi_2(2)|\Phi_2(2)\rangle = 1$$

and

$$\langle \Phi_1(1)|\Phi_2(1)\rangle = \langle \Phi_1(2)|\Phi_2(2)\rangle = 0$$

we get,

$$\langle \nabla_1^2 \rangle = \frac{1}{2}\left[\langle \Phi_1(1)|\nabla_1^2|\Phi_1(1)\rangle + \langle \Phi_2(1)|\nabla_1^2|\Phi_2(1)\rangle\right] \tag{2.4.9}$$

Similarly,

$$\langle \nabla_2^2 \rangle = \frac{1}{2}\left[\langle \Phi_2(2)|\nabla_2^2|\Phi_2(2)\rangle + \langle \Phi_1(2)|\nabla_2^2|\Phi_1(2)\rangle\right] \tag{2.4.10}$$

$$\langle 1/r_1 \rangle = \frac{1}{2}\left[\langle \Phi_1(1)|1/r_1|\Phi_1(1)\rangle + \langle \Phi_2(1)|1/r_1|\Phi_2(1)\rangle\right] \tag{2.4.11}$$

$$\langle 1/r_2 \rangle = \frac{1}{2}\left[\langle \Phi_2(2)|1/r_2|\Phi_2(2)\rangle + \langle \Phi_1(2)|1/r_2|\Phi_1(2)\rangle\right] \tag{2.4.12}$$

It may be noted that Eqs (2.4.9) and (2.4.10) are equivalent due to indistinguishability of electrons. For the same reason, the Eqs (2.4.11) and (2.4.12) are also equivalent. This would simplify Eq. (2.4.7).

Finally,

$$\begin{aligned}\langle 1/r_{12} \rangle = \frac{1}{2}&\left[\langle \Phi_1(1)\Phi_2(2)|1/r_{12}|\Phi_1(1)\Phi_2(2)\rangle + \langle \Phi_2(1)\Phi_1(2)|1/r_{12}|\Phi_2(1)\Phi_1(2)\rangle\right.\\ &\left. - \langle \Phi_1(1)\Phi_2(2)|1/r_{12}|\Phi_2(1)\Phi_1(2)\rangle - \langle \Phi_2(1)\Phi_1(2)|1/r_{12}|\Phi_1(1)\Phi_2(2)\rangle\right]\end{aligned} \tag{2.4.13}$$

Since, r_{12} depends on the position of both the electrons, nothing can be integrated out and the equation cannot be further simplified. However, the first two terms in this equation are equivalent and so are the last two.

Substituting Eqs (2.4.9)–(2.4.13) into Eq. (2.4.7), we get

$$\begin{aligned}E = \left\langle \hat{H} \right\rangle = &-\frac{1}{2}\langle \Phi_1|\nabla^2|\Phi_1\rangle - \frac{1}{2}\langle \Phi_2|\nabla^2|\Phi_2\rangle - Z\langle \Phi_1|1/r|\Phi_1\rangle - Z\langle \Phi_2|1/r|\Phi_2\rangle\\ &+ \langle \Phi_1\Phi_2|1/r_{12}|\Phi_1\Phi_2\rangle - \langle \Phi_1\Phi_2|1/r_{12}|\Phi_2\Phi_1\rangle\end{aligned} \tag{2.4.14}$$

Here, electron numbering for everything except r_{12} has been dropped.

A physical significance can be attributed to the different terms of Eq. (2.4.14).

The first two terms represent the kinetic energy of an electron moving in spin orbitals 1 and 2, respectively. The third and fourth terms are the attractive forces between these electrons and the nucleus. Thus, the sum of first and third term is the Hamiltonian of the first electron while the sum of second and fourth term is the Hamiltonian of the second electron. The fifth term represents repulsion between the electrons moving in orbitals 1 and 2, respectively. This term represents Coulomb integral and is abbreviated J_{12}. No adequate classical description is possible for the last term in Eq. (2.4.14). In this last integral the same electron appears in two different spin orbitals like $\Phi_1(1)$ and $\Phi_2(1)$ or $\Phi_1(2)$ and $\Phi_2(2)$. These integrals will obviously vanish if the two spin functions are different in accordance with the Pauli's exclusion Principle. It remains only if the two electrons have the same spin. This term is called the exchange integral and is abbreviated as K_{12}. This term may be treated as a correction to the repulsive energy of the two electrons if they have the same spin and happen to be simultaneously in the same region of space. It also accounts to some extent for the correlation effect of the two electrons.

2.4.1.1 *Generalization of the HF method to a many-electron atom*

We shall now consider generalization of Eq. (2.4.14) to a many-electron atom for which the Slater determinant form of the wavefunction is given by

$$\Psi(1,2,3\ldots N) = \frac{1}{\sqrt{N!}}|\Phi_1(1)\Phi_2(2)\Phi_3(3)\ldots\Phi_N(N)| \tag{2.4.15}$$

where Φ_1, Φ_2, etc., are spin orbitals which form an orthonormal set.

The Hamiltonian for the N-electron atom is given as

$$\begin{aligned}\hat{H} &= \left\{-\frac{1}{2}\sum_{\mu}\nabla_{\mu}^2 - \sum_{\mu}\frac{Z}{r_{\mu}}\right\} + \sum_{\mu<\nu} 1/r_{\mu\nu} \\ &= \sum_{\mu} h_{\mu} + \sum_{\mu<\nu} 1/r_{\mu\nu}\end{aligned} \tag{2.4.16}$$

where h_μ is the one-electron operator for electron μ and $\frac{1}{r_{\mu\nu}}$ is the electron repulsion term.

Using Eq. (2.4.16), the expectation value of $\hat{H}$ for the state Ψ can be obtained from the relation

$$\left\langle\Psi\middle|\hat{H}\middle|\Psi\right\rangle = \left\langle\Psi\middle|\sum_{\mu} h_{\mu} + \sum_{\mu<\nu} 1/r_{\mu\nu}\middle|\Psi\right\rangle \tag{2.4.17}$$

If the number of electrons is N, there will be N number of one-electron operators and $N(N-1)/2$ number of two-electron operators. Since all the electrons are equivalent, the expectation value $\langle\hat{H}\rangle$ may be given by the sum of N times the average of $\langle h_\mu\rangle$ and $N(N-1)/2$ times the average of $\langle 1/r_{\mu\nu}\rangle$.

Hence,

$$\left\langle\Psi\middle|\hat{H}\middle|\Psi\right\rangle = N\langle\Psi|h_\mu|\Psi\rangle + N(N-1)/2\cdot\langle\Psi|1/r_{\mu\nu}|\Psi\rangle \tag{2.4.18}$$

Various treatments are available in literature for evaluating integrals involving Slater determinants. Some clear treatments have been given by Pople and Beveridge [5], Szabo and Ostlund [6], Trindle and Shillady [7], and Chandra [8].

The first term in Eq. (2.4.18) is given by

$$\begin{aligned} N\langle\Psi|h_\mu|\Psi\rangle &= N\cdot\frac{1}{N}\cdot\{\langle\Phi_1|h_\mu|\Phi_1\rangle+\langle\Phi_2|h_\mu|\Phi_2\rangle+\ldots\langle\Phi_N|h_\mu|\Phi_N\rangle\} \\ &= \sum_i\langle\Phi_i|h_\mu|\Phi_i\rangle \end{aligned} \tag{2.4.19}$$

and the second term by

$$\begin{aligned} \frac{N(N-1)}{2}\cdot\langle\Psi|1/r_{\mu\nu}|\Psi\rangle &= \frac{N(N-1)}{2}\cdot\frac{2}{N(N-1)}\left\langle\frac{1}{2}\frac{[\Phi_i(\mu)\Phi_j(\nu)-\Phi_i(\nu)\Phi_j(\mu)]^2}{r_{\mu\nu}}\right\rangle \\ &= \left\langle\frac{\Phi_i^2(\mu)\Phi_j^2(\nu)}{r_{\mu\nu}}\right\rangle-\left\langle\frac{\Phi_i(\mu)\Phi_j(\nu)\Phi_i(\nu)\Phi_j(\mu)}{r_{\mu\nu}}\right\rangle \end{aligned} \tag{2.4.20}$$

As in the case of two-electron system, the first term of Eq. (2.4.20) is called Coulomb integral denoted by J_{ij}.

$$\begin{aligned} J_{ij} &= \iint\frac{\Phi_i^2(\mu)\Phi_j^2(\nu)}{r_{\mu\nu}}dr_\mu dr_\nu = \left\langle\Phi_i(\mu)\Phi_i(\mu)\left|\frac{1}{r_{\mu\nu}}\right|\Phi_j(\nu)\Phi_j(\nu)\right\rangle \\ &= \langle\Phi_i\Phi_i|\Phi_j\Phi_j\rangle \end{aligned} \tag{2.4.21}$$

In the second term in Eq. (2.4.20) the same electron appears in two different spin orbitals and shall have a nonzero value only if both of them have the same spin (α or β). This term arises from exchange of electrons between two orbitals and is the exchange integral denoted by K_{ij}.

$$\begin{aligned} K_{ij} &= \iint\frac{\Phi_i(\mu)\Phi_j(\nu)\Phi_i(\nu)\Phi_j(\mu)}{r_{\mu\nu}}dr_\mu dr_\nu = \left\langle\phi_i(\mu)\phi_j(\mu)\left|\frac{1}{r_{\mu\nu}}\right|\phi_i(\nu)\phi_j(\nu)\right\rangle \\ &= \langle\phi_i\phi_j|\phi_i\phi_j\rangle \end{aligned} \tag{2.4.22}$$

In Dirac representation of Eqs (2.4.21) and (2.4.22), the electrons μ and ν are written on different sides of the vertical line and the presence of operator $r_{\mu\nu}$ is implied.

Coulomb and exchange integrals are also sometimes written in a more convenient form. Thus

$$J_{ij} = \langle\Phi_j(\nu)|J_i|\Phi_j(\nu)\rangle \tag{2.4.23}$$

where

$$J_i = \left\langle\Phi_i(\mu)\left|\frac{1}{r_{\mu\nu}}\right|\Phi_i(\mu)\right\rangle \tag{2.4.24}$$

and

$$K_{ij} = \langle\Phi_i(\mu)|K_i|\Phi_j(\mu)\rangle \tag{2.4.25}$$

where

$$K_i = \left\langle\Phi_i(\nu)\left|\frac{1}{r_{\mu\nu}}\right|\Phi_j(\nu)\right\rangle \tag{2.4.26}$$

Thus using Eqs (2.4.19)–(2.4.22) in Eq. (2.4.17) we get,

$$\langle\Psi|H|\Psi\rangle = \sum_i \langle\Phi_i|h_\mu|\Phi_i\rangle + \sum_{i<j} J_{ij} - \sum\nolimits'_{i<j} K_{ij} \tag{2.4.27}$$

where $\sum'_{i<j}$ shows that the summation extends over all pairs of orbitals Φ_i and Φ_j having the same spin (α or β).

Given a set of wavefunctions Φ_i, Φ_j,...etc., we have now to find the best of them which may minimize the expectation value of $\widehat{H}$. Hartree–Fock method determines the set of spin orbitals which minimize the energy subject to the constraints of orthonormality $\langle\Phi_i|\Phi_j\rangle = \delta_{ij}$ of the spin orbitals. The Hartree–Fock strategy to find self-consistent solution to the N-electron Schrödinger equation lies in the use of the variational principle which would make the expectation value of the Hamiltonian an extremum. The constraint of orthogonality is taken into account by using the Lagrange's method of undetermined multipliers, where a function is defined as

$$\mathscr{L} = \langle\Psi|H|\Psi\rangle - \sum_{i,j} \varepsilon_{ij}(\langle\Phi_i|\Phi_j\rangle - \delta_{ij}) \tag{2.4.28}$$

where ε_{ij} are the undetermined multipliers and $\langle\Phi_i|\Phi_j\rangle$ is the overlap between the spin orbitals.

Since we need to minimize the Hartree–Fock energy expression with respect to changes in the orbitals $\Phi_i \to \Phi_i + \delta\Phi_i$, we require that $\delta\mathscr{L} = 0$ for such changes in Φ_i. Using Eqs (2.4.27) and (2.4.28), we therefore get

$$\delta\mathscr{L} = \delta\left[\sum_i \langle\Phi_i|h_\mu|\Phi_i\rangle + \sum_{i<j} J_{ij} - \sum\nolimits'_{i<j} K_{ij} - \sum_{ij} \varepsilon_j(\langle\Phi_i|\Phi_j\rangle - \delta_{ij})\right] = 0 \tag{2.4.29}$$

Working through some mathematical steps, we finally arrive at the Hartree–Fock equations defining the orbitals

$$\left[\widehat{h}_\mu + \sum_i (J_i - K'_i)\right]\Phi_j = \sum_i \varepsilon_{ij}\Phi_i \tag{2.4.30}$$

where ε_{ij} is the undetermined multiplier and J_i and K_i are given by Eqs (2.4.24) and (2.4.26).

The bracketed term on the left-hand side of Eq. (2.4.30) is known as Fock operator $\widehat{F}$

$$\widehat{F} = \widehat{h}_\mu + \sum_i (J_i - K'_i) \tag{2.4.31}$$

The Hartree–Fock Eq. (2.4.30) is now written as

$$\widehat{F}\Phi_j = \sum_i \varepsilon_{ij}\Phi_i \tag{2.4.32}$$

There are several different solutions to this equation each corresponding to a different set of ε_{ij}. We have the freedom to concentrate upon those ε_{ij} which satisfy

$$\varepsilon_{ij} = \delta_{ij}\varepsilon_j \tag{2.4.33}$$

Here ε_j is essentially a new name for the Lagrange's multiplier as it can also be obtained by diagonalizing $||\varepsilon_{ij}||$ matrix by a unitary transformation.

Hence, Hartree–Fock Eq. (2.4.32) can be simplified to give Eq. (2.4.34)

$$\widehat{F}\Phi_j = \varepsilon_j \Phi_j \tag{2.4.34}$$

In this form Eq. (2.4.34) is a traditional eigenvalue equation. For each j there is an equivalent equation defining a system of Schrödinger-like, one particle equations. In matrix representation, it may be written as

$$F\Phi = \varepsilon\Phi \tag{2.4.35}$$

where ε is the eigenvalue matrix or the energies associated with the Hartree–Fock orbitals Φ_i, Φ_j, etc.

The total electronic energy cannot be given as a simple sum over the eigenvalues of the Fock operator due to the discrepancy that stems from the description of a true many-body system in terms of single-particle orbitals. The total energy is given as

$$E = \sum_j \varepsilon_j - \left(\sum_{i>j} J_{ij} - \sum\nolimits'_{i>j} K_{ij} \right) \tag{2.4.36}$$

2.4.2 Interpretation of the Eigenvalues of the Fock Operator

The eigenvalues of the Fock operator do not directly represent the energy of the system. They individually represent approximate ionization potentials of the electrons in the various spin orbitals. These eigenvalues are the energies required to remove an electron from the individual spin orbitals, assuming that the remaining system can be adequately described by the remaining electrons in their unaltered spin orbitals. This statement is in conformity with the Koopmann's theorem according to which: if adding or subtracting an electron does not change the energy levels then each eigenvalue of the Fock operator gives the energy required to remove an electron from the corresponding single-electron state.

Hartree–Fock SCF theory has been able to predict quite accurate values of total electronic energy and first ionization potentials of many-electron atomic and molecular systems [9]. The error between the calculated values and experimental values, however, increases with the number of electrons in the system. The error primarily arises due to the neglect of correlation effect and the relativistic effect. Though, to some extent, the correlation effect is taken care of by the inclusion of Pauli's exclusion principle but it is not enough. The relativistic effect arises from the fact that in large atoms the average velocity of inner-shell electrons can be comparable to the speed of light resulting in increase in their effective mass (For greater details, see Ratner and Schatz [10]).

2.5 Electronic Configuration and Electronic States

The solution of Schrödinger equation for an electronic system results in spin orbitals, each associated with a discrete orbital energy. Two cases may arise: the total number of electrons is either an even number, say $2N$, or an odd number, say $2N+1$. If we consider the ground state of the $2N$-electron system, each spatial orbital can occupy up to two electrons—one with α spin ($\uparrow$) and the other with β spin ($\downarrow$). Thus a total of N spin orbitals will be occupied resulting in an electronic configuration $\Psi_1^2\Psi_2^2\ldots\Psi_N^2$. Such a configuration is said to be a closed-shell configuration. Since there are equal numbers of α and β electrons, such a ground state will have total spin $S=0$, resulting in a state of multiplicity $2S+1=1$. The closed-shell configuration gives rise to a singlet state (Figure 2.1(a)).

If the number of electrons is odd, the ground state configuration will be $\Psi_1^2\Psi_2^2\ldots\Psi_N^2\Psi_{N+1}^1$. The system shall now have $N+1$ orbitals; N spin orbitals will be doubly occupied while the last orbital shall have an unpaired electron (Figure 2.1(b)). This type of configuration is said to be an open-shell configuration. With odd number of electrons $M_s=+\frac{1}{2}$ or $M_s=-\frac{1}{2}$ the multiplicity of the configuration shall be $2S+1=2$, giving rise to a doublet state.

In either case, the orbital wavefunctions are constructed as Slater determinants or by choosing an appropriate combination of these determinants. Thus, the spin-correct eigenfunctions for a 4-electron system Beryllium (Be) having configuration $1s^2 2s^2$ is

$$ {}^1\Psi(\mathrm{Be}) = \frac{1}{\sqrt{4!}} \begin{vmatrix} \Phi_{1s}(1) & \overline{\Phi}_{1s}(1) & \Phi_{2s}(1) & \overline{\Phi}_{2s}(1) \\ \Phi_{1s}(2) & \overline{\Phi}_{1s}(2) & \Phi_{2s}(2) & \overline{\Phi}_{2s}(2) \\ \Phi_{1s}(3) & \overline{\Phi}_{1s}(3) & \Phi_{1s}(3) & \overline{\Phi}_{2s}(3) \\ \Phi_{1s}(4) & \overline{\Phi}_{1s}(4) & \Phi_{1s}(4) & \overline{\Phi}_{2s}(4) \end{vmatrix} \tag{2.5.1} $$

where unbarred terms correspond to α spin and barred to β spin.

In contracted notation, the orbital function is written as

$$ {}^1\Psi(\mathrm{Be}) = \left|\Phi_{1s}(1)\overline{\Phi}_{1s}(2)\Phi_{2s}(3)\overline{\Phi}_{2s}(4)\right| \tag{2.5.2} $$

For the $(2N+1)$ system, the spin-correct form of orbital wavefunctions for the two components of the doublet state $M_s=+\frac{1}{2}$ and $M_s=-\frac{1}{2}$ may be written as

$$ {}^2\Psi = \left|\Phi_1(1)\overline{\Phi}_1(2)\ldots\Phi_n(2N-1)\overline{\Phi}_n(2N)\Phi_{n+1}(2N+1)\right| \quad \text{for } M_s = \frac{1}{2}(\alpha) \tag{2.5.3} $$

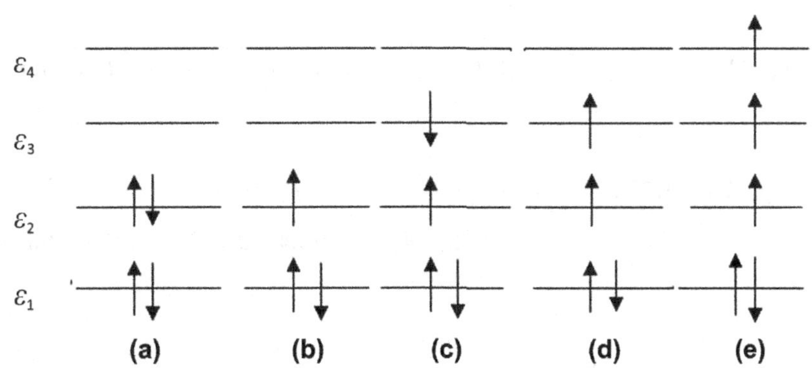

FIGURE 2.1 Orbital energy-level diagrams for the ground state. Closed-shell (a) and open-shell (b-e) electronic configurations.

and

$$^2\Psi = |\Phi_1(1)\overline{\Phi}_1(2)\ldots\Phi_n(2N-1)\overline{\Phi}_n(2N)\overline{\Phi}_{n+1}(2N+1)| \quad \text{for } M_s = -\frac{1}{2}(\beta) \tag{2.5.4}$$

Electronic configurations with more than one unpaired electrons may arise when systems of closed configuration are excited so that one of the paired electrons may rise to a higher energy state and occupy a previously unoccupied orbitals (Figure 2.1(c–d)). The contribution of doubly occupied states to magnetic quantum number M_s is zero. Two electrons in singly occupied states may either have parallel spin $\alpha\alpha$ or $\beta\beta$ giving rise to $M_s = 1$ or -1, or have opposite spin $\alpha\beta$ or $\beta\alpha$ giving rise to $M_s = 0$, 0. The state with $M_s = 0$ gives rise to a singlet state while that with $M_s = 1, 0, -1$ gives rise to a triplet state. The singlet and triplet states have different energies due to different electron–electron repulsion terms. According to Hund's rule, a triplet state lies deeper (has lower energy) than a singlet state due to the fact that electrons with parallel spin are kept apart due to antisymmetry condition.

Thus, in the singlet-excited state of four-electron system Beryllium (Be), the orbital Φ_1 is occupied by two electrons with spins α and β, while the other two orbitals Φ_2 and Φ_3 will be occupied by one electron each with spins either $\alpha\beta$ or $\beta\alpha$. The antisymmetric spin orbital in this case is given by

$$^1\Psi(\text{Be}) = \frac{1}{\sqrt{4!}}\left\{\begin{vmatrix} \Phi_1(1) & \overline{\Phi}_1(1) & \Phi_2(1) & \overline{\Phi}_3(1) \\ \Phi_1(2) & \overline{\Phi}_1(2) & \Phi_2(2) & \overline{\Phi}_3(2) \\ \Phi_1(3) & \overline{\Phi}_1(3) & \Phi_2(3) & \overline{\Phi}_3(3) \\ \Phi_1(4) & \overline{\Phi}_1(4) & \Phi_2(4) & \overline{\Phi}_3(4) \end{vmatrix} - \begin{vmatrix} \Phi_1(1) & \overline{\Phi}_1(1) & \overline{\Phi}_2(1) & \Phi_3(1) \\ \Phi_1(2) & \overline{\Phi}_1(2) & \overline{\Phi}_2(2) & \Phi_3(2) \\ \Phi_1(3) & \overline{\Phi}_1(3) & \overline{\Phi}_2(3) & \Phi_3(3) \\ \Phi_1(4) & \overline{\Phi}_1(4) & \overline{\Phi}_2(4) & \Phi_3(4) \end{vmatrix}\right\} \tag{2.5.5}$$

In general, for a $2N$-electron system, the possible singlet and triplet states, in contracted notation, can have the following corrected spin orbitals.

$$\begin{aligned} ^1\Psi(1,2,\ldots i,j\ldots 2N-1,2N) = \frac{1}{\sqrt{2}}\{&|\Phi_1(1)\overline{\Phi}_1(2)\ldots\Phi_k(i)\overline{\Phi}_l(j)\ldots\Phi_N(2N-1)\overline{\Phi}_N(2N)| \\ &- |\Phi_1(1)\overline{\Phi}_1(2)\ldots\overline{\Phi}_k(i)\Phi_l(j)\ldots\Phi_N(2N-1)\overline{\Phi}_N(2N)|\} \quad \text{for } M_s = 0 \end{aligned} \tag{2.5.6a}$$

$$^3\Psi(1,2,\ldots i,j\ldots 2N-1,2N) = \frac{1}{\sqrt{2}}|\Phi_1(1)\overline{\Phi}_1(2)\ldots\Phi_k(i)\Phi_l(j)\ldots\Phi_N(2N-1)\overline{\Phi}_N(2N)| \quad \text{for } M_s = 1 \tag{2.5.6b}$$

$$\begin{aligned} ^3\Psi(1,2,\ldots i,j\ldots 2N-1,2N) = \frac{1}{\sqrt{2}}\{&|\Phi_1(1)\overline{\Phi}_1(2)\ldots\Phi_k(i)\overline{\Phi}_l(j)\ldots\Phi_N(2N-1)\overline{\Phi}_N(2N)| \\ &- |\Phi_1(1)\overline{\Phi}_1(2)\ldots\overline{\Phi}_k(i)\Phi_l(j)\ldots\Phi_N(2N-1)\overline{\Phi}_N(2N)|\} \quad \text{for } M_s = 0 \end{aligned} \tag{2.5.6c}$$

and

$$^3\Psi(1,2,\ldots i,j\ldots 2N-1,2N) = \frac{1}{\sqrt{2}}\{|\Phi_1(1)\overline{\Phi}_1(2)\ldots\overline{\Phi}_k(i)\overline{\Phi}_l(j)\ldots\Phi_N(2N-1)\overline{\Phi}_N(2N)|\} \quad \text{for } M_s = -1 \tag{2.5.6d}$$

In other excited open-shell configurations the multiplicity of states is determined by the vector addition of electron spins. Thus for a system having three open shells (Figure 2.1(e)), these may be doublet and quartet states, all of which may have different energy. Following Hunds' rule the state with highest multiplicity shall lie deepest. The observables of a system are always referred to the states of the system and not to their configuration as there may be several states for the same configuration.

2.6 Restricted and Unrestricted Wavefunctions

In the Hartree–Fock description of a closed-shell system each spin orbital Φ_i is occupied by two electrons with opposite spins. Thus, we have

$$\Phi_i = \Phi_i(r)\alpha(s)$$

$$\overline{\Phi}_i = \Phi_i(r)\beta(s) \tag{2.6.1}$$

where Φ_i are the spatial orbitals. Such orbitals construct a wavefunction given by Slater determinant which is called restricted wavefunction. Thus, for a two-electron system the wavefunction

$$\Psi = \frac{1}{\sqrt{2}}\begin{vmatrix} \Phi_1(1)\alpha(1) & \Phi_1(1)\beta(1) \\ \Phi_1(2)\alpha(2) & \Phi_1(2)\beta(2) \end{vmatrix}$$

$$= \frac{1}{\sqrt{2}}[\Phi_1(1)\alpha(1)\Phi_1(2)\beta(2) - \Phi_1(1)\beta(1)\Phi_1(2)\alpha(2)] \tag{2.6.2}$$

The Hartree–Fock theory in this case is called the Restricted Hartree–Fock theory (RHF). The advantage of restricted functions is that the Slater determinant is an eigenfunction of S^2. However, the disadvantage is that the constraint of two electrons with α and β spins to occupy the same spatial orbital ignores the correlation between the electrons with opposite spins.

In another description called the unrestricted wavefunction, no restrictions are imposed on the spatial nature of the spatial orbitals Φ_i. The electron of α spin may be described by a set of orbitals $\{\Phi_i\}$ and electrons of β spin by a different set of spatial orbitals $\{\Phi_i'\}$. Thus

$$\Phi_i = \Phi(r)\alpha(s)$$

$$\Phi_i' = \Phi_i'(r)\beta(s) \tag{2.6.3}$$

The theory using the unrestricted wavefunction is known as Unrestricted Hartree–Fock (UHF) theory. This method has also been called as different orbitals for different spins (DODS) method. Such a wavefunction for a two-electron system is

$$\Psi = \frac{1}{\sqrt{2}}[\Phi_1(1)\alpha(1)\Phi_1'(2)\beta(2) - \Phi_1'(1)\beta(1)\Phi_1(2)\alpha(2)]$$

$$= \frac{1}{\sqrt{2}}\begin{vmatrix} \Phi_1(1)\alpha(1) & \Phi_1'(1)\beta(1) \\ \Phi_1(2)\alpha(2) & \Phi_1'(2)\beta(2) \end{vmatrix} \tag{2.6.4}$$

In UHF theory one assumes that the basic one-electron functions would be of pure α and β type. This theory is particularly useful for open-shell systems as it takes into account correlation between electrons of different spins. The disadvantage, however, is that the unrestricted wavefunction is not an eigenfunction of S^2.

References

[1] J.C. Slater, Phys. Rev. 35 (1930) 509, 34 (1959) 1293.

[2] J.E. Lennard-Jones, Proc. Roy. Soc. (London), A198 1 (1949) 14.

[3] D.R. Hartree, Proc. Cambridge Phil. Soc. 24 (1928) 89,111,426.

[4] V. Fock, Z. Physik. 31 (1926) 126.

[5] J.A. Pople, D.L. Beveridge, Approximate Molecular Orbital Theory, McGraw Hill Book Co, 1970.

[6] A. Szabo, N.S. Ostlund, Modern Quantum Chemistry: Introduction to Advanced Electronic Structure Theory, Dover Books on Chemistry, 1996.

[7] C. Trindle, D. Shillady, Electronic Structure Modeling, CRC Press, Boca Raton, USA, 2008.

[8] A.K. Chandra, Introductory Quantum Chemistry, McGraw Hill Publishing Co, New Delhi, 2003.

[9] [a] F. Herman, S. Skillman, Atomic Structure Calculations, Prentice – Hall, New Jersey, 1963; [b] E. Clementi, Tables of Atomic Functions, I.B.M., Corporation, San Jose, CA, 1965.

[10] M.A. Ratner, G.C. Schatz, Introduction to Quantum Mechanics in Chemistry, Prentice Hall, New Jersey, USA, 2002.

Further Reading

[1] I.N. Levine, Quantum Chemistry, fourth ed., Prentice Hall, Englewood Cliffs, NJ, 1991.

[2] R.L. Flurry, Molecular Orbital Theories of Bonding in Organic Molecules, Marcell Dekker, Inc., New York, 1968.

[3] A.K. Chandra, Introductory Quantum Chemistry, McGraw Hill Publishing Co, New Delhi, 2003.

[4] A. Szabo, N.S. Ostlund, Modern Quantum Chemistry: Introduction to Advanced Electronic Structure Theory, Dover Books on Chemistry, 1996.

3

Self-consistent Field Molecular Orbital Theory

CHAPTER OUTLINE

Principles and Applications of Quantum Chemistry. http://dx.doi.org/10.1016/B978-0-12-803478-1.00003-0

3.1 Introduction

The Hartree–Fock (HF) self-consistent field (SCF) theory of atoms can be extended to molecules with certain modifications. The electron is now no more confined to individual atoms but is influenced by all the electrons and nuclei in the molecule. The molecule contains not only the moving electrons but also the moving nuclei. As such, while constructing the Hamiltonian one needs to take into account electron nuclear attraction, nuclear–nuclear repulsion, and electron–electron repulsion terms. The Hamiltonian now becomes very complex and in most cases the Schrödinger equation cannot be solved even by analytical methods. Exact solutions of the Schrödinger equation have been worked out only for small molecules like hydrogen molecular ion (H_2^+) and hydrogen molecule (H_2). Drastic assumption, therefore, have to be made for solving Schrödinger equation for polyatomic molecules. The two approaches that have historically emerged to solve the Schrödinger equation of molecules are (1) molecular orbital (MO) approximation and (2) valence bond (VB) approximation.

In view of the fact that nuclear mass is very large in comparison with the electron mass (proton mass is 1836 times the electron mass), nuclei are expected to move much more slowly than the electrons; the electrons in a molecule may have an average distribution over a period of time during which the nuclei hardly move. It may therefore be presumed that in a given nuclear configuration the electronic motion is independent of the nuclear motion. The argument was put forth by Born and Oppenheimer in 1927 and is known as the Born–Oppenheimer approximation.

3.2 Born–Oppenheimer Approximation

If we assume the nuclei and electrons in a molecule to be point masses and neglect spin–orbit and other relativistic corrections, then the molecular Hamiltonian in atomic units can be written as

$$\hat{H} = -\frac{1}{2}\sum_{\alpha}\frac{1}{M_\alpha}\nabla_\alpha^2 - \frac{1}{2}\sum_i \nabla_i^2 + \sum_\alpha\sum_{\beta>\alpha}\frac{Z_\alpha Z_\beta}{R_{\alpha\beta}} - \sum_\alpha\sum_i\frac{Z_\alpha}{r_{i\alpha}} + \sum_j\sum_{i>j}\frac{1}{r_{ij}} \tag{3.2.1}$$

or more compactly as,

$$\hat{H} = \hat{T}_n + \hat{T}_e + \hat{V}_{nn} + \hat{V}_{en} + \hat{V}_{ee} \tag{3.2.2}$$

where symbols α and β refer to nuclei and i and j to electrons, $r_{i\alpha}$ is the distance between the electron i and nucleus α, $R_{\alpha\beta}$ is the distance between the nuclei α and β with atomic numbers Z_α and Z_β, and r_{ij} is the distance between electrons i and j. The first two terms in Eq. (3.2.1) are the kinetic energy operators of the nuclei ($\hat{T}_n$) and electrons ($\hat{T}_e$), respectively, the third term is the potential energy of repulsion between the nuclei (V_{nn}), the fourth term is the potential energy of attraction between the electrons and the nuclei (V_{en}), and the fifth term is the potential energy of repulsion between the electrons (V_{ee}).

The molecular Schrödinger equation is

$$\hat{H}\Psi(q_i, q_\alpha) = E\Psi(q_i, q_\alpha) \tag{3.2.3}$$

where q_i and q_α symbolize the electronic and nuclear coordinates, respectively.

Born and Oppenheimer introduced an approximation in Eq. (3.2.1) to account for the large difference in the masses of the electrons and the nuclei because of which the electrons are capable of instantaneously adjusting to any change in the position of the nuclei. Due to these reasons, in the Born–Oppenheimer approximation, the kinetic energy term for the nuclei is ignored and $R_{\alpha\beta}$ is taken to be a constant quantity so that the electron motion could be determined for a fixed position of the nuclei.

We try an approximate solution in the form

$$\Psi = \Psi_e(q_i, q_\alpha)\Psi_n(q_\alpha) \tag{3.2.4}$$

where Ψ_e is a function of the electronic coordinates q_i and Ψ_n is a function of the nuclear coordinates q_α. Thus the total wavefunction can be attempted to be broken into electronic wavefunction Ψ_e and the vibrational rotational wavefunction dependent on nuclear coordinates (translation, rotation, vibration) Ψ_n. After substituting Eq. (3.2.4) in Eq. (3.2.3), where $\hat{H}$ is given by Eq. (3.2.1), and neglecting terms of the type $\frac{\partial\Psi_e}{\partial x_\alpha}, \frac{\partial^2\Psi_e}{\partial x_\alpha^2}, \ldots$ which represent the variation of the electronic wavefunction Ψ_e with internuclear distance, we get

$$\left(\hat{H}_e + V_{nn}\right)\Psi_e = U\Psi_e \tag{3.2.5}$$

$$\text{where, } \hat{H}_e = -\frac{1}{2}\sum_i \nabla_i^2 - \sum_\alpha\sum_i\frac{Z_\alpha}{r_{i\alpha}} + \sum_j\sum_{i>j}\frac{1}{r_{ij}}$$

$$\text{and } V_{nn} = \sum_\alpha\sum_{\beta>\alpha}\frac{Z_\alpha Z_\beta}{R_{\alpha\beta}} \tag{3.2.6}$$

The energy U in Eq. (3.2.5) is the sum of electronic energy E_e and the internuclear repulsion (V_{nn}),

$$U = E_e + V_{nn}$$

It may be seen from Eq. (3.2.5) that for a given electronic state, the internuclear distances $R_{\alpha\beta}$ is fixed and so V_{nn} is a constant quantity. There may be an infinite number of nuclear conformations with different values of $R_{\alpha\beta}$ and for each conformation the Schrödinger Eq. (3.2.5) is to be solved to get a set of electronic wavefunctions Ψ_e and electronic energies E_e. These, therefore, parametrically depend upon the nuclear configuration

$$\Psi_e = \Psi_{e,n}(q_i, q_\alpha) \tag{3.2.7}$$

$$\text{and } E_e = E_n(q_\alpha) \tag{3.2.8}$$

where n is an electronic quantum number defining different electronic states.

In order to get the total electronic energy including nuclear repulsion U, we may first solve the equation

$$\widehat{H}_e \Psi_e = E_e \Psi_e \tag{3.2.9}$$

to get the purely electronic energy E_e and then add to it the constant value V_{nn}.

Having solved for the electronic wavefunctions and energies, we use the electronic energy including nuclear repulsion U as the potential energy in the Schrödinger equation for nuclear motion:

$$\widehat{H}_n \Psi_n = E \Psi_n \tag{3.2.10}$$

$$\widehat{H}_n = -\frac{1}{2}\sum_\alpha \frac{1}{M_\alpha}\nabla_\alpha^2 + U(q_\alpha) \tag{3.2.11}$$

The constant E in Eq. (3.2.10) is the total energy of the molecule, since $\widehat{H}_n$ includes operators for both nuclear and electronic energies. Each electronic state has a different electronic energy $U(q_\alpha)$; hence we must solve a different nuclear Schrödinger equation for each electronic state of the molecule.

In a given electronic state, different fixed arrangements of the nuclei may then be adopted and the calculation of total energy is repeated. The set of solutions for different internuclear distances can then be used to construct a potential energy curve for a diatomic molecule or a potential energy surface for a polyatomic molecule. Thus, if we plot the total energy of a diatomic molecule against the interatomic distance for a bound state, we expect to get a potential energy curve as shown in Figure 3.1.

The minimum of the potential energy curve or surface is identified as the equilibrium geometry of the molecule in the given electronic state. The wavefunctions that result from calculations are called molecular orbitals (MOs). In the case of a diatomic molecule, the nuclear configuration is a function of only one variable, the internuclear distance R. At $R \to 0$, the V_{nn} term causes U to tend to infinity. At $R \to \infty$, U approaches the sum of energies of the separated atoms. At $R = R_e$, which is the equilibrium internuclear

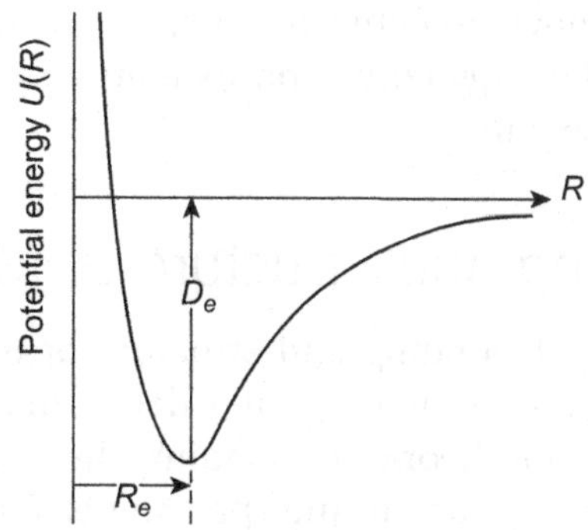

FIGURE 3.1 Potential energy curve of a diatomic molecule. The minimum energy point corresponds to equilibrium bond length of the molecule.

distance or bond length of the molecule, U has its minimum value. The quantity $U(\infty) - U(R_e)$ is the equilibrium dissociation energy D_e of the molecule. Although no exact expression for the potential energy curves of all molecules is known, a simple and often convenient expression that yields curves of the shape of Figure 3.1 has been given by P.M. Morse which is also known as Morse relation

$$U(q) = D_e\left(1 - e^{-\beta q}\right)^2 \tag{3.2.12}$$

where $q = (R - R_e)$ measures the distortion of the bond from its equilibrium length, D_e is the equilibrium dissociation energy, and β is a constant for any given molecule and determines the narrowness or curvature of the potential energy curve. The use of this expression in the Schrödinger equation however, causes some mathematical difficulties and a simpler and more fruitful procedure is followed.

Since the vibrations of a molecule result only in small distortions of the bond from the equilibrium length, we are particularly interested in an expression for the potential energy near the minimum of the potential energy curve. This is obtained by writing a Taylor series expansion of U about $q = 0$, where q is the displacement from the mean position.

$$U(q) = U_0 + \left(\frac{\partial U}{\partial q}\right)_0 q + \frac{1}{2!}\left(\frac{\partial^2 U}{\partial q^2}\right)_0 q^2 + \frac{1}{3!}\left(\frac{\partial^3 U}{\partial q^3}\right)_0 q^3 + \ldots \tag{3.2.13}$$

Taking $U_0 = 0$ at equilibrium and remembering that the potential energy at $q = 0$ is a minimum, that is $\left(\frac{\partial U}{\partial q}\right)_0 = 0$ at the minimum, the first two terms in Eq. (3.2.13) disappear. The equation therefore reduces to

$$U_0(q) = \frac{1}{2!}\left(\frac{\partial^2 U}{\partial q^2}\right)_0 q^2 + \frac{1}{3!}\left(\frac{\partial^3 U}{\partial q^3}\right)_0 q^3 + \ldots = \frac{1}{2}kq^2 + aq^3 + bq^4 + \ldots \tag{3.2.14}$$

Limiting the above equation to the quadratic term we get for the potential energy a curve in the shape of a parabola, whereas inclusion of higher order terms leads to a curve as shown in Figure 3.1. When dealing with an oscillator, $k = \left(\frac{\partial^2 U}{\partial q^2}\right)_0$ is called the force constant which is identical to the spring constant for a harmonic oscillator following

Hooke's law and *a, b,...* are known as cubic, quartic, etc., anharmonicity constants. For a chemical bond, the potential energy curve approximates a parabola only in the vicinity of the minimum of the Morse curve.

3.3 Chemical Bonding and Structure of Molecules

To obtain an understanding of bonding and structure of molecules and to study the electronic states of molecular systems in greater detail, quantum mechanics has provided chemistry with two general theories of bonding: the MO theory and the VB theory. MO theory was developed as a means to interpret the electronic spectra of molecules, and has canonical MOs delocalized over the entire molecule. This theory bears little relationship to the familiar language of chemists in terms of localized bonds. The nuclei in the MO theory are held at a fixed separation and the electrons assigned to MOs, which are one-electron wavefunctions similar to atomic orbitals, but now extending overall the nuclei. The "building-up" procedure is analogous to that used to obtain electron configurations for polyelectronic atoms. VB theory is essentially a quantum mechanical formulation of the classical concept of chemical bond wherein the molecule is regarded as a set of atoms held together by local bonds. This model is deeply rooted in chemistry, such as Lewis structural formula, chemical valency, hybrid orbitals and resonance. In this theory, the individual atoms complete with all their electrons are brought together from an infinite separation. The vector model is then used to correlate adiabatically the terms of the resultant molecule with the terms of the separate atoms. We shall go into further details of these theories in the following sections.

3.4 Molecular Orbitals as Linear Contribution of Atomic Orbitals (LCAO)

The most generally applicable and useful representation of a MO is a linear combination of atomic orbitals (LCAOs). An atomic orbital is understood to be any convenient one-centre basis function which is normally, but not invariably, centered on a nucleus of the molecule. In the precomputational era of quantum chemistry, the atomic orbitals were often taken as the HF orbitals of separate atoms or as simple, usually Slater-type orbitals (STOs). Two types of basis functions, each with its own advantages and disadvantages, are in common use for the construction of MOs. These are Slater-type functions and the Gaussian-type functions (GTFs). Each atom may contribute several atomic orbitals in the construction of MOs. Thus, if $\Phi_1, \Phi_2, \ldots \Phi_n$ are the atomic orbitals, then the MO shall have the form

$$\Psi_m = \sum_{i=1}^{n} c_{mi}\Phi_i \tag{3.4.1}$$

The functions $\Phi_1, \Phi_2, \ldots \Phi_n$ are said to form a basis set of functions.

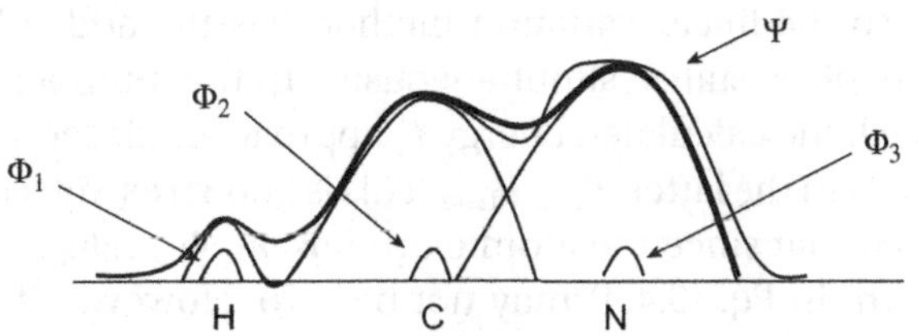

FIGURE 3.2 Molecular orbital as linear combination of atomic orbitals.

A graphical representation of the MO for triatomic molecules HCN in terms of the basis set functions is given in Figure 3.2. If Φ_1, Φ_2, Φ_3 are the basis functions each represented by a parabola, then the MO can be approximated by

$$\Psi = \sum_{i=1}^{3} c_i\Phi_i = 1.2\,\Phi_1 + 10\,\Phi_2 + 12\,\Phi_3$$

and represented as the contour of the parabolas shown by bold line [1].

In the orbital approximation, the coefficients c_i are calculated by SCF procedure. The optimal values of these coefficients are determined using the variational method (Chapter 1) so as to minimize the energy E of the system

$$E = \frac{\int \Psi^* H \Psi \, d\tau}{\int \Psi^* \Psi \, d\tau}$$

The variational method leads to a set of equations equivalent to Eq. (1.13.23)

$$\begin{pmatrix} H_{11} - ES_{11} & H_{12} - ES_{12} & \ldots & H_{1n} - ES_{1n} \\ H_{21} - ES_{21} & H_{22} - ES_{22} & \ldots & H_{2n} - ES_{2n} \\ \vdots & \vdots & \vdots & \vdots \\ H_{n1} - ES_{n1} & H_{n2} - ES_{n2} & \ldots & H_{nn} - ES_{nn} \end{pmatrix} \begin{pmatrix} c_1 \\ c_2 \\ \vdots \\ c_n \end{pmatrix} = 0 \qquad (3.4.2)$$

where,

$$H_{ij} = \int \Phi_i^* H \Phi_j d\tau \qquad (3.4.3)$$

$$\text{and } S_{ij} = \int \Phi_i^* \Phi_j d\tau \qquad (3.4.4)$$

In shorthand notation, Eq. (3.4.2) may be written as

$$HC = ESC \qquad (3.4.5)$$

If this set of equations is to have a nontrivial solution, the determinant $|H - ES|$ must be zero, which shall provide n-values of energy $E_1, E_2, \ldots E_n$. Substitution of these energies back into Eq. (3.4.2) shall provide the corresponding values of coefficients $(c_{11}, c_{12}, \ldots c_{1n})$, $(c_{21}, c_{22}, \ldots c_{2n}) \ldots (c_{n1}, c_{n2}, \ldots c_{nn})$ and hence a set of wavefunctions from Eq. (3.4.1). It turns out that each value of energy E_i calculated from Eq. (3.4.2) is greater than or equal to the true energy of each state, so the linear variation method has the added bonus of giving approximations to each of n states simultaneously. If the number of basis functions $\Phi_1, \Phi_2, \ldots \Phi_n$ is increased, the calculated energy E_i approaches closer to the true energy but shall always be greater than the latter, $E_i \geq E_{true}$. This is known as MacDonald's theorem [1].

It may also be noted that since the atomic orbitals $\Phi_1, \Phi_2, \ldots \Phi_n$ are based on different atoms, the integral term in Eq. (3.4.4) may not be zero. However, if a set of normalized atomic functions are used then

$$S_{ii} = \int \Phi_i^* \Phi_i d\tau = 1$$

$$\text{but,} \quad S_{ij} = \int \Phi_i^* \Phi_j d\tau \neq 0 \tag{3.4.6}$$

The term S_{ij} is known as the overlap integral and is indicative of the extent to which the two atomic orbitals overlap. It is a function of internuclear distance. For maximum overlap, it is essential that the two orbitals have the same symmetry relative to the molecular axis. A judicious choice of atomic orbitals is made for the construction of the MOs so that they may be able to reflect the real ground situation for the molecule. Various types of orbitals with diffuse and polarization functions are used in quantum chemistry. These shall be described separately. The conceptual basis of the MO theory shall be clearer, from the examples of the H_2^+ ion and the H_2 molecule that are being discussed.

3.4.1 Molecular Orbital Treatment of H_2^+ Molecule

The hydrogen molecule ion consists of two protons and an electron as shown in the Figure 3.3. If we take the two nuclei to be fixed with internuclear distance R_{AB} then within the Born–Oppenheimer approximation the Hamiltonian can be written as

$$\hat{H} = -\frac{1}{2}\nabla^2 - \frac{1}{r_A} - \frac{1}{r_B} + \frac{1}{R_{AB}} \tag{3.4.7}$$

The term $\frac{1}{R_{AB}}$ is a constant for a particular nuclear configuration. Since the Hamiltonian is written in atomic units, the internuclear distance is in dimensionless unit in terms of a_0, Bohr's radius. The Schrödinger equation can be written as

$$\hat{H}\Psi = E\Psi \tag{3.4.8}$$

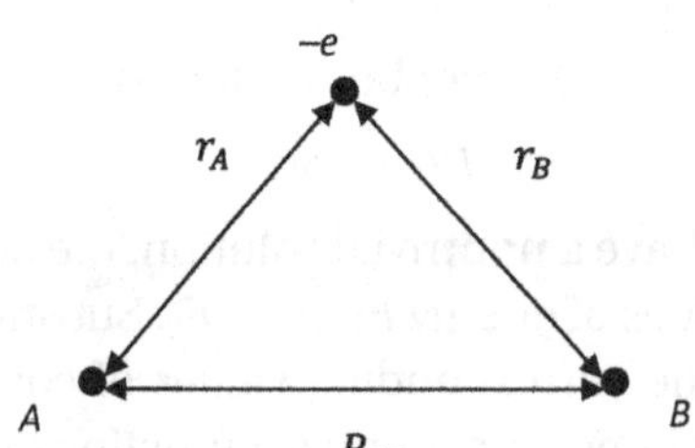

FIGURE 3.3 Relative positions of particles in H_2^+ molecule.

Let us now choose the atomic orbitals centered on each of the nuclei to construct MOs. When the electron is close to nucleus A, its distance from nucleus B (r_B) is very large and the potential due to nucleus B is negligible. In this case, the atomic orbital is very similar to that of hydrogen-like orbital around nucleus A. Let this orbital be named Φ_A. Similarly when electron is close to nucleus B, r_A is very large and the atomic orbital Φ_B is similar to the hydrogen-like orbital around the nucleus B. We may therefore construct a MO as a LCAOs Φ_A and Φ_B.

$$\Psi = c_A\Phi_A + c_B\Phi_B \tag{3.4.9}$$

where c_A and c_B are variable parameters which indicate the magnitude of contribution of each atomic orbital to the MO. The optimal value of the coefficients c_A and c_B are obtained by using the variational theory.

Multiplying Eq. (3.4.8) by Ψ^* on the left and integrating, we get

$$\int \Psi^* H \Psi d\tau = E \int \Psi^* \Psi d\tau \tag{3.4.10}$$

and on using Eq. (3.4.9), we get

$$E\left[\int (c_A\Phi_A + c_B\Phi_B)^*(c_A\Phi_A + c_B\Phi_B)d\tau\right] = \int \left[(c_A\Phi_A + c_B\Phi_B)^*\hat{H}(c_A\Phi_A + c_B\Phi_B)\right]d\tau \tag{3.4.11}$$

If we use the notations

$$\int \Phi_A^*\Phi_A d\tau = S_{AA}, \quad \int \Phi_A^*\Phi_B d\tau = S_{AB} \quad \text{and} \quad \int \Phi_A^* H \Phi_B \, d\tau = H_{AB} \tag{3.4.12}$$

Equation (3.4.11) simplifies to

$$E\left[c_A^2 S_{AA} + c_A c_B(S_{AB} + S_{BA}) + c_B^2 S_{BB}\right] = c_A^2 H_{AA} + c_A c_B H_{AB} + c_B c_A H_{BA} + c_B^2 H_{BB} \tag{3.4.13}$$

According to the variation principle, if the coefficients c_A, c_B, etc., can be chosen, so as to minimize the energy E, i.e., $\frac{\partial E}{\partial c_A} = \frac{\partial E}{\partial c_B} = 0$ then the so-obtained minimum energy shall give the upper limit of the actual energy of the system.

Differentiating E with respect to c_A and c_B respectively, and remembering that $S_{AB} = S_{BA}$ and $H_{AB} = H_{BA}$ for real wavefunctions, we get

$$\frac{\partial E}{\partial c_A}\left[c_A^2 S_{AA} + 2c_A c_B S_{AB} + c_B^2 S_{BB}\right] + E[2c_A S_{AA} + 2c_B S_{AB}] = 2c_A H_{AA} + 2c_B H_{AB} \tag{3.4.14}$$

Since $\frac{\partial E}{\partial c_A} = 0$, Eq. (3.4.14) reduces to

$$c_A(H_{AA} - ES_{AA}) + c_B(H_{AB} - ES_{AB}) = 0 \tag{3.4.15}$$

Similarly for $\frac{\partial E}{\partial c_B} = 0$, we get

$$c_A(H_{BA} - ES_{BA}) + c_B(H_{BB} - ES_{BB}) = 0 \tag{3.4.16}$$

If the atomic orbitals Φ_A and Φ_B are normalized $S_{AA} = S_{BB} = 1$, but since Φ_A and Φ_B are centered on different atoms $S_{AB} = S_{BA} \neq 0$.

Further, if both the atomic orbitals are of the same type, e.g., both are 1s, then $\Phi_A = \Phi_B$, $H_{AA} = H_{BA}$, $H_{AA} = H_{BB}$ and $S_{AB} = S_{BA} = S$.

Equations (3.4.15) and (3.4.16) can therefore be written as

$$c_A(H_{AA} - E) + c_B(H_{AB} - ES) = 0 \tag{3.4.17}$$

$$c_A(H_{AB} - ES) + c_B(H_{AA} - E) = 0 \tag{3.4.18}$$

The coefficients c_A and c_B are determined by solving the secular determinant

$$\begin{vmatrix} H_{AA} - E & H_{AB} - ES \\ H_{AB} - ES & H_{AA} - E \end{vmatrix} = 0 \tag{3.4.19}$$

which gives a quadratic equation in E and hence two values of energy

$$E_+ = \frac{H_{AA} + H_{AB}}{1 + S} \tag{3.4.20}$$

and

$$E_- = \frac{H_{AA} - H_{AB}}{1 - S} \tag{3.4.21}$$

The term E_+ represents symmetric bonding mode and E_- represents antisymmetric or antibonding mode.

Substituting the values of E_+ and E_- in either Eq. (3.4.17) or (3.4.18), we get $c_A = \pm c_B$ and hence two possible MOs

$$\Psi_+ = c_A(\Phi_A + \Phi_B) \tag{3.4.22}$$

and

$$\Psi_- = c_A(\Phi_A - \Phi_B) \tag{3.4.23}$$

The coefficients c_A can be determined by normalizing the two MOs. The normalization constant for Ψ_+ is found to be $\frac{1}{\sqrt{2+2S}}$ and for Ψ_- to be $\frac{1}{\sqrt{2-2S}}$.

Thus, the normalized MOs Ψ_+ and Ψ_- are given as

$$\Psi_+ = \frac{\Phi_A + \Phi_B}{\sqrt{2+2S}} \tag{3.4.24}$$

$$\Psi_- = \frac{\Phi_A - \Phi_B}{\sqrt{2-2S}} \tag{3.4.25}$$

The lowest energy atomic orbitals 1*s* of the hydrogen atom can be used to determine the integrals, H_{AA}, H_{AB}, and S and hence the energies E_+ and E_- of the symmetric and antisymmetric states of the $H_2{}^+$ molecule defined by the MO wavefunctions Ψ_+ and Ψ_-.

Using Eq. (3.4.7) for the Hamiltonian of the hydrogen ion, we get

$$\begin{aligned} H_{AA} &= \int \Phi_A^* H \Phi_A d\tau = \int \Phi_A^* \left(-\frac{1}{2}\nabla^2 - \frac{1}{r_A} - \frac{1}{r_B} + \frac{1}{R_{AB}} \right) \Phi_A d\tau \\ &= \int \Phi_A \left(-\frac{1}{2}\nabla^2 - \frac{1}{r_A} \right) \Phi_A d\tau - \int \Phi_A^* \frac{1}{r_B} \Phi_A d\tau + \int \frac{\Phi_A^* \Phi_A}{R_{AB}} d\tau \end{aligned} \tag{3.4.26}$$

Since Φ_A is the solution of the Schrödinger equation for the hydrogen atom,

$$\left(-\frac{1}{2}\nabla^2 - \frac{1}{r_A} \right) \Phi_A = E_H \Phi_A \tag{3.4.27}$$

we get

$$H_{AA} = E_H \int \Phi_A^* \Phi_A d\tau - \int \Phi_A^* \frac{1}{r_B} \Phi_A d\tau + \frac{1}{R_{AB}} \int \Phi_A^* \Phi_A d\tau$$

For normalized wavefunction Φ_A, this gives

$$H_{AA} = E_H + J + \frac{1}{R_{AB}} \quad (3.4.28)$$

where,

$$J = -\int \Phi_A^* \frac{1}{r_B} \Phi_A d\tau \quad (3.4.29)$$

J is called the Coulomb integral and represents the energy due to electrostatic attraction of one nucleus (charge +1 in au) for the electron cloud of charge density Φ_A^2 located on the other nucleus.

Similarly, we get

$$H_{AB} = \int \Phi_A^* H \Phi_B d\tau = E_H \int \Phi_A^* \Phi_B d\tau - \int \Phi_A^* \frac{1}{r_B} \Phi_B d\tau + \frac{1}{R_{AB}} \int \Phi_A^* \Phi_B d\tau = E_H S + K + \frac{S}{R_{AB}} \quad (3.4.30)$$

where,

$$S = \int \Phi_A^* \Phi_B d\tau \quad (3.4.31)$$

is the overlap integral for the two $1s$ orbitals.

$$K = -\int \Phi_A^* \frac{1}{r_B} \Phi_B d\tau \quad (3.4.32)$$

K is called a resonance or exchange integral, since both functions Φ_A and Φ_B occur in it. Unlike J it has no simple interpretation.

Substituting the values of H_{AA}, H_{AB}, and S in Eqs (3.4.20) and (3.4.21), we get

$$E_+ = E_H + \frac{J+K}{1+S} + \frac{1}{R_{AB}} \quad (3.4.33)$$

$$\text{and } E_- = E_H + \frac{J-K}{1-S} + \frac{1}{R_{AB}} \quad (3.4.34)$$

As seen from Eq. (3.4.32), K has a negative value and so $E_+ < E_-$. The atomic energies E_H which were equal when the two hydrogen atoms were infinitely apart, become separate and take values E_+ and E_- in the molecule (Figure 3.4).

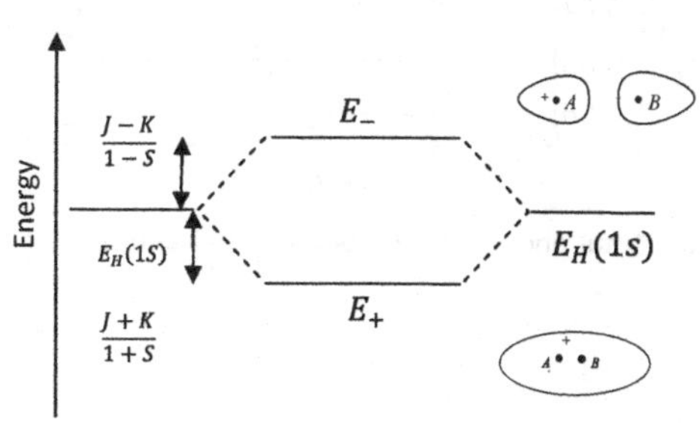

FIGURE 3.4 Energies and orbitals for the bonding and antibonding states of H_2^+.

For 1*s* orbitals of hydrogen with $\Phi_A = \Phi_B = 1s$, the integrals *J*, *S*, *K* have the following values:

$$-J = \frac{1}{a_0}\left[\frac{1}{\rho} - e^{-2\rho}\left(\frac{1}{\rho}+1\right)\right] \tag{3.4.35}$$

$$S = e^{-\rho}\left(1+\rho+\frac{1}{3}\rho^2\right) \tag{3.4.36}$$

$$\text{and } -K = \frac{1}{a_0}e^{-\rho}(1+\rho) \tag{3.4.37}$$

where $\rho = \frac{R_{AB}}{a_0}$. They depend upon the internuclear distance. The dependency is given in Figure 3.5.

Because of the dependence of *J*, *S*, and *K* on the internuclear distances, the energies E_+ and E_- also depend on R_{AB}. Curves showing energies E_+ and E_- as functions of the internuclear distance R_{AB} are given in Figure 3.6.

The curve for E_+ refers to the ground state of the molecule where a minimum energy is found for a nuclear distance of approximately $2a_0$ (1.32 Å). Thus the bond length of

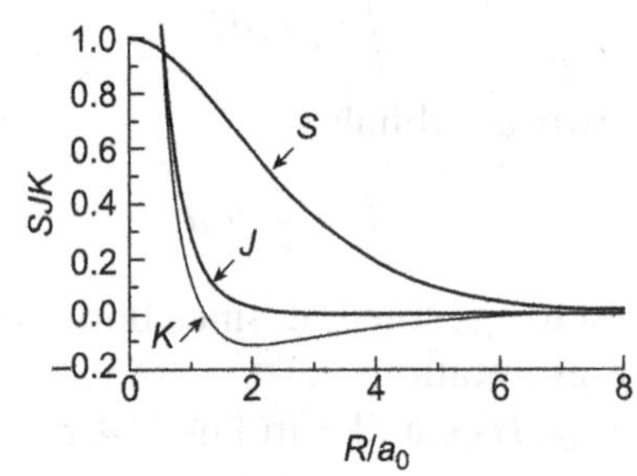

FIGURE 3.5 Dependence of Overlap integral *S*, Coulomb integral *J*, and resonance integral *K* on the internuclear distance *R*.

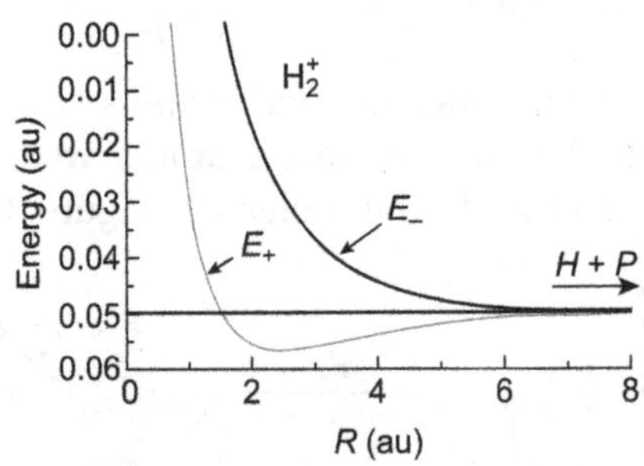

FIGURE 3.6 Curves representing the total energy for the bonding (+) and the antibonding (−) MOs as a function of the internuclear distance *R*.

$H_2{}^+$ molecule is 1.320 Å. This value is higher than the experimental value of 1.06 Å. Since the $H_2{}^+$ has a potential energy minimum, it should exist as a stable molecule. The calculated bonding energy is 1.77 eV, which is lower than the experimental value of 2.77 eV. This result is however satisfying if we consider the fact that we have made an assumption that the molecular wavefunction can be approximated using the individual atom ground states. This assumption is valid only for very large distances between the hydrogen atoms and is not accurate if they are 1.32 Å apart.

The curve for E_- shows no minimum. The molecule in this state is instable and with emission of energy, it splits simultaneously in one hydrogen atom and a proton $P(H^+)$. The energy curves E_+ and E_- are called "attractive" and "repulsive" and the respective MOs as "bonding" and "antibonding."

Using Eqs (3.4.24) and (3.4.25), the electron distributions in states Ψ_+ and Ψ_- are given as

$$\rho_+ = \Psi_+^2 = \frac{1}{(2+2S)}(\Phi_A + \Phi_B)^2 = \frac{1}{(2+2S)}\left[1s_A^2 + 1s_B^2 + 2.1s_A \cdot 1s_B\right]$$

$$\rho_- = \Psi_-^2 = \frac{1}{(2-2S)}(\Phi_A - \Phi_B)^2 = \frac{1}{(2-2S)}\left[1s_A^2 + 1s_B^2 - 2.1s_A \cdot 1s_B\right] \quad (3.4.38)$$

At a midpoint between the nuclei A and B, we have $1s_A = 1s_B$.

Hence, $\rho_+ = \frac{4 \cdot 1s_A^2}{(2+2S)}$ and $\rho_- = 0$

Thus, in the bonding state Ψ_+ there is more electron density between the two nuclei than what would be obtained from the sum of two separate atomic densities or in other words there is an accumulation of charge between the two nuclei. The attraction between the accumulated charge and the two protons leads to the stability of $H_2{}^+$ in the Ψ_+ state (Figure 3.7). It is for this reason, Ψ_+ is called bonding orbital.

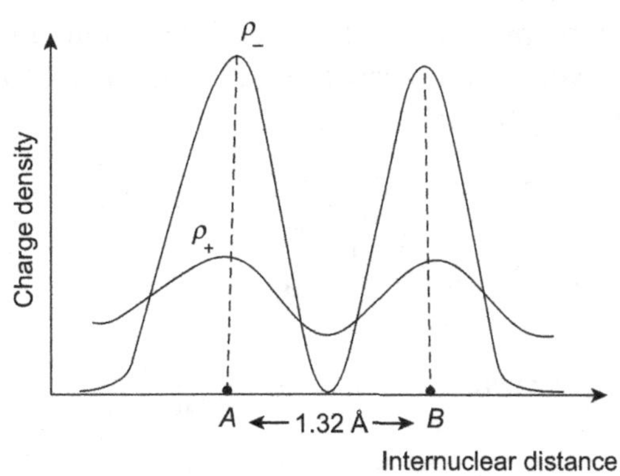

FIGURE 3.7 Distribution of electron density along the molecular axis in $H_2{}^+$.

On the other hand, in the Ψ_- state, the charge density at the center of the bond is zero and is therefore less than sum of the electron densities of separate atoms. The electron has maximum probability of existing near the two protons leading to structures $(H^+ + H)$ and $(H + H^+)$.

As we have seen, the calculated energies and bond lengths are in large error from their experimental values. This may be explained in terms of the approximations used in the calculations. Besides the fact that the variation technique gives only an upper limit of the energy, the use of spherically symmetric functions in the LCAO treatment is the main cause of error in the determination of energy and bond length of the H_2^+ molecule. The spherical symmetry in electron distribution is destroyed on molecule formation and the atomic orbitals get polarized or distorted. Thus, a spherical orbital $1s$ may be mixed with a $2p$ orbital to polarize it. Atomic orbitals of the type $(1s_A(\alpha) + c2p_A(\alpha'))$ and $(1s_B(\alpha) + c2p_B(\alpha'))$ were used by Dickenson [2] to calculate the energies of the H_2^+ molecule. The functions $1s_A(\alpha)$ and $2p_A(\alpha')$ had the form

$$1s_A(\alpha) = e^{-\alpha r_a}$$

$$2p_A(\alpha') = re^{-\alpha' r_a}$$

where, α and α' are the orbital exponents that may be changed to minimize the energy and c is an additional variational parameter. The $2p$ orbitals were oriented along the internuclear axis. Using such polarization functions in the variational technique, it was found that for $\alpha = 1.247$, $\alpha' = 2.868$ and $c = 0.145$. The bonding energy at bond length $r = 1.06$ Å was found to be 2.77 eV in close agreement with the experimental values. Various types of atomic orbitals are now in use for solving the energy value problems of diatomic and polyatomic molecules.

3.4.2 LCAO-MO Theory for Hydrogen Molecule

The treatment of hydrogen molecule by MO theory is essentially similar to that for the H_2^+ ion except that now the presence of two electrons results in several new integrals. The Hamiltonian for the hydrogen molecule (Figure 3.8) having two electrons moving in the field of fixed protons may be written as.

$$\begin{aligned} H &= -\frac{1}{2}\nabla_1^2 - \frac{1}{2}\nabla_2^2 - \frac{1}{r_{A1}} - \frac{1}{r_{B1}} - \frac{1}{r_{A2}} - \frac{1}{r_{B2}} + \frac{1}{r_{12}} + \frac{1}{R_{AB}} \\ &= H_e(1) + H_e(2) + \frac{1}{r_{12}} + \frac{1}{R_{AB}} \end{aligned} \tag{3.4.39}$$

$$\text{where,}\quad H_e(1) = -\frac{1}{2}\nabla_1^2 - \frac{1}{r_{A1}} - \frac{1}{r_{B1}}$$

$$\text{and,}\quad H_e(2) = -\frac{1}{2}\nabla_1^2 - \frac{1}{r_{B2}} - \frac{1}{r_{A2}} \tag{3.4.40}$$

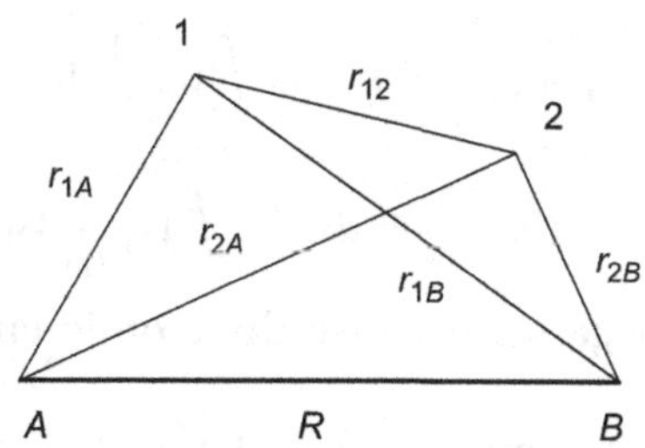

FIGURE 3.8 Coordinates used for hydrogen molecule.

These are similar to the Hamiltonian of the electrons 1 and 2 of the H_2^+ molecule. Since the internuclear distance R_{AB} is fixed for a given configuration, the correction due to nuclear–nuclear repulsion to the total energy can be made, as before, by adding it to the electron energy E_e at the end of the calculations

$$E = E_e + \frac{1}{R_{AB}}$$

If Φ_A and Φ_B are the two atomic orbitals corresponding to the hydrogen atoms A and B, both of which are 1s orbitals to be named as $1s_A$ and $1s_B$ then, proceeding in the same manner as in H_2^+ molecule, it is possible to construct two MOs

$$\Psi_1 = \frac{1}{\sqrt{2(1+s)}}(\Phi_A + \Phi_B) = \frac{1}{\sqrt{2(1+s)}}(1s_A + 1s_B) \tag{3.4.41}$$

$$\text{and,}\quad \Psi_2 = \frac{1}{\sqrt{2(1-s)}}(\Phi_A - \Phi_B) = \frac{1}{\sqrt{2(1-s)}}(1s_A - 1s_B) \tag{3.4.42}$$

Corresponding to energies

$$E_1 = \frac{H_{AA} + H_{AB}}{1+s}$$

$$\text{and,}\quad E_2 = \frac{H_{AA} - H_{AB}}{1-s} \tag{3.4.43}$$

where,

$$H_{AA} = \int \Phi_A^*(i) H_e(i) \Phi_A(i) d\tau_i$$

$$H_{AB} = \int \Phi_A^*(i) H_e(i) \Phi_B(i) d\tau_i$$

$$S = \int \Phi_A^*(i) \Phi_B(i) d\tau_i \tag{3.4.44}$$

as in Eqs (3.4.28) and (3.4.30)

$$H_{AA} = E_H + J$$

$$\text{and,}\quad H_{AB} = E_H S + K \tag{3.4.45}$$

$$J = -\int \Phi_A^* \frac{1}{r_{B1}} \Phi_A d\tau = -\int 1s_A^* \frac{1}{r_{B1}} 1s_A d\tau_1$$

$$\text{and, } K = -\int \Phi_A^* \frac{1}{r_{A1}} \Phi_B d\tau = -\int 1s_A^* \frac{1}{r_{A1}} 1s_B d\tau_1 d\tau_2 \quad (3.4.46)$$

Both these integrals are negative because the atomic energy E_H is negative, hence $E_1 < E_2$.

Substituting the values of H_{AA} and H_{AB} from Eq. (3.4.45) into Eq. (3.4.43), we get

$$E_1 = \frac{E_H + J + E_H s + K}{1+s} = E_H + \frac{J+K}{1+s}$$

$$E_2 = \frac{E_H + J - E_H s - K}{1-s} = E_H + \frac{J-K}{1+s} \quad (3.4.47)$$

E_1 and E_2 which are the energies of the ground and first excited states of the hydrogen molecule.

The hydrogen molecule has two electrons and both of them can occupy the lower energy MO Ψ_1 with opposite spins. The Slater determinant form of the ground state wavefunction can be written as

$$\Psi_0 = \frac{1}{\sqrt{2}} \begin{vmatrix} \Psi_1(1)\alpha(1) & \Psi_1(1)\beta(1) \\ \Psi_1(2)\alpha(2) & \Psi_1(2)\beta(2) \end{vmatrix} = \Psi_1(1)\Psi_1(2)\frac{1}{\sqrt{2}}[(\alpha(1)\beta(2) - \alpha(2)\beta(1)] \quad (3.4.48)$$

Ignoring the spin part of the wavefunction and using Eqn (3.4.41), we get

$$\Psi_0 = \Psi_1(1)\Psi_1(2) = \frac{1}{\sqrt{2(1+s)}}[1s_A(1)1s_A(2) + 1s_B(1)1s_B(2) + 1s_A(1)1s_B(2) + 1s_B(1)1s_A(2)] \quad (3.4.49)$$

Using Eq. (3.4.40) and the Slater determinant-type wavefunction; the electronic energy E_e can be written as

$$E_e = \int \Psi_0 H \Psi_0 d\tau = \int \Psi_0 \left[H_e(1) + H_e(2) + \frac{1}{r_{12}} \right] \Psi_0 d\tau = 2E_1 + J_{11} \quad (3.4.50)$$

$$\text{where, } J_{11} = \int \Psi_1(1)\Psi_1(1)\frac{1}{r_{12}}\Psi_1(2)\Psi_1(2) d\tau_1 d\tau_2 \quad (3.4.51)$$

Using Eq. (3.4.41), we get,

$$\begin{aligned} J_{11} = \frac{1}{4(1+s)^2} \bigg[& 2\iint 1s_A(1)1s_A(1)\frac{1}{r_{12}}1s_A(2)1s_A(2)d\tau_1 d\tau_2 \\ & + 2\iint 1s_A(1)1s_A(1)\frac{1}{r_{12}}1s_B(2)1s_B(2)d\tau_1 d\tau_2 \\ & + 8\iint 1s_A(1)1s_A(1)\frac{1}{r_{12}}1s_A(2)1s_B(2)d\tau_1 d\tau_2 \\ & + 4\iint 1s_A(1)1s_B(1)\frac{1}{r_{12}}1s_A(2)1s_B(2)d\tau_1 d\tau_2 \bigg] \\ = \frac{1}{4(1+s)^2} & [2\langle AA|AA\rangle + 2\langle AA|BB\rangle + 8\langle AA|AB\rangle + 4\langle AB|AB\rangle] \end{aligned} \quad (3.4.52)$$

The last equation being written in Dirac notations.

The integral represented by $\langle AA|AA\rangle$ is called a one-center integral, the integral $\langle AA|BB\rangle$ is called the two-center Coulomb integral, the integral $\langle AA|AB\rangle$ is called the two-center hybrid integral, and the integral $\langle AB|AB\rangle$ is called the two-center exchange integral. While the first integral is independent of the internuclear distance R_{AB}, all others depend on it.

Using Eqs (3.4.43), (3.4.44), and (3.4.45) in Eq. (3.4.50), we get

$$E_e = 2E_H + \frac{2(J+K)}{1+s} + J_{11} \tag{3.4.53}$$

After including the nuclear repulsion energy we get for the total energy of the ground state of the hydrogen molecule.

$$E_0 = 2E_H + \frac{2(J+K)}{1+s} + J_{11} + \frac{1}{R_{AB}} \tag{3.4.54}$$

It follows from Eq. (3.4.54) that the total energy depends upon the internuclear distance. The difference between the energy of the hydrogen molecule (E_0) and the two hydrogen atoms (E_H), that is (E_0–$2E$), is called the interaction energy or bond energy of the molecule. The plot of E_0 versus R is called the potential energy curve (Figure 3.9). It has a minimum at 0.84 Å showing that the equilibrium bond length of hydrogen molecule is 0.84 Å. This is larger than the experimental value of 0.74 Å. The calculated bond energy is found to be 2.68 eV against the experimental value of 4.75 eV. The MO theory is thus only an approximate method which is unable to give accurate values of bond length and binding energy due to some of its shortcomings.

The two 1*s* atomic orbitals used in the above treatment give rise to two MOs Ψ_1 and Ψ_2, also called σ_s and σ_s^*, which are symmetric and antisymmetric, respectively, in the spatial coordinates. According to Pauli's exclusion principle, the complete wavefunction must be antisymmetric in electrons and hence the complete wavefunctions for the σ_s state of the hydrogen molecule shall be

$$\Psi_{\sigma_s} = \Psi_i \Psi_{\text{spin}}(\uparrow\downarrow) = \Psi_1[\alpha(1)\beta(2) - \beta(1)\alpha(2)]$$

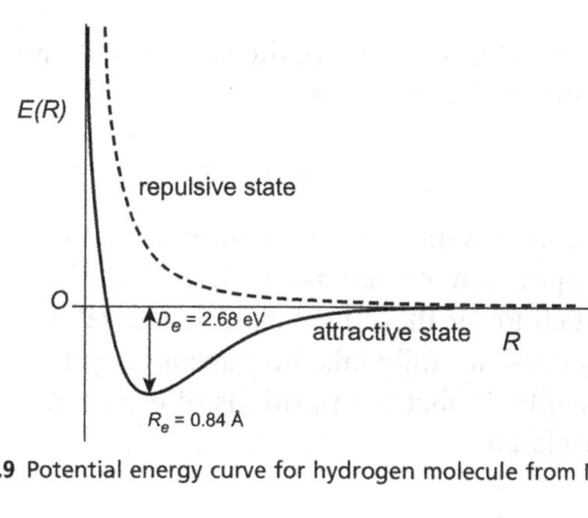

FIGURE 3.9 Potential energy curve for hydrogen molecule from MO theory.

with the two electrons with opposite spin occupying the same orbital Ψ_1. The total spin S being zero, this attractive ground state is a singlet state.

Likewise, since Ψ_2 is antisymmetric in spatial coordinates, the complete wavefunction for the σ_s^* state shall be

$$\Psi_{\sigma^*} = \Psi_2 \cdot \Psi_{spin}(\uparrow\uparrow)$$

Since the spins of the two electrons in this state are parallel they cannot occupy the same MO. The total spin in this case is $S = 1$ and so this state is a triplet state. This is the first excited state of the hydrogen molecule. On excitation, one of the electrons jumps from the electronic state E_1 to E_2.

3.4.2.1 Shortcoming of MO Wavefunctions

1. It follows from Eq. (3.4.54) that even at $R_{AB}\rightarrow\infty$ the total energy E_0 does not reduce to $2E_H$ because while at $R_{AB}=\infty$ the terms J, K, and S vanish, the term J_{11} does not vanish. In fact all other terms in Eq. (3.4.52) vanish except the first term $\langle AA|AA\rangle$ and hence

$$E_0(\infty) = 2E_H + \frac{1}{2}\langle AA|AA\rangle$$

2. The ground state wavefunction Ψ_0 given by Eq. (3.4.49) has four components. The first two components namely $1s_A(1)1s_A(2)$ and $1s_B(1)1s_B(2)$ show that both the electrons are lying on the same atomic orbital and hence correspond to configurations $H_A^- - H_B^+$ and $H_A^+ - H_B^-$. These therefore correspond to the ionic structures for the H_2 molecule. The other two terms $1s_A(1)1s_B(2)$ and $1s_B(1)1s_A(2)$ correspond to the situation where the electrons are equally shared by both the nuclei A and B and hence represent the covalent structure of the molecule. The wavefunction Ψ_0 may, therefore, be written as $\Psi_0 = \Psi_{cov} + \Psi_{ion}$. Thus in Eq. (3.4.49), the covalent and ionic structures have the same weightage which is not true; in fact the ionic structure has an extremely small weightage in the hydrogen molecule. This is one of the major reasons for disagreement between the calculated and observed results.

   An improvement in the MO description of the H_2 molecule was made by Coulson [3] by taking the atomic orbitals in the form

$$1s_A(1) = \sqrt{\frac{\alpha^3}{\pi}}e^{-\alpha r_{A1}} \text{ and } 1s_B(2) = \sqrt{\frac{\alpha^3}{\pi}}e^{-\alpha r_{B1}}$$

   and varying the adjustable parameter α to minimize the energy by the variation method. While this approach considerably improved the dissociation energy to 3.49 eV, it was still much lower than the experimental value 4.75 eV.

3. MO wavefunction Ψ_0 does not fully take into account spin correlation within the Pauli's exclusion principle. In fact the positions of electrons in a many-electron system should be correlated.

In later development of molecule orbital theory, this shortcoming was removed by introducing configuration interaction (CI) between different states of the molecule as well as by incorporating horizontal correlation.

The MO treatment however, has an advantage that it is easily extendable to polyatomic molecules.

3.5 VB Theory for Hydrogen Molecule—Heitler–London Model

The hydrogen molecule is described by an equation which considers the motion of two electrons in a field of two fixed protons separated by a distance say R_{AB} where 1 and 2 refer to electrons and A and B to protons (Figure 3.8)

$$H\Psi = E\Psi \tag{3.5.1}$$

where,

$$H = -\frac{1}{2}\nabla_1^2 - \frac{1}{2}\nabla_2^2 - \frac{1}{r_{A1}} - \frac{1}{r_{B1}} - \frac{1}{r_{A2}} - \frac{1}{r_{B2}} + \frac{1}{r_{12}} + \frac{1}{R_{AB}} \tag{3.5.2}$$

If the two hydrogen atoms were isolated, that is $R_{AB} = \infty$, then the Schrödinger equation for the first atom can be written as,

$$\left(-\frac{1}{2}\nabla_1^2 - \frac{1}{r_{A1}}\right)\Phi_A(1) = E_A\Phi_A(1), \tag{3.5.3}$$

where $\Phi_A(1)$ is the atomic orbital for atom A, and for the second atom as

$$\left(-\frac{1}{2}\nabla_2^2 - \frac{1}{r_{B2}}\right)\Phi_B(2) = E_B\Phi_B(2) \tag{3.5.4}$$

where $\Phi_A(1)$ and $\Phi_B(2)$ show that electrons 1 and 2 are in atomic orbitals Φ_A and Φ_B, respectively.

When the two atoms are brought together to form a molecule, due to the indistinguishability of the electrons, two molecular wavefunctions can be formed:

$$\Psi_1 = \Phi_A(1)\Phi_B(2) \tag{3.5.5}$$

$$\text{and } \Psi_2 = \Phi_A(2)\Phi_B(1) \tag{3.5.6}$$

By the principle of superposition, the trial wavefunction can be a combination of Ψ_1 and Ψ_2.

$$\Psi_{MO} = a_1\Psi_1 + a_2\Psi_2 \tag{3.5.7}$$

If we now use variation technique to solve Eq. (3.5.1), by choosing for the trial wavefunction Eq. (3.5.7), we shall get the secular equations

$$(H_{11} - S_{11}E)a_1 + (H_{12} - S_{12}E)a_2 = 0$$

$$(H_{21} - S_{21}E)a_1 + (H_{22} - S_{22}E)a_2 = 0 \tag{3.5.8}$$

that given nontrivial solutions, if

$$\begin{vmatrix} H_{11} - S_{11}E & H_{12} - S_{12}E \\ H_{21} - S_{21}E & H_{22} - S_{22}E \end{vmatrix} = 0 \tag{3.5.9}$$

where $H_{11} = \int \Psi_1^* H \Psi_1 d\tau,\ \ H_{22} = \int \Psi_2^* H \Psi_2 d\tau,\ \ H_{12} = \int \Psi_1^* H \Psi_2 d\tau, \ldots$

$$\text{and}\quad S_{11} = \int \Psi_1^* \Psi_1 d\tau,\ \ S_{12} = \int \Psi_1^* \Psi_2 d\tau \ \text{etc.} \tag{3.5.10}$$

If we use normalized atomic orbitals, then $S_{11} = S_{22} = 1$.

Also, since the two atoms are identical, $H_{11} = H_{22}$ and

$$S_{12} = S_{21} = \int \Psi_1^* \Psi_2 d\tau = \int \Phi_A^*(1)\Phi_B^*(2)\Phi_A(2)\Phi_B(1) d\tau = S^2 \tag{3.5.11}$$

where $S = \int \Phi_A^*(1)\Phi_B(1) d\tau$

Equation (3.5.9), therefore, reduces to

$$\begin{vmatrix} H_{11} - E & H_{12} - S^2E \\ H_{12} - S^2E & H_{11} - E \end{vmatrix} = 0$$

and gives two solutions,

$$E_S = \frac{H_{11} + H_{12}}{1 + S^2} \tag{3.5.12}$$

$$E_A = \frac{H_{11} - H_{12}}{1 - S^2} \tag{3.5.13}$$

By substituting these two eigenvalues in secular Eq. (3.5.8), we get $a_1 = \pm a_2$, and hence

$$\Psi_s = a_1(\Psi_1 + \Psi_2) \tag{3.5.14}$$

$$\text{and}\quad \Psi_A = a_1(\Psi_1 - \Psi_2) \tag{3.5.15}$$

If we choose normalized MO, then

$$\int \Psi_{\text{MO}}^* \Psi_{\text{MO}} d\tau = 1 \tag{3.5.16}$$

we get, $a_1 = \frac{1}{\sqrt{2+2s^2}}$ for $a_1 = a_2$ and $a_1 = \frac{1}{\sqrt{2-2s^2}}$ for $a_1 = -a_2$

On substituting the values of a_1 and a_2 in Eqs (3.5.14) and (3.5.15), we get,

$$\Psi_s = \frac{\Psi_1 + \Psi_2}{\sqrt{2 + 2s^2}} = \frac{1}{\sqrt{2 + 2s^2}}[\Phi_A(1)\Phi_B(2) + \Phi_B(1)\Phi_A(2)] \tag{3.5.17}$$

$$\Psi_A = \frac{\Psi_1 - \Psi_2}{\sqrt{2 - 2s^2}} = \frac{1}{\sqrt{2 + 2s^2}}[\Phi_A(1)\Phi_B(2) - \Phi_B(1)\Phi_A(2)], \tag{3.5.18}$$

where Ψ_s is symmetrical in spatial coordinates and Ψ_A is antisymmetrical.

We can now evaluate the various integrals involved in Eqs (3.5.12) and (3.5.13) for E_S and E_A.

$$\text{(a) } H_{11} = \int \Psi_1^* H \Psi_1 d\tau$$

$$= \int \Phi_A^*(1)\Phi_B^*(2)\left[-\frac{1}{2}(\nabla_1^2+\nabla_2^2) - \frac{1}{r_{A1}} - \frac{1}{r_{B1}} - \frac{1}{r_{A2}} - \frac{1}{r_{B2}} + \frac{1}{r_{12}} + \frac{1}{r_{AB}}\right]\Phi_A(1)\Phi_B(2)d\tau_1 d\tau_2 \tag{3.5.19}$$

Using Eqs (3.5.3) and (3.5.4) we get,

$$H_{11} = \int \Phi_A^*(1)\Phi_B^*(2)\left[E_A + E_B + \frac{1}{r_{AB}} + \frac{1}{r_{12}} - \frac{1}{r_{A2}} - \frac{1}{r_{B1}}\right]\Phi_A(1)\Phi_B(2)d\tau_1 d\tau_2$$

Since $E_A = E_B = E_0$ (say), and the nuclei are fixed in position, that is, $R_{AB} =$ constant, we get

$$H_{11} = 2E_0 + \frac{1}{r_{AB}} + \int \Phi_A^*(1)\Phi_B^*(2)\frac{1}{r_{12}}\Phi_A(1)\Phi_B(2)d\tau_1 d\tau_2$$

$$\int \Phi_A^*(1)\Phi_B^*(2)\left(\frac{1}{r_{A2}} + \frac{1}{r_{B1}}\right)\Phi_A(1)\Phi_B(2)d\tau_1 d\tau_2 = 2E_0 + \frac{1}{r_{AB}} + J_1 - 2J_2 \tag{3.5.20}$$

where,

$$J_1 = \int \Phi_A^*(1)\Phi_B^*(2)\frac{1}{r_{12}}\Phi_A(1)\Phi_B(2)d\tau_1 d\tau_2 \tag{3.5.21}$$

$$\text{and } J_2 = \int \Phi_A^*(1)\Phi_B^*(2)\frac{1}{r_{A2}}\Phi_A(1)\Phi_B(2)d\tau_1 d\tau_2 = \int \Phi_A^*(1)\Phi_B^*(2)\frac{1}{r_{B1}}\Phi_A(1)\Phi_B(2)d\tau_1 d\tau_2 \tag{3.5.22}$$

J_1 and J_2 are Coulomb integrals and have negative values.

Similarly,

$$H_{12} = \int \Phi_A^*(1)\Phi_B^*(2)\left[-\frac{1}{2}(\nabla_1^2+\nabla_2^2) - \frac{1}{r_{A1}} - \frac{1}{r_{B1}} - \frac{1}{r_{A2}} - \frac{1}{r_{B2}} + \frac{1}{r_{12}} + \frac{1}{r_{AB}}\right]\Phi_A(2)\Phi_B(1)d\tau_1 d\tau_2$$

$$= \int \Phi_A^*(1)\Phi_B^*(2)\left\{2E_0 + \frac{1}{r_{AB}} + \frac{1}{r_{12}} - \frac{1}{r_{A1}} - \frac{1}{r_{B2}}\right\}\Phi_A(2)\Phi_B(1)d\tau_1 d\tau_2 \tag{3.5.23}$$

$$= 2E_0 s^2 + \frac{1}{r_{AB}}s^2 + K_1 - 2K_2$$

where,

$$K_1 = \int \Phi_A^*(1)\Phi_B^*(2)\frac{1}{r_{12}}\Phi_A(2)\Phi_B(1)d\tau \tag{3.5.24}$$

$$K_2 = \int \Phi_A^*(1)\Phi_B^*(2)\frac{1}{r_{A1}}\Phi_A(2)\Phi_B(1)d\tau = \int \Phi_A^*(1)\Phi_B^*(2)\frac{1}{r_{B2}}\Phi_A(2)\Phi_B(1)d\tau \tag{3.5.25}$$

K_1 and K_2 are known as the exchange integrals.

Substituting the values of H_{11} and H_{12} from Eqs (3.5.20) and (3.5.23) into Eq. (3.5.12) and (3.5.13) we get,

$$E_s = 2E_0 + \frac{1}{r_{AB}} + \frac{J_1 - 2J_2 + K_1 - 2K_2}{1 + s^2} \tag{3.5.26}$$

$$E_A = 2E_0 + \frac{1}{r_{AB}} + \frac{J_1 - 2J_2 - K_1 + 2K_2}{1 - s^2} \tag{3.5.27}$$

Curves representing E_s and E_A as functions of r_{AB} are similar to Figure 3.9 in the LCAO-MO theory. As before, the curve E_s corresponds to attraction of the two hydrogen atoms with the formation of a stable molecule. The value of r_{AB} at the potential energy minimum corresponding to the equilibrium bond length which works out to 0.80 Å. This is in rough agreement with the experimental value 0.740 Å. The energy of dissociation of the molecule into atoms (neglecting vibrational energy of the nuclei) is calculated to be 3.14 eV, which is somewhat less than the experimental value 4.72 eV. E_A corresponds to repulsion at all distances, there being no equilibrium position of the nuclei.

So far we have not included electron spin in our treatment. According to Pauli's exclusion principle, the complete wavefunction must be antisymmetric in electrons. Since the wavefunction Ψ_s is symmetric in the spatial part, it must be antisymmetric in the spin part so that the complete wavefunction becomes antisymmetric. Thus,

$$\begin{aligned}\Psi = \Psi_s \Psi_{\text{spin}} &= \frac{1}{\sqrt{2+2s^2}}[\Phi_A(1)\Phi_B(2) + \Phi_B(1)\Phi_A(2)]\Psi(\uparrow\downarrow) \\ &= \frac{1}{\sqrt{2+2s^2}}[\Phi_A(1)\Phi_B(2) + \Phi_B(1)\Phi_A(2)][\alpha(1)\beta(2) - \beta(1)\alpha(2)]\end{aligned} \tag{3.5.28}$$

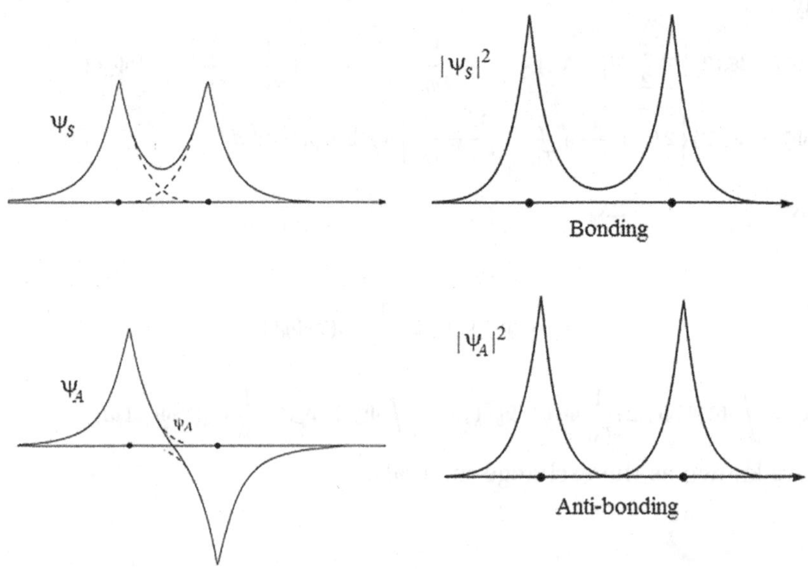

FIGURE 3.10 Wavefunctions and probability densities for bonding and antibonding states of H_2.

This state has total spin $S = 0$, corresponding to a singlet state. Thus, the normal state of hydrogen molecule is a singlet state.

In contrast, the wavefunction Ψ_A is antisymmetric and must be multiplied with a symmetric spin function, which may either be $\alpha(1)\alpha(2)$, $\beta(1)\beta(2)$ or $[\alpha(1)\beta(2) + \beta(1)\alpha(2)]$.

The complete wavefunction in this case

$$\Psi = \Psi_A \Psi_{\text{spin}} = \frac{1}{\sqrt{2 - 2s^2}} [\Phi_A(1)\Phi_B(2) - \Phi_B(1)\Phi_A(2)]\Psi(\uparrow\uparrow) \quad (3.5.29)$$

can therefore, be

$$\Psi_A \alpha(1)\beta(2)$$

$$\Psi_A \frac{1}{\sqrt{2}} [\alpha(1)\beta(2) + \beta(1)\alpha(2)] \quad (3.5.30)$$

$$\Psi_A \beta(1)\beta(2)$$

The total spin in this case is $S = 1$, and so this state shall be a triplet state (Figure 3.10).

3.5.1 Shortcoming of VB Theory

In the VB theory, as in the case of the MO theory, the distortion in the spherically symmetric $1s$ orbitals of the hydrogen atom as they form a molecule is neglected. This shortcoming is removed by Wang [4] by changing the value of the orbital coefficient α in the exponential term $e^{-\alpha r}$ from its original value 1.0. By this method he obtained binding energy of 3.78 eV which is a considerable improvement over the previous value of 3.14 eV. Weinbaum [5] further improved upon these results by including ionic terms $\Phi_A(1)\Phi_A(2)$ and $\Phi_B(1)\Phi_B(2)$ in the original Heitler–London wavefunction Ψ_0. Thus, he used an improved wavefunction

$$\Psi = \Phi_A(1)\Phi_B(2) + \Phi_B(1)\Phi_A(2) + \lambda[\Phi_A(1)\Phi_A(2) + \Phi_B(1)\Phi_B(2)]$$

and with $\alpha = 1.193$ and $\lambda = 0.256$ obtained the binding energy of 4.02 eV and equilibrium bond length of 0.75 Å. Results of some of the calculations on hydrogen molecule using VB theory are listed in Table 3.1 and are compared with the values obtained by MO calculations.

3.6 One-Electron Density Function and Charge Distribution in Hydrogen Molecule

If we consider a single electron in the normalized orbital $\Psi(\mathbf{r})$, then the probability of finding the electron in the volume element $d\tau$ is

$$\Psi(\mathbf{r})\Psi^*(\mathbf{r})d\tau = |\Psi(\mathbf{r})|^2 d\tau \quad (3.6.1)$$

The quantity $|\Psi(\mathbf{r})|^2$ is a charge density function which describes the spatial distribution of electronic charge in the system: $-e|\Psi(r)|^2 d\tau$ is the amount of electronic

Table 3.1 Bond energies and equilibrium bond lengths of hydrogen molecule by different methods

Type of wavefunction	Bond energy (eV)	R_e (Å)
Heitler–London	3.14	0.87
Wang [4]	3.78	0.744
Weinbaum [5]	4.02	0.74
Molecular orbital (simple)	2.68	0.84
Molecular orbital with optimized orbital exponent (Coulson [3])	3.49	0.730
Experimental	4.746	0.741

charge to be found in volume element $d\tau$ at position $\mathbf{r}$. We can similarly use spin orbital $\Psi(\mathbf{r},\sigma)$ instead of $\Psi(\mathbf{r})$. Likewise, for a two-electron system in the state with normalized wavefunction $\Psi(\mathbf{r}_1,\sigma_1,\mathbf{r}_2,\sigma_2)$ the quantity $|\Psi(\mathbf{r}_1,\sigma_1,\mathbf{r}_2,\sigma_2)|^2 d\tau_1 d\tau_2$ is interpreted as the simultaneous probability of finding the one electron (labeled 1) with coordinates in the space-spin volume element $d\tau_1$ and the second electron with coordinates in the element $d\tau_2$. The probability of finding electron 1 in element $d\tau_1$ irrespective of position and spin of electron 2 is then found by integrating over the coordinates of electron 2:

$$\left[\int |\Psi(\mathbf{r}_1,\sigma_1,\mathbf{r}_2,\sigma_2)|^2 d\tau_2\right] d\tau_1 \tag{3.6.2}$$

Since the wavefunction Ψ is antisymmetric with respect to exchange of coordinates, the two-electron density $|\Psi|^2$ is symmetric in the coordinates of the two electrons. The probability of finding either electron in $d\tau_1$ is therefore twice of this value.

Thus, if $\rho(\mathbf{r}_1,\sigma_1)d\tau_1$ is the probability of finding either electron in the volume element $d\tau_1$, independent of the position and spin of the other electron

$$\rho(\mathbf{r}_1,\sigma_1)d\tau_1 = \left[2\int |\Psi(\mathbf{r}_1,\sigma_1,\mathbf{r}_2,\sigma_2)|^2 d\tau_2\right] d\tau_1 = 2\int |\Psi(\mathbf{r}_1,\sigma_1,\ \mathbf{r}_2,\sigma_2)|^2 d\tau_2 \tag{3.6.3}$$

$\rho(\mathbf{r}_1,\sigma_1)$ is the one-electron density function of the two-electron system.

If we generalize it to the case of an N-electron wavefunction $\Psi(\mathbf{r}_1,\sigma_1,\mathbf{r}_2,\sigma_2 \ldots \mathbf{r}_N,\sigma_N)$, the one-electron density function will be

$$\rho(\mathbf{r}_1,\sigma_1) = N\int |\Psi(\mathbf{r}_1,\sigma_1,\mathbf{r}_2,\sigma_2 \ldots \mathbf{r}_N,\sigma_N)|^2 d\tau_2 \ldots . d\tau_N \tag{3.6.4}$$

$\rho(\mathbf{r}_1,\ \sigma_1)d\tau$ is then the probability of finding any one of the N electrons in the volume element $d\tau$, independent of the positions and spins of the other electrons.

The one-electron density function has a particular simple form when the wavefunction is a single Slater determinant and may be written as the sum of the individual spin–orbit density functions.

$$\rho(\mathbf{r}) = \sum_{i=1}^{N} \sum_{\sigma_1=-\frac{1}{2},+\frac{1}{2}} |\Psi_i(\mathbf{r}, \sigma_1)|^2 \tag{3.6.5}$$

$\sigma_1 = 1/2$ corresponds to $\alpha(\uparrow)$ and $\sigma_1 = -1/2$ corresponds to $\beta(\downarrow)$ electron. Thus, for example, in the case of the hydrogen molecule,

$$\rho(1)d\tau_1 = \left| \int \Psi_0^2(1,2)d\tau_2 \right| d\tau_1 \tag{3.6.6}$$

where $\Psi_0(1,2)$ is the ground state MO wavefunction given by equation

$$\Psi_0(1,2) = \frac{1}{\sqrt{2}}\left[\Psi_1(1)\overline{\Psi}_1(2) - \overline{\Psi}_1(1)\Psi_1(2)\right] \tag{3.6.7}$$

$$\begin{aligned}\rho(1) &= \frac{1}{2}\left[\Psi_1^2(1)\int \overline{\Psi}_1^2(2)d\tau_2 + \overline{\Psi}_1^2(1)\int \Psi_1^2(2)d\tau_2 - 2\Psi_1(1)\overline{\Psi}_1(1)\int \overline{\Psi}_1(2)\Psi_1(2)d\tau_2\right] \\ &= \frac{1}{2}\left[\Psi_1^2(1) + \overline{\Psi}_1^2(1)\right]\end{aligned} \tag{3.6.8}$$

The integral $\int \overline{\Psi}_1(2)\Psi_1(2)d\tau_2$ vanishes due to the orthogonality of the spin functions α and β as electrons 1 and 2 are indistinguishable. The density function $\rho(2)$ shall also have the same value. The total one-electron density function is therefore,

$$\rho = \rho(1) + \rho(2) = 2\rho(1) \tag{3.6.9}$$

Similarly, for an *N*-electron wavefunction, assuming double occupancy of molecular orbitals, we have from Eq. (3.6.5)

$$\rho(\mathbf{r}) = \sum_{i=1}^{N} \sum_{\sigma_1=\frac{1}{2},-\frac{1}{2}} |\Psi_i(\mathbf{r}, \sigma_1)|^2 = \sum_{i=1}^{N/2}\sum_{\sigma_1} |\Psi_i(\mathbf{r})\alpha(\sigma_1)|^2 + \sum_{i=1}^{N/2}\sum_{\sigma_1} |\Psi_i(\mathbf{r})\beta(\sigma_1)|^2 = \sum_{i=1}^{N/2} 2|\Psi_i(\mathbf{r})|^2 \tag{3.6.10}$$

In the case of open-shell systems, $\rho(\mathbf{r})$ may be written as

$$\rho(\mathbf{r}) = \sum_i n_i |\Psi_i(\mathbf{r})|^2 \tag{3.6.11}$$

with $n_i = 1$ or 2 as the occupation number of the orbital Ψ_i.

Proceeding from Eq. (3.6.9) and using the wavefunctions of the ground state, we can calculate the one-electron density function $\rho(1)$ in the MO and VB theories. Thus, using Eq. (3.4.41) for the MO wavefunction, we get

$$\rho(1) = \Psi_1^2(1) = \frac{1}{2(1+s)}[1s_A(1) + 1s_B(1)]^2 = \frac{1}{2(1+s)}\left[1s_A^2(1) + 1s_B^2(1) + 2.1s_A(1)1s_B(1)\right] \tag{3.6.12}$$

Since electrons 1 and 2 are indistinguishable, $\rho(2)$ shall also have the same value. Hence,

$$\rho = \rho(1) + \rho(2) = \frac{1}{(1+s)}\left[1s_A^2(1) + 1s_B^2(1) + 2.1s_A(1)1s_B(1)\right] \tag{3.6.13}$$

If we integrate ρ over all space for electron 1, we get

$$\begin{aligned}\int \rho d\tau_1 &= \frac{1}{(1+s)}\left[\int 1s_A^2(1)d\tau_1 + \int 1s_B^2(1)d\tau_1 + 2\int 1s_A(1)1s_B(1)d\tau_1\right] \\ &= \frac{1}{(1+s)}[1+1+2s] = 2\end{aligned} \tag{3.6.14}$$

Thus, the total electron density is 2, which is also the total number of electrons.

For the case of an N-electron system, ρ represents density of the electron cloud carrying N electrons. Hence, integrating over all the three-dimensional space,

$$\int \rho(\mathbf{r})d\tau = N \tag{3.6.15}$$

We may analyze Eq. (3.6.14) to say that, of the two electrons of the hydrogen molecule $\frac{1}{1+s}$ are associated with each of the two hydrogen atoms and $\frac{2s}{1+s}$ electrons remain in the internuclear region. Since $\frac{2s}{1+s} > 0$, it immediately follows that the H–H bond has an electron population greater than zero, i.e., the atom–atom interaction is bonding and the electronic charge is concentrated in the internuclear region (Figure 3.11).

If we now consider the MO Ψ_2 from Eq. (3.4.42), we get

$$\int \rho d\tau_1 = \frac{1}{(1-s)}[1+1-2s]$$

In this case, $\frac{1}{(1-s)}$ electrons are associated with each hydrogen atom and $-\frac{2s}{(1-s)}$ with the internuclear region. Since the last term is negative, it shows that the atoms are interacting in an antibonding manner and the electronic charge is concentrated near the two hydrogen atoms (Figure 3.11).

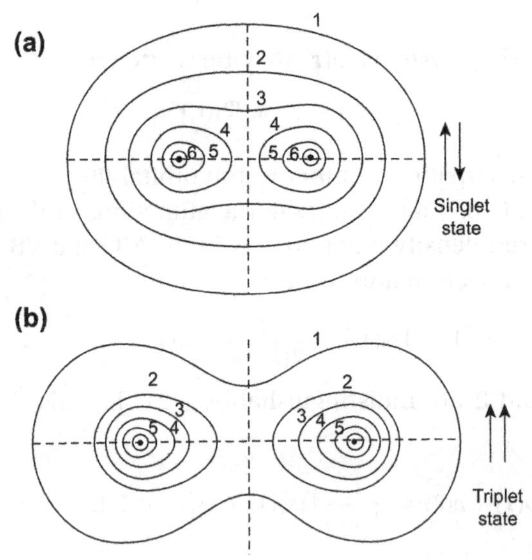

FIGURE 3.11 A contour map of the electron density distribution for homonuclear diatomic molecules like H_2 in (a) stable and (b) unstable state.

3.7 Formation of Molecular Quantum Numbers for Diatomic Molecules

Formation of molecular quantum numbers follows the same pattern as the atomic quantum numbers. The orbital angular momenta l_1, l_2,... in an atom couple together to give the total orbital angular momentum L and the spin angular momenta s_1, s_2,... give total spin angular momentum S.

$$\text{Magnitude of the total orbital angular momentum } L^* = \sqrt{L(L+1)}\hbar \text{ and} \quad (3.7.1)$$

$$\text{Magnitude of the total spin angular momentum } S^* = \sqrt{S(S+1)}\hbar. \quad (3.7.2)$$

When two atoms, one characterized by angular momenta L_1 and S_1 and the other with angular momenta L_2 and S_2, respectively, form a molecule, then the possible values of L and S for the molecule are given as

$$L = (L_1 + L_2),\ (L_1 + L_2 - 1), \ldots\ldots\ldots\ldots |L_1 - L_2|,$$

and

$$S = (S_1 + S_2),\ (S_1 + S_2 - 1), \ldots\ldots |S_1 - S_2| \quad (3.7.3)$$

A diatomic molecule, however, possesses an axial symmetry the strong internuclear field in the molecule may break coupling between the angular momenta L and S. L therefore precesses around the internuclear axis giving components $M_L = L, L-1, \ldots, -L$ (Figure 3.12).

If the direction of motion of the electrons is reversed the energy of the molecule remains unchanged though a positive M_L value changes into negative one. So the states differing only in the sign of M_L have the same energy. States with different M_L have in general widely different energies. These different energy states are characterized by a quantum number Λ which represents the component of the resultant angular momentum $\sqrt{L(L+1)}\hbar$ for each state along the internuclear axis.

$$\Lambda = |M_L| = 0, 1, 2, \ldots L \quad (3.7.4)$$

An electronic state with $\Lambda = 0$ is called a Σ state, the one with $\Lambda = 1$, a Π state, etc. In general, states with Λ values of 0, 1, 2, 3, 4,... are represented by code letters Σ, Π, Δ, Φ, Λ,...

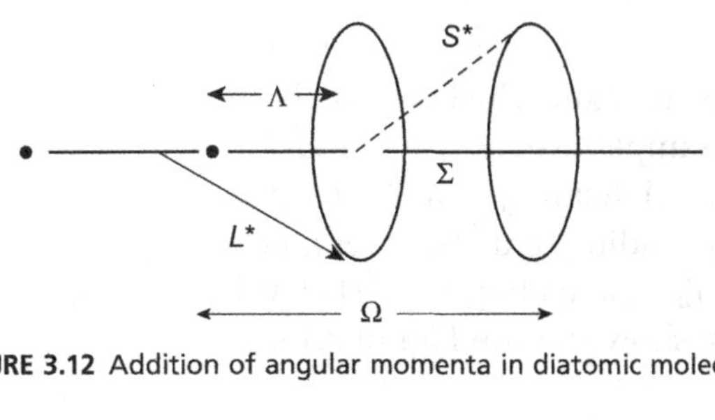

FIGURE 3.12 Addition of angular momenta in diatomic molecules.

A Σ MO having $\Lambda = 0$ is nondegenerate. For each nonzero Λ value, there are two M_L values: $+\Lambda$ and $-\Lambda$. It is found that the electronic energy depends on M_L^2 and therefore a twofold degeneracy is associated with Λ, Δ, Φ... states. Π (or Δ or...) MOs hold four electrons. A set of degenerate MOs constitutes a molecular subshell; specification of the number of electrons in each subshell defines the electronic configuration.

The orbital motion of the electrons in states with $\Lambda > 0$ produces a magnetic field along internuclear axis which causes S to precesses about this axis. S has therefore quantized components M_S with magnitude $M_S\hbar$ along the axis, and the quantum number M_S is identified by the symbol Σ

$$\Sigma = M_S = S, S-1, \dots, -S \tag{3.7.5}$$

where Σ can have (2S+1) possible values. (Σ here may not be confused with Σ for $\Lambda = 0$.)

The axial components of electron spin and angular momenta add up to give the total electronic angular momentum $(\Lambda + \Sigma)\hbar$ along the internuclear axis. The magnitude of the component of total electronic angular momentum about the axis is written as $\Omega\,\hbar$, where

$$\Omega = |\Lambda + \Sigma| \tag{3.7.6}$$

Ω can be an integer or half integer. If $\Lambda \neq 0$, then there are $(2S+1)$ different values of $\Lambda + \Sigma$ for a given Λ. As per international convention energy levels are specified by adding the value of $\Lambda + \Sigma$ as right subscript to the term symbol (like, Π_2) and the multiplicity $2S+1$ is affixed as a left superscript to the Λ code letter. The terminology singlet, doublet, triplet, quadruplet, etc., is used for atomic and molecular terms according to whether $2S+1$ is 1, 2, 3, 4,....

Thus, for example, if we consider a state for which $\Lambda = 1$, and $S = 1$, then from Eqs (3.7.5) and (3.7.6)

$\Sigma = M_S = 1, 0, -1$ and,

$\Omega = |\Lambda + \Sigma| = 2, 1, 0$

In this case, the three states shall have term representation $^3\Pi_2$, $^3\Pi_1$, $^3\Pi_0$.

Diatomic electronic states having the same electronic configuration and the same values of Λ and S have the same electronic term. As in the case of atoms, if spin–orbit interaction is neglected, states of the same term have the same energy.

Spin–orbit interaction splits a molecular term into closely spaced energy levels. The splitting of energy levels in a multiplet state may be related to the values of Λ and Σ. For example, the term value T_e of a multiplet term is given approximately by

$$T_e = T_o + \mathrm{A}\,\Lambda\,\Sigma$$

where A is the coupling constant which can be both positive or negative.

Thus in the above example, where $\Lambda = 1$ and $S = 1$

$\Lambda = 1, \Sigma = 1, T_e = T_o + A$, leading to a $^3\Pi_2$ level

$\Lambda = 1, \Sigma = 0, T_e = T_o$, leading to a $^3\Pi_1$ level and

$\Lambda = 1, \Sigma = -1, T_e, = T_o - A$, leading to a $^3\Pi_0$ level.

This is schematically shown in the Figure 3.13.

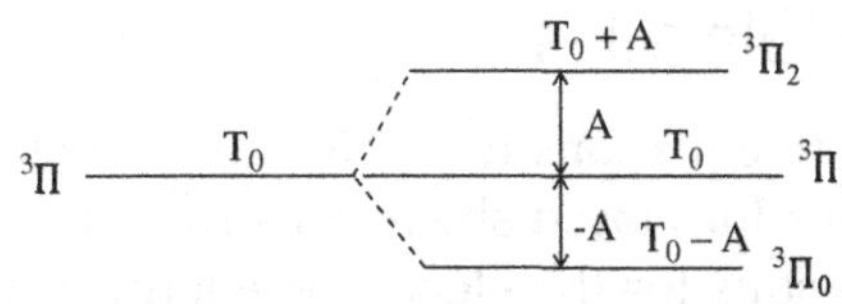

FIGURE 3.13 Term splitting due to spin–orbit interaction.

The ground term of a molecule is prefixed by the letter X. Excited terms that have the same multiplicity as the ground term are usually prefixed by A, B, C,... The letters *a, b, c*,... usually denote excited terms of different multiplicity than the ground term.

3.7.1 Scripts Giving Information on the Wavefunction Symmetry

If the electron eigenfunction of a diatomic molecule does not change sign on reflection of the coordinates to the other side of plane through the two nuclei, then a plus (+) superscript is given and if the sign is changed then the electronic state is characterized by a minus (−) sign. This is particularly true for the Σ terms ($\Lambda = 0$). Thus, we may have Σ^+ and Σ^- terms. As mentioned, non-Σ terms ($\Lambda > 0$) have a twofold spatial degeneracy, and we can form linear combinations of the two wavefunctions. Since these linear combinations have the same energy, one does not ordinarily use the plus or minus superscripts for the non-Σ terms.

Homonuclear molecules possess a center of inversion. If the electron wavefunction does not change sign on replacing the coordinates of the electrons from (x,y,z) to (−x,−y,−z), a symbol *g* (gerade, even) is attached to the Λ code letter as a subscript. If the sign changes, then the symbol (ungerade, odd) is attached. Thus, for example, in H_2^+ molecule which has only one electron, the multiplicity $2S + 1 = 2$ and the possible terms are $^2\Sigma_g^+$ and $^2\Sigma_u^-$ for the bonding and anti-bonding orbitals, respectively as shown in Figure 3.14. The notations *g* and *u* are also valid for polyatomic molecules having center of symmetry.

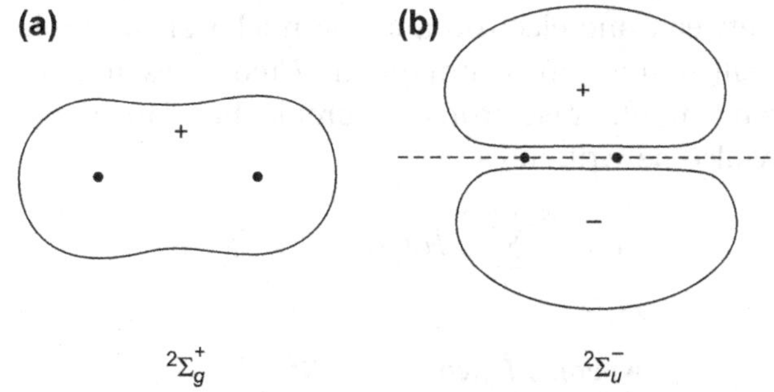

FIGURE 3.14 (a) Bonding ($^2\Sigma_g^+$) and (b) antibonding ($^2\Sigma_u^-$) orbitals of H_2^+.

3.8 HF Theory of Molecules

In Chapter 2, we obtained HF equations for atoms and the Slater determinant form for many-electron wavefunctions for closed shells constructed from spin orbitals. We also obtained convenient expressions for the electronic energy. However, the HF equations for molecules are more difficult to solve because of the dependence of MO functions on two or more coordinates. Fock first proposed the extension of Hartree's SCF to Slater determinantal wavefunction. Just like the Hartree's product orbitals described earlier, the HF MOs can be individually determined as eigenfunctions of a set of one-electron operators. However, now the interaction of each electron with average field of all other electrons should also include the Coulomb repulsion terms and the exchange effects as described in the case of the H_2^+ and hydrogen molecules. In 1951, Roothan developed [6] a SCF model for molecules which has been critical to further development of practical computations based on matrix formalism. The following treatment is restricted to closed-shell systems.

3.8.1 HF Formalism

As discussed under Born–Oppenheimer approximation, the Hamiltonian for an n-electron molecule is given as,

$$H = H_e + V_{nn} = -\frac{1}{2}\sum_i \nabla_i^2 - \sum_\alpha \sum_i \frac{Z_\alpha}{r_{i\alpha}} + \sum_j \sum_{i>j} \frac{1}{r_{ij}} + \sum_\alpha \sum_{\beta>\alpha} \frac{Z_\alpha Z_\beta}{R_{\alpha\beta}} \tag{3.8.1}$$

where i refers to electrons and α to nuclei. H_e is the Hamiltonian for the electronic motion and V_{nn}, the last term in the equation, is the internuclear repulsion energy.

The internuclear distance $R_{\alpha\beta}$ is constant for a particular nuclear configuration and so the total energy E of the molecule is obtained by adding the internuclear repulsion energy (V_{nn}) to the electronic energy E_e.

$$E = E_e + \sum_\alpha \sum_{\beta>\alpha} \frac{Z_\alpha Z_\beta}{R_{\alpha\beta}} = E_e + V_{nn}$$

The first two terms in Eq. (3.8.1) correspond to kinetic energy operator of the electron and the attraction between one electron and the nuclei of the molecule. These terms constitute what is called the core Hamiltonian H(core) as it does not contain any interaction of electron with other electrons. H(core) is therefore a one-electron operator. Hamiltonian H_e may also be written as

$$\widehat{H}_e = \sum_i^{\text{all electrons}} \widehat{H}_i(\text{core}) + \sum_j \sum_{i>j} \frac{1}{r_{ij}} \tag{3.8.2}$$

$$\text{where,}\quad \widehat{H}_i(\text{core}) = -\frac{1}{2}\nabla_i^2 - \sum_\alpha \frac{Z_\alpha}{r_{i\alpha}} \tag{3.8.3}$$

Since the $\frac{1}{r_{ij}}$ term in Eq. (3.8.2) involve coordinates of both electrons simultaneously, it cannot be evaluated exactly. In order to obtain contribution of this term to the

Hamiltonian, we consider two electrons: an electron, say 1, which is subjected to the average field of another electron, say 2, which occupies a spin MO, say Ψ_i.

This average field is

$$\int \frac{|\Psi_i(2)|^2}{r_{12}} d\tau_2$$

and is represented by an operator $\hat{J}_i(1)$—the Coulomb operator. The Coulomb operator thus accounts for the smeared-out electron potential with an electron density of $|\Psi_i(2)|^2$.

$$\hat{J}_i(1) = \int \frac{|\Psi_i(2)|^2}{r_{12}} d\tau_2$$

$$\text{or } \hat{J}_i(1)\Psi_j(1) = \Psi_j(1) \int \frac{|\Psi_i(2)|^2}{r_{12}} d\tau_2 \tag{3.8.4}$$

Also, since electrons can exchange between MOs, this may be considered by including the term $\int \frac{\Psi_i^*(2)\Psi_j(2)}{r_{12}} d\tau_2$, called the exchange operator $\hat{K}$.

$$\hat{K}_i(1) = \int \frac{\Psi_i^*(2)\Psi_j(2)}{r_{12}} d\tau_2 \tag{3.8.5a}$$

When $\hat{K}_i(1)$ operates on an orbital $\Psi_j(1)$, it gives

$$\hat{K}_i(1)\Psi_j(1) = \Psi_i(1) \int \frac{\Psi_i^*(2)\Psi_j(2)}{r_{12}} d\tau_2 \tag{3.8.5b}$$

As is evident from the definition, the exchange operator leads to an exchange of the variables in the orbitals. Both the terms given by Eqs (3.8.4) and (3.8.5) ought to summed over all occupied orbitals. Thus, the equation to be solved for electron 1 in MO say, $\Psi_j(1)$ may be given as

$$\left[\sum_i^{\text{all electrons}} \hat{H}_i(\text{core}) + \sum_i^{\text{occ MOs}} \left(2\hat{J}_i - \hat{K}_i\right)\right] \Psi_j(1) = \varepsilon_j \Psi_j(1) \tag{3.8.6}$$

If the number of electrons is n, the number of occupied MOs is $n/2$. The factor of 2 in the Coulomb operator arises because there are two electrons in each MO. However, the exchange operator will be a single term as it would become zero unless the MOs Ψ_i and Ψ_j refer to the same spin functions. The exchange operator has no physical interpretation. It takes into account the effects of spin correlation.

The operator on the left side of Eq. (3.8.6) is called Effective HF operator $\hat{F}$

$$\hat{F}(1) = \sum_i^{\text{all electrons}} \hat{H}_i(\text{core}) + \sum_i^{\text{occ MOs}} \left(2\hat{J}_i - \hat{K}_i\right) \tag{3.8.7}$$

Eq. (3.8.6) may be written as

$$\hat{F}\Psi_i = \varepsilon_i \Psi_i \tag{3.8.8}$$

The HF operator $\hat{F}$ is a one-electron operator, as sometimes indicated by writing it as dependent on electron 1.

The MO Ψ_i in Eq. (3.8.8) are eigenfunctions of the HF operator $\widehat{F}$ and can be chosen to be orthogonal. This may cause many integrals in the expression to vanish. The operator $\widehat{F}$ includes the coordinates of only one electron and is a differential equation in terms of only one electron. As against this, the true Hamiltonian and wavefunction of a molecule includes coordinates of all n electrons.

Like many-electron atoms, the solution of HF equation must be done by an iterative process. The energy of the closed-shell molecule in terms of the HF SCF method is determined from the expression

$$E = 2\sum_{i}^{\text{occ MO}} \varepsilon_i + \sum_{i,j}^{\text{occ MOs}} (2J_{ij} - K_{ij}) + \sum_{\alpha>\beta}\sum_{\beta} \frac{Z_\alpha Z_\beta}{R_{\alpha\beta}} \tag{3.8.9}$$

The first term is the summation over the energies of all the occupied molecule orbitals, each having two electrons. The orbital energies are given by the expression

$$\varepsilon_i = H_{ii}(\text{core}) + \sum_{j}^{\text{occ MOs}} (2J_{ij} - K_{ij}) \tag{3.8.10}$$

$$\text{where,}\quad H_{ii}(\text{core}) = \left\langle \Psi_i(1) \middle| -\frac{1}{2}\nabla_1^2 - \sum_{\alpha}\frac{Z_\alpha}{r_{1\alpha}} \middle| \Psi_i(1) \right\rangle \tag{3.8.11}$$

$$J_{ij} = \left\langle \Psi_i(1)\Psi_j(2) \middle| \frac{1}{r_{12}} \middle| \Psi_i(1)\Psi_j(2) \right\rangle \tag{3.8.12}$$

$$K_{ij} = \left\langle \Psi_i(1)\Psi_j(2) \middle| \frac{1}{r_{12}} \middle| \Psi_j(1)\Psi_i(2) \right\rangle \tag{3.8.13}$$

J_{ij} is obtained by operating the Coulomb operator J_i Eq. (3.8.4) on $\Psi_i(1)$ and multiplying the result by $\Psi_i^*(1)$ followed by integration over all space. Similarly, K_{ij} is obtained by operating the exchange operator Eq. (3.8.5) over $\Psi_j(1)$and multiplying by $\Psi_i^*(1)$ followed by integration over all space. The last term in Eq. (3.8.9) is the internuclear repulsion potential for a given nuclear configuration. From Eq. (3.8.9) we note that the total energy is not simply the sum of orbital energies. The terms J_{ij} and K_{ij} are similar to those in many-electron atoms.

3.8.2 Roothan Formalism

Roothan proposed an orbital expansion method for the calculation of accurate SCF wavefunctions for molecules from the HF Eq. (3.8.8). The spatial one-electron wavefunction Ψ_i can be expressed as a LCAOs which form a basis set $\{\Phi_i\}$.

$$\Psi_i = \sum_{\nu}^{\text{all } AO} c_{i\nu}\Phi_\nu \tag{3.8.14}$$

The LCAO method has been described in Section 3.4. As per convention, the Greek letters shall be used as suffixes for the atomic orbitals in expansions such as Eq. (3.8.14) and the Roman letters as suffixes for the MOs.

To exactly represent MOs, the basis functions should form a complete set. For best representation, an infinite number of atomic orbitals is needed but in practice only a finite number of them is used. MOs can be obtained to essentially any accuracy desired by appropriately adjusting the number of base functions in the LCAO expansion.

The coefficient $c_{i\nu}$ correspond to the contribution of each atomic orbital to the corresponding MO. These coefficients are determined by using the variation technique by minimizing the energy of the molecular system.

Using Eq. (3.8.14) in Eq. (3.8.8), we get

$$\widehat{F}\sum_{\nu} c_{i\nu}\Phi_{\nu} = \varepsilon_i \sum_{\nu} c_{i\nu}\Phi_{\nu} \tag{3.8.15}$$

Multiplying by Φ_{μ}^* and integrating gives

$$\sum_{\nu} c_{i\nu} \int \Phi_{\mu}^* \widehat{F}\Phi_{\nu} d\tau = \varepsilon_i \sum_{\nu} c_{i\nu} \int \Phi_{\mu}^* \Phi_{\nu} d\tau$$

$$\text{or } \sum_{\nu} c_{i\nu}(F_{\mu\nu} - \varepsilon_i S_{\mu\nu}) = 0 \tag{3.8.16}$$

$$\text{where } F_{\mu\nu} = \int \Phi_{\mu}^* \widehat{F}\Phi_{\nu} d\tau \tag{3.8.17}$$

are the elements of the Fock matrix $\widehat{F}$ and

$$S_{\mu\nu} = \int \Phi_{\mu}^* \Phi_{\nu} d\tau \tag{3.8.18}$$

are the elements of the overlap matrix.

The set of Eq. (3.8.16) are known as Hall and Roothan equations or more generally as Roothan equations. These equations are algebraic equations instead of the differential equations. In the matrix form it may be written as

$$FC = SCE \tag{3.8.19}$$

where E is the diagonal matrix of ε_i.

Using variational technique, the coefficients $c_{i\nu}$ are optimized by taking the derivative of ε_i with respect to each coefficient and setting it equal to zero.

This would result in a set of equations similar to those obtained for the H_2^+ molecule (Eq. 3.4.19)

$$\begin{pmatrix} F_{11} - \varepsilon & F_{12} - S_{12}\varepsilon & \ldots & F_{1n} - S_{1n}\varepsilon \\ F_{21} - S_{12}\varepsilon & F_{22} - \varepsilon & \ldots & F_{2n} - S_{2n}\varepsilon \\ \vdots & \vdots & \vdots & \vdots \\ F_{n1} - S_{n1}\varepsilon & F_{n2} - S_{n2}\varepsilon & \ldots & F_{nn} - \varepsilon \end{pmatrix} \begin{pmatrix} c_1 \\ c_2 \\ \vdots \\ c_n \end{pmatrix} = 0 \tag{3.8.20}$$

For a nontrivial solution, we must have

$$\det(F_{\mu\nu} - \varepsilon_i S_{\mu\nu}) = 0 \tag{3.8.21}$$

$$\text{where, } F_{\mu\nu} = \int \Phi_{\mu}^* \widehat{F}\Phi_{\nu} d\tau \tag{3.8.22}$$

This is a secular equation whose roots give the orbital energies ε_i.
Using Eq. (3.8.6), we get for $F_{\mu\nu}$,

$$F_{\mu\nu} = \langle \Phi_\mu | H(\text{core}) | \Phi_\nu \rangle + \sum_i^{\text{occ MO}} \left[\langle \Phi_\mu | 2\widehat{J}_i | \Phi_\nu \rangle - \langle \Phi_\mu | \widehat{K}_i | \Phi_\nu \rangle \right] \tag{3.8.23}$$

If we neglect the differential overlap $S_{\mu\nu}$, that is $\Phi_\mu \Phi_\nu = 0$, the calculations show that

$$\langle \Phi_\mu | J_i | \Phi_\nu \rangle = \sum_\lambda^{AO} \sum_\sigma^{AO} c_{i\lambda} c_{i\sigma} \langle \mu\nu | \lambda\sigma \rangle \tag{3.8.24}$$

$$\text{and } \langle \Phi_\mu | K_i | \Phi_\nu \rangle = \sum_\lambda \sum_\sigma c_{i\lambda} c_{i\sigma} (\mu\lambda | \nu\sigma) \tag{3.8.25}$$

where,

$$\langle \mu\nu | \lambda\sigma \rangle = \iint \Phi_\mu(1) \Phi_\nu(1) \frac{1}{r_{12}} \Phi_\lambda(2) \Phi_\sigma(2) d\tau_1 d\tau_2$$

$$\text{and } \langle \mu\nu | \lambda\sigma \rangle = \iint \Phi_\mu(1) \Phi_\lambda(1) \frac{1}{r_{12}} \Phi_\nu(2) \Phi_\sigma(2) \tag{3.8.26}$$

are the two-electron Coulomb and exchange integrals.

Hence for an n-electron closed-shell molecule, Eq. (3.8.23) can be written as

$$\begin{aligned} F_{\mu\nu} &= H_{\mu\nu}(\text{core}) + \sum_{i=1}^{n/2} \sum_\lambda^{AO} \sum_\sigma^{AO} c_{i\lambda} c_{i\sigma} [2(\mu\nu|\lambda\sigma) - (\mu\lambda|\nu\sigma)] \\ &= H_{\mu\nu}(\text{core}) + \sum_\lambda \sum_\sigma P_{\lambda\sigma} \left[(\mu\nu|\lambda\sigma) - \frac{1}{2} (\mu\lambda|\nu\sigma) \right] \end{aligned} \tag{3.8.27}$$

$$\text{where, } P_{\lambda\sigma} = 2 \sum_{i=1}^{n/2} c_{i\lambda} c_{i\sigma} \tag{3.8.28}$$

$P_{\lambda\sigma}$ are the elements of electron density matrix or charge-density-bond-order matrix. The density matrix describes the degree to which individual basis functions contribute to the many-electron wavefunctions. The factor of 2 appears because in closed-shell system we are considering only singlet wavefunctions in which all orbitals are doubly occupied. Also, it may be noted that in Eq. (3.8.27) the exchange integrals $(\mu\lambda \mid \nu\sigma)$ are preceded by a factor of 1/2 because they are limited to electrons of the same spin while Coulomb interactions are present for any combination of spins.

The HF Roothan Eq. (3.8.16) is solved by an iterative process because the $F_{\mu\nu}$ integrals depend on the orbitals Ψ_i which in turn depend on the unknown coefficients $c_{i\nu}$. In actual practice, one starts with an initial guess for the occupied MOs as LCAOs (Eq. 3.8.14). This is usually done by semiempirical methods. The initial set of MOs is then used to compute an approximate Fock operator $\widehat{F}$. The matrix elements $F_{\mu\nu}$ are then computed and secular Eq. (3.8.20) is solved to give an initial set of orbital energies ε_i. These values of ε_i are then used to solve Eq. (3.8.16) to get improved set of coefficients $c_{i\nu}$ which provide improved set of MOs. These are then again used to compute improved $\widehat{F}$. This process is repeated and the calculated coefficients and energies for each cycle are compared with those of the previous cycle. If there is a difference, the computation is repeated with new optimized coefficients. If there is no significant difference the computation is terminated. This iterative process, as in the case of atoms, is called a SCF method.

3.9 Closed-Shell and Open-Shell Molecules

Like atoms, the molecules can also be classified as closed-shell and open-shell molecules. In case of closed-shell molecules, all the electrons in the occupied MOs are paired. In the case of open-shell molecules not all the electrons in the MOs are paired. Most of the ions and radicals and excited electronic states belong to this class.

Thus, for an n-electron closed-shell molecule, $n/2$ MOs will be doubly occupied and may be represented by a wavefunction

$$\Psi = \Psi_1(1)\alpha(1)\Psi_1(2)\beta(2)\ldots\Psi_{\frac{n}{2}}(n-1)\alpha(n-1)\Psi_{\frac{n}{2}}(n)\beta(n) \tag{3.9.1}$$

where α and β are the spin functions.

The HF approximation results in separation of electron motions resulting in the ordering of the electrons into MOs. Thus, a nitrogen molecule is formed through the combination of two nitrogen atoms each having configuration $1s^2$, $2s^2$, $2p^3$. The orbitals of the same type and of equal or approximately equal configurations combine to produce MOs. Thus, in this case, the $1s$, $2s$, $2p_x$, $2p_y$ and $2p_z$ orbitals of the two nitrogen atoms combine with each other resulting in 10 MOs (Figure 3.15). The 14 electrons of the

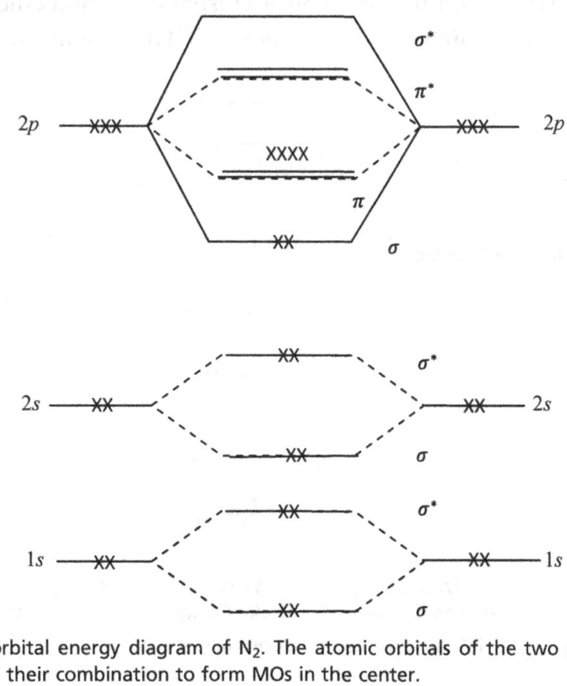

FIGURE 3.15 Molecular orbital energy diagram of N_2. The atomic orbitals of the two nitrogen atoms are shown on the left and right and their combination to form MOs in the center.

nitrogen molecule are then filled up in an increasing sequence of MO energy. Thus, the total 14 electrons will distribute themselves into 7 of the 10 MOs.

Similarly, the MOs of water molecule can be constructed from a linear combination of two 1*s*-type orbitals of the two hydrogen atoms (H_1 1*s* and H_2 1*s* for hydrogen atoms 1 and 2) and five atomic orbitals of the oxygen atom namely 1*s*, 2*s*, $2p_x$, $2p_y$ and $2p_z$. Water molecule has 10 electrons which may distributed over 5 MOs $\Psi_1, \Psi_2, \Psi_3, \Psi_4$ and Ψ_5. The ground state wavefunction for the water molecule will then be an SCF wavefunction represented as a Slater determinant.

$$|\Psi_1\alpha\ \Psi_1\beta\ \Psi_2\alpha\ \Psi_2\beta\ \Psi_3\alpha\ \Psi_3\beta\ \Psi_4\alpha\ \Psi_4\beta\ \Psi_5\alpha\ \Psi_5\beta|$$

$$\text{or}\quad |\Psi_1\overline{\Psi}_1\ \Psi_2\overline{\Psi}_2\ \Psi_3\overline{\Psi}_3\ \Psi_4\overline{\Psi}_4\ \Psi_5\overline{\Psi}_5|$$

For a closed-shell configuration the MO wavefunction is a single determinant; open-shell wavefunctions are usually linear combination of a few Slater determinants.

The above five MOs ($\Psi_1, \Psi_2, \Psi_3, \Psi_4, \Psi_5$) are delocalized but it is also possible to form an appropriate description in which their linear combination is localized. This is done by addition of multiples of columns in the Slater determinant. Thus, we can get five occupied MOs for water that are localized one inner shell MO, two equivalent localized lone pair MOs and two equivalent localized bond MOs (Figure 3.16). Thus, there is some degree of freedom in the MO description that we use. Of the many ways to choose localized MOs, the most widely used is to choose them to minimize interorbital repulsions, thereby giving the energy-localized MOs.

Two common procedures are used for computations on open-shell molecules. One procedure is to use RHF wavefunction as in a closed-shell molecule. This approach is called the restricted open-shell Hartree–Fock (ROHF) method. Thus, the ROHF

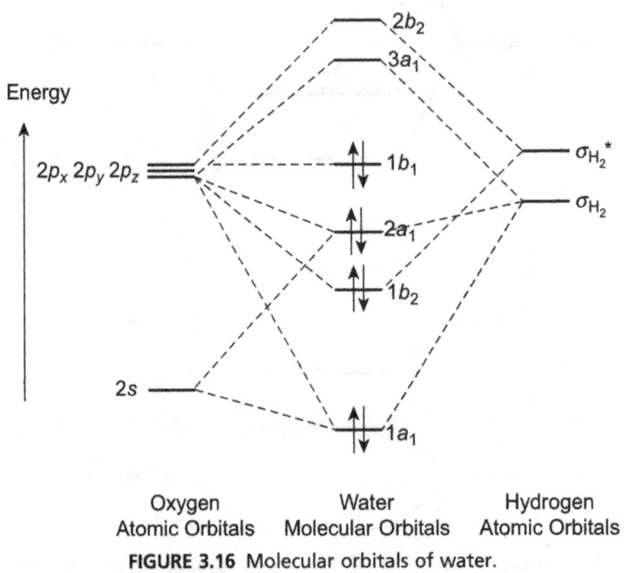

FIGURE 3.16 Molecular orbitals of water.

wavefunction of the Li ground state is $|1s\,1\bar{s}\,2s|$, where the two $1s$ electrons occupy the same spatial MO. The $2s$ electron in this ROHF function has been given spin α. The difficulty in this approach is that the lone electron in the MO shall interact only with the other electrons in the molecule with the same spin. Since, according to Pauli's principle, electrons with the same spin tend to keep away from each other, the interaction between the $2s\,\alpha$ and $1s\,\alpha$ electrons in the case of Li differ from the interaction between $2s\,\alpha$ and $1s\,\beta$ electrons. To relax this constraint on the solution, each electron in the MO is given a different spatial function. Thus, the two $1s$ orbitals may be called $1s$ and $1s'$, where $1s \neq 1s'$ and the wavefunction for the ground state of Li is written as $|1s\,1s'\,2s|$. This approach is called the unrestricted Hartree–Fock (UHF) method. Thus, in the UHF approach, the spatial orbitals of spin $-\alpha$ electrons are allowed to differ from spin $-\beta$ electrons. Unlike RHF wavefunctions, in the open-shell excited states, the MO is a linear combination of a few determinants.

Thus, for a system of n-electrons and a basis set of M-spin orbitals, it is possible to form a total of $K = \frac{M!}{n!(M-n)!}$ independent Slater determinants Ψ_I and the most general wavefunction can be written as,

$$\Psi = \sum_{I=1}^{K} c_I \Psi_I \tag{3.9.2}$$

For a large class of open-shell states, containing a single open shell, the energy can be expressed in the form

$$E = \left\{ 2\sum_{i}^{K} \varepsilon_i + \sum_{i}\sum_{j} (2J_{ij} - K_{ij}) \right\} + \gamma \left\{ 2\sum_{l} \varepsilon_l + \gamma \sum_{l,m} (2aJ_{lm} - bK_{lm}) \right\}$$
$$+ 2\gamma \sum_{i}\sum_{l} (2J_{il} - K_{il}) + \sum_{\alpha>\beta}\sum_{\beta} \frac{Z_\alpha Z_\beta}{R_{\alpha\beta}} \tag{3.9.3}$$

where i and j refer to the closed-shell orbitals, and l and m to the open-shell orbitals. γ is the fractional occupation number, $0 < \gamma < 1$ of the open shell and is calculated by dividing the number of electrons in the shell by the total number of spin orbitals which make up the shell. The two constants, a and b, depend on the coefficient in Eq. (3.9.2) for the particular state being considered. The first group (shown in curly brackets) of terms in Eq. (3.9.3) is the energy of closed shell, the second group is the energy of the open shell, and the third represents interaction between the two.

Application of the variation principle to the energy Eq. (3.9.3), gives two sets of RHF equations.

$$\widehat{F}_c \Psi_i = \varepsilon_i \Psi_i \tag{3.9.4a}$$

$$\text{and } \widehat{F}_o \Psi_l = \varepsilon_l \Psi_l \tag{3.9.4b}$$

where F_c is the HF operator for the closed-shell orbitals and F_o for the open-shell orbitals. Using the procedure suggested by McWeeny and Sutcliffe [7] these operators, which are complicated functions of all the occupied orbitals, and the constants γ, a, and b may be combined to give a single HF operator whose eigenfunctions include both the closed-shell and open-shell orbitals.

The open-shell UHF wavefunctions may not be the eigenfunctions of the total squared spin angular momentum operator $\hat{S}^2$. This leads to impure spin states for the molecule. In contrast the true wavefunctions and ROHF wavefunctions are eigenfunctions of $\hat{S}^2$.

Energies given by UHF wavefunctions are slightly lower than those by ROHF. UHF wavefunctions are much more useful in the prediction of electron spin resonance spectra and in the treatment of electron correlation affects in open-shell systems.

3.10 Atomic Orbitals—Their Types and Properties

In LCAO calculations, it is necessary to choose a set of functions to represent atomic orbitals. Atomic orbitals can be hydrogen-like orbitals which are exact solutions of the Schrödinger equation for a hydrogen-like atom or even Hartree–Fock (HF) orbitals for atoms. However neither of these are often used because of their complicated functional forms that make evaluation of integrals very difficult. The coordinate system chosen for atomic orbitals in polyatomic molecules is usually the spherical coordinate system. The advantage of spherical coordinates for atoms is that an orbital wavefunction is a product of three factors, each dependent on a single coordinate:

$$\Psi(r, \theta, \phi) = R(r)\,\Theta(\theta)\,\Phi(\phi)$$

The angular factors of atomic orbitals $\Theta(\theta)\,\Phi(\phi)$ generate s, p, d,... functions as real combinations of spherical harmonics $Y_{lm}(\theta, \phi)$. Three mathematical forms have been tried to describe the radial functions $R(r)$ as a starting point for the calculation of the properties of atoms and molecules with many electrons. These are:

1. *Hydrogen-like atomic functions* derived from the exact solution of Schrödinger equation. In this case the function decays as $e^{-\alpha r}$, where α is a constant and r is the distance from the nucleus. These types of functions are usually not used because of their complicated form, presence of several nodes, and difficulty in the evaluation of integrals.
2. *Slater-type orbitals (STOs)* which have the form $\Psi = Ar^{n-1}e^{-\zeta r}Y_{l,m}(\theta, \phi)$, where ζ is the orbital exponent and A is the normalization constant. The set of all STOs with integral values of quantum numbers n, l, and m and all possible values of ζ form a complete set.
   For constructing molecular orbital (MOs), the STOs can be of different types, like:
   - **(i)** SCF orbitals calculated for separate atoms
   - **(ii)** Individual STOs obtained from atomic calculations

(iii) STOs of type (ii) but with reoptimized values of exponents ζ for the molecule under consideration and

(iv) STOs of type (ii) or (iii) supplemented by additional functions of lower symmetries to describe the distortion or polarization of atomic orbitals in the molecular environment. Thus, p-type functions are used to describe polarization of *s*-type orbitals, d-type functions for the polarization of *p*-orbitals, etc.

Accuracy of results and flexibility increases from type (i) to (iv) and so also the computational labor. A large number of calculations of all these types are available for small molecules in the literature [8]. Some broad conclusions have been drawn on the basis of calculations using STOs, three of which are:

a. The accuracy of the total energy does not in general represent a valid criterion for the accuracies of other properties like dipole moment, dissociation energy, etc.

b. With the sole exception of the 1*s* orbital of hydrogen, the reoptimization of the exponent ζ in the molecule does not normally lead to a significant improvement of the energy or other properties of the molecule. In the case of hydrogen the exponent ζ for 1*s* orbital increases from its value 1.0 in isolated atom to 1.2 in molecule. This may be interpreted as contraction of the orbital in molecular environment and has been found to be important for correct description of charge distribution near a proton in the molecule.

c. It is important to have at least double-zeta (DZ) basis plus some polarization functions for getting values of properties comparable with accurate RHF values.

3. The Gaussian-type orbitals (GTOs)

Although the Slater functions form a satisfactory basis for the representation of MOs, they have been used for SCF calculations only for diatomic and small linear polyatomic molecules. The reason for the limited use lies in the great difficulty with the evaluation of multiple center electron-interaction integrals of the type

$$(\mu\nu|\sigma\lambda) = \iint \frac{\Phi_\mu^*(1)\Phi_\nu(1)\Phi_\sigma^*(2)\Phi_\lambda(2)d\tau_1 d\tau_2}{r_{12}}$$

which appear in the matrix elements of the SCF Hamiltonian matrix *H*. The difficulty in evaluation is particularly severe for the three- and four-center integrals where the STOs are centered on non-collinear atoms in a molecule.

To simply the molecular integral evaluation, Boys proposed in 1950 the use of Gaussian-type orbitals (GTOs) instead of STOs for the atomic orbitals. Wavefunctions for polyatomic molecules are now almost exclusively treated in this way.

GTOs are one-center functions having form similar to STOs, but the radial term $R_n(r)$ is now taken as

$$R_n(r) = r^{n-1}e^{-\zeta r^2}(n > l) \tag{3.10.1}$$

The Gaussian orbitals have no nodes and decay as $e^{-\zeta r^2}$. When $(n - l - 1)$ is restricted to zero or even integer values, it can easily be shown that the product of two Gaussians on different centers becomes equivalent to a single Gaussian on a new center. An integral like $(\mu\nu|\lambda\sigma)$ in which the four Φ functions are Gaussians on different centers then reduces to a two-center integral which can be explicitly evaluated in terms of known functions (for details, see Ref. [8]).

A Cartesian Gaussian centered on an atom is defined as

$$\Psi_{ijk} = Ax^i y^j z^k e^{-\zeta r^2} \tag{3.10.2}$$

where i, j, k are nonnegative integers and ζ is a positive orbital exponent. When $i = j = k = 0$, the GTO is called an s-type Gaussian. Similarly, when $i + j + k = 1$ (for any combination of the three-integer), we have a p-type Gaussian and when $i + j + k = 2$, we have d-type of Gaussian. The angular dependence of GTO with $l > 0$ is simulated by the superposition of a number of symmetrically distributed spherical Gaussians. Thus, a linear combination of Cartesian Gaussians can be constructed to have the form $Ae^{-\zeta r^2}(Y^*_{lm} \pm Y_{lm})/\sqrt{2}$. Thus, any s atomic orbital may be represented as combination of several Gaussians with different orbital expansion coefficients ζ. The p_x atomic orbital may be represented by a linear combination of Gaussians ($i = 1, j = 0, k = 0$), each of the form $xe^{-\zeta r^2}$, the p_y AO as combination of $ye^{-\zeta r^2}$ and p_z as combination of $ze^{-\zeta r^2}$ Gaussians.

Although not as accurate as STOs, the combination of many Gaussians can attain the accuracy of hydrogen-like orbitals (Figure 3.17). Thus, for example, 10 Gaussians can produce 6-figure accuracy for the energy of hydrogen atom.

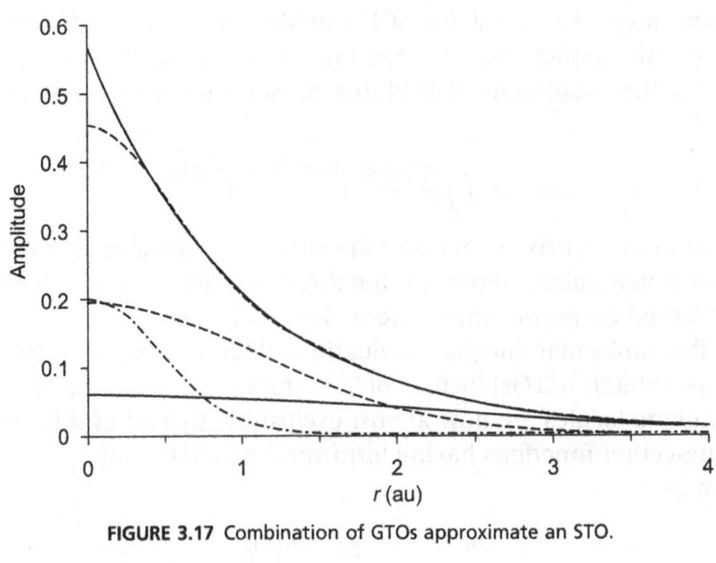

FIGURE 3.17 Combination of GTOs approximate an STO.

Gaussian functions have several drawbacks but the main drawback is its incorrect behavior at and near a nucleus. While a Slater type function $e^{-\zeta r}$ has a nonzero derivates at $r = 0$

$$\left.\frac{d}{dr}e^{-\zeta r}\right|_{r=0} = -\zeta \neq 0$$

and this cusp property is essential for the accurate description of an atomic or molecular wavefunction at or near a nucleus, the GTO $e^{-\zeta r^2}$ has a zero derivative at $r = 0$. To get an accurate representation of an atomic orbital at or near the nucleus, combination of a large number of GTOs is used as they have no cusp in the region. Thus, for example, of the 10 Gaussians required to describe the 1*s* and 2*s* orbitals of the first row atoms, 5 or 6 are used for the region very close to the nucleus and a very small number is used for the intermediate to large distances from the nucleus. Also, a Gaussian function $e^{-\zeta r^2}$ falls off too rapidly with *r* and so larger number of Gaussian functions than the Slater functions are needed to describe the correct behavior.

Although the use of large number of Gaussian functions than STOs significantly increases the number of integrals to be evaluated (number of two-electron integrals is proportional to the fourth power of the number of basis functions), the advantage of using GTOs is that it takes much lesser computational time to evaluate the Gaussian functions than the Slater functions.

A very large number of Gaussian atomic orbitals are available and have been compiled by Poirer et al. [9] and Huzinaga et al. [10].

3.11 Classification of Basis Sets

In this section we shall discuss the terminology used to describe the STO and GTO basis sets.

3.11.1 Slater-Type Basis Sets

3.11.1.1 Minimal Basis Sets

The smallest basis set used to describe an atomic orbital is called the "*minimal*" or "*minimum*" basis set. In this basis set one STO is used for each inner shell and valence shell atomic orbital. Thus, for example, for methane (CH_4) a minimal basis set consists of 1*s*, 2*s*, $2p_x$, $2p_y$, and $2p_z$ atomic orbital on the carbon atom and 1*s* STO on each hydrogen. Thus, there are five STOs on the carbon and one on each hydrogen for a total of nine basis functions. Since this set contains two s-type STOs and one set of p-type STO on each carbon and one s-type STO on each hydrogen, this is denoted by (2*s*1*p*) for the carbon functions and (1*s*) for the hydrogen functions. This notation is further abbreviated as (2*s*1*p*/1*s*).

The minimal basis sets are not flexible enough for accurate representation of orbitals and to describe electron distribution between the nuclei to form chemical bonds. To bring greater flexibility DZ basis sets are used.

3.11.1.2 Split-Valence Basis Sets

To bring greater flexibility, the valence atomic orbitals are split into two parts: an inner compact orbital and an outer diffuse one. One such split-valence basis set is a double-zeta (DZ) basis set which is obtained by replacing each STO of minimal basis set by two STOs that differ in their orbital exponents. For example, the 2*s* atomic orbital in the STO representation may be expressed as

$$\Phi_{2s}(r) = \Phi_{2s}^{\mathrm{STO}}(r,\zeta_1) + \lambda\Phi_{2s}^{\mathrm{STO}}(r,\zeta_2) \tag{3.11.1}$$

The value of ζ accounts for the size of the orbital and the constant λ determines the contribution of each STO toward the final orbital. Thus a DZ set of methane will consist of two 1*s* STO on each hydrogen and two 1*s* STOs, two 2*s* STOs, two $2p_x$, two $2p_y$, and two $2p_z$ STOs on the carbon for a total of 18 basis functions. Triple-zeta (TZ) and quadruple-zeta basis sets work in the same manner, except that they use three and four STOs instead of two.

Often it takes more effort to calculate a DZ for every orbital hence for simplification DZ is calculated only for the valence electrons. Since the inner shell electrons are not as vital to the calculation they are described by a single STO. Thus, in double ζ valence basis set, inner or core atomic orbitals are represented by one basis function and the valence atomic orbitals by two basis functions. In other words it is minimal for the inner shell atomic orbitals and DZ for the valence orbitals. Hence, the carbon atom in CH_4 is represented by 1*s* inner orbital and $2(2s, 2p_x, 2p_y, 2p_z) = 8$ valence orbital. Each hydrogen atom is represented by two valence orbitals. Thus, a total of 17 orbitals shall be used for CH_4.

3.11.1.3 Polarization Basis Sets

The next step beyond the DZ and SV basis sets involves adding polarization functions, that is, adding d-type functions to the first row atoms Li-F and p-type functions to hydrogen H. This is sometimes necessary to account for the distortion in the atomic orbitals when they enter a molecule. Thus, for example, when the 1*s* orbital of hydrogen atom enters a molecule it shall experience a nonuniform electric field arising from its nonspherical environment. This effect is accommodated by adding polarization function to a basis set of H. A common kind of basis set is a double-zeta plus polarization set (DZP) which adds a set of five 3D functions on each first row and each second row atom and a set of three 2*p* functions ($2p_x$, $2p_y$, $2p_z$) on each hydrogen atom.

3.11.2 Gaussian-Type Basis Set

Since a single GTF is not suitable to describe an atomic orbital, instead of using individual Gaussian function as basis functions, each basis function is taken as a linear combination of a small number of Gaussian functions. Such functions are called contracted Gaussian-type functions. Thus,

$$\Psi_r = \sum_l c_{lr} g_l \tag{3.11.2}$$

where g_l are the Cartesian Gaussians (Eq. (3.10.2)) centered on the same atom and having the same *i*, *j*, *k* values but different exponents ζ, c_{lr} are the coefficients and g_l are the primitive Gaussians.

3.11.2.1 Minimal and Extended Basis Sets

The smallest GTF is called minimal basis set. It composed of minimum number of primitive Gaussian functions required to represent all the electrons on each atom. The most common minimal basis set is STO-*n*G, where *n* is an integer and represents the number of Gaussian primitive functions contracted to give a single STO. In these basis sets the same number of Gaussian primitives are used for both the core and valence orbitals. Thus, an STO-3G shows that 3GTF have been used to construct an STO. The minimal basis set describes only the most basic aspects of the orbitals. To describe the orbitals in greater details, extended basis sets are used. These sets include multiple functions for each angular momentum component in the electronic configuration description. A number of basis functions per orbital are used to increase the size of the basis set as in the case of DZ and split-valence basis sets.

3.11.2.2 Split-Valence Basis Sets

Split-valence basis sets use sums of Gaussians such that there is more than one set of basis functions for each subshell. Pople and coworkers [13–18] have suggested split-valence sets of the type N-*ij*G and N-*ijk*G. In N-*ij*G, N represents the number of primitive Gaussians for the inner (core) shells. The *ij* notation describes sets of DZ quality for the valence shell. Thus, each valence orbital is composed of two basis functions—the first one is composed of a linear combination of *i* primitive Gaussian functions and the other composed of a linear combination of *j* primitive Gaussian functions. Similarly, in N-*ijk*, the *ijk* notation describes sets of valence TZ quality, that is, the valence orbitals are composed of three basis functions having linear combination of *i*, *j*, and *k* primitive Gaussian functions, respectively. Some of the commonly used split-valence basis sets are 3-21G, 4-21G, 4-31G, 6-21G, 6-31G, 6-311G with or without polarization and diffuse functions. As an example, in 6-31G basis, the inner shell of carbon is represented by six primitives and the valence shell orbitals are represented by two contracted orbitals each consisting of four primitives: three of these are contracted and one uncontracted. So, for the $1s^2 2s^2 2p^2$ configuration of C we get $6 + 4 \times 4 = 22$ primitives. The single hydrogen valence shells are represented by two orbitals of two primitives each. Thus, for a methane (CH_4) molecule we shall have $6 + 4(4) + 4(2 \times 2) = 38$ primitives.

It may be noted that while the basis sets can be improved by going to TZ, quadruple zeta, etc., in some cases this may however lead to incorrect results unless functions of higher angular momentum are included. A typical example is the equilibrium geometry of ammonia which actually becomes planer if only large numbers of s and p functions are used without including d and f functions.

3.11.2.3 Polarization Basis Sets

As in the case of STOs, in order to account for the distortions in atomic orbitals when they enter a molecule, it is important to use polarized Gaussian basis functions with higher angular momentum.

Polarized basis sets are constructed by adding orbitals having angular momentum l one higher than the primitive Gaussian. Thus, for describing a $2p$ orbital on an atom, a polarization function shall have 3d character. Adding a single set of polarization functions to the DZ basis forms a DZP-type basis. Frequently, polarization functions are added only to nonhydrogen atoms partly due to the fact that hydrogen sits at the end of bonds that do not take active part in the property of interest. However, in those cases where hydrogen plays an important role in the property of interest, polarization functions on hydrogen must be included. In using Gaussian orbitals, basis sets that allow for polarized functions are denoted by an asterisk (*) after G. Thus, the basis 6-31G* is 6-31G basis plus a set of six d-type Cartesian Gaussian polarization functions on each nonhydrogen atom and the basis 6-31G** has, in addition, a set of three p-type Gaussians on each hydrogen atom. The 6-31G* and 6-31G** sets are sometimes denoted as 6-31G(d) and 6-31G(d, p), respectively. A more general representation is 6-31G(kp, ld), where k and l are integers that indicate the number of p (for hydrogen) and d (for other atoms) polarization functions included in the basis.

3.11.2.4 Diffuse Function Basis Sets

Another set of basis functions that are commonly used are diffuse functions. These are used to take into account the effect of electrons when they are away from the nucleus. Thus, anions, hydrogen-bonded dimers and loose supermolecular complexes, compounds with lone pairs, etc., have significant electron density at large distances from the nucleus. Similarly, highly excited electronic states tend to be much more spatially diffuse than ground state MOs. The diffuse functions have very small ζ value (typically, 0.01–0.1) and so they allow the wavefunction to extend far from the nucleus. In the general case, a rough rule of thumb is that diffuse functions should have an exponent about a factor of four smaller than the smallest valence exponent. Diffuse functions are normally s and p functions and in Pople's style of basis sets are denoted by + or ++ before G; the first + indicates one set of diffuse s and p functions (s, p_x, p_y, p_z)on heavy atoms, and the second + indicates that a diffuse s function is added also to hydrogen. Thus, 6-31+G(d) or 6-31+G* indicates that heavy atoms have been augmented with an additional one s and one set of p functions having small exponents and a single d-type polarization function. A second plus indicates the presence of diffuse s functions on hydrogen also.

The inclusion of diffuse functions, where needed may significantly affect the calculated energies and molecular properties.

3.11.2.5 Correlation Consistent Basis Sets

Dunning pointed out that basis sets optimized at the HF level might not be ideal for correlated computations. They developed Gaussian orbitals, expressed using Eq. (3.9.2), in which the coefficients are determined by solving the Schrödinger equation for atoms with high level of correlation included, such as MP2, MP3, MP4, CISD, etc. The use of Dunning's *correlation consistent* basis sets recovers the correlation energy of the valence electrons. For first- and second-row atoms, the basis sets are *cc–pVNZ* where N = D, T, Q,… (D = double, T = triples, Q = quadruples, etc.). The "cc," stands for "correlation consistent," "p" for polarized and the "V" indicates they are valence-only basis sets. They include successively larger shells of polarization (correlating) functions (*d, f, g,* etc.). Several different sizes of cc basis sets are available in terms of final number of contracted functions. These are known by their acronyms: *cc–*p*VDZ, cc–*p*VTZ, cc–*p*VQZ, cc–*p*V5Z,* and *cc–*p*V6Z* (*correlation consistent* polarized valence double/triple/quadruple/quintuple/sextuple zeta). In the Dunning family of *cc–pVNZ* basis sets, diffuse functions on all atoms are indicated by prefixing with aug-. Moreover, one set of diffuse functions is added for each angular momentum already present. Thus, *cc–pVDZ* for C atom consists of 3s 2p 1d functions and *aug–cc–*p*VTZ* has diffuse f, d, p, and s functions (4s 3p 2d 1f) on heavy atoms and diffuse d, p, and s functions (3s 2p 1d) on H and He.

3.11.3 Some Other Basis Sets

The basis set considered so far have all been made up of nuclear-centered functions; that is functions whose centers are fixed on nuclei of a molecule. In addition there are some other kinds of basis functions that are occasionally used. Some of these are the following:

1. Floating spherical Gaussian in which the position of the center of the orbital and the Gaussian width are variationally optimized.
2. Plane waves which are completely delocalized.
3. Numerical atomic HF orbitals which are products of spherical harmonics and numerically tabulated radial functions that are obtained from atomic HF calculations.

The first two of these shall be described here.

3.11.3.1 Floating Spherical Gaussian Orbitals

Frost [19a,b] provided flexibility to the GTOs, which are nuclear-centered functions by treating the centers of Gaussians as variational parameters. The functions were now allowed to "float" during the variational optimization of the energy. Such orbitals are known as floating spherical Gaussian orbitals (FSGOs). In this construction, a MO which is expressed as a linear combination of n 1s-type Gaussians has $5n - 1$ variational parameters. These include $(n - 1)$ independent coefficients, n Gaussian exponents, and $3n$ position coordinates of the centers. Because of this flexibility, the orbitals no longer remain bound to a given center and all reference to the atomic orbitals of the constituent atoms of a molecule is lost. Hence, there is no need of polarization functions to describe the distortion of atomic orbitals in the molecular environment. Frost also suggested a model in which the electronic structure of a molecule in a closed-shell state is described

Table 3.2 Experimental and FSGO (floating spherical Gaussian orbital)-based ground state geometries of some molecules. Bond length in Å

Molecule	Parameter	FSGO	Exp.
CH_4 [a]	R(C–H)	1.115	1.085
C_2H_2 [b]	R(C–H)	1.079	1.061
	R(C–C)	1.214	1.203
C_2H_4 [b]	R(C–H)	1.101	1.076
	R(C–C)	1.351	1.330
	θ(HCH)	118.7	116.6
C_2H_6 [b]	R(C–H)	1.120	1.096
	R(C–C)	1.501	1.531
	θ(HCH)	108.2	107.8
H_2O [a]	R(O–H)	0.881	0.957
	θ(HOH)	88.4	104.5
LiH [c]	R(Li–H)	1.707	1.595

[a] A.A. Frost, J. Phys. Chem. 72 (1968) 1289.
[b] A.A. Frost, R.A. Rouse, J. Am. Chem. Soc. 90 (1968) 1965.
[c] A.A. Frost, J. Chem. Phys. 47 (1967) 3707.

by $n/2$ double occupied localized bond and lone-pair orbitals, each of which is represented by a single normalized spherical Gaussian (1*s*), of the type

$$\Psi = \left(2/\pi\rho^2\right)^{3/4}\exp\left(-|r-R|^2/\rho^2\right)$$

whose effective radius ρ and the position R are obtained by minimizing the energy. Thus, for a diatomic molecule like LiH having four electrons only two spherical Gaussian orbitals (SGOs) having their centers on the molecular axis are needed. Ethane which has 18 electrons can be described by 9 SGOs—each carbon inner shell is represented by 1 SGO, each of the 6 equivalent CH bonds by a single SGO and CC bond by a single SGO centered on a bond axis. Due to unsatisfactory description of the inner shell electrons, FSGO, though simple, does not allow a very accurate calculation of the total or orbital energies (Table 3.2).

The FSGOs provide simple nonempirical method for the description of the shapes and electronic structures of molecules. This method is superior to the semiempirical methods and has been widely used for small- and medium-sized molecules as: (1) It can be applied to any configuration of a molecule for which experimental information may or may not be available, (2) Unlike the semiempirical theories it does not require empirical parameterization and calibration, and (3) It is, in principle, capable of improvement to any desired accuracy.

3.11.3.2 Plane Wave Basis Sets

In addition to the common type of basis sets localized on the nuclei, another type of basis sets known as plane wave basis sets are used for quantum chemical simulations.

These are helpful for extended (infinite) systems involving periodic boundary conditions such as in crystals, metals, polymers, etc. Certain integrals and operations are easier to carry out with plane wave basis functions rather than with their localized counter parts.

The plane wavefunctions were first introduced for understanding the behavior of free electrons in metals. The outer valence electrons in metals behave almost like free electrons for which solutions to one-dimensional Schrödinger equation can be written as

$$\Psi(x) = Ae^{ikx} + Be^{-ikx} \tag{3.11.3}$$

For infinite systems, the energy difference between the MOs almost vanishes and they coalesce into bands. The electrons in a band can be described by orbitals expanded in a basis set of plane waves, which in three-dimension may be written as

$$\Psi_k(\mathbf{r}) = e^{i\mathbf{k}\cdot\mathbf{r}} \tag{3.11.4}$$

where $\mathbf{k}\left(=\frac{2\pi}{\lambda}\right)$ is the wave vector and is related to the energy (in eV) by the relation

$$E = \frac{1}{2}k^2 \tag{3.11.5}$$

$\mathbf{k}$ being proportional to frequency may also be thought of as frequency factor; high $\mathbf{k}$ values indicate rapid oscillation.

A comparison with a STO shows that $\mathbf{k}$ plays the same role as the exponent ζ. If we consider boundary conditions in a periodic system having a unit cell translational vector ($\mathbf{T}$), then $\mathbf{k}\cdot\mathbf{T} = 2\pi n$ where n is a positive integer. This leads to a typical spacing between $\mathbf{k}$ vectors of ~0.01 eV. The size of a plane wave basis set is characterized by the highest value of energy corresponding to $\mathbf{k}$. Thus if a system having highest cutoff energy of 100 eV then about 10,000 plane wavefunctions will be involved. Thus, the size of this type of basis set is very much larger than that of the typical Slater-type or Gaussian-type basis set. It is also to be noted that the size of a plane wave basis set depends only on the size of the periodic cell and not on the actual system described within the cell. This is in contrast to the GTF where the number of basis function increases with the system size. For this reason the plane wave basis sets become more favorable for larger systems and have been primarily used for periodic system.

In periodic molecular systems, the core electrons are concentrated very close to the atomic nuclei. This results in large wavefunction and density gradients near the nuclei which are not easily described by the plane wave basis sets unless very high energy cutoffs are involved. This obviously drastically increases size of the basis sets. In practice, these basis sets are used in combination with an effective core potential also called pseudopotential. So that the plane waves are used to describe only the valence electrons.

Some attempts have also been made to use plane wavefunctions for molecular species by using a super cell approach, where the molecule is placed in a sufficiently large unit cell such that it has no interaction with its repetitive molecules in the neighboring cells. The use of large super cells requires a very large basis set.

Since problems related to molecular species can more conveniently be solved by using Gaussian basis set, the plane wave basis sets are primarily used for describing a three-dimensional periodic system.

Besides the ease with which they can be handled, the plane wave basis sets have three main advantages over the Gaussian basis sets:

1. Since all the functions in the basis are mutually orthogonal and are not associated with any particular atom, they do not show any basis set superposition error (BSSE).
2. It is possible to work with them also in reciprocal space in which all integrals and the derivatives of function can be easily calculated.
3. They converge in a smooth, monotonic manner to the target function. This is in contrast to GTFs where there is no guarantee of smooth convergence, except in variational calculations.

3.11.4 Basis Set Superposition Error

Most common basis sets used for molecular applications are centered on the nuclei. For getting accurate results, in principle, complete basis sets should be used for calculations. However, due to enormous computational efforts involved in such calculations, usually a limited number of basis sets, having a few hundred to a few thousand basis functions are used. This limitation may cause large errors of several atomic units or thousands of kcal/mol. Since in most of applications, the basic interest lies in the computation of relative energies, obviously, it is necessary that the errors are constant so that they cancel out when relative energies are calculated. Thus, for example, if enthalpy difference between two isomers is to be calculated, it will be meaningless to compare their energies if calculations are performed using different basis sets, although they may be of the same quality like DZ or TZ. Further the nuclei-fixed basis sets introduce an error when comparing energies at different geometries. The quality of the basis set is not the same for all geometries as the electron densities around one nucleus has contributions from functions centered at another nucleus. The error is especially large when calculating small effects such as energies of hydrogen bonds or van der Walls complexes. Thus, in the dimer of water where two water molecules join together, the energy gain is obtained by subtracting the energies of two water monomers from that of the dimer. There is however an inconsistency involved here. When calculating the energy of a dimer, the total number of basis sets becomes double of what is used for a monomer. This will lower down the energy of the dimer with the result that the strength of the hydrogen bond, as calculated by the above prescription, is overestimated. This effect is known as the BSSE. If however a complete basis set were used for all the calculations, this error would be zero and adding more basis function would not have given any improvement. In order to eliminate BSSE the conceptually simplest approach is to add more and more basis functions until the change in interaction energy is zero. This approach is however not practicable, particularly in larger systems, because it requires very large basis sets and hence an enormous increase in the computational efforts.

No completely reliable scheme for either eliminating or estimating superposition errors is available. An approximate method of accessing the BSSEs is the counterpoise correction as described by van Duijneveldt et al. [20]. In this method the BSSE is estimated as the difference between energies of monomers with the regular basis and the energy for the whole complex calculated with the full set of basis functions.

3.12 Quality of HF Results

HF theory is fundamental to much of electronic structure theory. Methods of calculating molecular electronic structure have reached chemical accuracy of roughly 1 kcal/mol for small- and medium-sized organic molecules and for compounds containing elements in the first and second rows of the periodic table. The HF calculations are usually done for a variety of basis functions of increasing complexity in order to study the convergence behavior of the properties of interest like energy structure and reactivity. By using large enough basis, one recovers the exact solution of the HF equations which is known as the HF limit. This limit is however, not enough for exact description of the electronic structure and for many properties. Some of the results from HF calculations using modest basis sets are given below:

3.12.1 Energetic Predictions

Among the most chemically important energetic quantities are conformational energy changes, heats of reaction, dissociation energies, and activation energies or barrier heights. In general only the first or to some extent the second quantity is reliably predicted at the HF level of theory. The energetic quantities are often sensitive to the effect of electron correlation. Generally speaking, the prediction of exothermicities or heat of reaction is often unreliable except for systems having closed-shell reactants and products such as, isodesmic reactions where the number of bonds of each type is conserved.

The dissociation energies of covalent molecules are generally predicted poorly by single configuration SCF method and may be typically low by a substantial fraction like 50%. Thus as against the experimental value of 9.91 eV, the calculated dissociation energy for N_2 is 5.27 eV [21].

Another frequent failing of the HF method lies in the prediction of the barrier heights or activation energies of chemical reactions. In many cases these barriers may be very high, up to several hundred percent of the experimental value. However, in many classes of reactions, the HF theory does yield meaningful barrier heights. Much improved results are obtained when polarization functions and correlation effects are included.

The unimolecular reactions seem to be treated more reliably by HF methods than bimolecular reactions. Thus, the predicted barrier heights for geometrical isomerization of cyclopropane is 53 kcal/mol, in reasonable agreement with the experimental value 64 kcal/mol.

3.12.2 Structural Predictions

Most molecular structures appear to be reliably predicted at the HF level of theory. While carbon–carbon bond distances differ typically by 0.02 Å from the experimental value, in some cases the error may be up to 0.05 Å. The error in bond angle predictions may be between 2 and 5°. Transition state structures are more sensitive to basis sets than the equilibrium geometries. This is true because potential energy surfaces are often rather flat in the vicinity of a saddle point. Typical structural variations at the transition state are 0.02–0.05 Å in internuclear separations and 2°–5° in angles. In general it may be said that as long as we consider molecules near their equilibrium geometry, HF theory often provides a good starting point for more elaborate theoretical methods which take into account the electron correlation effects.

3.12.3 Vibrational Frequencies

The vibrational frequencies of molecules calculated by HF theory are high by 10–15% on an average. The inclusion of electron correlation does improve the frequency to some extent but they still remain on the higher side of the experimental values. It has become customary to multiply the calculated frequencies by a scaling factor between 0.89 and 0.95 before comparing with the experiment.

3.13 Beyond HF Theory

The HF method makes some major simplifications in the study of electronic structure of molecules. It assumes that each energy eigenfunction can be described by a single Slater determinant and each electron moves in an average electronic field. It completely neglects electron correlation for the electrons of opposite spin but it does take into account to some extent correlation between electrons of parallel spin. For an HF function, there is little probability of finding electrons of the same spin in the same region of space. This explains correlation of the motion of electrons with the same spin. The correlation energy E_c is conveniently defined as

$$E_c = E_{\text{exact}} - E_{\text{HF}}$$

that is, the difference between the nonrelativistic exact energy and the HF-limit energy. The quantum chemical calculations are sometimes judged by the percentage of correlation energy they can account for. This may be between 80% and 95% depending on the size of the molecule.

The neglect of electron correlation can lead to large deviations from experimental results. A number of approaches, collectively called post HF method, have been devised to include electron correlation to the multielectron wavefunction. They help to include repulsions between electrons in a more accurate way than the HF method where the repulsions are only averaged. Some of these post HF methods are the following:

1. Perturbation methods or Møller–Plesset (MP) theory (MP2, MP3, MP4, etc.) which treats correlation as a perturbation of the HF operator.
2. Configuration interaction (CI) which is a linear variational method for solving the nonrelativistic Schrödinger equation within the Born–Oppenheimer Approximation.

3. Multiconfiguration SCF (MCSCF) with/without CI method which involves a wavefunction that consists of a sum of determinants. The coefficients in front of each determinant and the orbitals in each determinant are optimized variationally.
4. Coupled-Cluster (CC) theory which is a numerical technique used to describe many body systems. It is one of the most prevalent methods in quantum chemistry that includes electronic correlation.
5. Complete active space SCF (CASSCF) method. Like MCSCF and CI theories it expands the true multielectron wavefunction in terms of a linear combination of Slater determinants.

3.13.1 CI Method

This is the oldest and perhaps the easiest method based on the variational method. Basically, the method is based upon the idea that if we can obtain a number of different solutions to our Schrödinger equation corresponding to different configurations (different orbitals occupancies), then linear combinations of these individual solutions will also be solutions. If our solutions were exact solutions, then the linear combination of solutions would give us the same energy values as the individual solutions (generally the expansion coefficient for a given configuration would turn out to be 1 and all others 0). However, since we cannot directly obtain the exact solution of the Schrödinger equation, the use of a combination of configurations would amount to expanding the set of functions to obtain correct wavefunction. Further, since, the calculated energies in variation technique are larger than the true energy, one can expect an improved energy from a wavefunction made up of a linear combination of the configurational functions. Calculations taking advantage of this approach are referred to as CI calculations.

Frequently, a distinction has been made between two types of CI: first-order CI, where the only configurations used are those which are energetically degenerate, and second-order CI, where many other configurations are included.

The concept of configuration has already been discussed in Section (2.5). An example of the various configurations of a six electrons molecular system is given in Figure 3.18.

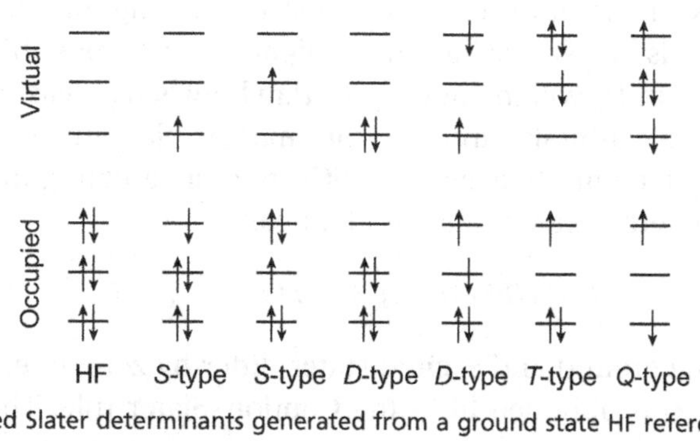

FIGURE 3.18 Excited Slater determinants generated from a ground state HF reference configuration.

Since an electronic configuration is defined in terms of Slater determinants composed of spin orbitals, a whole series of determinants may be generated by replacing the occupied MOs by the unoccupied MOs. These can be denoted according as how many occupied MOs have been replaced by unoccupied MOs, i.e., Slater determinants that are singly, doubly, triply, quadruple, etc., excited relative to the ground state HF determinant, up to a maximum of N-excited electrons. These determinants are often referred to as singles (S), doubles (D), triples (T), quadruples (Q), etc. The trial wavefunction is written as a linear combination of determinants with the expansion coefficients determined by requiring that the energy should be a minimum (or at least stationary). Subscripts S, D, T, etc., indicate determinants that are singly, doubly, triply, etc., excited relative to the HF configuration.

$$\Psi_{CI} = c_0\Phi_{\mathrm{HF}} + \sum_S c_S\Phi_S + \sum_D c_D\Phi_D + \sum_T c_T\Phi_T + \sum_Q c_Q\Phi_Q \ldots = \sum_i c_i\Phi_i \tag{3.13.1}$$

When all the configurations are taken in to account, the calculations are called "full configuration interaction or full CI calculations." Such calculations cannot be conducted but for the small molecules.

The MOs used for building the excited Slater determinants are taken from a HF calculation and held fixed. The energy is then minimized under the constraint that the total CI wavefunction is normalized. The coefficients c_i, also known as CI coefficients, are determined by solving the eigenvalue problem

$$HC = EC \tag{3.13.2}$$

Solving the secular Eq. (3.13.2) is equivalent to diagonalizing the CI matrix. The CI energy can be obtained as the lowest eigenvalue of CI matrix and the corresponding eigenvector contains the c_i coefficients in Eq. (3.13.1). The second lowest eigenvalue corresponds to the first excited state, etc.

The Hamiltonian operator in Eq. (3.13.2) is composed of only one- and two-electron operators and the Hamiltonian matrix H is composed of elements

$$H_{ij} = \int \Phi_i H \Phi_j d\tau \tag{3.13.3}$$

which are usually taken as linear combination of only one- and two-electron integrals to simplify calculations. It therefore follows that if two configurations differ by three or more occupied orbitals (triple excitation or higher), every possible integral over the electronic coordinates in H_{ij} will include an overlap between at least one pair of different and hence orthogonal orbitals and so the matrix element shall be zero. Thus, Hamiltonian matrix elements between the SCF reference configuration and all triple, quadruple, quintuple, etc., excitations are identically zero.

$$\int \Phi_{\mathrm{HF}} H \Phi_T d\tau = \int \Phi_{\mathrm{HF}} H \Phi_Q d\tau = \ldots = 0 \tag{3.13.4}$$

For cases when the ground and excited states differ by zero, one, or two orbitals, the solution of Eq. (3.13.2) is obtained from the Condon–Slater rule. This rule dictates that the matrix element between the HF ground state (I) and the singly excited states (S) when a transition is taking place between an occupied orbital Φ_i and the virtual orbital, say, Φ_r, is

$$H_{IS} = \int \Phi_{HF}\, H\, \Phi_S^{ir}\, d\tau = \int \phi_r F \phi_i\, d\tau \tag{3.13.5}$$

where F is the Fock operator. Since these orbitals ϕ_i *and* ϕ_r are eigenfunctions of the Fock operator, we have

$$\int \phi_r F \phi_i\, d\tau = \varepsilon_i \int \phi_r \phi_i\, d\tau = \varepsilon_i \delta_{ir} \tag{3.13.6}$$

where ε_i is the MO eigenvalue. Thus, all matrix elements between the HF determinant and singly excited determinants are zero as r is different from i. So, there will be no interaction between the ground and singly excited states and the inclusion of singly excited state in Eq. (3.13.1) shall have no affect on the ground state energy. This result is known as Brillouin theorem and the method is known as CI-singles (CIS). Thus, CIS method is not useful for the ground states though it can be useful for excited states as shall be seen later in Chapter 9. We should, therefore, consider double excitation. After taking into account CI the total energy is reduced relative to the HF energy. This difference is called the correlation energy.

In order to illustrate the effect of double excitation we may consider the case of hydrogen molecule which has two MOs σ and σ^* corresponding to the electronic ground and excited states. In this case, the doubly excited electronic state will be σ^{*2}.

The energies resulting from the interaction between the ground (I) and doubly excited (II) states can be found by solving the determinant

$$\begin{vmatrix} H_{I\ I} - E & H_{I\ II} \\ H_{I\ II} & H_{II\ II} - E \end{vmatrix} = 0 \tag{3.13.7}$$

which results in

$$E = \frac{1}{2}[(H_{I\ I} + H_{II\ II}) \pm \sqrt{(H_{II\ II} - H_{I\ I})^2 + 4H_{I\ II}}] \tag{3.13.8}$$

$H_{I\ I}$ and $H_{II\ II}$ are the energies of the ground and the doubly excited states, respectively, where as $H_{I\ II}$ is interaction energy of the two states which, as per the Condon–Slater rule, is the exchange integral ($K_{I\ II}$) and has a positive value. So the value of the second term in Eq. (3.13.8) shall be slightly larger than the difference between the states I and II. So,

$$E_+ = \frac{1}{2}\left[(H_{I\ I} + H_{II\ II}) + \sqrt{(H_{II\ II} - H_{I\ I})^2 + 4H_{I\ II}}\right] \text{ and}$$

$$E_- = \frac{1}{2}\left[(H_{I\ I} + H_{II\ II}) - \sqrt{(H_{II\ II} - H_{I\ I})^2 + 4H_{I\ II}}\right] \tag{3.13.9}$$

Since $\frac{1}{2}[H_{I\ I} + H_{II\ II}]$ is the average of the energies of the ground and doubly excited state, it is clear that the energy E_{-} will be below the average energy. (So, in this case the total energy is reduced relative to the HF energy. This difference is called the correlation energy.)

The correlation energy in H_2 is found to be about 13 kcal/mol using STO-3G basis set. It may also be noted that while singly excited states do not interact with the ground states, they do interact with the doubly excited states. Since the latter interacts with the ground states, the singly excited states may indirectly influence the ground state. The method which includes both single and double excited states is called CISD.

The triply excited state does not directly interact with the ground state as the matrix element between the two is zero. However, they interact with the doubly excited states and through them influence the ground state. Methods based on this approach are known as CISDT. Since, there may be a lot of triply excited states, their inclusion makes the calculations practically difficult. Similarly, taking into account also quadruple excited determinants yields the CISDTQ method. This method, in general, gives results close to the full-CI limit. As the number of triple and higher excitations is very large, they are usually not included in truncated CI calculations of the excited states.

For small molecules, single and double (S + D) excitations give rise to almost all of the correlation energy. Thus, in the case of water molecule, CISD accounts for ~95–96% of the correlation energy. In larger molecules the CISD method recovers less and less of the correlation energy and so the higher excitations become more important.

3.13.2 MP Perturbation Theory

Perturbation theory described in Chapter 1 has been used since the early days of quantum chemistry to obtain descriptions of electronic structures of atoms and molecules free from electron correlation effects. Møller and Plesset described how the HF method can be corrected for electron pair correlation by using second-order perturbative theory. This approach is now known as MP perturbation theory and referred to as MPn, where n is the order at which the perturbation theory is truncated, for example, MP2, MP3, MP4, etc. It can be seen as below that the HF energy is correct up to first order of MP perturbation (MP1).

In the MP theory the energy corrections are obtained with a perturbating operator $\widehat{H}^{(1)}$ given by

$$\widehat{H}^{(1)} = \widehat{H} - \widehat{H}^{(0)} \tag{3.13.10}$$

where $\widehat{H}$ and $\widehat{H}^{(0)}$ are the exact and the unperturbed or zeroth-order HF Hamiltonians. We shall be limiting our treatment to closed-shell ground state molecules.

The MP approach takes $\widehat{H}^{(0)}$ to be the sum of one-electron Fock operators, that is, the sum of noninteracting operators $\widehat{F}_i$ given by

$$\widehat{H}^{(0)} = \sum_{i=1}^{n} \widehat{F}_i \tag{3.13.11}$$

$$\widehat{F}_i = -\frac{1}{2}\nabla_i^2 - \sum_{\alpha}\frac{Z_\alpha}{r_{i\alpha}} + \sum_{i}(J_i - K_i) \tag{3.13.12}$$

where n is the number of basis functions and $\widehat{F}_i$ is defined by an equation identical to Eq. (2.4.31). Taking $\widehat{H}^{(0)}$ as the starting point for perturbation expansion we may define a zero-order problem

$$\widehat{H}^{(0)}\Psi^{(0)} = E^{(0)}\Psi^{(0)} \tag{3.13.13}$$

$$\text{or} \quad E^{(0)} = \int \Psi^{*(0)}\widehat{H}^{(0)}\Psi^{(0)}d\tau \tag{3.13.14}$$

where $\widehat{H}^{(0)}$ and $\Psi^{(0)}$ are the zeroth-order Hamiltonian and wavefunction, respectively. $\Psi^{(0)}$ is taken to be the HF wavefunction which is a Slater determinant formed from the occupied orbitals by analogy to Eq. (2.1.8). It can easily be shown that the eigenvalue of $\widehat{H}^{(0)}$ when applied to the HF wavefunction is the sum of the occupied orbital energies, i.e.,

$$\widehat{H}^{(0)}\Psi^{(0)} = \sum_{i}^{\text{occ}}\varepsilon_i\Psi^{(0)} \tag{3.13.15}$$

where the orbital energies are the eigenvalues of the specific one-electron Fock operators.

Hence,

$$E^{(0)} = \sum_{i}^{\text{occ}}\varepsilon_i \tag{3.13.16}$$

which defines the eigenvalue of the operator $\widehat{H}^{(0)}$. The definition of $\widehat{H}^{(0)}$ neglects all electron repulsion terms. In actual molecule the energy is determined by using the correct Hamiltonian and the HF wavefunctions. Thus, the error in Eq. (3.13.15) is that each orbital energy includes the repulsion of the occupying electron(s) with all other electrons. Thus, each electron–electron repulsion is counted twice (once in each orbital corresponding to each pair of electrons). In exact Hamiltonian, the electron repulsion is counted only once. Hence, the perturbative Hamiltonian should be such that it will lead to the correct Hamiltonian and allow to improve the HF wavefunctions and eigenvalues by the perturbation theory. The correction $\widehat{H}^{(1)}$ must therefore include the electron repulsion term as it appears in the exact Hamiltonian $\widehat{H}$ and removes the effect of counting electron repulsion twice. Thus,

$$\widehat{H}^{(1)} = \sum_{i}^{\text{occ}}\sum_{j>i}^{\text{occ}}\frac{1}{r_{ij}} - \sum_{i}^{\text{occ}}\sum_{j}^{\text{occ}}\left(J_{ij} - \frac{1}{2}K_{ij}\right) \tag{3.13.17}$$

Here, the first term is the electron repulsion term as appearing in Eq. (3.8.1) and the second term is half of the term appearing in Eq. (3.8.9). J_{ij} and K_{ij} are the Coulomb and exchange operators, respectively, given by Eqs (3.8.12) and (3.8.13). The first-order perturbation correction $E^{(1)}$ to the energy can now be found by using Eq. (1.13.6).

$$E^{(1)} = \int \Psi^{*(0)}\widehat{H}^{(1)}\Psi^{(0)}d\tau \tag{3.13.18}$$

Thus, from Eqs (3.13.14) and (3.13.18), we get

$$\begin{aligned} E^{(0)} + E^{(1)} &= \int \Psi^{(0)*} \widehat{H}^{(0)} \Psi^{(0)} d\tau + \int \Psi^{(0)*} \widehat{H}^{(1)} \Psi^{(0)} d\tau \\ &= \int \Psi^{(0)*} \left(\widehat{H}^{(0)} + \widehat{H}^{(1)}\right) \Psi^{(0)} d\tau = \int \Psi^{(0)*} \widehat{H} \Psi^{(0)} d\tau = E_{\mathrm{HF}} \end{aligned} \tag{3.13.19}$$

that is, the HF energy is correct up to the first order of the MP perturbation theory.

We now, take into account the second-order MP perturbation (MP2) and calculate the second-order energy correction. For this purpose, we must go beyond the closed-shell system and consider all configurations resulting from the excitation of one or more electrons. We must therefore construct more than one Slater determinant by replacing one or more occupied orbitals within the HF determinant by a virtual orbit. Thus, in a single substitution, a virtual orbital, say, Ψ_a replaces an occupied orbital Ψ_i in the determinant. This is equivalent to exciting an electron to a higher energy orbital. Similarly, in a double substitution, two occupied orbitals are replaced by two virtual orbitals say, Ψ_a replacing Ψ_i and Ψ_b replacing Ψ_j. Likewise, triple substitution replaces three occupied orbitals Ψ_i, Ψ_j, Ψ_k by three virtual orbitals Ψ_a, Ψ_b, Ψ_c and so on. These are equivalent to the S, D, T,... configurations described in the Section 3.13.1.

A linear combination of all such determinants is then used for the calculations. The second-order correction to the correlation energy can be obtained by using Eq. (1.13.9) in the form

$$E^{(2)} = \sum_{j>0} \frac{\left| \int \Psi_j^{(0)} \widehat{H}^{(1)} \Psi_0^{(0)} d\tau \right|^2}{E_j^{(0)} - E_0^{(0)}} \tag{3.13.20}$$

where $\Psi_j^{(0)}$ is the appropriate normalized ground state eigenfunction of $\widehat{H}^{(0)}$. We must evaluate Eq. (3.13.20) using the set of all possible excited state eigenfunctions and eigenvalues of the operator $\widehat{H}^{(0)}$.

From Brillouin's theorem we know that the integrals involving the singly excited determinants will all be zero. Triple and higher substitutions also result in zero values as the Hamiltonian contains only one- and two-electron terms meaning that only pair-wise electron interactions are considered. Thus, only doubly excited determinants give nonzero values. Evaluation of the numerator in Eq. (3.13.20), therefore, gives

$$\sum_{j>0} < \Psi_j^{(0)} | \widehat{H}^{(1)} | \Psi_0^0 \geq \sum_{i}^{\text{occ}} \sum_{j>i}^{\text{occ}} \sum_{a}^{\text{vir}} \sum_{b>a}^{\text{vir}} [(ij|ab) - (ia|jb)] \tag{3.13.21}$$

where the two-electron integrals are defined by Eq. (3.8.26).

As for the denominator of Eq. (3.13.20), $E^{(0)}$ for each doubly excited determinant will differ from that for the ground state $E_0^{(0)}$ only by excluding in the sum the energies of the two occupied orbitals from which the excitation has taken place and instead including the energies of the two virtual orbitals into which the excitation has occurred. The second-order energy correction can therefore be written as

$$E^{(2)} = \sum_{i}^{\text{occ}} \sum_{j>i}^{\text{occ}} \sum_{a}^{\text{vir}} \sum_{b>a}^{\text{vir}} \frac{[(ij|ab) - (ia|jb)]^2}{\varepsilon_i + \varepsilon_j - \varepsilon_a - \varepsilon_b} \tag{3.13.22}$$

Thus, the total MP2 energy will be

$$E(\text{MP2}) = E^{(0)} + E^{(1)} + E^{(2)} \tag{3.13.23}$$

In a similar process, the third- and fourth-order energy corrections can also be derived.

The second-order perturbation energy will always be negative so as to lower down the HF energy. A serious drawback in the perturbation theory may here be noted. A perturbation theory works best when the perturbation is small but in the case of MP2 theory the perturbation is the full electron–electron repulsion energy which is a very substantial contributor to the total energy. Thus, it is possible that the MP2 estimate for the correlation energy may be too large. Also since the MP theory is not variational, the higher order corrections may even be positive.

The third-order MP3 calculations again involve doubly excited state but tend to offer rather little improvement over MP2. The MP4 calculations involve triply and quadruple excited determinants. Though very costly, the MP corrections can together account for more than 95% of the correlational energy. Though one expects convergent behavior for higher orders (MPn, $n > 4$), in practice only an oscillatory behavior is observed.

3.13.3 Multiconfiguration SCF Method (MCSCF-CI)

The idea behind a MCSCF calculation is to use a wavefunction that consists of a sum of determinants (about hundreds to thousands) and to optimize variationally both the coefficients in front of each determinant and the orbitals in each determinant.

$$\Psi = c_0 \Psi_{\text{HF}} + c_1 \Psi_1 + c_2 \Psi_2 + \ldots \tag{3.13.24}$$

where the coefficients c reflect the weight of each determinant in the expansion and also ensure normalization. MCSCF theory is designed to handle both multiple configurations and the multideterminantal character of individual configurations. In this regard MCSCF is a generalization of ROHF theory which can handle multiple determinants but is not capable of handling multiple configuration state functions (CSF). A CSF refers to a molecular spin state and for closed-shell systems can be represented as single determinants.

As an example, consider the case of ground state of helium. In this case we may construct a wavefunction as

$$\Psi = c_1 \Psi_1 + c_2 \Psi_2 = c_1 \left| \Psi_1 \overline{\Psi}_1 \right| + c_2 \left| \Psi_2 \overline{\Psi}_2 \right| \tag{3.13.25}$$

If we write the MOs as

$$\Psi_i = \sum_j a_{ji}\Phi_j, \tag{3.13.26}$$

We get, $$\Psi = c_1\left|(a_{11}\Phi_1 + a_{21}\Phi_2)\left(\overline{a_{11}\Phi_1 + a_{21}\Phi_2}\right)\right| + c_2\left|(a_{12}\Phi_1 + a_{22}\Phi_2)\left(\overline{a_{12}\Phi_1 + a_{22}\Phi_2}\right)\right| \tag{3.13.27}$$

In order to get an MCSCF function, we may then have to vary all the six coefficients c_1, c_2, a_{11}, a_{21}, a_{12}, and a_{22} to minimize the energy. The optimized MCSCF orbitals are obtained by an iterative procedure like the SCF. Since in this case the orbitals are also optimized, one can get good results by using relatively lesser number of CSF. However, since the number of MCSCF iterations required for achieving convergence tends to increase with the number of configurations included, the size of MCSCF wavefunction that can be treated is somewhat smaller than for CI methods. MCSCF calculations are especially useful for describing bond breaking and forming.

In fact, one type of MCSCF calculation is based on summing determinants as in the VB theory and is also called the generalized VB theory. MCSCF methods are rarely used for calculating large fractions of the correlation energy because the orbital relaxation usually does not recover most of the electron correlation. The major problem with MCSCF methods is selecting which configurations are necessary to be included for the property of interest. A type of commonly used MCSCF method is the CASSCF described below.

3.13.3.1 CASSCF Method

In this method the configurations are divided into active and inactive orbitals. The orbitals involved in the CI are known as active space. The active MOs are usually taken to be some of the highest occupied and some of the lowest unoccupied MOs from an RHF calculation. The inactive orbitals are always either doubly occupied or empty. The wavefunction is then written as a linear combination of all symmetry-adapted configurations by distributing the active electrons among the active orbitals in all possible ways. The choice of the configurations must be decided manually by considering the problem at hand and the amount of computational effort involved. Thus, for example, if several points on the energy surface are desired the CASSCF calculations are done by including all those orbitals that change significantly or for which the electron correlation is expected to change. After choosing the configurations, the coefficients c_i and a_{ij} involved in Eqs (3.13.25) and (3.13.26) are optimized to obtain CASSCF wavefunctions. A common notation for CASSCF calculations is CASSCF [n,m] which indicates that n electrons are distributed in all possible ways in m orbitals. This method is included in most of the quantum chemistry softwares with useful examples [24]. The CASSCF method becomes unmanageably large even for quite small active spaces.

3.13.4 Coupled-Cluster Method

Several methods have been used to account for the correlation energies in molecules. The more thoroughly developed and widely adopted approaches focus on the systematic treatment of correlation by introducing refinements in the HF approximation either by rigorous methods to obtain chemical accuracy or by semiempirical method such as Monte Carlo method. The CC method belongs to the first category. It essentially takes the basic HF MO method and constructs wavefunctions using the exponential cluster operator to account for the electron correlation. Although this method is quite demanding on computer resources and time, it is still widely regarded as the best available method to treat difficult systems. Some of the most accurate calculations for small- to medium-sized molecules have used this method [23–25].

CC theory provides an exact solution to time-independent Schrödinger equation

$$\widehat{H}|\Psi\rangle = E|\Psi\rangle \tag{3.13.28}$$

where $\widehat{H}$ is the Hamiltonian of the system, $|\Psi\rangle$ is the exact wavefunction, and E is the exact energy of the ground state.

The starting point for the CC expansion is the expression of the exact wavefunction in terms of a reference function which is written in the form

$$|\Psi\rangle = e^{\widehat{T}}|\Phi_0\rangle \tag{3.13.29}$$

where $e^{\widehat{T}}$ is the CC operator and $|\Phi_0\rangle$ is the reference function which is usually a Slater determinant constructed from HF MOs. Other wavefunctions such as CI-SCF, MCSCF, or Brueckner orbitals can also be used. The cluster operator $\widehat{T}$ when acting on $|\Phi_0\rangle$ produces a linear combination of excited determinants from the reference wavefunction.

The operator $\widehat{T}$ can be written as

$$\widehat{T} = \widehat{T}_1 + \widehat{T}_2 + \widehat{T}_3 + \ldots \widehat{T}_n \tag{3.13.30}$$

$\widehat{T}_1$ represents single excitation from the ground state $|\Phi_0\rangle$, $\widehat{T}_2$ double excitation, $\widehat{T}_3$ triple excitation, etc., up to the excitation of all electrons in the system.

The advantage of applying cluster exponential operator to the wavefunction is that one can generate more than doubly excited determinants due to the various powers of $\widehat{T}_1$, $\widehat{T}_2$, etc., in the expansion of $e^{\widehat{T}}$. In CC doubles (CCD) theory the quadruple excitations are produced by the operator $\widehat{T}_2^2$ and so the coefficients of the quadruple-substituted determinants are determined as products of the coefficients of the doubly substituted determinants rather than being determined independently. Thus,

$$e^{\widehat{T}} = 1 + \widehat{T} + \frac{\widehat{T}^2}{2!} + \frac{\widehat{T}^3}{3!} + \ldots = \sum_{k=0}^{\infty} \frac{\widehat{T}^k}{k!} \tag{3.13.31}$$

In case we consider only single and double excitations and take $\widehat{T} = \widehat{T}_1 + \widehat{T}_2$, this equation reduces to

$$e^{\widehat{T}} = e^{\widehat{T}_1 + \widehat{T}_2} = 1 + \widehat{T}_1 + \widehat{T}_2 + \frac{1}{2}\widehat{T}_1^2 + \widehat{T}_1\widehat{T}_2 + \frac{1}{2}\widehat{T}_2 + \ldots \tag{3.13.32}$$

The one- and two-electron excitation operators $\widehat{T}_1$ and $\widehat{T}_2$ are defined by

$$\widehat{T}_1|\Phi_0\rangle = \sum_{i=1}^{n} \sum_{a=n+1}^{\infty} t_i^a |\Phi_i^a\rangle \tag{3.13.33}$$

and

$$\widehat{T}_2|\Phi_0\rangle = \sum_{i=1}^{n-1}\sum_{j=i+1}^{n}\sum_{a=n+1}^{\infty}\sum_{b=a+1}^{\infty} t_{ij}^{ab}\left|\Phi_{ij}^{ab}\right\rangle \tag{3.13.34}$$

where $i, j, k,\ldots$ refer to the occupied MOs and $a, b, c,\ldots$ refer to the vacant or virtual MOs. $|\Phi_i^a\rangle$ refers to a singly excited state determinant in which one of the occupied spin orbitals, say χ_i, is replaced by the virtual orbitals χ_a and $|\Phi_{ij}^{ab}\rangle$ is a doubly excited Slater determinant in which two occupied orbitals χ_i and χ_j are replaced by two virtual orbitals χ_a and χ_b. t_i^a and t_{ij}^{ab} are the coefficients, also called amplitudes, whose values are determined by solving the Schrödinger Eq. (3.13.28) written as

$$\widehat{H}e^{\widehat{T}}|\Phi_0\rangle = Ee^{\widehat{T}}|\Phi_0\rangle \tag{3.13.35}$$

Multiplying the two sides by $\langle\Phi_0|e^{-\widehat{T}}$, we get

$$\left\langle\Phi_0|e^{-\widehat{T}}\widehat{H}e^{\widehat{T}}|\Phi_0\right\rangle = E\langle\Phi_0|\Phi_0\rangle = E \tag{3.13.36}$$

$$\left\langle\Phi^*|e^{-\widehat{T}}\widehat{H}e^{\widehat{T}}|\Phi_0\right\rangle = E\langle\Phi^*|\Phi_0\rangle = 0 \tag{3.13.37}$$

Use has been made of the identity operator $e^{-\widehat{T}}e^{\widehat{T}} = 1$. Also, it is assumed that $|\Phi_0\rangle$ is a normal orbital. $|\Phi^*\rangle$ refers to the excited states.

Since the excited state Slater determinants are orthogonal to $|\Phi_0\rangle$, we have $\langle\Phi^*|\Phi_0\rangle = 0$. This may however, not necessarily be true, for example, in the case of VB orbitals. In such cases Eq. (3.13.37) need not be equal to zero.

The nomenclature for the CC method is based on the highest number of excitations allowed in the definition of $\widehat{T}$. Thus in the case of single (*S*) excitation it is named as CCS, in case of double (*D*) it is named as CCD, and analogously for triple (*T*) and quadruple (*Q*) excitations. Thus CCSD and CCSD-*T* refer to $\widehat{T} = \widehat{T}_1 + \widehat{T}_2$ and $\widehat{T} = \widehat{T}_1 + \widehat{T}_2 + \widehat{T}_3$, respectively. Terms in parenthesis such as in CCSD (*T*) indicate that these terms are calculated from perturbation theory. Thus, CCSD (*T*) refers to a case where a full treatment of singles and doubles has been made in the CC method and an estimate of the three-electron excitation is calculated noniteratively by perturbation theory.

In order to understand the implications of Eqs (3.13.36) and (3.13.37), let us consider the cases of CCSD and CCD.

In case of CCSD these equations can be written as

$$\left\langle\Phi_0|e^{-(\widehat{T}_1+\widehat{T}_2)}\widehat{H}e^{(\widehat{T}_1+\widehat{T}_2)}|\Phi_0\right\rangle = E \tag{3.13.38}$$

$$\left\langle\Phi_i^a|e^{-(\widehat{T}_1+\widehat{T}_2)}\widehat{H}e^{(\widehat{T}_1+\widehat{T}_2)}|\Phi_0\right\rangle = 0 \tag{3.13.39}$$

$$\text{and}\quad \left\langle\Phi_{ij}^{ab}|e^{-(\widehat{T}_1+\widehat{T}_2)}\widehat{H}e^{(\widehat{T}_1+\widehat{T}_2)}|\Phi_0\right\rangle = 0 \tag{3.13.40}$$

If we consider the case of double excitation, i.e., CCD, we need to consider an equation of the form

$$\left\langle \Phi_{ij}^{ab} \middle| e^{-\hat{T}_2} \hat{H} e^{\hat{T}_2} \middle| \Phi_0 \right\rangle = 0 \qquad (3.13.41)$$

to determine the coefficients t_{ij}^{ab}. This equation may be expanded as

$$\begin{aligned} &\left\langle \Phi_{ij}^{ab} \middle| \left(1 - \hat{T}_2 + \frac{1}{2}T_2^2 + \ldots\right) \hat{H} \left(1 + \hat{T}_2 + \frac{1}{2}T_2^2 + \ldots\right) \middle| \Phi_0 \right\rangle \\ &= \left\langle \Phi_{ij}^{ab} | \hat{H} | \Phi_0 \right\rangle + \left\langle \Phi_{ij}^{ab} | \hat{H} \hat{T}_2 | \Phi_0 \right\rangle - \left\langle \Phi_{ij}^{ab} | \hat{T}_2 \hat{H} | \Phi_0 \right\rangle + \ldots \qquad (3.13.42) \\ &= \left\langle \Phi_{ij}^{ab} | \hat{H} | \Phi_0 \right\rangle + \Phi_{ij}^{ab} \left| \left[\hat{H}, \hat{T}_2\right] \right| \Phi_0 + \ldots \end{aligned}$$

Using Hausdorff expansion, the similarity transformed Hamiltonian $\hat{H}$ can be written as

$$\hat{H} = e^{-\hat{T}} \hat{H} e^{\hat{T}} = \hat{H} + \left[\hat{H}, \hat{T}\right] + \frac{1}{2!}\left[\left[\hat{H}, \hat{T}\right], \hat{T}\right] + \frac{1}{3!}\left[\left[\hat{H}, \hat{T}\right], \hat{T}\right] \ldots \qquad (3.13.43)$$

In case of CCSD this expansion will be truncated at the fourth level commutator because H refers to only two electrons. The commutator $[\hat{H}, \hat{T}] = \hat{H}\hat{T} - \hat{T}\hat{H}$. The resulting similarity transformed Hamiltonian is non-Hermitian resulting in different left- and right-handed wavefunctions for the same state of interest. The set of equations resulting from similarity transformed Hamiltonian are nonlinear and are solved by an iterative method (for detailed treatment, see Refs [26,28]). The most commonly used CC method is CCSD in which single and double excitations are included. This method has two very desirable features, namely,

1. Exactness, for two-electron system and
2. Size consistency, that is, E_{CCSD} (N-isolated systems) $= N\, E_{\text{CCSD}}$ (single systems).

For closed-shell systems, CCSD method gives excellent values for structures (bond length accuracy about 0.002 Å), vibrational frequencies (within about 2%), and energies of chemical reactions (less than 2 kcal/mol for bond energies). The CCSD method is more time-consuming than the HF method but is much less expensive than the MCSCF–CI method. For closed-shell molecules the inclusion of triple excitation as in the CCSD (T) method provides a good description near the equilibrium geometry. In more complicated situations as in diradicals or dissociation, this method does not provide good results. More complicated CC method such as CCSDT and CCSD (TQ) are used only for highly accurate calculation of small molecules where the quadruple contribution is estimated by fifth-order perturbation theory.

Two new intermediate CC methods known as CC2 and CC3 methods have been developed by Christiansen et al. [28a,b], which are modifications of the CCSD and CCSDT methods. Though not as accurate as the CCSD or CCSDT methods they are computationally more economical.

Based on the performance in terms of the computational cost and accuracy, the computational methods described above may roughly be ordered as below:

$$\mathrm{HF} < \mathrm{MP2} \sim \mathrm{MP3} \sim \mathrm{CCD} < \mathrm{CISD} < \mathrm{MP4SDQ} \sim \mathrm{QCISD} < \mathrm{CC2} < \mathrm{CCSD} < \mathrm{CC3} < \mathrm{CCSD(T)} < \mathrm{CCSDT}$$

CC theory has also been developed for the open-shell systems and is based on a UHF reference wavefunction. It is known as the unrestricted coupled cluster (UCC) theory. UCC, like UHF also suffers from spin contamination though it is comparatively much less in UCC due to the infinite nature of the method.

References

[1] J.K.L. MacDonald, Phys. Rev. 43 (1933) 830.

[2] B.N. Dickenson, J. Chem. Phys. 1 (1933) 317.

[3] C.A. Coulson, Trans. Faraday Soc. 33 (1937) 1479.

[4] S. Wang, Phys. Rev. 31 (1928) 579.

[5] S. Wienbaum, J. Chem. Phys. 1 (1933) 593.

[6] C.C.J. Roothan, Rev. Mod. Phys. 23 (1951) 69.

[7] R. McWeeny, B.T. Sutcliffe, Methods of Molecular Quantum Mechanics, Academic Press, NY, 1969.

[8] E. Steiner, The Determination and Interpretation of Molecular Wavefunctions, Cambridge University press, London, 1976.

[9] I.R. Poirer, R. Kari, I.G. Czismadia, Handbook of Gaussian Basis Sets, Elsevier, Amsterdam, 1985, http://dx.doi.org/10.1002/qua560320414. Online October 2004.

[10] S. Huzinaga, J. Andzelm, Gaussian Basis Sets for Molecular Calculations, Elsevier, 1984.

[11] W.J. Hehre, R.F. Stewart, J.A. Pople, J. Chem. Phys. 51 (1969) 2657.

[12] E. Davidson, D. Feller, Chem. Rev. 86 (4) (1986) 681.

[13] J.A. Pople, M. Head-Gordon, D.J. Fox, K. Raghavachari, L.A. Curtiss, J. Chem. Phys. 90 (1989) 5622.

[14] R. Ditchfield, W.J. Hehrc, J.A. Pople, Chem. Phys. 54 (1971) 724.

[15] W.J. Hehre, R. Ditchfield, J.A. Pople, J. Chem. Phys. 56 (1971) 2257.

[16] J.A. Pople, W.J. Hehre, J. Compt. Phys. 27 (1978) 161.

[17] S. Binkley, J.A. Pople, W.J. Hehre, J. Am. Chem. Soc. 102 (1979) 939.

[18] R. Krishnan, J.S. Binkley, R. Seeger, J.A. Pople, J. Chem. Phys. 72 (1980) 650.

[19] [a] A.A. Frost, J. Chem. Phys. 47 (1967) 3707;
[b] A. Frost, J. Phys. Chem. 72 (1968) 1289.

[20] F.B. van Duijneveldt, J.G.C.M. van Duijneveldt-van de Rijdt, J.H. van Lenthe, Chem. Rev. 94 (1994) 1873.

[21] P.E. Cade, K.D. Sales, A.C. Wahl, J. Chem. Phys. 44 (1966) 1973.

[22] T.B. Foresman, E. Frisch, Exploring Chemistry with Electronic Structure Method, Gaussian, Inc., Pittsburgh USA, 1996.

[23] H.G. Kümmel, A biography of the coupled cluster method, in: R.F. Bishop, T. Brandes, K.A. Gernoth, N.R. Walet, Y. Xian (Eds.), Recent Progress in Many-Body Theories, Proceedings of the 11th International Conference, World Scientific Publishing, Singapore, 2002, p. 334.

[24] C.J. Cramer, Essentials of Computational Chemistry, John Wiley & Sons, Ltd, Chichester, 2002, 191.

[25] I. Shavitt, R.J. Bartlett, Many-Body Methods in Chemistry and Physics: MBPT and Coupled-Cluster Theory, Cambridge University Press, 2009.

[26] C. Trindle, D. Shillady, Electronic Structure Modelling: Connections between Theory and Software, CRC Press, NY, 2008.

[27] P. Carskyand, M. Urban, *Ab Initio* Calculations, Springer Verlag, NY, 1980.

[28] [a] O. Christiansen, H. Koch, P. Jørgensen, Chem. Phys. Lett. 243 (1995) 409;
[b] O. Christiansen, H. Koch, P. Jørgensen, J. Chem. Phys. 103 (1995) 7429.

Further Reading

[1] A.K. Chandra, Introductory Quantum Chemistry, Tata-McGraw Hill Publishing Co., India, 2006.

[2] I. Shavitt, R.J. Bartlett, Many-Body Methods in Chemistry and Physics: MBPT and Coupled-Cluster Theory, Cambridge University Press, 2009.

[3] M.A. Ratner, G.C. Schatz, Introduction to Quantum Mechanics in Chemistry, Dover, 2000.

[4] C.J. Cramer, Essentials of Computational Chemistry, John Wiley & Sons, Ltd, Chichester, 2002.

4 Approximate Molecular Orbital Theories

CHAPTER OUTLINE

4.1 Introduction

Availability of high speed computers and easy to use programs have made the *ab initio* methods a widely used computational tool. Nonetheless the *ab initio* calculations can be extremely expensive and time-consuming and pose their limitation in very large polyatomic systems such as biomolecules and polymeric systems. For large polyatomic molecules a variety of approximations have been made to the Hartree–Fock–Roothan equations discussed in Chapter 3 so as to make them more computationally viable and

Principles and Applications of Quantum Chemistry. http://dx.doi.org/10.1016/B978-0-12-803478-1.00004-2

applicable for routine work. These approximate methods require significantly less computational resources. Due to incorporation of parameters derived from experimental data, some of the approximate methods can calculate certain properties more accurately than even the highest level of *ab initio* methods. These approximate methods are also known as semiempirical methods. The semiempirical methods can be classified as:

1. Methods restricted to π electrons.
2. Methods restricted to all valence electrons.

In this chapter we shall indicate the relationship between the approximate and *ab initio* methods and will focus on some of the more popular methods developed in the research groups of Pople, such as complete neglect of differential overlap (CNDO), intermediate neglect of differential overlap (INDO), and neglect of diatomic differential overlap (NDDO) methods, and Dewar like modified INDO (MINDO), modified neglect of diatomic overlap (MNDO), Austin Model-1 (AM1), and Parameterization Method 3 (PM3), etc. Some of the practical aspects related to these semiempirical methods and their successes and problems will also be discussed. The discussion shall be based on the Roothan formulation developed in Chapter 3.

4.2 Semiempirical Methods

All commonly used approximate molecular orbital methods start with the Roothan form of equations discussed earlier in Chapter 3. Some of these are,

$$FC = SCE \tag{4.2.1}$$

$$F_{\mu\nu} = H_{\mu\nu}(core) + \sum_{\lambda}\sum_{\sigma} P_{\lambda\sigma}\left[(\mu\nu|\lambda\sigma) - \frac{1}{2}\langle\mu\lambda|\nu\sigma\rangle\right] \tag{4.2.2}$$

$$P_{\lambda\sigma} = 2\sum_{i=1}^{occ\ MOs} C_{i\lambda}C_{i\sigma} \tag{4.2.3}$$

and

$$H_{\mu\nu}(core) = \int \phi_\mu(1)\left(-\frac{1}{2}\nabla^2 - \sum_{A}\frac{Z_A}{r_{1A}}\right)\phi_\nu(1)d\tau_1 = \int \phi_\mu(1)\left(-\frac{1}{2}\nabla^2 - \sum_{A} V_A(r)\right)\phi_\nu(1)d\tau_1 \tag{4.2.4}$$

where $\sum_A \frac{Z_A}{r_{1A}} = \sum_A V_A(r)$ is the electrostatic field of the core written as a sum of the potentials for the various atoms in the molecule. $H_{\mu\nu}$ is the matrix element of one-electron Hamiltonian.

Also, the two-electron integrals

$$\begin{aligned}\langle\mu\nu|\lambda\sigma\rangle &= \iint \phi_\mu^*(1)\phi_\nu(1)\frac{1}{r_{12}}\phi_\lambda^*(2)\phi_\sigma(2)d\tau_1 d\tau_2 \\ \langle\mu\lambda|\nu\sigma\rangle &= \iint \phi_\mu^*(1)\phi_\lambda(1)\frac{1}{r_{12}}\phi_\nu^*(2)\phi_\sigma(2)d\tau_1 d\tau_2\end{aligned} \tag{4.2.5}$$

and

$$S_{\mu\nu} = \int \phi_\mu^*\phi_\nu d\tau \tag{4.2.6}$$

Since the greatest proportion of time in *ab initio* HF calculations is spent in calculating and manipulating integrals, the various approximate methods make increasingly drastic assumptions about the integrals which appear in the above expressions. Some of these are set equal to zero and the others replaced by semiempirical estimates. Semiempirical methods explicitly consider only the valence electrons of the system; the core electrons are subsumed into the nuclear core. The rationale behind this approximation is that only the valence electrons are involved in chemical bonding and in determining molecular properties.

To discuss semiempirical methods, it is useful to consider the Fock matrix elements into three groups:

1. $F_{\mu\mu}$ (the diagonal elements),
2. $F_{\mu\nu}$, where orbitals Φ_μ and Φ_ν lie on the same atom, and
3. $F_{\mu\nu}$, where Φ_μ and Φ_ν are on different atoms.

It is also important to note that the overlap matrix S (Eq. (4.2.6)) is set equal to a unit matrix I. This is a feature common to all the semiempirical methods. Thus, all the diagonal elements of the overlap matrix are equal to 1 and the nondiagonal elements are set equal to 0. Roothan Eq. (4.2.1) then reduces to

$$FC = CE, \tag{4.2.7}$$

which is the standard matrix form.

Setting S equal to a unit matrix does not mean that all overlap integrals are set to zero in the calculation of the Fock matrix elements. In fact, it is important to include some of the overlap integrals even in the simplest of the semiempirical methods to account for chemical bonding. The semiempirical calculations invariably use basis sets comprising of Slater-type *s*, *p*, and *d* orbitals. The orthogonality of such orbitals sometimes allows further simplifications to these equations.

Finally, in making approximations, it is necessary to ensure that the results are invariant to simple transformation of the atomic orbital basis sets, such as rotation of axes or replacing simple *s*, *p*, *d* orbitals by their hybrids.

4.2.1 Zero-Differential Overlap (ZDO)

This is the most basic approximation in all semiempirical theories. In this approximation, the overlap between pairs of different orbitals is set to zero for all volume elements $d\tau$;

$$\Phi_\mu \Phi_\nu d\tau = 0, \quad \text{for} \quad \mu \neq \nu \tag{4.2.8}$$

This will set $S_{\mu\nu}$ equal to 0 unless $\mu = \nu$ in which case $S_{\mu\nu} = 1$ or $S_{\mu\nu} = \delta_{\mu\nu}$.

If the two atomic orbitals Φ_μ and Φ_ν are located on the same atom then the differential overlap is referred to as monoatomic differential overlap and if they are located on different atoms then it is said to be diatomic differential overlap.

In the zero-differential overlap (ZDO) approximation, the two-electron repulsion integral $(\mu\nu|\lambda\sigma)$ (Eq. (4.2.5)) will be zero unless μ and ν or λ and σ are identical. This may be written as

$$(\mu\nu|\lambda\sigma) = (\mu\mu|\lambda\lambda)\delta_{\mu\nu}\delta_{\lambda\sigma} \tag{4.2.9}$$

where $\delta_{\mu\nu}$ and $\delta_{\lambda\sigma}$ are Kronecker delta. This means that the overlapping charge densities of the basis orbitals on different atoms are neglected.

All the three are four-center integrals are therefore set equal to zero under the ZDO approximation. Hence, under the ZDO, Eq. (4.2.2) for closed-shell molecules shall simplify considerably to give the following equation for $\mu = \nu$

$$F_{\mu\mu} = H_{\mu\mu}(core) + \sum_{\lambda} P_{\lambda\lambda}(\mu\mu|\lambda\lambda) - \frac{1}{2}P_{\mu\mu}(\mu\mu|\mu\mu) \tag{4.2.10}$$

This may also be written as

$$F_{\mu\mu} = H_{\mu\mu}(core) + \frac{1}{2}P_{\mu\mu}(\mu\mu|\mu\mu) + \sum_{\lambda\neq\mu} P_{\lambda\lambda}(\mu\mu|\lambda\lambda) \tag{4.2.11}$$

If $\mu \neq \nu$, in the ZDO approximation, Eq. (4.2.2) reduces to

$$F_{\mu\nu} = H_{\mu\nu}(core) - \frac{1}{2}P_{\mu\nu}(\mu\mu|\nu\nu) \tag{4.2.12}$$

Correct results cannot be obtained by applying the ZDO approximation across the board. Thus, for example, the major contribution towards bond formation is made by electron–core interaction between pairs of orbitals and the nuclear core ($H_{\mu\nu}(core)$). These interactions cannot therefore be subjected to the ZDO approximation. Also, it must be ensured that the system is invariant under local rotation and hybridization of the basis sets. The basic features of all-valence-electron ZDO self-consistent field (SCF) molecular orbital (MO) theories have been thoroughly analyzed over the years. A crucial point to express their successes and failures is the way the $H(core)$ matrix elements between atomic orbitals Φ_μ and Φ_ν on different atoms A and B are evaluated. These matrix elements $H_{\mu\nu}$, also called resonance integrals $\beta_{\mu\nu}$, are assumed to be overlap dependent according to the relationship,

$$\beta_{\mu\nu} = \beta^0_{AB}S_{\mu\nu}, \tag{4.2.13}$$

where $\beta^0_{AB} = \frac{1}{2}(\beta^0_A + \beta^0_B)$ is a bonding parameter depending only on the nature of atoms A and B. The use of averaged β^0_{AB} bonding parameters (i.e., they are equal for all atomic orbital pairs of atoms A and B) ensures its invariance with both rotation and hybridization of the basis atomic orbitals. Only certain types of approximations fulfill these conditions of invariance. Two of these are:

1. Complete neglect of differential overlap (CNDO) in which product of two different atomic orbitals $\Phi_\mu(1)\Phi_\nu(1)$ associated with electron 1 is always neglected in electron interaction integrals.
2. Neglect of diatomic differential overlap (NDDO) in which this product $\Phi_\mu(1)\Phi_\nu(1)$ is neglected only if $\Phi_\mu(1)$ and $\Phi_\nu(1)$ are on separate centers.

4.2.2 Complete Neglect of Differential Overlap

The CNDO method uses the ZDO approximation for all pairs of atomic orbitals. In the CNDO approximation, there is a complete neglect of differential overlap that is

$$S_{\mu\nu} = \langle \mu | \nu \rangle$$

$S_{\mu\nu}$ is taken to be zero even when different atomic orbitals belong to the same atom. In the approach developed by Pople et al. [1], all two-electron integrals which depend on the overlapping of charge densities of different atomic basis orbitals are neglected. If orbitals μ and ν lie on atom A and λ and σ on atom B, then

$$(\mu\nu|\lambda\sigma) = 0, \text{ unless } \mu = \nu \text{ and } \lambda = \sigma \tag{4.2.14}$$

The two-electron integrals $(\mu\mu|\lambda\lambda)$ are set equal to a parameter γ_{AB} which depends only on the nature of atoms A and B and on the internuclear distance and not on the type of the orbital. The parameter γ_{AB} can be interpreted as the average electrostatic repulsion between two electrons—a valence electron on atom A and another valence electron on atom B. When both the atomic orbitals μ and λ are on the same atom, the parameter is written as γ_{AA} and interpreted as the average electron–electron repulsion between the two valence electrons on atom A. CNDO is a fairly drastic approximation as some one-center integrals like $\langle 2s_A 2p_{xA} | 2s_A 2p_{xA} \rangle$ are ignored.

Using the CNDO approximation and replacing the two-electron integrals $\langle \mu\mu|\lambda\lambda \rangle$ by γ_{AB} when μ and λ are on different atoms A and B and by γ_{AA} when they are on the same atom, Eqs (4.2.10) and (4.2.12) reduce to

$$F_{\mu\mu} = H_{\mu\mu}(core) + \sum_{\lambda}^{A} P_{\lambda\lambda}\gamma_{AA} + \sum_{\lambda}^{B} P_{\lambda\lambda}\gamma_{AB} - \frac{1}{2}P_{\mu\mu}\gamma_{AA};\ B \neq A \tag{4.2.15}$$

$$F_{\mu\nu} = H_{\mu\nu}(core) - \frac{1}{2}P_{\mu\nu}\gamma_{AA}, \tag{4.2.16}$$

when μ and ν are both of atom A, and

$$F_{\mu\nu} = H_{\mu\nu}(core) - \frac{1}{2}P_{\mu\nu}\gamma_{AB}, \tag{4.2.17}$$

when μ and ν are on different atoms A and B.

However, when λ is on atom A,

$$\sum_{\lambda}^{A} P_{\lambda\lambda} = P_{AA}, \tag{4.2.18}$$

where P_{AA} is the total electron density on atom A, and when λ is not on atom A,

$$\sum_{\lambda}^{B} P_{\lambda\lambda} = P_{BB}, \text{ where } B \neq A \tag{4.2.19}$$

Hence, Eq. 4.2.15 may be written as

$$F_{\mu\mu} = H_{\mu\mu}(core) + \left(P_{AA} - \frac{1}{2}P_{\mu\mu}\right)\gamma_{AA} + \sum_{B \neq A} P_{BB}\gamma_{AB} \tag{4.2.20}$$

The core Hamiltonians $H_{\mu\mu}$ and $H_{\mu\nu}$ correspond to electrons moving in the field of the parent nucleus and the other nuclei. The integral $H_{\mu\nu}(core)$ may be separated into two parts: an integral involving the atom A on which μ is situated, i.e., ($H_{\mu\mu}(core)$) and all the other atoms B, i.e., ($H_{\mu\nu}(core)$),

$$H_{\mu\mu}(core) = U_{\mu\mu} - \sum_{B \neq A} V_{AB}, \tag{4.2.21}$$

where $U_{\mu\mu}$ is given by Eq. (4.2.4) as

$$U_{\mu\mu} = \left\langle \mu \middle| -\frac{1}{2}\nabla^2 - \frac{z_A}{r_{1A}} \middle| \mu \right\rangle \tag{4.2.22}$$

and represents the energy of the orbital Φ_μ in the field of its own nucleus A and the core electrons. The energy of the electron in the field of another nucleus B is represented by - V_{AB}.

$$V_{AB} = \left\langle \mu \middle| \frac{z_B}{r_{1B}} \middle| \mu \right\rangle \tag{4.2.23}$$

Using Eqs (4.2.21) and (4.2.22) into Eq. (4.2.20), we get

$$F_{\mu\mu} = U_{\mu\mu} - \sum_{B \neq A} V_{AB} + \left(P_{AA} - \frac{1}{2} P_{\mu\mu} \right) \gamma_{AA} + \sum_{B \neq A} P_{BB} \gamma_{AB} \tag{4.2.24}$$

From similar considerations, if the orbitals Φ_μ and Φ_ν are on the same atom, $H_{\mu\nu}(core)$ may be written as

$$H_{\mu\nu}(core) = U_{\mu\nu} - \sum_{B \neq A} \left\langle \mu \middle| \frac{z_B}{r_{1B}} \middle| \nu \right\rangle = U_{\mu\nu} - \sum_{B \neq A} V_{AB}, \tag{4.2.25}$$

where

$$U_{\mu\nu} = \left\langle \mu \middle| -\frac{1}{2}\nabla^2 - \frac{Z_A}{r_{1A}} \middle| \nu \right\rangle \tag{4.2.26}$$

Since Φ_μ and Φ_ν are on the same atom, $U_{\mu\nu} = 0$ due to orthogonality of the atomic orbitals. The second term in Eq. (4.2.25) is zero due to the ZDO approximation. Hence, in this case

$$H_{\mu\nu}(core) = 0 \tag{4.2.27}$$

Finally, if μ and ν are on two different atoms A and B, $H_{\mu\nu}(core)$ is written as $\beta_{\mu\nu}$ and given by

$$H_{\mu\nu}(core) = \beta_{\mu\nu} = \beta^0_{AB} S_{\mu\nu}, \tag{4.2.28}$$

where β^0_{AB} is a parameter which depends only on atoms A and B and is known as resonance integral.

By substituting Eqs (4.2.21), (4.2.27), and (4.2.28) into Eqs (4.2.15–4.2.17), respectively, we get the following expressions for the Fock operators.

$$F_{\mu\mu} = U_{\mu\mu} - \sum_{B \neq A} V_{AB} + \left(P_{AA} - \frac{1}{2} P_{\mu\mu} \right) \gamma_{AA} + \sum_{B \neq A} P_{BB} \gamma_{AB} \tag{4.2.29}$$

$$F_{\mu\nu} = -\frac{1}{2} P_{\mu\nu} \gamma_{AA}, \tag{4.2.30}$$

when μ and ν are on the same atom A and

$$F_{\mu\nu} = \beta^0_{AB} S_{\mu\nu} - \frac{1}{2} P_{\mu\nu} \gamma_{AB}, \tag{4.2.31}$$

when μ and ν are on different atoms A and B.

The SCF calculations in the CNDO approximation can be carried out by calculating overlap integrals $S_{\mu\nu}$, core Hamiltonian $U_{\mu\mu}$, electron–core interactions V_{AB}, repulsion integrals γ_{AA} and γ_{AB}, and the bonding parameters β^0_{AB}. The parameters β^0_{AB} are calculated empirically either to fit the experimental data or to reproduce *ab initio* results. It is usually written in terms of single atom values as

$$\beta^0_{AB} = \frac{1}{2} \left(\beta^0_A + \beta^0_B \right), \tag{4.2.32}$$

where β^0 values on right-hand side of the equation are determined from *ab initio* calculations for diatomic molecules. The term γ can either be computed explicitly from s-type STOs or it can be treated as a parameter. One popular parametric form for the one-center integral γ_{AA} involves the use of Pariser–Parr approximation [2,2a]

$$\gamma_{AA} = (IP)_A - (EA)_A, \tag{4.2.33}$$

where *IP* and *EA* are the atomic ionization potential and electron affinity, respectively.

For the two-center term γ_{AB}, the Mataga–Nishimoto formalism is adopted

$$\gamma_{AB} = \frac{\gamma_{AA} + \gamma_{BB}}{2 + R_{AB}(\gamma_{AA} + \gamma_{BB})}, \tag{4.2.34}$$

where R_{AB} is the interatomic distance [3,4]. Thus for large distance R_{AB}, γ_{AB} tends to R_{AB}^{-1} as expected for widely separated charge clouds, while for short distances it approaches $\frac{\gamma_{AA}+\gamma_{BB}}{2}$, i.e., it approaches the average of the two one-center parameters.

The CNDO approximation, also called CNDO/1, does not give good energies or spectral information but it does predict correct conformations of molecules. The bond lengths and vibrational frequencies are also predicted very poorly. The predicted equilibrium distances for diatomic molecules are also too short and dissociation energies too large. The CNDO theory suffers from the defect that the nuclear attraction due to penetration of electrons of one atom in the valence shell of another atom has not been accounted for. This penetration effect has been accounted for in the modified version of the theory called CNDO/2 by introducing penetration integrals, which are two-center integrals of the type $\langle B|\mu\mu\rangle$ representing the interaction of an electron in orbital μ on atom A with the core of atom B. Pople suggests that this can be calculated using *S* orbitals on atom A and a point charge of Z_{eff} on atom B. This follows the fact that the net charge on atom B equals the difference between the nuclear charge Z_B and the total electron density P_{BB}

$$Q_B = Z_B - P_{BB}.$$

$$\text{or } P_{BB} = Z_B - Q_B$$

Table 4.1 Some parameters for CNDO/2 calculations in electron volt

Atom (A)	H	C	N	O
Z'	1.2	1.625	1.950	2.275
Z_{eff}	1.0	4.0	5.0	6.0
β_A	−9.0	−21.0	−25.0	−31.0[a]
γ_{AA}	12.85	10.33	11.31	13.91
$1/2(I_S + A_S)$	−7.176	−14.051	−19.316	−25.390
$1/2(I_P + A_P)$	–	−5.572	−7.275	−9.111
$U_{\mu\mu}(s)$	−13.6	−50.69	−70.09	−101.31
$U_{\mu\mu}(p)$	–	−41.50	−57.85	−84.28

[a]−45.0 eV when the atom is a part of a long-conjugated system.

Substitution of the value of P_{BB} into Eq. (4.2.24) gives

$$F_{\mu\mu} = U_{\mu\mu} + \left(P_{AA} - \frac{1}{2}P_{\mu\mu}\right)\gamma_{AA} + \sum_{B \neq A}[-Q_B\gamma_{AB} + (Z_B\gamma_{AB} - V_{AB})] \tag{4.2.35}$$

The term $(Z_B\gamma_{AB} - V_{AB})$ is called the penetration integral.

The core Hamiltonian $U_{\mu\mu}$ in CNDO/2 theory is defined in terms of ionization energies and electron affinities. It is chosen as the energy of the orbital Φ_μ for the valence state of the atom on which it is centered,

$$U_{\mu\mu} = -(\text{VSIP}) - (N' - 1)\gamma_{AA}, \tag{4.2.36}$$

where N' is the total number of valence electrons on atom A, and VSIP refers to valence state ionization potential. The integral $U_{\mu\mu}$ is determined from the relationship

$$-\frac{1}{2}(I_\mu + A_\mu) = U_{\mu\mu} + \left(Z_A - \frac{1}{2}\right)\gamma_{AA} \tag{4.2.37}$$

where I_μ and A_μ are the ionization potential and electron affinity, respectively. Some of the parameters used in CNDO/2 calculations for a few atoms frequently involved in organic reactions are listed in Table 4.1.

4.2.3 Neglect of Diatomic Differential Overlap

In this approximation, the product of two different atomic orbitals $\Phi_\mu(1)\Phi_\nu(1)$ is neglected in electron–interaction integrals only if $\Phi_\mu(1)$ and $\Phi_\nu(1)$ are on separate atoms. In this approximation also the atomic orbitals are considered to be orthonormal

$$S_{\mu\nu} = 1, \quad \text{if} \quad \mu = \nu \text{ and}$$

$$S_{\mu\nu} = 0, \quad \text{if} \quad \mu \neq \nu$$

Also, $\langle\mu\nu|\lambda\sigma\rangle = 0$, except when μ, ν belong to the same atom and λ, σ belong to the same atom.

This is in contrast to CNDO, where $(\mu\nu|\lambda\sigma)$ is always zero except when $\mu = \nu$ and $\lambda = \sigma$.

As in the case of CNDO/2, the matrix elements $H_{\mu\nu}$, where μ and ν are on different atoms, are calculated using the relationship

$$H_{\mu\nu} = \beta_{\mu\nu} = \beta^0_{AB} S_{\mu\nu}, \tag{4.2.38}$$

and the parameter β^0_{AB}, the resonance integral, is calculated empirically either to fit experimental data or to reproduce *ab initio* results. The number of two-electron two-center integrals in NDDO approximation is very large, may be 100 times than in CNDO calculations for each pair of heavy atom.

The Fock matrix elements in this case are

$$F_{\mu\mu} = H_{\mu\mu}(core) + \sum_{\nu}^{A} \left| P_{\nu\nu}\langle\mu\mu|\nu\nu\rangle - \frac{1}{2} P_{\nu\nu}\langle\mu\nu|\mu\nu\rangle \right| + \sum_{B\neq A} \sum_{\lambda}^{B} \sum_{\sigma}^{B} P_{\lambda\sigma}\langle\mu\mu|\lambda\sigma\rangle \tag{4.2.39}$$

$$F_{\mu\nu} = H_{\mu\nu}(core) + \frac{3}{2} P_{\mu\nu}\langle\mu\nu|\mu\nu\rangle - \frac{1}{2} P_{\mu\nu}\langle\mu\mu|\nu\nu\rangle + \sum_{B\neq A} \sum_{\lambda}^{B} \sum_{\sigma}^{B} P_{\lambda\sigma}\langle\mu\nu|\lambda\sigma\rangle, \tag{4.2.40}$$

when μ and ν are both on A, and

$$F_{\mu\nu} = H_{\mu\nu}(core) - \frac{1}{2} \sum_{\lambda}^{B} \sum_{\sigma}^{A} P_{\lambda\sigma}\langle\mu\sigma|\nu\lambda\rangle, \tag{4.2.41}$$

when μ on A and ν on B. The summation notation represents orbital λ on atom B.

4.2.4 INDO, INDO/S, CS-INDO

4.2.4.1 INDO (Intermediate Neglect of Differential Overlap)

This method is an improvement over CNDO where no allowance was made for the fact that the interaction between two electrons depends upon their relative spin. This interaction may be very large for electrons on the same atom, for example, in the case of methylene CH_2 (Figure 4.1(a)). Likewise, in two-center cases, CNDO fails to distinguish either between two orbitals or different orbital orientations Figure 4.1(b).

Unlike CNDO, in case of INDO, the differential overlap between atomic orbitals on the same atom is not neglected in one-center electron repulsion integrals but is still neglected in two-center electron repulsion integrals. As a result, the interaction between electrons on the same atom with parallel spins leads to lower energy as compared to interaction between electrons with paired spin. Except that a fewer two-center integrals are neglected in INDO, the CNDO and INDO methods are the same.

Fock matrix elements in INDO are usually written by explicitly specifying the electron spin α or β

$$F^{\alpha}_{\mu\mu} = U_{\mu\mu} + \sum_{\lambda}^{A} \sum_{\sigma}^{A} \left[P_{\lambda\sigma}(\mu\mu|\lambda\sigma) - P^{\alpha}_{\lambda\sigma}(\mu\lambda|\mu\sigma) \right] + \sum_{B\neq A} (P_{BB} - Z_B)\gamma_{AB} \tag{4.2.42}$$

$$F^{\alpha}_{\mu\nu} = U_{\mu\nu} + \sum_{\lambda}^{A} \sum_{\sigma}^{A} \left[P_{\lambda\sigma}(\mu\nu|\lambda\sigma) - P^{\alpha}_{\lambda\sigma}(\mu\lambda|\nu\sigma) \right], \tag{4.2.43}$$

μ and ν are both on atom A.

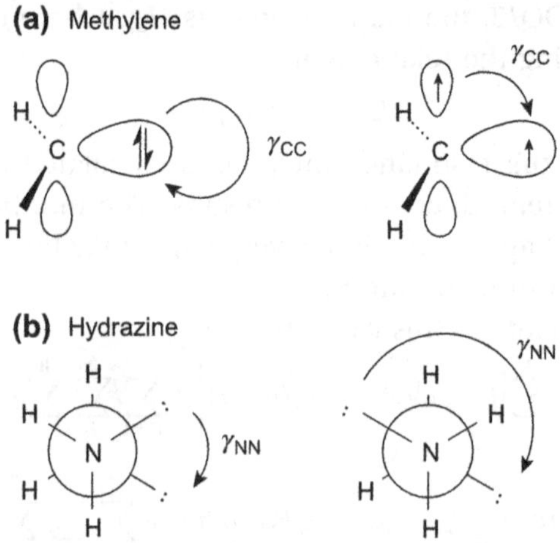

FIGURE 4.1 One- and two-center cases where CNDO fails in correctly estimating repulsive two-electron interactions.

For a closed system $P^{\alpha}_{\mu\nu} = P^{\beta}_{\mu\nu} = \frac{1}{2}P_{\mu\nu}$. When using a basis set comprising of s and p orbitals, several one-center integrals are equal to zero and so also the core elements $U_{\mu\nu}$. $U_{\mu\mu}$ is determined from ionization potential.

The INDO method gives fairly good bond lengths and bond angles, but not very good values for dipole moments and dissociation energies. It has the key advantage that states of different multiplicities can be distinguished as, for example, the singlet and triplet configurations $1s^2\ 2s^2p^2$ of carbon.

4.2.4.2 INDO/S

The success of a semiempirical method depends upon the use of parameters which replace some or all of the integrals. These parameters are assigned values based on fitting to many molecular experimental data. Ridley and Zerner [5] first described a careful parameterization of INDO specifically for spectroscopic problems and named it as INDO/S. Subsequently, parameters for most of the elements in the periodic table were calculated [6]. The INDO/S is especially designed for calculating electronic spectra of large molecules or systems involving heavy atoms. INDO/S is a modified version of INDO in which the one-center core integrals $U_{\mu\mu}$ are obtained from ionization potentials only. By choosing one-center core integrals from ionization potentials, rather than from ionization potentials and electron affinities, improvement in spectroscopic results are obtained.

Coulomb integrals in INDO/S are evaluated by a modification of the Mataga–Nishimoto relation

$$\gamma_{AB} = \frac{f_{\gamma}(\gamma_{AA} + \gamma_{BB})}{2f_{\gamma} + (\gamma_{AA} + \gamma_{BB})R_{AB}}, \tag{4.2.44}$$

where R_{AB} is the distance between the two centers in Bohr radii, and $\gamma_{AA} = I_A - A_A$. I_A is the ionization potential and A_A is the electron affinity of atom A. The parameter f_{γ} is set equal to 1.2 to reproduce the benzene spectrum.

The INDO/S method is very successful for predicting spectroscopic transitions that are not localized to a single center, for example, in extended π-electron systems, metal-to-ligand or ligand-to-metal excitation, and also for *d–d* transitions within transition metal complexes. The INDO/S method also exhibits good accuracy for the prediction of ionization potentials and oscillator strengths for weak electronic transitions. However, oscillator strengths for strong transitions are overestimated by this method.

4.2.4.3 CS-INDO

A semiempirical all-valence-electron method capable of correctly describing both ground and excited state properties of large conjugated molecules, especially molecular conformations, by modifying INDO method was developed by Momicchioli et al. [7]. The method was named as CS-INDO, where C and S stand for conformation and spectra, respectively. This method is especially suitable for conjugated molecules and is capable of making reasonably correct predictions about (1) barriers to internal rotation, conformations, and energy differences between geometrical isomers, (2) electronic transition energies and ordering of the lowest lying excited states, and (3) bond lengths which can be used to describe nuclear motions and simple internal rotation about single bonds.

CS-INDO uses a basis set of hybrid atomic orbitals which conserve their σ or π (local) symmetry in nonplanar geometries. In this method, the nondiagonal elements of the core matrix $H_{\mu\nu}(core)$ correspond to chemically well-defined interactions (σ–σ, π–π, σ–π) which can be differently screened by introducing specific screening factors ($K_{\sigma\sigma}$, $K_{\sigma\pi}$, $K_{\pi\pi}$) to correctly describe internal rotation around single bonds or to correctly reproduce *cis–trans* energy differences. By spectroscopic parameterization of resonance integrals β_{AB} and electron repulsion integrals γ_{AB}, and a newly designed core–core repulsion formula that ensures correct balance between attractive and repulsive interactions, they handled both the ground and lowest excited states of conjugated nonrigid molecules with a satisfactory degree of accuracy.

The basic idea lies in replacing the usual Slater AO valence set $\{\chi\}$ by a basis set $\{\chi'\}$ of atomic hybrid orbitals directed along the bonds. In general, a hybrid orbital χ'_α of atom A can be written as

$$\chi'_\alpha = \sum_{\mu}^{A} t_{\mu\alpha}\chi_\mu, \tag{4.2.45}$$

where χ_μ are AOs of the original basis set and $t_{\mu\alpha}$ are proper transformation coefficients.

The resonance integrals are determined from hybridized orbitals,

$$\beta'_{\alpha\beta} = \frac{1}{2}\left(\beta_A^0 + \beta_B^0\right)K_{i(\alpha)j(\beta)}S'_{\alpha\beta}, \quad i,j = \sigma, \pi, \tag{4.2.46}$$

where $K_{i(\alpha)j(\beta)}$ is the screening constant and $S'_{\alpha\beta}$ is the overlap integral between a hybrid orbital χ'_α of atom A and a hybrid orbital χ'_β of atom B. On using Eq. (4.2.45), we get

$$S'_{\alpha\beta} = \sum_{\mu}^{A}\sum_{\nu}^{B} t_{\mu\alpha}t_{\nu\beta}S_{\mu\nu} \tag{4.2.47}$$

The electron repulsion integrals γ_{AB} vary with the internuclear distance R_{AB} and are evaluated by the semiempirical Ohno–Klopman formula [8,9]

$$\gamma_{AB} = e^2\left[R_{AB}^2 + 0.25\left(e^2/\gamma_{AA} + e^2/\gamma_{BB}\right)^2\right]^{-1/2} \tag{4.2.48}$$

The one-center integrals γ_{AA} are given the semiempirical values adopted by Del Bene and Jaffe [10,10a]. Thus, for example,

$$\gamma_{HH} = 12.85\ \text{eV}, \quad \gamma_{CC} = 10.93\ \text{eV}$$

In CS-INDO method, both $\beta'_{\alpha\beta}$ and γ_{AB} integrals have been reparameterized in order to improve spectral predictions. The core–core repulsion is determined by adopting the expression

$$E_{AB}(core) = Z_A Z_B\left[e^2/R_{AB} - \left(\gamma_{AB}^T - \gamma_{AB}\right)\right]\left[1 + a_0^{-1}\left(R_{AB}^0 - R_{AB}\right)\exp(-\alpha_{AB}R_{AB})\right], \tag{4.2.49}$$

where Z_A and Z_B are the formal charges of the two atomic cores, a_0 is the Bohr radius, while R_{AB}^0 and α_{AB} are parameters characteristic of the atomic pair AB. γ_{AB}^T is the theoretical Zener–Slater, electron repulsion integral.

Extensive calculations on conjugated hydrocarbon and their simpler substituents conducted by Momicchioli [7] and Gupta et al. [11,12,12a] using Ohno–Klopman γ integrals and complete configuration interaction of the excited states resulted in good description of the electronic spectra, both for the lowest-lying singlet states and triplet states and also for the potential energy curve for rotation about the single bonds.

As an example of application of the CS-INDO method, the conformational isomerization and electronic spectra of acrolein may be considered [11].

The molecular geometry of acrolein with numbering of atoms and the spatial orientation of the two lone pairs of electrons (L_9 and L_{10}) of the oxygen atom are shown in the Figure 4.2. Following Momicchioli et al. [13], the geometrical parameters were given values: $L_9O_4 = L_{10}O_4 = 1.0$ Å, $\angle L_9O_4C_2 = \angle L_{10}O_4C_2 = 120°$, $\phi(L_9O_4C_2C_1) = 180°$, and $\phi(L_{10}O_4C_2C_1) = 0°$. The screening constants K_{nn}, $K_{n\pi}$, $K_{n\sigma}$, $K_{\pi\pi}$, and $K_{\pi\sigma}$ were optimized to obtain best fit of the transitions of the *trans*-conformer of acrolein with the experimental values. The optimized values of these constants are 0.74, 0.68, 0.80, 0.60, and 0.75, respectively.

The potential energy curves of acrolein in the ground state (S_0) and the first excited state $S_1(n\pi^*)$ based on CS-INDO calculations are given in Figure 4.3 and the

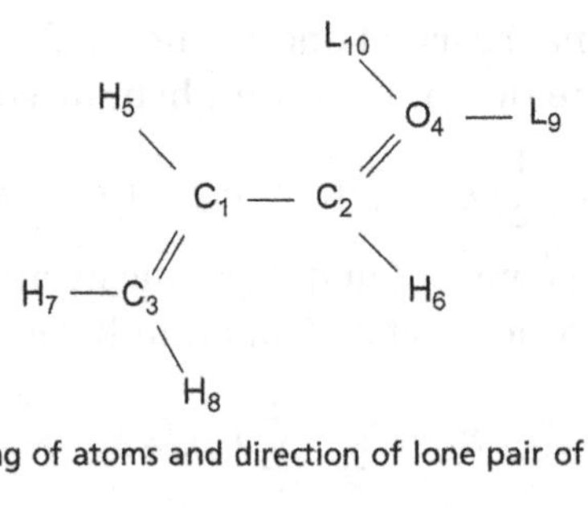

FIGURE 4.2 Numbering of atoms and direction of lone pair of electrons of acrolein.

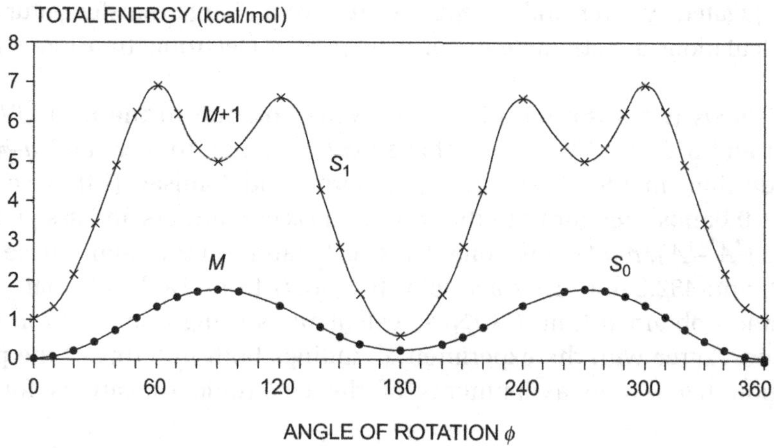

FIGURE 4.3 Potential energy curves of acrolein in ground (S_0) and first excited (S_1)($n\pi^*$) electronic states. ϕ is the angle of rotation about the C_1–C_2 bond relative to the *trans* conformation ($\phi(C_3C_1C_2O_4) = 180°$). The ordinate for S_1 is shifted by 1.00 kcal/mol relative to the S_0 state.

Table 4.2 Electronic transitions of *trans* and *cis* acrolein

Experimental		*Trans* Calculated	Assignment	Experimental[b]	*cis* Calculated	Assignment
Vapor[a] 0–0 band λ nm	**Water λ nm**	**CS-INDO λ nm**		**Vapor[a] 0–0 band λ nm**	**CS-INDO λ nm**	
Singlet						
386.5	315	389.6	*n*–*π**	406.0	412.9	*n*–*π**
	210	216.1	*π*–*π**		218.7	*π*–*π**
	–	187.6	*π*–*π**		163.5	*π*–*π**
	–	150.4	*π*–*π**		153.9	*π*–*π**
Triplet						
412.2	–	415.7	*n*–*π**	432.2	415.7	*n*–*π**

[a]Ref. [14].

[b]No experimental value in solution available.

experimental and calculated electronic transitions for its *cis* and *trans* conformers are given in Table 4.2.

In agreement with experimental measurements, the CS-INDO calculations predict that in the S_0 state the *trans* conformer is more stable than the *cis* conformer while the reverse is true in the S_1 state (Figure 4.3). In between the two most stable conformers (*cis* and *trans*), in the S_1 state, the calculations also predict two other higher energy stable conformers with dihedral angle $\phi = \pm 90°$ having enthalpy difference of about 4.40 kcal/mol with respect to the stable *cis* conformer and the potential well depth of about 1.8 kcal/mol. The existence of intramolecular perpendicular minimum in the first excited state is found in many conjugated systems and is reported to have a charge transfer nature [13]. The CS-INDO calculations are also able to correctly predict electronic transitions in acrolein (Table 4.2).

Acrolein shows two absorption bands in water solution in the near-UV region at 210 nm (VS) and 315 nm (VW), which have been assigned to π–π^* and n–π^*, respectively. In addition, in the vapor phase, Osborne and Ramsay [14] have identified two sets of 0–0 bands each for both the *trans* and *cis* conformers and assigned them to ($^1A''$–$^1A'$) and ($^3A''$–$^1A'$), n–π^* transitions. These 0–0 bands, in the vapor phase, appear at 386.5, 412.2, and 432.2 nm in *cis* acrolein. It follows from Table 4.2 that singlet and triplet transition obtained from CS-INDO calculations using multielectron configuration interaction agree with the experimental findings both in terms of the position of the absorption bands and assignments of the electronic transitions for both the conformers.

4.2.5 Modified INDO (MINDO/3)

The CNDO, INDO, and NDDO methods described above are of considerable importance in showing that a systematic series of approximations could be used to develop methods of real practical value. These methods, however, did not produce very accurate results primarily because they were parameterized upon the results of low-level *ab initio* calculations which themselves did not agree well with the experimental values and were limited to small classes of molecules. Dewar and colleagues developed a series of methods such as MINDO, MNDO, AM1, PM3, etc., where the parameterization was based on the experimental values on heats of formation, dipole moments, ionization potential, and geometries. Thus, the MINDO (modified INDO) method does not differ significantly from INDO in theory except in the value of the parameters which are based on closeness to experimental data. Some of the parameters that have fixed values in CNDO, INDO methods are allowed to vary during MINDO/3 parameterization procedures, Thus, MINDO/3 uses an s, p basis set and the exponents of the Slater atomic orbitals can be varied. The core Hamiltonian $U_{\mu\mu}$ and the resonance integral are also regarded as variable parameters. The core–core interaction between pairs of nuclei can also be changed in MINDO/3 from the form used in CNDO/2. The core–core interaction is no more taken as a simple Coulombic interaction $\left(\frac{Z_A Z_B}{R_{AB}}\right)$, but is given a modified expression which also accounts for the decrease in the screening by the core electrons. Thus, in MINDO/3, the core–core interaction is taken as a function of the electron–electron repulsion integral and given by the expression,

$$E_{AB} = Z_A Z_B \left[\gamma_{AB} + \left\{ \frac{e^2}{R_{AB}} - \gamma_{AB} \right\} e^{-\alpha_{AB} R_{AB}} \right], \tag{4.2.50}$$

where α_{AB} is a parameter which depends on the nature of atoms A and B.

4.2.6 MNDO (Modified Neglect of Diatomic Overlap)

The MINDO/3 method, despite its remarkable success in predicting experimental data, has some significant limitations. These limitations are due to the use of INDO approximation which fails to deal with systems containing lone pairs of electrons. Thus, the heats of formation for molecules having adjacent atoms with lone pairs of electrons were found to be too negative by MINDO/3 calculations. Dewar and Thiel [15] developed a method which was based on NDDO instead of INDO and is known as MNDO (modified neglected of diatomic differential overlap). Since MNDO is based on NDDO, Eqs (4.2.39–4.2.41) for the Fock matrix elements $F_{\mu\mu}$ and $F_{\mu\nu}$ are still valid in this case. However, difference lies in the expressions for $H_{\mu\mu}(core)$ and $H_{\mu\nu}(core)$, which in MNDO are:

$$H_{\mu\mu}(core) = U_{\mu\mu} - \sum_{B \neq A} V_{\mu\mu B} \tag{4.2.51}$$

$$H_{\mu\nu}(core) = -\sum_{B \neq A} V_{\mu\nu B}, \text{ where } \mu \text{ and } \nu \text{ both are on atom A} \tag{4.2.52}$$

$$H_{\mu\nu}(core) = \frac{1}{2} S_{\mu\nu}(\beta_\mu + \beta_\nu) \tag{4.2.53}$$

where β_μ and β_ν are the bonding parameters.

$V_{\mu\mu B}$ and $V_{\mu\nu B}$ are two-cenetr one-electron attractions between an electron distribution $\Phi_\mu \Phi_\mu$ and $\Phi_\mu \Phi_\nu$, respectively, on atom A and the core of atom B. These are given as:

$$V_{\mu\mu B} = -Z_B \langle \mu_A \mu_A | S_B S_B \rangle \tag{4.2.54}$$

and

$$V_{\mu\nu B} = -Z_B \langle \mu_A \nu_A | S_B S_B \rangle \tag{4.2.55}$$

In its original implementation, the one-center terms were derived from atomic spectroscopic data with the refinement that slight adjustments of one-electron one-center energies are allowed in parameterization.

The total energy E_{Tot} of a molecule is the sum of its electronic energy E_{el} and the core–core repulsion energy $E_{AB}(core)$; the latter is composed of an electrostatic term E_{AB}^{coul} and an effective term E_{AB}^{eff}. This last term with an essentially exponential repulsion attempts to account for Pauli exchange repulsions and compensates for the errors introduced by other assumptions. The core–core repulsion terms in MNDO are different from those in MINDO/3. In MNDO

$$E_{AB}(core) = Z_A Z_B (s_A s_A | s_B s_B)\{1 + e^{-\alpha_A R_{AB}} + e^{-\alpha_B R_{AB}}\} \tag{4.2.56}$$

where α is a parameter to be optimized for each atom. Thus, bonds like OH and NH are treated separately.

The parameterization of MNDO is mainly based on ground state properties like heats of formation and geometries; the ionization potentials and dipole moments are used as additional reference data. MNDO, like MINDO/3, also uses (s, p) basis set making it unsuitable for the calculations on transition metals which require a basis set containing *d* orbitals. Thiel and Voityuk [16,17] developed an improved version of MNDO by explicitly including *d* orbitals known as MNDO(D).

Though MNDO method gave much improved results than MINDO/3 for molecular ionization potentials, dipole moments, heats of formation of hydrocarbons, and nitrogen- and oxygen-containing molecules, it was unable to make accurate predictions for intermolecular system involving hydrogen bonds or conjugated systems. Thus, for example, in nitrobenzene the nitro group is predicted to be perpendicular to the aromatic ring instead of its being conjugated to it. The main source of error is the tendency of MNDO to overestimate repulsions between atoms separated by distances approximately equal to the sum of their van der Waal's radii. Several of these problems were eliminated in AM1 developed by Dewar et al. [18].

4.2.7 AM1 (Austin Model-1)

AM1 method [18] is based on exactly the same model as MNDO and differs from it only in respect of the technique to remove the overestimation of nonbonded repulsion. In both these methods, the effective atom-pair term E_{AB}^{eff} in the core–core repulsion function is represented by a more flexible function with several adjustable parameters. Also the bonding parameters like β_s and β_p and the Slater orbital exponents ζ_s and ζ_p, which were taken to be equal ($\beta_s = \beta_p$ and $\zeta_s = \zeta_p$), in MNDO were allowed to be unequal and flexible in AM1. A number of attractive as well as repulsive Gaussian functions were used to modify E_{AB}^{eff} by centering the attractive functions in the region where the repulsions were too large and the repulsive functions at smaller internuclear distances. The core–core term in AM1 takes the form:

$$E_{AB} = E_{MNDO} + \frac{Z_A Z_B}{R_{AB}}\left[\sum_i K_{A_i}\exp\left\{-L_{A_i}(R_{AB} - M_{A_i})^2\right\} + \sum_j K_{B_j}\exp\left\{-L_{B_j}(R_{AB} - M_{B_j})^2\right\}\right] \quad (4.2.57)$$

M and K in Eq. (4.2.57) and α in Eq. (4.2.56) in MNDO are parameters which are optimized for each atom. The values of these parameters are not critical and many of them may have the same value. The parameter L determines the width of the spherical Gaussian functions. All these additional parameters in E_{AB}^{eff} are determined empirically for reducing the overestimated nonbonded repulsions in MNDO. The addition of Gaussian functions in AM1 significantly increases the number of parameters to be parameterized (almost double of MNDO), but AM1 represents a very real improvement over MNDO with no increase in the computational time. AM1 is able to reproduce hydrogen bonds and gives better estimation of activation energies for reactions. AM1 has been used very widely because of its performance over the other semiempirical methods, especially for modeling organic molecules. However, it has several limitations such as (1) the steric effects are also overestimated, (2) the predicted rotational barriers are too low and the five-membered rings are predicted to be too stable, and (3) the predicted heat of formation tends to be inaccurate for molecules with a large amount of charge localization.

4.2.8 PM3 (Parameterization Method 3)

Developed by Stewart [20], PM3 derives its name from the fact that it is third parameterization of MNDO; AM1 is considered to be second parameterization. Like AM1, it is based on the same model as MNDO. PM3 uses an automated parameterization procedure, as against AM1 where the parameters were adjusted manually with the help of chemical knowledge or intuition. PM3 differs from AM1 in the number of Gaussian terms in the core repulsion function. Thus, PM3 uses only two Gaussian terms per atom instead of upto four used by AM1. Both AM1 and PM3 predict various structural and thermodynamic parameters to almost the same level of accuracy. Statistically, PM3 is more accurate than the other semiempirical methods. The overall heats of formation are more accurate than with MNDO or AM1. Hypervalent molecules are predicted more accurately. It is more accurate than AM1 for hydrogen bond angles but AM1 is more accurate for hydrogen bond energies. On average, PM3 predicts energies and bond lengths more accurately than AM1. However, it has several deficiencies that limit its usefulness. For example, the rotational barrier of the amide bond predicted by PM3 is too low or even nonexistent. PM3 has a very strong tendency to make the environment around nitrogen pyramidal and so is unsuitable for use where the state of hybridization of nitrogen is important.

A comparison of MNDO, AM1, and PM3 shows that on average, AM1 predicts energies and geometries better than MNDO, but not as well as PM3. Thus, Dewar et al. [19] evaluated bond lengths and valence angles in a large number of organic molecules having H, C, N, O, F, Cl, Br, and I by using AM1 and PM3. They found that the average absolute errors in AM1 and PM3 were 0.027 and 0.022 Å, respectively for the bond lengths and 2.3° and 2.8°, respectively, for the angles. Against this, calculations on a larger set of molecules by MNDO, AM1, and PM3, conducted by Stewart [20,21] showed that the absolute errors for the bond lengths by these methods were 0.054, 0.050, and 0.036 Å, respectively where as for the valence angles they were 4.3°, 3.3°, and 3.9°, respectively. The mean unsigned errors for the heats of formation from MNDO, AM1, and PM3 methods were found to be 7.35, 5.80, and 4.71 kcal/mol for molecules/radicals containing lighter elements and 29.2, 15.3, and 10.0 kcal/mol for those containing heavier elements (Al-Hg), respectively. Thus, the results from AM1 were found to be superior to MNDO but not better than PM3.

In a more recent study Stewart [22,23] compared the accuracy of some semiempirical and some density functional theory (DFT) methods for predicting heats of formation. It was noted that the PM3 and PM5 methods gave much improved values of the heats of formation than all other methods for a set of 1276 molecules (Table 4.3).

4.3 Semiempirical Methods for Planar-Conjugated Systems

These methods treat the π electrons separately from the σ electrons. The σ–π separability is based on different symmetry of the σ and π orbitals. The π electrons are more polarizable than the σ electrons due to which they are more susceptible to

Table 4.3 Average errors in heats of formation (kcal/mol) for various methods

Method	Median and good error	Average unsigned error	Root mean square error	Largest error
MNDO	6.91	15.38	31.41	+223.0
AM1	6.69	10.31	16.05	+150.5
PM3	4.74	6.54	10.74	+153.8
PM5	3.75	5.57	9.79	+155.1
B88-LYP(D)[a]	4.31	6.69	11.34	+155.6
B88-PW91(D)[a]	4.08	6.77	11.98	+155.8

[a]Heats of formation estimated using a simple atomic additivity expression.

perturbations as in chemical reactions. These methods are limited to conjugate systems and were originally devised to explain their nonadditive properties and to calculate the electronically excited states of polyenes—both linear and cyclic. These methods belong to two categories:

1. Those in which electrons are being treated as independent and the Hamiltonian is independent of the r_{12}^{-1} terms, e.g., Hückel theory and free electron theory.
2. Those in which electron interactions are taken into account, i.e., the r_{12}^{-1} term is not neglected such as the Pariser–Parr–Pople (PPP) method.

4.3.1 Hückel Theory

Hückel theory, formulated in early 1930s, can be considered as the "grandfather" of approximate molecular orbital methods. Its remarkability can be judged by the fact that it still survives in the present age of computer revolution when sophisticated calculations can be performed on molecules to yield results in favorable agreement with experimental values. Hückel theory separates the π-electron system from the σ-electron framework and considers the π electrons to be moving in a field created by the nuclei and the core of σ electrons which results in an effective Hamiltonian H. It is a LCAO-MO (Linear Combination of Atomic Orbitals–Molecular Orbital) method in which the molecular orbitals are constructed from a linear combination of $p\pi$ atomic orbitals. The basis set consists of one $p\pi$ orbital on each aromatic C, N, O, S, etc. Thus, in a minimal basis set calculation of a planar conjugated hydrocarbon, the only atomic orbitals of symmetry are the $2p$ orbitals that are perpendicular to the molecular plane. So,

$$\Psi_i = \sum_{r=1}^{N} C_{ir}\Phi_r, \tag{4.3.1}$$

where Φ_r is a $p\pi$ orbital on the rth atom, C_{ir} are the coefficients that will be optimized variationally, and N is the number of atoms contributing π electrons.

Since the electron–electron repulsion is neglected and hence the Hamiltonian for the π electrons is treated as independent of r_{12}^{-1} term, the π Hamiltonian is simply the sum of Hamiltonians for each π electron.

$$H_{\pi}(1,2,3\ldots\ldots n) = \sum_{i} H^{eff}(i) \quad i = 1,2,3,\ldots,n \tag{4.3.2}$$

where n is the total number of π-electrons. This equation implies that the Schrödinger equation may be written as

$$H_{\pi}\psi = E\psi \tag{4.3.3}$$

where ψ is the product of n spatial orbitals

$$\psi = \psi_1\psi_2\ldots\psi_n \tag{4.3.4}$$

and the total energy is the sum of energy ε_i of each electron i

$$E = \sum_{i=1}^{n} \varepsilon_i \tag{4.3.5}$$

Since the Hückel π-electron Hamiltonian is the sum of one-electron Hamiltonians, a separation of variables is possible and the Hückel molecular orbitals (HMOs) satisfy

$$H^{eff}(i)\,\Psi_i = \varepsilon_i\Psi_i \tag{4.3.6}$$

Instead of product of spatial orbitals in Eq. (4.3.4) we could have also used Slater determinants, but in the absence of electron–electron repulsion, there is no difference between the energies of spatial product and the Slater determinant.

Equation (4.3.6) is solved by substituting the value of Ψ_i from Eq. (4.3.1) and using the variation technique described earlier for obtaining the coefficients C_{ir}. The optimum values of the coefficients for the lowest energy $p\pi$ MOs satisfy the equation

$$\sum_{s=1}^{N} \left[\left(H_{rs}^{eff} - S_{rs}\varepsilon_i\right)C_{is}\right] = 0, \quad r = 1,2,\ldots,N, \tag{4.3.7}$$

or in matrix representation

$$(H^{eff} - S\varepsilon)C = 0, \tag{4.3.8}$$

where H^{eff} is the effective Hamiltonian matrix and S is the overlap matrix.

The significance of H^{eff} term is better understood if we consider Hückel theory in terms of CNDO approximation which involves ZDO. Thus, if we consider Eq. (4.2.35) from which the penetration effects have been eliminated, we get

$$F_{\mu\mu} = U_{\mu\mu} + \left(P_{AA} - \frac{1}{2}P_{\mu\mu}\right)\gamma_{AA} \tag{4.3.9}$$

If each nucleus (A) in the π system is the same as, for example, in hydrocarbons, then $F_{\mu\mu}$ shall be approximately constant for all the nuclei that are considered. The matrix elements H_{rr}^{eff} in Eq. (4.3.7) are equivalent to $F_{\mu\mu}$ in Eq. (4.3.9) and are often called Coulomb integral and assigned a symbol α. Similarly H_{rs}^{eff} in Eq. (4.3.7) are equivalent to $F_{\mu\nu}$ in Eq. (4.2.30), and are called bond integral or resonance integral. It is assigned symbol β.

Looking back to Eq. (4.3.7), Hückel theory makes the following assumptions:

1. $H_{rs} = < \Phi_r|H|\Phi_s > = \alpha,$ for $r = s$, the two $p\pi$-orbitals lying on the same atom

$$\begin{aligned} &= \beta \quad \text{for atom } r \text{ boned to atom } s \\ &= 0 \quad \text{otherwise} \end{aligned} \tag{4.3.10}$$

2. All overlap integrals between the $p\pi$-atomic orbital pair,

$$\begin{aligned} S_{rs} &= < p_r|p_s > = \delta_{rs} \\ &= 1 \text{ for } r = s \\ &= 0 \text{ otherwise} \end{aligned} \tag{4.3.11}$$

The parameters α and β are taken to be constants but may be adjusted to match experimental values. The values of α and β are given the same value for all carbon atoms but if heteroatoms are present in the molecule, then different values of α and β may be associated with different atoms. Empirically, the following values are suggested for atom X and Y

$$\alpha_X = \alpha_c + h_X\beta_{cc} \text{ and } \beta_{XY} = K_{XY}\beta_{cc}$$

Atom	-N=		-O-	F	Cl	Br
h_X	0.5		2.0	3	2	1.5
Bond	C-N-	C=N-	C≡N	C–F	C–Cl	C–Br
K_{XY}	0.8	1.0	1	0.7	0.4	0.3

In order to normalize the molecular orbitals, the normalization condition for the *i*th molecular orbital is written as

$$< \psi_i|\psi_i > = 1,$$

$$\text{or } \sum_{r=1}^{N} |C_{ir}|^2 = 1 \tag{4.3.12}$$

Also, it may be noted from Eq. (4.3.1) that the total number of MOs is equal to the total number of atoms contributing the π electrons.

As, an example, we may consider the case of butadiene having four carbon atoms each of which contributes one π electron to the π-electron system.

$$\underset{1}{CH_2}{=}\underset{2}{CH}{-}\underset{3}{CH}{=}\underset{4}{CH_2}$$

Using Hückel's two assumptions as mentioned above, the determinant given by Eq. (4.3.8) may be written as

$$\begin{vmatrix} \alpha-\varepsilon & \beta & 0 & 0 \\ \beta & \alpha-\varepsilon & \beta & 0 \\ 0 & 0 & \alpha-\varepsilon & \beta \\ 0 & 0 & \beta & \alpha-\varepsilon \end{vmatrix} = 0 \tag{4.3.13}$$

If we divide this equation by β and put $\frac{\alpha-\varepsilon}{\beta} = x$, we get

$$\begin{vmatrix} x & 1 & 0 & 0 \\ 1 & x & 1 & 0 \\ 0 & 1 & x & 1 \\ 0 & 0 & 1 & x \end{vmatrix} = 0,$$

which gives $x^4 - 3x^2 + 1 = 0$

Solutions to this equation are

$$x = \pm 1.61804 \text{ and } \pm 0.61804 \tag{4.3.14}$$

These values of x give four values of energy

$$\varepsilon = \alpha - \beta x$$

Using these energies, Eq. (4.3.8) can be used to calculate the normalized MOs,

$$\Psi_1 = 0.372\Phi_1 + 0.602\Phi_2 + 0.602\Phi_3 + 0.372\Phi_4$$

$$\Psi_2 = 0.602\Phi_1 + 0.372\Phi_2 - 0.372\Phi_3 - 0.602\Phi_4$$

$$\Psi_3 = 0.602\Phi_1 - 0.372\Phi_2 - 0.372\Phi_3 + 0.602\Phi_4$$

$$\Psi_4 = 0.372\Phi_1 - 0.602\Phi_2 + 0.602\Phi_3 - 0.372\Phi_4 \tag{4.3.15}$$

The energy levels and molecular orbitals for butadiene are given in Figure 4.4.

From Eq. (4.3.15) it follows that Ψ_1 has no node. Maximum overlap between the atoms leads to maximum charge building between the atoms. This state, therefore, must have lowest energy. The resonance integral must therefore be negative.

Since butadiene has four π electrons, both the Ψ_1 and Ψ_2 states would be doubly occupied and the total π electron energy will be

$$E = 2(\alpha + 1.618\beta) + 2(\alpha + 0.618\beta) = 4\alpha + 4.472\beta \tag{4.3.16}$$

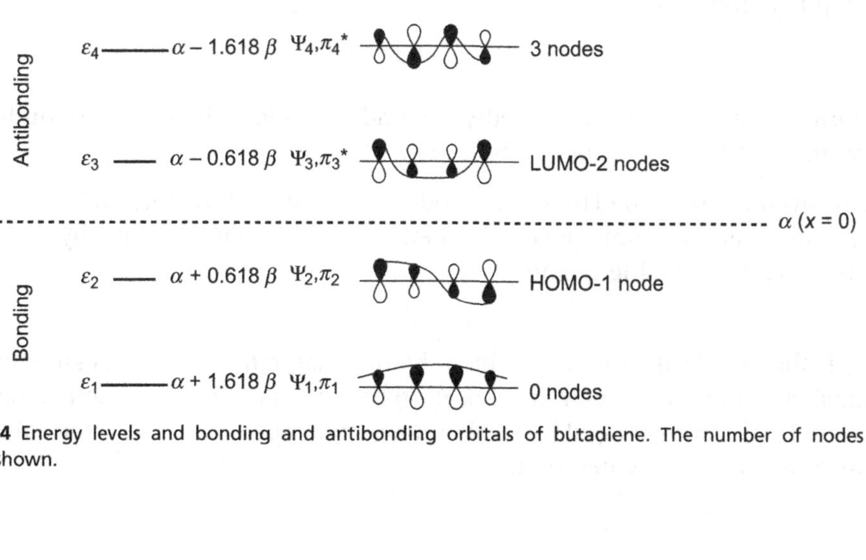

FIGURE 4.4 Energy levels and bonding and antibonding orbitals of butadiene. The number of nodes in each orbital is shown.

Some other properties that may be obtained from the Hückel model are as follows:

1. **The π charge density at atoms**
   It is obtained by the relation

$$q_r = \sum_i n_i |C_{ir}|^2, \tag{4.3.17}$$

where n_i is the occupation number of each MO (i.e., $n_i = 0$, 1, or 2 depending on whether the orbital is vacant, singly occupied, or doubly occupied) and C_{ir} are the orbital coefficients. The net charge, f_r, on each atom is simply the difference between the net nuclear charge (excluding the σ electrons) and the π charge density. Thus,

$$f_r = 1 - q_r \tag{4.3.18}$$

2. **π bond order between atoms r and s**

$$p_{rs} = \sum_i n_i C_{ir}^* C_{is} \tag{4.3.19}$$

The total π bond order at atom r is $\sum_s p_{rs}$. The maximum possible value of this sum for a molecule with no heteroatom is $\sqrt{3}$ (e.g., at the trigonal carbon in $C(CH_2)_3$).

3. **Free valence**
   Free valence is a measure of the ability of each atom (carbon) to form additional π bonds.

   The difference between $\sqrt{3}$ and $\sum_s p_{rs}$ is called the free valence F_r of atom r

$$F_r = \sqrt{3} - \sum_s p_{rs} \tag{4.3.20}$$

4. **Bond length**
   The π bond order can be used to make rough estimate of C–C bond lengths using the empirical formula

$$R_{rs} = 1.52\ \text{Å} - 0.186\ \text{Å} \cdot p_{rs} \tag{4.3.21}$$

The formula gives $R_{rs} = 1.33$ Å for ethylene and 1.52 Å for ethane, corresponding to π bond orders of 1 and 0, respectively.

An improvement over the Hückel method was introduced by Wheland–Mann which allows the inclusion of atomic polarization effect by neighboring atoms by making the Coulomb integrals a variable quantity

$$\alpha_r = \alpha_r^0 + f_r|\beta|\omega, \tag{4.3.22}$$

where α_r^0 is the initial value of α_r for which the net charge $f_r = 0$ and ω is an empirically determined constant that is typically chosen to be 1.4. Since the net charge depends on the MOs, and these are obtained by solving the secular equations, the values of α_r must therefore be determined by iteration.

4.3.2 Extended Hückel Theory

While the Hückel theory is limited to π electron systems, the extended Hückel theory (EHT) developed by Hoffmann [24,25,25a,25b,26] takes into account all the valence electrons in the molecule. The valence electrons are treated as independent of each other and the Hamiltonian is taken as the sum of one electron Hamiltonians.

$$\widehat{H}_{val} = \sum_i \widehat{H}_{eff}(i)$$

Hence, as in the case of Hückel theory, the many problem separates into several one-electron problems. Equation (4.3.7) is now again valid except that it is extended to all valence electrons. The atomic orbitals used in EHT are Slater orbitals with fixed orbital exponents. Thus, 1s orbitals are used for hydrogen, $2s\ 2p_x 2p_y 2p_z$ for elements from carbon to fluorine and $3s\ 3p_x 3p_y 3p_z\ (3d)$ for elements from silicon to chlorine. So, in the EHT treatment of nonplanar hydrocarbons, each MO has contributions from four AOs on each carbon atom (one $2s$ and three $2p$ orbitals) and one $1s$ orbital on hydrogen atom. As against Hückel theory, where only π electron overlaps between neighboring atoms are included, EHT does not neglect overlap and all overlap integrals (S_{rs}) are included. Since $S_{rs} \neq \delta_{ij}$, the nondiagonal elements in Eq. (4.3.7) are present in the secular determinant. In this case, the diagonal elements H_{rr} are considered as average energy for an electron in the atomic orbital centered on atom r and so, using Koopmann's theorem, they are taken to be equal to minus of valence state ionization potential (VSIP).

$$H_{rr}^{eff} = -(\text{VSIP})_r \tag{4.3.23}$$

For the nondiagonal elements H_{rs}, Wolfsberg–Helmholtz approximation [27] is used

$$H_{rs} = \frac{K\left(H_{rr}^{eff} - H_{ss}^{eff}\right)}{2} S_{rs} \tag{4.3.24}$$

Another approximation was suggested by Ballhausen and Gray for the evaluation of nondiagonal elements H_{rs}

$$H_{rs} = -K\left(H_{rr}^{eff} \cdot H_{ss}^{eff}\right)^{1/2} S_{rs} \tag{4.3.25}$$

EHT ignores both the electron–electron and nuclear–nuclear repulsions and so is not very useful for geometry predictions. It also gives poor predictions for several molecular properties such as bond lengths, dipole moments, energies, and rotational barriers. The theory has however been successful in providing qualitative results.

4.3.3 Pariser–Parr–Pople Method

PPP method considers only the π orbitals of molecules. It is the first semiempirical method which included the effect of electron repulsion between the valence electrons within the π-electron framework and is therefore an improvement over the HMO theory. Both HMO and PPP methods are only applied to planar conjugate molecules. Like CNDO, the PPP method is also based on ZDO approximation. The two-electron

Coulomb integral is treated as an empirical parameter and several two-electron integrals are ignored. Thus,

$$\begin{aligned}(\mu\nu|\lambda\sigma) &= \delta_{\mu\nu}\delta_{\lambda\sigma}(\mu\mu|\lambda\lambda) \\ &= \delta_{\mu\nu}\delta_{\lambda\sigma}\gamma_{\mu\lambda}\end{aligned} \tag{4.3.26}$$

The parameter $\gamma_{\mu\lambda}$ is calculated empirically.

The Hamiltonian for each electron is similar to the Hartree Hamiltonian except that the exchange integral in neglected. We can, therefore, write

$$H_{\mathrm{PPP}} = H_{\mathrm{Hückel}} + \text{Electron Repulsion term} \tag{4.3.27}$$

Roothan Eqs (4.2.1–4.2.5) are again used except that now only the π electrons are included. The π-electron Hamiltonian H_π has the form

$$\widehat{H}_\pi = \sum_{i=1}^{n_\pi} H_\pi^{core}(i) + \sum_{i=1}^{n_\pi}\sum_{j>i}\frac{1}{r_{ij}} \tag{4.3.28}$$

$$H_\pi^{core}(i) = -\frac{1}{2}\nabla_i^2 - \sum_\alpha \frac{Z_\alpha}{r_{i\alpha}} \tag{4.3.29}$$

where i represents the electron, n_π the number of π electrons, and α the nuclei. The summation is overall the π electrons.

The Roothan Eq. (4.2.1) takes the form

$$FC = SCE \tag{4.3.30}$$

$$\text{or} \quad \sum_s (F_{rs} - \varepsilon_i S_{rs})C_{is} = 0, \tag{4.3.31}$$

where F_{rs} is the equivalent of Eq. (4.2.2) for the π electrons

$$F_{rs} = H_{rs}(core) + \sum_t\sum_u P_{tu}\left[\langle rs|tu\rangle - \frac{1}{2}\langle rt|su\rangle\right], \tag{4.3.32}$$

where atomic orbitals μ, ν, λ, and σ are located on atoms r, s, t, and u, respectively.

Within the ZDO approximation $S_{rs} = \delta_{rs}$ and Eq. (4.3.32) reduces to equations similar to Eqs (4.2.10) and (4.2.12)

$$F_{rr} = H_{rr}(core) + \sum_{t\neq r} P_{tt}\gamma_{rt} - \frac{1}{2}P_{rr}\gamma_{rr} \tag{4.3.33}$$

and

$$F_{rs} = H_{rs}(core) - \frac{1}{2}P_{rs}\gamma_{rs}, \quad r \neq s \tag{4.3.34}$$

In the PPP approximation, $H_{rs}(core)$ is taken as zero for nonbonded atoms, but for the bonded atoms it is treated as an empirical parameter $\beta_{rs}(core)$ whose value depends up on the nature of the two atoms. The electron repulsion integrals γ_{rs} can either be calculated from the STOs or by using semiempirical formulas. All three- and four-center electron repulsion integrals are ignored.

For solving Roothan equations under PPP method, an iterative procedure is adopted. One usually starts with Hückel's method to calculate the initial values of the coefficients C_{is} in Eq. (4.3.31) and the density matrix elements P_{rs} to generate the initial F_{rs} matrix elements. Equation (4.3.31) is then solved to obtain π-electron energies and improved set of coefficients C_{is} and density matrix elements P_{rs}. This process is repeated till convergence in energies and charge densities (P_{rr}) is achieved within a preset limit.

PPP method is often used for developing simple parameterized analytic expressions for molecular properties. It gives a good account of electronic spectra of many, but not all, aromatic hydrocarbons.

4.4 Comparative Study of the Performance of Semiempirical Methods

The relative merits and demerits of the various semiempirical methods and their limitations have been outlined during the course of discussion of these methods. The performance of successive semiempirical methods has gradually improved from one method to another but anomalous results have also been obtained for certain types of systems. The parameterization for different semiempirical methods is usually based on geometrical variables, dipole moments, ionization energies, and heats of formation. Thus, parameters for MNDO, AM1, and PM3 have been parameterized to reproduce experimental equilibrium geometries and heats of formation of organic compounds. The parameters for PM3 have been obtained by parameterization on organic molecules and on a number of transition metals. Accordingly, the suitability of a particular method and the quality of the results obtained depend upon the properties, physical variables, and group of molecules used for their parameterization. Thus, for example, if the parameters of CNDO, INDO, CS-INDO are determined on the basis of systems which have no conformational flexibility then, clearly, the rotational barriers obtained from these parameters will not have the same accuracy as the heats of formation. Summary of the performance of MINDO/3, MNDO, AM1, and PM3 based on the publications of the research group of Dewar [15,18,19] is given in Table 4.4.

Table 4.4 Comparison of quantities calculated with various semiempirical methods

	MINDO/3	MNDO	AM1	PM3
Heats of formation (kcal/mol) [15]	11.0	6.3	8.82	7.12
Bond lengths (Å)	0.022	0.014	0.026	0.015
Angles	5.6°	2.8°		
Dipole moments (D)	0.54	0.32	0.35	0.40
Heats of formation of hydrocarbons (kcal/mol) [18]	9.7	5.87	5.07	
Heats of formation for species with N and/or O (kcal/mol)	11.69	6.64	5.88	
Ionization energies (eV)	0.31	0.39	0.29	

It is found that while geometries calculated from semiempirical methods are in good agreement with the experimental geometries, the same is not true of the thermochemical parameters. The thermochemical computations from semiempirical methods are however useful for comparison purposes to explain or predict trends.

References

[1] J.A. Pople, G.A. Segal, Chem. Phys. 43 (1965) 5136.

[2] R. Pariser, R.G. Parr, J. Chem. Phys. 21 (1953) 466–471;
[a] R. Pariser, R.G. Parr, J. Chem. Phys. 21 (1953) 767–776.

[3] K. Nishimoto, N. Mataga, Z. Phys. Chem. N. F. 12 (1957) 335.

[4] N. Mataga, K. Nishimoto, Z. Phys. Chem. N. F. 13 (1957) 140.

[5] J.E. Ridley, M.C. Zerner, Theor. Chim. Acta 32 (1973) 111.

[6] M. Kotzian, N. Rösch, M.C. Zerner, Theor. Chim. Acta 81 (1992) 201.

[7] F. Momicchioli, I. Baraldi, M.C. Bruni, Chem. Phys. 70 (1982) 161.

[8] K. Ohno, Theoret Chim. Acta 2 (1964) 219.

[9] G. Klopmann, J. Am. Chem. Soc. 86 (1964) 4550.

[10] J. Del Bene, H.H. Jaffe, J. Chem. Phys. 48 (1968) 1807;
[a] J. Del Bene, H.H. Jaffe, J. Chem. Phys. 48 (1968) 4050.

[11] S. Thakur, V.P. Gupta, B. Ram, Spectrochim. Acta A 53 (1997) 749.

[12] S. Thakur, V.P. Gupta, Indian J. Pure Appl. Phys. 36 (1998) 177;
[a] S. Thakur, V.P. Gupta, Indian J. Pure Appl. Phys. 36 (1998) 567.

[13] F. Momicchioli, I. Baraldi, G. Berthier, Chem. Phys. 123 (1988) 103.

[14] G.A. Osborne, D.A. Ramsay, Can. J. Phys. 51 (1973) 1170.

[15] M.J.S. Dewar, W. Thiel, J. Am. Chem. Soc. 99 (1977) 4899.

[16] W. Thiel, A.A. Voityuk, J. Mol. Struct. 313 (1994) 141.

[17] W. Theil, A.A. Voityuk, Int. J. Quantum Chem. 44 (1994) 807.

[18] M.J. S Dewar, E.G. Zoebisch, E.F. Healy, J.J.P. Stewart, J. Amer. Chem. Soc. 107 (1985) 3902.

[19] M.J.S. Dewar, C. Jie, J. Yu, Tetrahedron 49 (1993) 5003.

[20] J.J.P. Stewart, J.P. James, J. Comput. Chem. 10 (1989) 221.

[21] J.J.P. Stewart, J.P. James, J. Comput. Chem. 12 (1991) 320.

[22] J.J.P. Stewart, J. Mol. Model 10 (2004) 6.

[23] J.J.P. Stewart, J. Comput. Chem. 11 (1990) 543.

[24] R. Hoffmann, J. Chem. Phys. 39 (1963) 1397.

[25] R. Hoffmann, J. Chem. Phys. 40 (1964) 2745;
[a] R. Hoffmann, J. Chem. Phys. 40 (1964) 2474;
[b] R. Hoffmann, J. Chem. Phys. 40 (1964) 2480.

[26] R. Hoffmann, J. Chem. Phys. Tetrahedron 22 (1966) 521;
[a] R. Hoffmann, J. Chem. Phys. Tetrahedron 22 (1966) 539.

[27] M. Wolfsberg, L. Hemholtz, J. Chem. Phys. 20 (1952) 837.

Further Reading

[1] J.A. Pople, D.L. Beveridge, Approximate Molecular Orbital Theory, McGraw Hill Book Co., 1970.

[2] A. Szabo, N.S. Ostlund, Modern Quantum Chemistry: Introduction to Advanced Electronic Structure Theory, Dover Books on Chemistry, 1996.

[3] T. Veszprémi and, M. Fehér, Quantum Chemistry: Fundamentals to Applications, Springer, 1999.

[4] A.R. Leach, Molecular Modelling: Principles and Applications, second ed., Pearson, 2001.

[5] W. Thiel, Semi-empirical quantum chemical methods in computational chemistry, in: C. Dykstra, et al. (Eds.), Theory and Applications of Computational Chemistry: The First 40 Years, Elsevier, 2005.

[6] G.A. Segal, in: W. Miller, et al. (Eds.), Semi-empirical Methods of Electronic Structure Calculations, Part A & B, vols 7 and 8 of Modern Theoretical Chemistry, Plenum, New York, 1977.

5

Density Functional Theory (DFT) and Time Dependent DFT (TDDFT)

CHAPTER OUTLINE

Principles and Applications of Quantum Chemistry. http://dx.doi.org/10.1016/B978-0-12-803478-1.00005-4

5.1 Introduction

Density functional theory (DFT) is a successful theory to calculate the electronic structure of materials and to provide a quantitative understanding of their properties from the fundamental laws of quantum mechanics. Its applicability ranges from atoms, molecules, and solids to nuclei and quantum and classical fluids. In its original formulation, the DFT provides the ground state properties of a system and predicts a great variety of molecular properties: molecular structures, vibrational frequencies, atomization energies, ionization energies, electric and magnetic properties, reaction paths, etc. It has now been generalized to deal with many different situations: spin-polarized systems, multicomponent systems such as nuclei and electron–hole droplets, superconductors with electronic pairing mechanisms, time-dependent phenomena and excited states, molecular dynamics (MD), etc.

DFT differs from the wavefunction-based methods by using the physical observable electron density $n(\mathbf{r})$ as the central quantity for determining energy and molecular properties instead of wavefunction which, as mentioned earlier, is essentially uninterpretable and cannot be probed experimentally. An important advantage of using electron density over the wavefunction is a significant reduction in the dimension of the problem. Regardless of how many electrons one has in the system, the electron density is always three-dimensional. This enables DFT to be applied to much larger systems, even biomolecules and polymeric compounds having hundreds or even thousands of atoms. Partly for this reason, DFT has become the most widely used electronic structure method in condensed matter physics.

5.2 Theoretical Motivation—Thomas–Fermi Model

The first attempts to use the electron density rather than the wavefunction for obtaining information about atomic and molecular systems date back to the early work of Thomas and Fermi (1927) which is based on quantum statistical model of electrons. In its original formulation, this model takes into account only the kinetic energy while treating the nuclear–electron and electron–electron contributions in a completely classical way. Thomas and Fermi arrived at the following, very simple expression for the kinetic energy based on the uniform electron gas, which is a fictitious model system of constant electron density $n(\mathbf{r})$

$$T_{TF}[n(\mathbf{r})] = \frac{3}{10}(3\pi^2)^{2/3} \int n^{5/3}(\mathbf{r})d\mathbf{r} \tag{5.2.1}$$

When combined with the classical expression for the nuclear–electron attractive potential and the electron–electron repulsive potential, this gives the famous Thomas–Fermi expression for the energy of an atom,

$$E_{TF}[n(\mathbf{r})] = \frac{3}{10}(3\pi^2)^{2/3} \int n^{5/3}(\mathbf{r})d\mathbf{r} - Z\int \left(\frac{n(\mathbf{r})}{\mathbf{r}}\right)d\mathbf{r} + \frac{1}{2}\iint \left(\frac{n(\mathbf{r}_1)n(\mathbf{r}_2)}{\mathbf{r}_{12}}\right)d\mathbf{r}_1 d\mathbf{r}_2 \tag{5.2.2}$$

The importance of this equation lies in the fact that the energy is given completely in terms of the electron density $n(\mathbf{r})$. Though the energy T_{TF} so obtained is only a very coarse approximation to the true kinetic energy due to the complete neglect of exchange and correlation effects, Eq. (5.2.2) provides a method of calculating energy from density without any additional information. It is particularly to be noted that no use of wave-function has been made here. Thus, if an appropriate functional to express energy in terms of the density has been obtained, the next step will be to look for a strategy for identifying the correct density that may be inserted into Eq. (5.2.2).

5.3 Formalism of the DFT

The central statement of the DFT is the Hohenberg–Kohn (HK) theorem which for the nondegenerate ground states can be summarized in the following three statements:

Statement 1: The ground state density $n(\mathbf{r})$ of a bound state system of interacting electrons subjected to some external potential $V(\mathbf{r})$ can uniquely determine this potential. In other words there is a one-to-one correspondence between the external potential $V_{ext}(\mathbf{r})$ and the electron density $n(\mathbf{r})$. This is known as the first HK theorem.

The main question in DFT is therefore finding a way of calculating electron density in the ground state if the potential felt by this electron is known. Since the emphasis is on calculating electron density, our thinking has to be the reverse of what we had in the molecular orbital (MO) theory. In MO theory, we think of a molecule as a set of fixed nuclei with an electron placed around it. In DFT, the electron cloud is the origin of our model and the nuclei are considered as if they were immerged in this "electron gas." The effect of the nuclei on the electron gas is then regarded as if it was an external potential.

The relation between electron density and external potential was demonstrated by the first HK theorem, according to which the electron density uniquely determines the Hamiltonian operator $\widehat{H}$ and thus all the molecular electronic properties of the system. Formally speaking, this theorem states that, to within a constant, the external potential $V_{ext}(\mathbf{r})$ is a unique functional of electron density $n(\mathbf{r})$ and $\mathbf{r}$. Since $V_{ext}(\mathbf{r})$ in turn fixes $\widehat{H}$, the full many particle ground state is a unique functional of $n(\mathbf{r})$.

The electron density is defined as the integral over the spin coordinates of all electrons and over all but one of the spatial variables ($\mathbf{x} \equiv \mathbf{r},\sigma$)

$$n(\mathbf{r}) = N\int \cdots \int |\Psi(x_1, x_2, \cdots, x_N)|^2 dx_1 dx_2 \cdots dx_N \tag{5.3.1}$$

where $\{x_i\}$ represents both spatial and spin coordinates, $n(\mathbf{r})$ determines the probability of finding any of the N electrons within volume element $d\mathbf{r}$. Also,

$$\int n(\mathbf{r})d\mathbf{r} = N \tag{5.3.2}$$

Proof: Suppose we have two external potentials $V_{ext}(\mathbf{r})$ and $V'_{ext}(\mathbf{r})$ which differ by more than a constant but give the same electron density $n(\mathbf{r})$ for the ground state. We, therefore, have two different Hamiltonians $\widehat{H}$ and $\widehat{H}'$ whose ground state densities are the same although the normalized wavefunctions Ψ and Ψ' would be different. The Hamiltonian $\widehat{H}$ of a system of N-interacting electrons has the form

$$\begin{aligned}\widehat{H} &= -\frac{1}{2}\sum_{i=1}^{N}\nabla_i^2 - \sum_{i=1}^{N}\sum_{\alpha=1}^{M}\frac{Z_\alpha}{|\mathbf{r}_{i\alpha}|} + \sum_{i=1}^{N}\sum_{j>i}^{N}\frac{1}{|\mathbf{r}_{ij}|} \\ &= \widehat{T} + \widehat{V}_{ne} + \widehat{V}_{ee},\end{aligned} \tag{5.3.3}$$

where as usual i represents electrons and α represents nuclei. Here, T is the kinetic energy, V_{ne} is the N-electron potential energy from the external field,

$$\widehat{V}_{ne} = \int V_{ext}(\mathbf{r})\Psi^*(\mathbf{r})\Psi(\mathbf{r})d\mathbf{r} = \int V_{ext}(\mathbf{r})n(\mathbf{r})d\mathbf{r} \tag{5.3.4}$$

and V_{ee} is the electron repulsion energy.

Let E_0 and E'_0 be the ground state energies for $\widehat{H}$ and $\widehat{H}'$, then

$$E_0 = \langle\Psi|\widehat{H}|\Psi\rangle \quad \text{and} \quad E'_0 = \langle\Psi'|\widehat{H}'|\Psi'\rangle \tag{5.3.5}$$

Now, suppose, we take Ψ' as a trial wavefunction for the $\widehat{H}$ problem and Ψ as the trial function for the $\widehat{H}'$ problem. Then, in the first case, from variational theorem,

$$\begin{aligned}E_0 &= \langle\Psi|\widehat{H}|\Psi\rangle < \langle\Psi'|\widehat{H}|\Psi'\rangle = \langle\Psi'|\widehat{H} - \widehat{H}' + \widehat{H}'|\Psi'\rangle \\ &= \langle\Psi'|\widehat{H}'|\Psi'\rangle + \langle\Psi'|\widehat{H} - \widehat{H}'|\Psi'\rangle \\ &= E'_0 + \int n(\mathbf{r})\left[V_{ext}(\mathbf{r}) - V'_{ext}(\mathbf{r})\right]d\mathbf{r}\end{aligned} \tag{5.3.6}$$

and in the second case

$$\begin{aligned}E'_0 &= \langle\Psi'|\widehat{H}'|\Psi'\rangle < \langle\Psi|\widehat{H}'|\Psi\rangle = \langle\Psi|\widehat{H}|\Psi\rangle + \langle\Psi|\widehat{H}' - \widehat{H}|\Psi\rangle \\ &= E_0 + \int n(\mathbf{r})\left[V'_{ext}(\mathbf{r}) - V_{ext}(\mathbf{r})\right]d\mathbf{r}\end{aligned} \tag{5.3.7}$$

Adding Eqs (5.3.6) and (5.3.7), we get

$$E_0 + E'_0 < E'_0 + E_0 \tag{5.3.8}$$

which is a contradiction. It therefore follows that there cannot be two different $V_{ext}(\mathbf{r})$ that give the same $n(\mathbf{r})$ for the ground state. Thus, $n(\mathbf{r})$ determines N and $V_{ext}(\mathbf{r})$ and hence all properties of the ground state, like the kinetic energy T, the potential energy V, and the total energy E.

Statement 2: There exists a functional $F_{HK}[n]$ such that the energy functional for a given external potential can be written as

$$E[n] = F_{HK}[n] + \int V_{ext}(\mathbf{r})n(\mathbf{r})d\mathbf{r}$$

This statement follows simply from Eq. (5.3.3).

The expectation value of energy as a functional of electron density is

$$\begin{aligned} E[n] &= \left(\Psi[n]|\hat{H}|\Psi[n]\right) \\ &= (\Psi[n]|T|\Psi[n]) + (\Psi[n]|V_{ee}|\Psi[n]) + (\Psi[n]|V_{ne}|\Psi[n]) \\ &= T[n] + E_{ee}[n] + V_{ne}[n] = F_{HK}[n] + V_{ne}[n] \end{aligned} \tag{5.3.9}$$

$$\text{where, } F_{HK}[n] = T[n] + E_{ee}[n] \tag{5.3.10}$$

$F_{HK}[n]$ is universal in the sense that for a given particle–particle interaction (the Coulomb interaction in our case) it is independent of any external potential. It is valid for any number of particles. $F_{HK}[n]$ depends only on the number of electrons and is the same for different molecules because it is not a function of external potential. If it included V_{ne}, it would not have been universal because V_{ne} depends on the system. The energy contribution of the external potential V_{ne} can be expressed using electron density by the relation

$$V_{ne}[n] = \int V_{ext}(\mathbf{r})n(\mathbf{r})d\mathbf{r}, \tag{5.3.11}$$

where we take into account the fact that all electrons experience the same external potential. Thus, from Eqs (5.3.9–5.3.11), the energy functional can be written as

$$E[n] = F_{HK}[n] + \int V_{ext}(\mathbf{r})n(\mathbf{r})d\mathbf{r}, \tag{5.3.12}$$

as stated.

From Eq. (5.3.10), $F_{HK}[n]$ is the sum of two terms: the kinetic energy functional and the electron repulsion functional. The latter, in turn, is the sum of classical electron repulsion and a nonclassical exchange functional.

Thus,

$$\begin{aligned} E_{ee}[n] &= \frac{1}{2}\iint \frac{n(\mathbf{r}_1)n(\mathbf{r}_2)}{\mathbf{r}_{12}} d\mathbf{r}_1 d\mathbf{r}_2 + E_{XC}[n] \\ &= J[n] + E_{XC}[n] \end{aligned} \tag{5.3.13}$$

where the first term is the classical part, the Coulomb repulsion, and the second is the nonclassical part which involves self-interaction correction, exchange, and Coulomb correlation. E_{XC} is known as exchange–correlation energy.

Thus, from Eqs (5.3.9), (5.3.10), and (5.3.13), we get

$$F_{HK}[n] = T[n] + J[n] + E_{XC}[n] \tag{5.3.14}$$

$$\text{and} \quad E[n] = T[n] + J[n] + V_{ne}[n] + E_{XC}[n] \tag{5.3.15}$$

These terms are similar to those found in the Hartree–Fock (HF) theory.

Statement 3: The ground state energy E_0 and ground state density $n_0(\mathbf{r})$ of a system characterized by an external potential $V_0(\mathbf{r})$ can be obtained from variational principle which involves only the density. Thus, the ground state energy can be written as a functional of electron density $E_0[n]$.

According to this statement of the HK theorem, the functional $F_{HK}[n]$ delivers the lowest ground state energy E_0 if and only if the input density $n(\mathbf{r})$ is the true ground state density ($n_0(\mathbf{r})$). This is nothing but the variational principle according to which

$$E_0 \le E[n] = T[n] + E_{ee}[n] + V_{Ne}[n] = T[n] + J[n] + E_{XC}[n] + V_{ne}[n] \tag{5.3.16}$$

In other words, for any other trial density $n(\mathbf{r})$, the energy obtained from the functional $F_{HK}[n]$ represents an upper bound to the true ground state energy E_0. The explicit form of the functionals $T[n]$ and $E_{ee}[n]$ are the main challenges of DFT.

The ground state energy is therefore obtained by using the variational theorem subject to the constraint on the electron density as the number of electrons (N) is fixed in Eq. (5.3.2).

In order to minimize the energy, this constraint is introduced as a Lagrangian multiplier ($-\mu$) leading to

$$\frac{\delta}{\delta n(\mathbf{r})}\left[E[n(\mathbf{r})] - \mu \int n(\mathbf{r})d\mathbf{r}\right] = 0 \tag{5.3.17}$$

From this we can write,

$$\left(\frac{\delta E[n(\mathbf{r})]}{\delta n(\mathbf{r})}\right)_{V_{ext}} = -\mu \tag{5.3.18}$$

or using Eq. (5.3.12),

$$\frac{\delta F_{HK}[n(\mathbf{r})]}{\delta n(\mathbf{r})} + V_{ext}(\mathbf{r}) = -\mu \tag{5.3.19}$$

The subscript V_{ext} indicates that this is under the condition of constant external potential.

The Lagrangian multiplier μ can be identified with the chemical potential of an electron cloud for its nuclei which in turn is related to the electronegativity (χ).

Thus,

$$\chi = -\mu = \left(\frac{\partial E}{\partial N}\right)_{V_{ext}} \tag{5.3.20}$$

Equation (5.3.19) could be solved exactly if the functional $F_{HK}[n]$ in Eq. (5.3.12) were known. Such as solution would be very useful because it would have been universally applicable for any system having the same number of electrons. However $F_{HK}[n]$ is not known exactly for interacting systems. While $T[n]$ can still be calculated exactly, only approximate formula exist for $V_{ext}[n]$ and so only approximate solutions can be found.

5.4 Kohn–Sham Equations

Since a direct solution of Eq. (5.3.19) was not possible for interacting systems, Kohn and Sham replaced them by a system of noninteracting electrons having the same electron density ($n_S(\mathbf{r})$) as the system of interacting electrons ($n(\mathbf{r})$).

$$T_S = -\frac{1}{2}\sum_{i=1}^{N}(\Psi_i|\nabla^2|\Psi_i)$$

$$n_S(\mathbf{r}) = \sum_{i=1}^{N}\sum_{\sigma}|\Psi_0(\mathbf{r},\sigma)|^2 = n(\mathbf{r}), \tag{5.4.1}$$

where Ψ_i are the spin orbitals of the noninteracting system called the Kohn–Sham (KS) orbitals. Of course, T_S is not equal to the true kinetic energy of the system. The difference between two kinetic energies

$$T_c = T_S[n] - T[n] \tag{5.4.2}$$

is combined in the $E_{XC}[n]$ part given by Eq. (5.3.13). So, the exchange–correlation energy is defined as,

$$E_{XC}[n] = (T_S[n] - T[n]) + (E_{ee}[n] - J[n]) \tag{5.4.3}$$

E_{XC} is the functional that contains everything that is unknown.

The next problem is to define a potential V_S of the noninteracting system such that it may lead to a Slater determinant which is characterized by the same density as the real system. This problem is solved by writing down the energy of the interacting systems in a form similar to Eq. (5.3.15).

$$E[n] = T_S[n] + J[n] + E_{XC}[n] + V_{ne}[n] \tag{5.4.4}$$

$$\text{or, } E[n] = T_S[n] + \frac{1}{2}\iint \frac{n(\mathbf{r}_1)n(\mathbf{r}_2)}{\mathbf{r}_{12}} d\mathbf{r}_1 d\mathbf{r}_2 + E_{XC}[n] + V_{ne}$$

$$= -\frac{1}{2}\sum_{i=1}^{N}(\Psi_i|\nabla^2|\Psi_i) + \frac{1}{2}\sum_{i,j=1}^{N}\iint |\Psi_i(\mathbf{r}_1)|^2\frac{1}{r_{12}}|\Psi_j(\mathbf{r}_2)|^2 d\mathbf{r}_1 d\mathbf{r}_2 + E_{XC}[n] \tag{5.4.5}$$

$$-\sum_{i=1}^{N}\int\sum_{\alpha=1}^{M}\frac{Z_\alpha}{\mathbf{r}_{i\alpha}}|\Psi_i(\mathbf{r}_1)|^2 d\mathbf{r}_1,$$

where i and α refer to electrons and nuclei, respectively.

In this expression of energy $E[n]$, which is greater than the true energy $E_0[n]$, the only term with no explicit form is $E_{XC}[n]$. To minimize the energy, the variational principle is now used with the constraint that $(\Psi_i|\Psi_j) = \delta_{ij}$.

This results in the set of one-electron equations known as Kohn-Sham equations which are given as:

$$\left(-\frac{1}{2}\nabla_1^2 + V_S(\mathbf{r}_1)\right)\Psi_i^{KS} = \varepsilon_i\Psi_i^{KS} \tag{5.4.6}$$

where,

$$\begin{aligned} V_S(\mathbf{r}_1) &= \int \frac{n(\mathbf{r}_2)}{\mathbf{r}_{12}} d\mathbf{r}_2 + V_{XC}(\mathbf{r}_1) - \sum_{\alpha=1}^{M} \frac{Z_\alpha}{\mathbf{r}_{1\alpha}} \\ &= \sum_{j}^{N} \frac{|\Psi_j(\mathbf{r}_2)|^2}{\mathbf{r}_{12}} d\mathbf{r}_2 + V_{XC}(\mathbf{r}_1) - \sum_{\alpha}^{M} \frac{Z_\alpha}{\mathbf{r}_{1\alpha}} \end{aligned} \tag{5.4.7}$$

V_{XC} is known as the exchange–correlational potential and is defined as the functional derivative of E_{XC} with respect to the density n.

$$V_{XC} = \frac{\delta E_{XC}}{\delta n} \tag{5.4.8}$$

Ψ_i^{KS} are the KS orbitals which, unlike the HF orbitals, have no physical meaning. It follows that once we know the contributions of various terms, we can find V_S used in one-particle Eq. (5.4.6). Since V_S depends on the density, hence the KS equations have to be solved iteratively. If we know the exact forms of E_{XC} and V_{XC}, we can know the exact energy of the system. The exchange–correlational functional is therefore the key to the success or failure of the DFT. While DFT itself does not give a hint as how to calculate E_{XC} but it does tell that a true E_{XC} has the same functional form for all the systems and so is a universal functional of density. An approximate functional once constructed may be applied to any system of interest.

5.5 LCAO Ansatz in the KS Equations

Most of the applications in chemistry make use of the linear combination of atomic orbital (LCAO) expansion of the KS orbitals. In this approach, just as in the case of MO, a predefined set of l base functions $\{\Phi_\mu\}$ is used to construct KS orbitals,

$$\Psi^{KS} = \sum_{\mu=1}^{l} c_{\mu i}\Phi_\mu \tag{5.5.1}$$

In analogy to the MO theory we can write the one-electron KS Eq. (5.4.6), in a compact form as

$$\hat{f}^{KS}\Psi_i^{KS} = \varepsilon_i\Psi_i^{KS} \tag{5.5.2}$$

Substituting Eq. (5.5.1) into Eq. (5.5.2), we get

$$\hat{f}^{KS}(\mathbf{r}_1)\sum_{\nu=1}^{l} c_{\nu i}\Phi_\nu(\mathbf{r}_1) = \varepsilon_i \sum_{\mu=1}^{l} c_{\nu i}\Phi_\nu(\mathbf{r}_1) \tag{5.5.3}$$

Multiplying this equation from the left with an arbitrary function Φ_μ and integrating over space, we get l equations

$$\sum_{v=1}^{l} c_{vi} \int \Phi_\mu(r_1) f^{KS}(\mathbf{r}_1)\Phi_\nu(\mathbf{r}_1)d\mathbf{r}_1 = \varepsilon_i \sum_{\nu=1}^{l} C_{\nu_i} \int \Phi_\mu(\mathbf{r}_1)\Phi_\nu(\mathbf{r}_1)d\mathbf{r}_1 \quad \text{for } 1 \le i \le l \tag{5.5.4}$$

These may also be written as,

$$\sum_{\nu=1}^{l} C_{\nu_i} F_{\mu\nu}^{KS} = \varepsilon_i \sum_{\nu=1}^{l} c_{\nu i} S_{\mu\nu}^{KS} \tag{5.5.5}$$

$$\text{where, } F_{\mu\nu}^{KS} = \int \Phi_\mu(\mathbf{r}_1) f^{KS}(\mathbf{r}_1)\Phi_\mu(\mathbf{r}_1)d\mathbf{r}_1 \tag{5.5.6}$$

$$\text{and } S_{\mu\nu}^{KS} = \int \Phi_\mu(\mathbf{r}_1)\Phi_\nu(\mathbf{r}_1)d\mathbf{r}_1 \tag{5.5.7}$$

are the elements of the matrix F^{KS} called KS matrix, and matrix S called the overlap matrix, respectively. Both the matrices have $l \times l$ dimension. Equation (5.5.4) can be written in a compact form as

$$F^{KS}C = SCE \tag{5.5.8}$$

This equation has a form equivalent to the Hartree–Fock–Roothan equation. Using Eqs (5.4.6) and (5.4.7), the matrix element $F_{\mu\nu}^{KS}$ may be expanded as

$$F_{\mu\nu}^{KS} = \int \Phi_\mu(\mathbf{r}_1)\left[-\frac{1}{2}\nabla_1^2 - \sum_{\alpha}^{M} \frac{Z_\alpha}{\mathbf{r}_{1\alpha}} + \int \frac{n(\mathbf{r}_2)}{\mathbf{r}_{12}} d\mathbf{r}_2 + V_{XC}(\mathbf{r}_1)\right]\Phi_\nu(\mathbf{r}_1)d\mathbf{r}_1 \tag{5.5.9}$$

The first two terms in this equation are the kinetic energy and electron–nuclear attraction. They are combined as one-electron integral,

$$h_{\mu\nu} = \int \Phi_\mu(\mathbf{r}_1)\left(-\frac{1}{2}\nabla_1^2 - \sum_{\alpha}^{M} \frac{Z_\alpha}{\mathbf{r}_{1\alpha}}\right)\Phi_\nu(\mathbf{r}_1)d\mathbf{r}_1 \tag{5.5.10}$$

The third term in Eq. (5.5.9) is the Coulomb repulsion term $J_{\mu\nu}$ which involves electron density $n(\mathbf{r})$

$$J_{\mu\nu} = \int \frac{\Phi_\mu(\mathbf{r}_1)n(\mathbf{r}_2)\Phi_\nu(\mathbf{r}_1)}{\mathbf{r}_{12}} d\mathbf{r}_1 d\mathbf{r}_2 \tag{5.5.11}$$

$$n(\mathbf{r}) = \sum_{i=1}^{N} |\Psi_i(\mathbf{r})|^2 = \sum_{i=1}^{N}\sum_{\mu}^{l}\sum_{\nu}^{l} C_{\mu i}C_{\nu i}\Phi_\mu(\mathbf{r})\Phi_\nu(\mathbf{r}) \tag{5.5.12}$$

As before, we can collect the expansion coefficients in a density matrix $\widehat{P}$ with elements

$$P_{\mu\nu} = \sum_{i=1}^{N} C_{\mu_i}C_{\nu_i} \tag{5.5.13}$$

Using Eqs (5.5.12) and (5.5.13) in Eq. (5.5.11), we get

$$J_{\mu\nu} = \sum_{\lambda=1}^{l}\sum_{\sigma=1}^{l} P_{\lambda\sigma} \iint \Phi_\mu(\mathbf{r}_1)\Phi_\nu(\mathbf{r}_1)\frac{1}{\mathbf{r}_{12}}\Phi_\lambda(\mathbf{r}_2)\Phi_\sigma(\mathbf{r}_2)d\mathbf{r}_1 d\mathbf{r}_2 \tag{5.5.14}$$

It may be noted that so far the treatment is the same as in the HF theory, except in the exchange–correlation part (the fourth term in Eq. (5.5.9)). In DFT, it is given by the integral

$$V_{\mu\nu}^{\mathrm{XC}} = \int \Phi_\mu(\mathbf{r}_1)V_{\mathrm{XC}}(\mathbf{r}_1)\Phi_\nu(\mathbf{r}_1)d\mathbf{r}_1 \tag{5.5.15}$$

This differs from the HF theory (Eq. (3.8.13)) where the exchange integral $K_{\mu\nu}$ is

$$\begin{aligned} K_{\mu\nu} &= \sum_{\lambda}^{l}\sum_{\sigma}^{l} P_{\lambda\sigma} \iint \Phi_\mu(\mathbf{r}_1)\Phi_\lambda(\mathbf{r}_1)\frac{1}{\mathbf{r}_{12}}\Phi_\nu(\mathbf{r}_2)\Phi_\sigma(\mathbf{r}_2)d\mathbf{r}_1 d\mathbf{r}_2 \\ &= \sum_{\lambda}\sum_{\sigma} P_{\lambda\sigma}(\mu\lambda|\nu\sigma) \end{aligned} \tag{5.5.16}$$

5.5.1 Solution of KS Equations

In order to solve the KS equations, a self-consistent approach is adopted. An initial guess of density obtained from semiempirical or low-level HF calculations is fed into Eq. (5.4.6) and a set of orbitals is derived. These orbitals lead to an improved value for the density, which is then used in the second iteration. The process is repeated until convergence is achieved. The steps are summarized in flowchart given in Figure 5.1.

The approach is computationally very different from the direct solution of the Schrödinger equation where considerable time is spent in a search of the wavefunction. In DFT, with a given E_{XC}, the search is only for the three-dimensional density, which is a comparatively easier problem. In all $\frac{1}{2}l^2$ one-electron integrals are calculated in $h_{\mu\nu}$ and about l^4 two-electron integrals in the Coulomb term $J_{\mu\nu}$.

The exchange–correlation contribution to the KS matrix elements in Eq. (5.5.9) is invariably evaluated using a grid of points because of the complexity of the functionals employed in the method. The integration is performed using the grid directly or by finding a further auxiliary basis set expansion with which analytical integration can be used. Different types of basis functions may be used in DFT calculations including the basis functions which are not necessarily contracted. In most computer softwares employed for DFT calculations, the representation of the density in the classical electron–repulsion operator is carried out using KS orbital basis functions. If the density is represented using an auxiliary basis set, other options such as Pople-type basis sets and Slater-type functions, which have the advantage of having fewer functions, are available. The appearance of four-center integrals in the Coulomb repulsion term in Eq. (5.5.14) questions the advantage of the DFT approach with

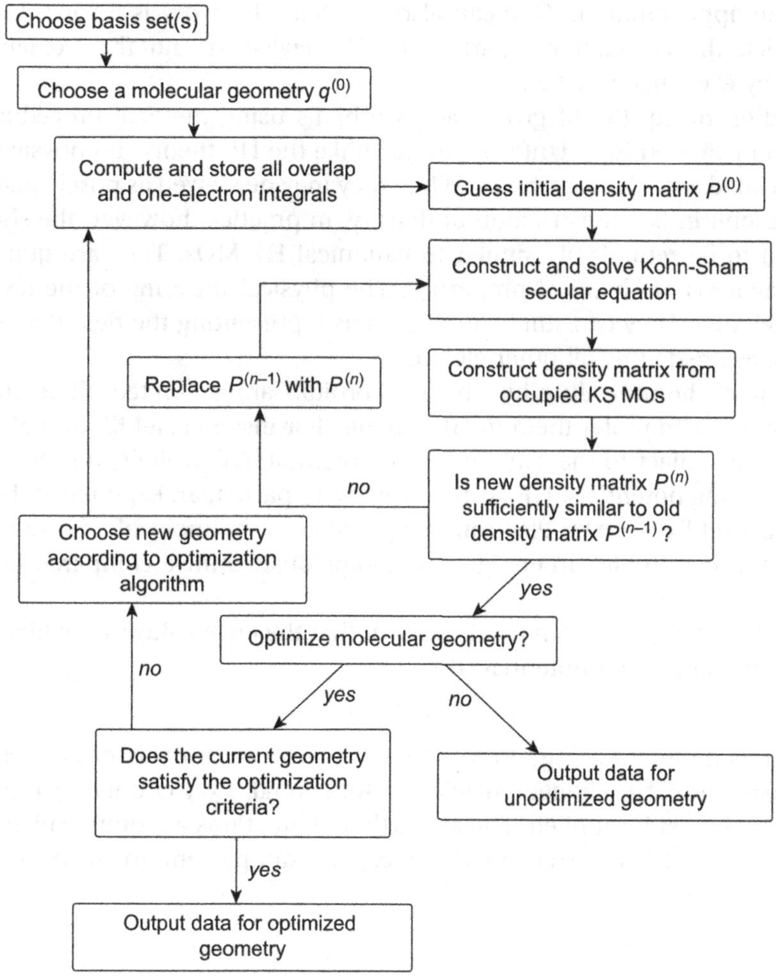

FIGURE 5.1 Flowchart of the KS self-consistent field procedure.

regards to computational efficiency. However in DFT they may be calculated by several other different methods which may reduce the four-center integrals to less demanding three-center two-electron integrals.

5.6 Comparison between HF and DFT

The KS Eq. (5.4.6) is identical to the HF Eq. (3.8.6) but the underlying concepts in the two cases are fundamentally different. While the potential V_S, as taken for noninteracting systems, is a simple multiplicative operator, the one in HF equations is a complicated integral. In contrast to the HF equations, the KS equations are exact and contain absolutely no approximation. One can also say that HF theory is a special case of the DFT in which the correlational part of E_{XC} is neglected and the exchange part is substituted by exchange integral.

The solution of Eq. (5.4.6) gives the KS orbitals using the SCF procedure; we can obtain optimum KS orbitals. Unfortunately, unlike the HF theory, no physical meaning can be associated with the KS orbitals. While they may be viewed as purely mathematical constructs useful in the construction of density, in practice, however, the shapes of KS orbitals tend to be remarkably similar to canonical HF MOs. They are quite useful in qualitative analysis of chemical properties. The physical meaning of the KS orbitals is still being debated. They can still be described as representing the field that an electron experiences, as created by all other electrons.

If we think of the procedure by which KS orbitals are generated, there are indeed a number of reasons to prefer them to HF orbitals. For instance, all KS orbitals, occupied and virtual, are subject to the same external potential. HF orbitals, on the other hand, experience varying potentials. HF virtual orbitals, in particular, experience the potential that would be felt by an extra electron being added to the molecule. Hence, HF virtual orbitals tend to be too high in energy and anomalously diffuse compared to KS virtual orbitals.

In exact DFT, if $V_{ne}(\infty) = 0$, it can be shown that the eigenvalue of the highest KS MO is the exact first ionization potential (IP)

$$\varepsilon_{\max} = -I$$

and so there is a direct analogy to Koopmann's theorem for this orbital. However, in practice, since correct exchange–correlation functional, E_{XC}, is unknown, an approximate functional must be applied. The approximate functions are quite bad at predicting *IPs* unless some sort of correction scheme, for example empirical linear scaling of eigenvalues, is applied.

5.7 Exchange–Correlation Functional

It follows from Eq. (5.4.3) that the exchange–correlation energy E_{XC} has contributions from various factors such as kinetic correlation energy (T_C), which is the difference between the kinetic energy of the real molecule and the reference system of noninteracting electrons, self-interaction correction, exchange energy, and Coulombic correlation energy (Eq. (5.3.13)). Thus, E_{XC} not only accounts for the difference between the classical and quantum mechanical electron–electron repulsion, but it also includes the kinetic energy between the fictitious noninteracting system and the real system. This last contribution (T_C) is very small as the kinetic energy of the noninteracting system (T_S) comes out to be close to T of the real system. The contribution of (T_C) to E_{XC} is however not negligible. Most modern functionals do not attempt to compute (T_C) explicitly. Instead, they either ignore the term or use a function that incorporates the kinetic energy reference between the interacting and noninteracting systems. In many functionals

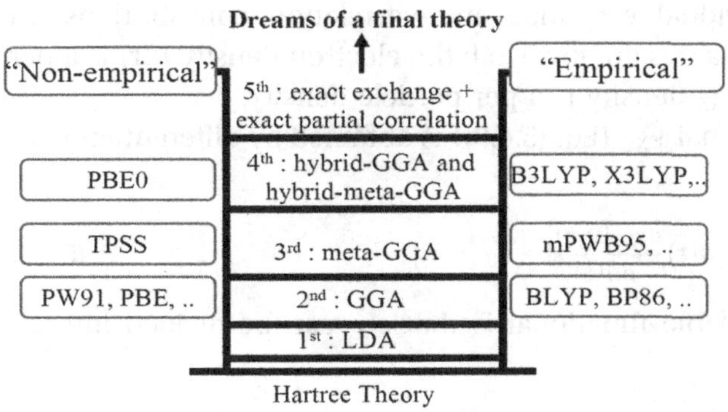

FIGURE 5.2 Schematic diagram of Jacob's Ladder of exchange–correlation functionals proposed by J.P. Perdew.

some empirical parameters appear which necessarily introduce some kinetic energy correction. Since the exact form of the exchange–correlation functional E_{XC} is not known, some sort of approximations for E_{XC} have been used right since the birth of DFT. There is an endless list of approximate functionals with varying level of complexity. A useful way of categorizing the many and varied E_{XC} functionals that exist has been proposed by Perdew and is known as "Jacob's Ladder" (Figure 5.2).

In this scheme, functionals are grouped according to their complexity on rungs of a ladder which lead from Hartree approximation to the exact exchange–correlation functional. Furthermore functionals can be categorized into nonempirical (formulated only by satisfying some physical rules) and empirical (made by fitting to the known results of atomic or molecular properties). A few of them shown on the Jacob's Ladder are being described here.

5.7.1 Local Density Approximation

The simplest method for obtaining the exchange–correlation functional (E_{XC}) uses the so-called local density approximation (LDA). In systems including spin polarization (e.g., open-shell systems), a spin-polarized formalism is used which is called local spin density approximation (LSDA). Both these formalisms are based upon a model called the uniform electron gas, in which the electron density is constant throughout the space. In this approach, a real inhomogeneous system is divided into infinitesimal volumes, and the electron density in each of the volumes is taken to be constant. In other words, the real electron density surrounding a volume element at position **r** is replaced by a constant electron density with the same value as at **r**. This constant electron density is, however, different for each point in space. The total exchange–correlation energy is then obtained by integrating over all space.

$$E_{XC}[n(\mathbf{r})] = \int n(\mathbf{r})\varepsilon_{XC}[n(\mathbf{r})]d\mathbf{r} \tag{5.7.1}$$

$\varepsilon_{XC}[n(\mathbf{r})]$ is the exchange–correlation energy per electron as a function of the density in the uniform electron gas. $\varepsilon_{XC}[n(\mathbf{r})]$ is also sometimes called energy density and is always treated as a sum of individual exchange and correlation contributions. In Eq. (5.7.1) two different kinds of densities are involved: the electron density $n(\mathbf{r})$ is a per unit volume density, while the energy density is a per particle density.

The exchange–correlation functional V_{XC} (Eq. (5.4.8)) is obtained by differentiation of this expression:

$$V_{XC}[\mathbf{r}] = n(\mathbf{r})\frac{\partial \varepsilon_{XC}[n(\mathbf{r})]}{\partial n(\mathbf{r})} + \varepsilon_{XC}[n(\mathbf{r})] \tag{5.7.2}$$

A common property of the available functional is that ε_{XC} can be divided into an exchange and a correlation part

$$\varepsilon_{XC}[n(\mathbf{r})] = \varepsilon_X[n(\mathbf{r})] + \varepsilon_C[n(\mathbf{r})] \ldots \tag{5.7.3}$$

The exchange part ε_X, which represents the exchange energy of an electron in a uniform electron gas of a particular density, as derived by Bloch and Dirac, is

$$\varepsilon_X[n(\mathbf{r})] = -\frac{3}{4}\left(\frac{3n(\mathbf{r})}{\pi}\right)^{1/3} \tag{5.7.4}$$

The analytical expression for the exchange energy is known exactly and is written as

$$E_X^{LDA}[n(\mathbf{r})] = \int n(\mathbf{r})\varepsilon_X d\mathbf{r} = -\frac{3}{4}\left(\frac{3}{\pi}\right)^{1/3}\int n^{4/3}(\mathbf{r})d\mathbf{r} \tag{5.7.5}$$

When this formula is used, the method is referred to as LDA. If it is used without any correlational functional, the method is called X_α.

No such explicit expression is known for the correlation part ε_C. The correlation energy is more complicated and generally obtained by fitting to the many-body equations of Gell-Mann and Brueckner [1] and Ceperley and Alder [2]. The LDA functionals are exceeding similar and differ only in how their correlation contributions have been fitted to the many-body free electron gas data. Some of the common LDA functionals are due to Gunnarsson and Lundquist [3], Perdew–Zunger (PZ) [4], Perdew–Wang (PW) [5], Slater and a complicated functional of Vosko–Wilk–Nusair (VWN) [6], for which the expressions are as below.

Gunnarsson and Lundquist analytical exchange–correlation functional

$$\varepsilon_{XC}[n(\mathbf{r})] = -\frac{0.458}{\mathbf{r}_s} - 0.0666G\left(\frac{\mathbf{r}_s}{11.4}\right)$$

$$\text{where, } r_s^3 = \frac{3}{4\pi n(\mathbf{r})} \text{ and } G(x) = \frac{1}{2}\left[(1+x)\log\left(1+\frac{1}{x}\right) - x^2 + \frac{x}{2} - \frac{1}{3}\right] \ldots$$

Slater's correlation functional

$$E_X[n_\alpha(\mathbf{r}), n_\beta(\mathbf{r})] = -\frac{3}{2}\left(\frac{3}{4\pi}\right)^{1/3}\int\left(n_\alpha^{4/3}(\mathbf{r}) + n_\beta^{4/3}(\mathbf{r})\right)d\mathbf{r}$$

where α and β represents spin-up and spin-down. The Slater functional was derived much before the KS formulation and as such $E_X \approx E_{XC}$.

Perdew and Zunger correlation functional

$$\varepsilon_C[n(\mathbf{r})] = \begin{cases} -0.1423/(1 + 1.9529 r_s^{1/2} + 0.3334 r_s), & \text{if } r_s \geq 1 \\ -0.0480 + 0.0311 \ln r_s - 0.0116 r_s + 0.0020 r_s \ln r_s, & \text{if } r_s < 1 \end{cases}$$

Vosko, Wilk, and Nusair (VWN) proposed a more accurate but less transparent representation

$$\varepsilon_C^{\mathrm{VWN}}(x) = A\left\{\ln\frac{x^2}{X(x)} + \frac{2b}{Q}\tan^{-1}\frac{Q}{2x+b} - \frac{bx_0}{X(x_0)}\left[\ln\frac{(x-x_0)^2}{X(x)} + \frac{2(2x_0+b)}{Q}\tan^{-1}\frac{Q}{2x+b}\right]\right\}$$

where $x = r_s^{1/2}$, $X(x) = x^2 + bx + c$, $Q = (4c - b^2)^{1/2}$ and A, b, c, and x_0 are parameters.

Strictly, LDA is valid only for slowly varying densities; Eq. (5.7.1) is applicable to atomic, molecular, and solid state systems. In fact, LDA also works surprisingly well for metals due to systematic error cancellation. In inhomogeneous systems LDA underestimates correlation but overestimates exchange, resulting in unexpectedly good values for $E_{\mathrm{XC}}^{\mathrm{LDA}}$. This error cancellation is systematic and is caused by the fact that for any density the LDA satisfies a number of so-called sum rules. The accuracy of LDA for the exchange energy is typically within 10%. In weakly bonded system such as van der Waals and H-bonded system like water dimer, the binding energy may be greater than 50% of the experimental value. Experience shows that LDA gives ionization energies of atoms, dissociation energies of molecules, and cohesive energies of within 10–20% accuracy. However, the bond lengths of molecules and solids are typically within an accuracy of 2%. LDA fails in systems like heavy Fermions having very large electron–electron interaction effect.

5.7.2 Local Spin Density Approximation

Instead of only one scalar external potential $V(\mathbf{r})$ (which is usually the electrostatic potential due to the nuclei) if we consider a system with a more generalized potential, e.g., a magnetic potential in addition to the usual scalar potential, then we need to describe the densities for electrons with different spin, i.e., $n_\alpha(\mathbf{r})$ and $n_\beta(\mathbf{r})$. The LSDA approximation allows electrons with opposite spins to have different spatial KS orbitals $\Psi_{i\alpha}^{\mathrm{KS}}$ and $\Psi_{i\beta}^{\mathrm{KS}}$. In this respect, LSDA is analogous to the open-shell unrestricted Hartree–Fock (UHF) method which allows different spatial HF orbitals for electrons with different spins. The DFT that allows different orbitals with different spin is also known as spin-DFT. When we consider electron spin, the exchange–correlation functional becomes functional of two quantities n_α and n_β for the spin-up and spin-down electrons:

$$E_{\mathrm{XC}} = E_{\mathrm{XC}}[n_\alpha, n_\beta]$$

In such a case, we shall have separate KS eigenvalue equations with different values of V_{XC}

$$V_{\mathrm{XC}}^{\alpha} = \frac{\partial E_{\mathrm{XC}}[n_\alpha, n_\beta]}{\partial n_\alpha}$$

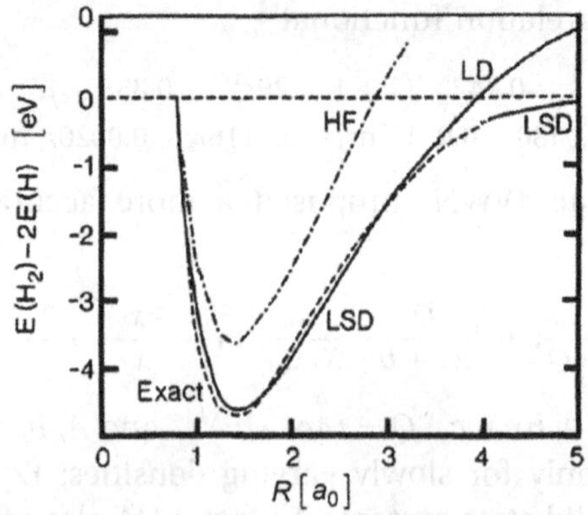

FIGURE 5.3 Potential energy curves and binding energy of H_2 by HF, LSDA, and LDA calculations. *Adapted with permission from Ref. [3]. Copyright @ 1976, American Physical Society.*

$$\text{and} \quad V_{XC}^{\beta} = \frac{\delta E_{XC}[n_\alpha, n_\beta]}{\delta n_\beta} \tag{5.7.6}$$

$$\text{Also, } E_{XC}[n_\alpha, n_\beta] = E_X[n_\alpha, n_\beta] + E_C[n_\alpha, n_\beta] \tag{5.7.7}$$

For species like CH_3 or O_2 triplet ground state, the electron density of α electrons differs from that of the β-electrons and hence the spin-DFT shall give different orbitals for electrons with different spin. For species with all electrons paired and with molecular

Table 5.1 Ionization potentials in electron volts of some light atoms calculated in the LSD, LDA, and HF approximations[a]

Atom	LSD	LDA	HF-ΔSCF[b]	Expt.
H	13.4	12.0	...	13.6
He	24.5	26.4	...	24.6
Li	5.7	5.4	5.3	5.4
Be	9.1	...	8.0	9.3
B	8.8	...	7.9	8.3
C	12.1	...	10.8	11.3
N	15.3	...	14.0	14.5
O	14.2	16.5	11.9	13.6
F	18.4	...	16.2	17.4
Ne	22.6	22.5	19.8	21.6
Na	5.6	5.3	4.9	5.1
Ar	16.2	16.1	14.8	15.8
K	4.7	4.5	4.0	4.3

Note: LSD, local spin-density method; LDA, local density approximation; HF, Hartree-Fock.

[a]IP, energy difference between neutral species and cations.

[b]HF-ΔSCF is a method in which the threshold energy is taken as the difference in total energies of two self-consistent-field (SCF) calculations (E. Clementi, J. Chem. Phys. 38 (1963) 996; 38 (1963) 1001; 41 (1964) 295,303)

Reprinted with permission from Ref. [3], Copyright @ 1976, American Physical Society.

geometries close to the equilibrium geometries we can expect $n_\alpha = n_\beta$ and so the spin-DFT will reduce to ordinary DFT.

Gunnarsson and Lundquist [3] did LSDA spin-DFT calculations of the H_2 molecule using a minimum basis set and showed that the LSDA gives good results for the energy curves for all separation studied in contrast to the spin-independent local approximation (LDA). The error in the binding energy is only 0.1 eV and bond breaking is properly described. It follows from Figure 5.3 that LSDA gives proper dissociation limit for the diatomic molecules while LDA fails. This parallels the performance of UHF versus HF. It also gives good values of bond length as compared to the LDA method.

LSDA methods provide more accurate values for the IP of some light elements than the LDA or HF methods as shown in Table 5.1. LSDA calculations are, however, known to underestimate exchange energy E_X by at least 10%, overestimate correlation energy by a factor of 2 or more and also cause electron self-interaction errors in approximate functionals.

5.7.3 Generalized Gradient Approximation

Despite its simplicity, the LDA has been found to be inadequate for some problems and for this reason extensions of LDA have been developed. The logical steps in this regard are the use of not only the information about the density $n(\mathbf{r})$ at a particular point $\mathbf{r}$, but also information about the gradient of the charge density, $\nabla n(\mathbf{r})$ so as to account for the nonhomogeneity of the true electron density distribution in a real system. Thus, we may write the exchange–correlation energy in a form known as Generalized Gradient Approximation (GGA).

$$E_{XC}^{GGA}[n(\mathbf{r})] = \int f^{GGA}[n(\mathbf{r}), \nabla n(\mathbf{r})] d\mathbf{r}$$

where f is some function of electron densities and their gradients. E_{XC}^{GGA} is usually split into exchange and correlation parts, which are modeled separately

$$E_{XC}^{GGA} = E_X^{GGA} + E_C^{GGA} \tag{5.7.8}$$

GGAs are often called "semilocal" functionals due to their dependence on $\nabla n(\mathbf{r})$. For many properties such as geometries and ground state energies of molecules and solids GGA can yield better results than the LDA; this is especially true for covalent bond and weakly bonded system. Several GGA functionals have been developed and are currently in use.

The functional form of f^{GGA} is taken as a correction to the LDA exchange and correlation while ensuring consistency with the known sum rules. Within GGA, the exchange energy E_X^{GGA} Eq. (5.7.8) takes the form

$$E_X^{GGA}[n] = \int n(\mathbf{r}) \varepsilon_X[n(\mathbf{r})] F_X^{GGA}(s) d\mathbf{r} \tag{5.7.9}$$

where ε_X corresponds to LDA, and $F_X^{GGA}(s)$ is the enhancement factor which tells how much exchange energy is enhanced over the LDA value for a given $n(\mathbf{r})$. The choice of F_X makes the difference between the different GGA exchange functionals reported in the literature. Here, s is a dimensionless reduced gradient

$$s = \frac{|\nabla n(\mathbf{r})|}{2(3\pi^2)^{1/3} n^{4/3}(\mathbf{r})} \tag{5.7.10}$$

Thus, in one of the most popular functionals, B88 or Becke 88 [7]

$$F_{\mathrm{X}}^{\mathrm{B88}}(s) = 1 + \frac{\beta x^2(s)}{c[1 + 6\beta x(s)\sinh^{-1}(x(s))]} \tag{5.7.11}$$

$$\text{where,}\ \ x(s) = 2(6\pi^2)^{1/3}\ s \tag{5.7.12}$$

and c and β are parameters obtained from empirical fitting. The exchange energy in this case is

$$E_{\mathrm{X}}^{\mathrm{B88}} = E_{\mathrm{X}}^{\mathrm{LDA}} - \beta \int \frac{n^{4/3} x^2}{(1 + 6\beta \sinh^{-1} x)} d\mathbf{r} \tag{5.7.13}$$

β has an empirical value of 0.0042 au. When the density gradient is zero, $F_{\mathrm{X}}^{\mathrm{GGA}}(s) = 1$ leading to the LDA exchange.

The well-known and popular correlational functionals for GGA are the Lee–Yang–Parr (LYP) functionals, the Perdew 1986 correlation functional (P86 or Pc86) and the Perdew–Wang 1991 parameter-free functional (PW91 or PWc91). In its original form the LYP correlation functional for a closed system is expressed as

$$E_{\mathrm{C}}[n(\mathbf{r})] = -a \int \frac{1}{1 + dn^{-1/3}} \left[\mathbf{r} + bn^{-2/3} \left\{ C_F n^{5/3} - 2t_w + \left(\frac{1}{9} t_w + \frac{1}{18} \nabla^2 n \right) e^{-cr^{-1/3}} \right\} \right] d\mathbf{r}$$

where,

$$t_w(\mathbf{r}) = \sum_{i=1}^{N} \frac{|\nabla n_i(\mathbf{r})|^2}{n_i(\mathbf{r})} - \frac{1}{8} \nabla^2 n;$$

$$C_F = \frac{3}{10}(3\pi^2)^{2/3} \tag{5.7.14}$$

and a, b, c, and d are constants with values 0.049, 0.132, 0.2533, and 0.3490, respectively. This expression has the advantage that it provides local and nonlocal components within a single expression and the gradient contribution to the second order. In the case of argon (Ar), the contributions of exchange and correlation energies are about 6% and 0.14%, respectively, of the total energy.

The different DFT methods are defined using a combination of an exchange and a correlation formula. For example, the Becke-Lee-Yang-Parr (BLYP) method uses LYP correlational and B88 exchange functionals, while BP86 uses Perdew's correlation functional with Becke 88 (B88) exchange functional.

5.7.4 Meta-GGA Functional

In order to improve accuracy, these functionals also use the second derivative of density $\nabla^2 n(\mathbf{r})$ and kinetic energy densities $(\nabla\Psi[n(\mathbf{r})])$, as additional degrees of freedom. Kinetic energy density is defined as $\tau[\mathbf{r}] = \frac{1}{2}\sum_i^{occ} |\nabla\Psi_i(\mathbf{r})|^2$. The form of the functional is typically

$E_{XC} \approx \int n(\mathbf{r})\varepsilon_{XC}(\rho, |\nabla\rho|, \nabla^2\rho, \tau)d\mathbf{r}$. In gas phase studies of molecular properties, meta-GGA such as Tao-Perdew-Staroverov-Scuseria (TPSS) [8] functionals have been shown to offer improved performance over LDAs and GGAs. The available meta-GGA functionals include B95, B98, 1SM, KC1S , PKZB, TPSS, and VSXC.

5.7.5 Hybrid Exchange Functionals

An exact connection exists between the noninteracting density functional system and the fully interacting system through the integration of workdone in gradually increasing electron–electron interactions. This approach is known as the adiabatic connection approach. While the noninteracting system corresponds identically to the HF approximation, fully interacting homogenous electron gas can be approximated by LDA and GGA functionals. In this background, Becke introduced hybrid functionals which are a mixture of HF exchange and DFT exchange–correlation functionals.

$$E_{XC}^{\text{hybrid}} \approx aE_X^{\text{HF}} + bE_{XC}^{\text{DFT}} \tag{5.7.15}$$

The coefficients are determined by fitting to the observed atomization energies, IPs, proton affinities, and total atomic energies for a number of small molecules for which exact results are known.

One such, three-parameter energy functional may be written as

$$E_{XC} = E_{XC}^{\text{LDA}} + 0.2\left(E_X^{\text{HF}} - E_X^{\text{LDA}}\right) + 0.72\Delta E_X^{\text{B88}} + 0.81\Delta E_C^{\text{PW91}} \tag{5.7.16}$$

where ΔE_X^{B88} and ΔE_C^{PW91} are the GGA corrections to LDA.

The most widely used functional in chemical applications is the Becke, three-parameter, Lee-Yang-Parr (B3LYP) functional [6,9,10].

$$E_{XC}^{\text{B3LYP}} = E_{XC}^{\text{LDA}} + a_0\left(E_X^{\text{HF}} - E_X^{\text{LDA}}\right) + E_C^{\text{LDA}} + a_C\left(E_C^{\text{GGA}} - E_C^{\text{LDA}}\right) + a_X\left(E_X^{\text{GGA}} - E_X^{\text{LDA}}\right) \tag{5.7.17}$$

where $a_0 = 0.20$, $a_X = 0.72$, and $a_C = 0.81$. All the three parameters are determined through fitting to the experiment so as to control mixing of the HF exchange and density-functional exchange and correlation. E_X^{GGA} and E_C^{GGA} are B88 exchange and LYP correlation functionals, respectively.

Reformulating this equation to eliminate two parameters leads to an equation of the form

$$E_{XC} = E_X^{\text{GGA}} + a\left(E_X^{\text{HF}} - E_X^{\text{GGA}}\right) \tag{5.7.18}$$

With $a = 0.25$, this leads to a simplified form belonging to the class of GGA functionals.

Several different forms of hybrid functionals have been constructed by varying the component functionals and the weighting factors. Another popular hybrid functional is BH&HLYP [11] which has 50% HF exchange and 50% GGA exchange. (The abbreviation H&H stands for half and half.) Such functionals have been shown to give much improved performance over LDA and GGA functionals for the calculation of gas phase properties of molecules and solids.

As an improvement over the B3LYP, B3PW91, and B1B95 hybrid functionals, Becke [12] introduced the B97 hybrid GGA functional in which the exchange parameter $[E_X^{GGA}]$ and correlation $[E_C^{GGA}]$ functionals contain three and six parameters, respectively.

$$E_{XC} = E_X^{GGA} + C_X E_X^{exact} + E_C^{GGA} \tag{5.7.19}$$

where C_X is a parameter. The systematic optimization based on least square fitting to accurate thermodynamic data led to very good agreement with the experimental data for the heat of formation with a mean absolute error (MAE) of 1.78 kcal/mol. Becke's 97 hybrid functional was reparameterized by Hamprecht et al. [13] that resulted in B97-1 functional. This functional gives best performance for van der Waals complex calculations. A further reoptimization of the empirical parameters of B97 resulted in B97-2. It incorporates 21% HF exchange. B97 hybrid GGA exchange–correlation functional was revised to create a hybrid meta-GGA called B98. It has 21.98% HF exchange.

Zhao et al. [14] developed hybrid meta-GGA functionals, called M05, M05-1X, and M05-2X for thermochemistry, thermochemical kinetics, and noncovalent interactions involving both metals and nonmetals. M05 functional has broad applicability to chemistry and performs well in main group thermochemistry and radical reaction barrier heights, and in transition-metal - transition-metal interactions. M05-2X functional has best performance for weak interactions, hydrogen bonding, $\pi\cdots\pi$ stacking, and interactions energies of nucleobases.

GGA, meta-GGA, hybrid-GGA, and hybrid–meta-GGA functionals give good equilibrium geometries, vibrational frequencies, dipole moments, and accurate molecular atomization energies. Some illustrative examples are given below. As of now, the list of exchange–correlation functionals is very large and they have been included in almost all quantum chemistry softwares for electronic structure calculations. A comparative study of some of these functionals in common use has been reported by Santra [15].

5.7.6 Selecting the Right Functional for Calculations

There is a vast choice of functionals available in the literature and it may be a daunting task to select a suitable functional. More so, the same functional which may be suitable for one task may not be suitable for another task. Some of the suggestions that may help in the selection process are:

1. Conduct a search of the literature for calculations on similar materials and properties of interest.
2. In the light of the type of calculation to be conducted, search for the strengths and weakness of the shortlisted functionals.
3. Evaluate the performance of the chosen functionals with different basis sets. Similarly parameterized functionals with empirical parameters sometimes produce better results with medium basis sets.

4. Before using for the actual system, carry out some calculations on similar smaller molecules to calibrate the results. This helps in comparing the accuracy of possible functionals and the basis sets for calculation on actual molecules.
5. It is also possible to create new functionals by mixing and matching available exchange and correlation functionals.
6. Since computational simulation of materials is a time-consuming job, selection of appropriate exchange and correlation functionals and basis sets can save a lot of efforts and provide better results.

5.8 Applications and Performance of DFT

DFT predicts a great variety of ground state molecular properties such as molecular structures and physicochemical properties, vibrational frequencies, thermodynamic properties, potential barriers and reaction paths, and hydrogen bonding, etc. It has been applied in different situations; for example, in the study of spin-polarized systems, multicomponent systems, free energy at finite temperatures, superconductors with electronic pairing mechanisms, relativistic electrons, time-dependent phenomena and excited states, MD, etc. In materials science it has found applications in the study of structural properties (such as lattice parameters, elastic constants, equilibrium geometry, and structural defects), lattice dynamics, electron density, electronic structure, and in different areas of spectroscopy such as Raman and Compton scatterings, photoemission spectroscopy, optical absorption spectroscopy, and magnetic resonance spectroscopy.

The performance of the DFT depends upon the choice of the exchange–correlation functional and hence utmost care is taken in its selection. The selection is made on the basis of their performance for a wide range of properties from energetics and geometries of molecules to reaction barriers and van der Waals interactions.

Cohen et al. [16] analyzed the performance of the various types of exchange–correlational functionals on the basis of thermochemical data sets (G3), geometries, kinetics and reaction barriers, and hydrogen bonding. The MAEs on thermochemistry, barriers, geometries, hydrogen bonding, and polarizabilities, all evaluated by post-B3LYP and post-PBE functionals are given in Table 5.2.

It follows from this table that approximate functionals generally perform relatively well for atomization energies but poorly for reaction barrier heights. However, functionals such as LDA and GGA can underestimate this barrier by 4–8 kcal/mol. Boese et al. [17] have carried out a careful evaluation of a large number of functionals for various energetic quantities (atomization enthalpies, ionization potentials, electron affinities, etc.) using polarized triple-ζ basis sets. They conclude that for pure GGA functionals the error ranking is HCTH < OLYP < BPW91 < BLYP < mPWPW91 < BP86 < PWPW91 < PBE (the lower, the better). They further conclude that for hybrid and meta-GGA functionals, which are overall better than pure functionals, an analogous error ranking is hybrid τ-HCTH < B97-1 < τ-HCTH < VSXC ~ B97-2 ~ B3LYP ~ B98 < PBE1PBE << MPW1K < PKZB. Zhao et al. [18] have also presented some meta-GGA exchange–correlation functionals like M06-L, M06, M06-HF, M05-2X, BMK, etc. (Table 5.3), having different

Table 5.2 Performance of post-B3LYP and post-PBE functionals in terms of mean absolute errors (MAE) on thermochemistry (G3), barriers, geometries, hydrogen bonding, and polarizabilities

	Post-B3LYP					Post-PBE				
Functional	G3 (kcal/mol)	Barriers (kcal/mol)	T96 (a_0)	H-bond (kcal/mol)	α_{iso} (au)	G3 (kcal/mol)	Barriers- (kcal/mol)	T96 (a_0)	H-bond (kcal/mol)	α_{iso} (au)
LDA	72.24	14.36	0.0107	3.02	0.78	73.08	14.95	0.0114	3.19	0.83
GGA and meta-GGA functionals										
BLYP	6.64	7.37	0.0205	1.46	0.79	6.77	7.58	0.0216	1.49	0.83
HCTH	5.59	4.15	0.0119	2.22	0.48	5.51	4.31	0.0216	2.25	0.44
HCTH407	5.72	4.69	0.0107	1.05	0.50	6.20	4.84	0.0115	1.10	0.48
PBE	15.99	8.29	0.0148	1.24	0.63	16.22	8.58	0.0157	1.32	0.66
BP86	15.71	8.49	0.0158	1.39	0.66	16.11	8.76	0.0169	1.43	0.67
BPBE	7.55	6.81	0.0155	1.67	0.53	7.81	7.08	0.0167	1.71	0.51
OLYP	5.22	5.36	0.0142	2.21	0.53	5.39	5.51	0.0152	2.24	0.53
OPBE	8.86	5.21	0.0121	2.55	0.31	9.48	5.38	0.0132	2.59	0.30
TPSS	7.85	8.03	0.0123	1.16	0.44	8.04	8.01	0.0131	1.20	0.43
M06-L	5.87	3.82	0.0056	0.58	0.40	7.67	3.72	0.0060	0.60	0.33
Hybrid functionals										
TPSSh	6.03	6.45	0.0082	0.98	0.30	6.05	6.19	0.0090	0.97	0.27
B3LYP	4.28	4.50	0.0097	1.01	0.37	4.38	4.22	0.0106	0.98	0.36
PBEO	6.37	4.11	0.0089	0.76	0.21	6.23	3.76	0.0096	0.71	0.19
B97-1	3.90	3.88	0.093	0.75	0.28	3.85	3.61	0.0100	0.67	0.26
B97-2	4.31	2.79	0.0087	0.97	0.19	4.49	2.58	0.0093	1.07	0.19
B97-3	3.70	2.22	0.0087	0.92	0.26	3.96	2.00	0.0095	1.07	0.24
M06	4.78	2.03	0.0088	0.47	0.39	5.48	1.79	0.0093	0.48	0.33
M06-2X	3.34	1.37	0.0110	0.34	0.35	3.67	1.59	0.0123	0.60	0.34
M06-HF	6.26	3.14	0.0167	0.88	0.73	8.13	4.23	0.0185	1.57	0.74
HF	132.38	15.12	0.0277	3.15	1.01	134.36	17.78	0.0307	4.11	1.03
HFLYP	35.39	9.18	0.0423	1.13	1.36	37.38	11.52	0.0444	1.87	1.55
Range-separated functionals										
CAMB3LYP	4.04	2.51	0.0119	0.69	0.23	4.22	2.40	0.0132	0.88	0.22
LCBLYP	16.91	3.73	0.0169	0.90	0.31	16.40	3.44	0.0182	0.90	0.31
rCAMB3LYP	5.50	2.76	0.0225	0.78	0.37	6.04	3.57	0.0240	1.20	0.42
LC-PBE	16.69	3.07	0.0245	0.75	0.53	16.34	3.50	0.0271	1.16	0.62
HSE	4.37	3.43	0.0082	0.77	0.21	4.50	3.09	0.0093	0.77	0.19

Table 5.3 Assessment of functionals for thermochemistry, kinetics, and noncovalent interactions based on mean unsigned error (MUE)

M06-2X	M05-2X	M06	M06-HF	M05	B97-3	M06-L	B98	B3LYP	PBE	BLYP	HFLYP	HF	Avg (DFT)	Avg (all)
1.32	1.63	1.82	2.29	2.57	2.80	2.94	2.59	3.59	5.20	4.80	12.73	39.18	3.45	5.55

percentages of exact (HF) exchange with an all-round good performance, where the forms were optimized to give both thermochemistry and kinetics. The composition of the various functionals is given as:

$$E_{XC} = y/100E_X^{HF} + (1 - y/100)E_X^{DFT} + E_C^{DFT}, \quad (5.8.1)$$

where y is the percentage of exact (HF) exchange; for M06-L, M06, M05, M06-2X, M05-2X, and M06-HF, y has values 0, 27, 28, 54, 58, and 100, respectively.

All these functionals surpass the performance of B3LYP in terms of the MAE. Thus MAE for M06-2X is 1.32 kcal/mol, for B98 it is 2.59 kcal/mol, while for B3LYP it is 3.59 kcal/mol. For thermochemistry and kinetics, the most accurate methods are M06-2X, M05-2X, BMK in that order. When noncovalent interactions such as hydrogen bonding are also considered, the best methods are in the sequence M06-2X, M05-2X, M06, BMK, M06-HF, and MK05. It was widely believed that one could not obtain high accuracy with full HF exchange. In this context, as follows from Table 5.3, the performance of M06-HF is noteworthy.

The prediction of geometries has been one of the great successes of DFT. Bond lengths computed using B3LYP and BP86 show an average absolute deviation from experiment of 0.013 and 0.010 Å compared to 0.026 and 0.022 Å for the pure GGA functionals BLYP and BP86, respectively. The choice of the basis set also affects the results. Thus, increasing the basis set from 6-31G(d) to 6-311+G(3df,2p) reduces the mean error further to 0.008 Å for B3LYP. In virtually all cases studied, hybrid functionals perform substantially better than LDA or GGA approaches in predicting molecular geometries, and in most comparative studies a 50% reduction in the mean errors for bond lengths is observed.

Gupta et al. [19] reported vibrationally averaged molecular geometry, dipole moment, and total energy of ketene using B97-1 and B3LYP functionals and aug-cc-pVTZ, 6-311++G*, and 6-31+G** basis sets and compared the calculated results with the experimental values of Cox et al. [20] and theoretical values of East et al. [21] (Table 5.4). It was found that the optimized bond lengths and bond angles of ketene were in good agreement with the experimental values as well as the CCSD/QZ(2d,2p)-based theoretical values of East et al. [21]. The vibrationally averaged bond lengths are longer than the optimized lengths by 0.001–0.003 Å. The averaged bond angles are shorter than the optimized bond angles by about 0.2°. The results of B97-1 are in better agreement with experiment than that of B3LYP.

The case of hydrogen bonds, where there is reasonable overlap of electron density as well as some electrostatic interactions between the fragments, has been a difficult challenge for functionals. Hydrogen bonds are of the order 1–10 kcal/mol and are much weaker than normal covalent and ionic bonds. Yet they are still much stronger that the weak dispersion and van der Waals interactions found between nonpolar closed-shell fragments. A wide difference in the performance of functionals that show good results for thermochemistry and geometries has been found in case of hydrogen bonds.

Table 5.4 Optimized and vibrationally averaged molecular geometries, dipole moment, and total energy of ketene

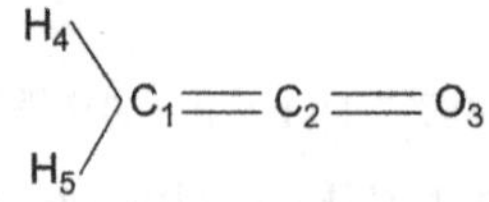

Bond length/ Bond angle	Exptl. values[a] Cox et al. [20]	East et al. [21] Opt.	B97-1/aug-cc-pVTZ Opt.	B97-1/aug-cc-pVTZ Aver.	B97-1/6-311++G** Opt.	B97-1/6-311++G** Aver.	B3LYP/6-31+G** Opt.	B3LYP/6-31+G** Aver.
$C_1{=}C_2$	1.314 ± 0.010	1.3142	1.3116	1.3153	1.3136	1.3173	1.3165	1.3201
$C_2{=}O_3$	1.161 ± 0.010	1.1609	1.1613	1.1627	1.1621	1.1636	1.1712	1.1726
$C_1{-}H_4$	1.083 ± 0.002	1.0753	1.0800	1.0825	1.0819	1.0844	1.0825	1.0854
$C_1{-}H_5$	1.083 ± 0.002	1.0753	1.0800	1.0825	1.0819	1.0845	1.0825	1.0855
$C_1C_2O_3$	–	180.00	180.00	180.00	179.95	179.96	179.98	179.98
$H_4C_1C_2$	–	–	119.42	119.22	119.69	119.50	119.71	119.51
$H_5C_1C_2$	–	–	119.42	119.22	119.68	119.49	119.70	119.50
$H_4C_1H_5$	122.6 ± 0.3	121.76	121.15	121.57	120.68	121.06	120.61	121.02
$H_5C_1C_2O_3$	0.0	0.0	0.0	0.0	0.0	0.0	0.0	0.0
Dipole-moment	1.41	–	1.4939	–	1.5333	–	1.6101	–
Energy			−152.6160790		−152.6024003		−152.6109023	

[a]Vibrationally averaged parameters at 0 K.

The performance of DFT has been evaluated for several standard sets related to response properties such as dipole polarizability which is given by the second derivatives of the energy with respect to an electric field F_x, $\alpha_{xx} = \partial^2 E/\partial F_x^2$. GGAs tend to overestimate the polarizability of standard small molecules, but the hybrid functionals generally give better results.

5.9 Challenges for DFT

Despite several successes of the DFT it still suffers from some limitations which need to be overcome before it can be treated as an exact theory rather than a semiempirical theory. One of the first great challenges for DFT in chemistry was to provide an accurate description of geometries and binding energies of simple molecules. This is largely due to the fact that though the simplest functional LDA gives good geometries, it massively overbinds molecules. Though this problem has been largely overcome by the introduction of B3LYP hybrid functionals, there are still problems associated with the transition state calculations, and inclusion of intermolecular interactions, hydrogen bonding, and van der Waals and dispersion forces. One of the main challenges for DFT is to maintain its simplicity and not to become as complicated as full configuration interaction (FCI) wavefunction theory. Some of the challenges before DFT continue to exist ever since its formulation. These relate to (1) the fundamental gaps in bulk solids which are underestimated (2) nonavailability of suitable functions for van der Waals interaction (3) poor treatment of strongly correlated systems and (4) nonavailability of a good scheme for excitations. During the last two decades a lot has been done to meet these challenges and some achievements have also been made but the progress has been slow. Two very informative review articles by Cohen et al. [16] and Burke [22] have been devoted to this issue and some suggestions have been made to meet these challenges. For an exhaustive account refer to these review articles and the references therein. We shall only be making a brief mention of the current challenges to DFT and the progress made in this regard.

5.9.1 To Develop a Functional of Nonempirical Nature

The practice of modern DFT suffers from a lack of understanding on how to approximate functionals. We begin from local approximations and then create more accurate, sophisticated versions. The present success of DFT is due to the excellent performance of these functionals. Thus, B3LYP is one of the most widely used of all the functionals so far. It has enjoyed a remarkable performance over a wide range of systems. However, it fails to correctly account for weak interactions, hydrogen bonding, $\pi\cdots\pi$ stacking, and radical reaction barrier heights. Although new ideas have been introduced into more recent functionals of different complexity, B3LYP is still the most popular. To remove the inadequacies of B3LYP, functionals like M05, M05-1X, and M05-2x have been developed for thermochemistry, thermochemical kinetics, and noncovalent interactions involving nonmetals. But the use of approximations gives DFT the stigma of a semiempirical method. It is important for DFT to fully connect to its roots as an exact theory.

5.9.2 Need to Improve Description of Transition States and Weak Interactions

In order to achieve a more complete description of chemistry, it is necessary to describe weakly interacting molecules as well as transition states in chemical reactions. Also, some very important chemical processes, which are much smaller in the energetic scale, may become significant for large systems, e.g., the weak but very important van der Waals force or London dispersion force. The performance of most popular functionals on simple weakly bound dimers is extremely poor. It may however be mentioned that the treatment of van der Waals interactions is a recent success story for DFT on which further work is in progress. Langreth and Lundquist [23] have developed a nonempirical nonlocal ground state density functional that has the right decay behavior of $-c/R^6$ (c—van der Waals coefficient and R—separation) for the dispersion force and can reasonably capture the effect of van der Waals interaction.

Two other methods like the DFT-D of Grimme [24], which provides empirical correction to DFT results, and the method proposed by Tkatchenko and Scheffler [25], which produces additive corrections for any functional with only slight empiricism and can be applied even to metals, have encouraged to remove semiempiricism. The correct and efficient description of the van der Waals attraction, covalent bonding, and transition states all remains a challenge particularly in the application of DFT to areas of biological importance where all these interactions can occur simultaneously.

5.9.3 Delocalization Error and Static Correlation Error

While a system with a single electron can be solved simply by using the wavefunction method, the DFT shows large errors in such cases. The problem is that DFT does not treat individual electrons in the same way as the wavefunction method, but rather only considers their total density, n. So one-electron systems play no special role in DFT; here, a single electron can unphysically interact with itself, in terms of self-interaction error. Of course the exact functional does not have any self-interaction; i.e., the exchange energy exactly cancels the Coulomb energy for one electron. Most currently used functionals are not completely free of self-interaction. Such and similar errors are the main cause of failures with the currently used approximations. So, it is important to seek out and understand practical and theoretical inconsistencies of DFT. It is known that most modern functionals can still have errors of 100 kcal/mol in extremely simple systems. In more complex systems these can be related to systematic errors such as delocalization error and static correlation errors.

5.9.4 Description of Strongly Correlated Systems

Except for FCI and the valence bond theory, most theories fail to correctly describe strongly correlated systems such as infinitely separated protons with varying numbers of electrons. Currently, all functionals used in DFT fail even for the simplest cases of infinitely stretched molecules like H_2, He_2^+, F_2^+, etc. In order to satisfy exact fundamental conditions and to remove systematic errors, the energy functionals must have the correct discontinuous behavior at integer numbers of electrons. This discontinuous behavior is important for correctly describing strong correlation.

5.9.5 Challenge of Larger Systems

While methods like molecular mechanics with classical force fields can handle millions of atoms, but their results regarding making or breaking of bonds are usually unreliable. As such, it may be necessary to combine DFT methods with some other method as is being done is the field of *ab initio* molecular dynamics (AIMD). This method is however able to handle only a few hundred atoms in each simulation. Use of DFT methods to deal with chemically active part in very large systems creates several problems, especially when the interface between the classical and DFT regions involves covalent bonds. This is a very active area of research and the integration of DFT with other methods is a challenging task.

5.9.6 Alternative View of DFT and beyond

An alternative view of DFT that uses the potential, instead of the electron density, as the basic variable has been invoked in potential functional theory [26]. The energy minimization can be carried out by minimizing the energy with respect to the potential. This provides the basis for the optimized effective potential (OEP) method and the understanding of orbital energies, which are functionals of the potential. Currently, there are important challenges in understanding exactly how most widely used approximations fit in with some known properties of the exact functional. Another approach known as density matrix functional theory (DMFT) has been developed by Donnelly and Parr [27] which uses the same variational principles as DFT. It has provided impressive results for equilibrium bond lengths and for total energies of closed-shell systems. Even energy gaps in insulators have also been predicted correctly. This method however still uses refinement of open-shell systems and size consistency remains an issue.

5.10 Time-Dependent DFT

The properties of excited states and excitation energies in particular are of interest in many respects and a number of strategies to approach this problem in the framework of the KS scheme have been put forward. One such effort was made by Grimme [28] who developed the method called density functional theory/single excitation configuration interaction (DFT/SCI), now extended to multireference configuration interactions. DFT/SCI, however, lacks a firm theoretical foundation but provides computed excitation energies for molecules, including fairly large hydrocarbons, within a few tenths of an electron volt of the experimental data.

Another completely different but a very promising approach to the calculation of excitation energies is based on time-dependent DFT (TDDFT). TDDFT is an extension of DFT having analogous conceptual and computational foundations. TDDFT approach uses the properties of the ground state—namely the ordinary KS orbitals and their corresponding orbital energies as obtained in a regular ground state calculation. Hence, excitation energies are expressed in terms of ground state properties. The TDDFT approach has even been extended from the mere prediction of excitation energies to the computational treatment of excited state surfaces including avoided crossings between states belonging to the same irreducible representation.

As of now, TDDFT is probably the most promising theory which provides a satisfactory treatment of the excited state within approximate DFT. From a practical point of view also, TDDFT has the important advantage that it has been implemented in many quantum chemical programs, such as Gaussian or Turbomole, and so can actually be used to solve physical problems. It is finding use in an ever-increasing range of applications to widely varying systems in chemistry, biology, and materials science. In the area of spectroscopy it is used to extract information about features like excitation energies, frequency-dependent response properties, and photoabsorption spectra of atoms, molecules, and solids. The excitation energies by TDDFT can be calculated to an accuracy of 0.1 eV but the actual value significantly depends on the system and type of excitation considered. Thus, for example, the errors for long-range charge-transfer excitations can be 10 times as large. TDDFT technique has a fairly involved theoretical background but we will confine our discussion to a very qualitative level.

5.10.1 Runge–Gross Theorem

The foundation of modern TDDFT was laid in 1984 by Runge and Gross [29], who derived a HK-like theorem for the time-dependent Schrödinger equation. This is the central theorem of TDDFT and is in many ways equivalent to the HK theorem in ground state DFT. It proves that there is a one-to-one correspondence between the external (time-dependent) potential, $V_{ext}(\mathbf{r}, t)$, and the electron density, $n(\mathbf{r}, t)$, for many-body systems evolving from a fixed initial state Φ^0. Thus, while in the HK theorem the ground state expectation value of any physical observable of a many-electron system is a unique functional of the electron density $n(\mathbf{r})$, in the Runge–Gross theorem, it is a unique functional of the time-dependent electron density $n(\mathbf{r}, t)$ and of the initial state Φ^0 corresponding to time $t = 0$, $\Phi^0 = \Phi(t = 0)$. Thus, if an operator $\widehat{O}$ represents the physical observable for the state Φ^0, then according to HK theorem

$$\langle \Phi^0 | \widehat{O} | \Phi^0 \rangle = O[n] \tag{5.10.1}$$

whereas according to Runge–Gross theorem

$$\langle \Phi(t) | \widehat{O}(t) | \Phi(t) \rangle = O[n, \Phi^0](t) \tag{5.10.2}$$

The density $n(\mathbf{r}, t)$ is the probability (normalized to the particle number N) of finding any one electron, of any spin σ, at position $\mathbf{r}$:

$$n(\mathbf{r}, t) = N \sum_{\sigma, \sigma_2 \ldots \sigma_N} \int d^3\mathbf{r}_2 \ldots \int d^3\mathbf{r}_N |\Phi(\mathbf{r}\sigma, \mathbf{r}_2\sigma_2 \ldots \mathbf{r}_N\sigma_N, t)|^2 \tag{5.10.3}$$

Since the time-dependent Schrödinger equation (Eq. (5.10.4)) is a first-order differential equation in time, a time-dependent problem in quantum mechanics is mathematically defined as an initial value problem. The wavefunction (or the density) thus depends on the initial state, which implies that the Runge–Gross theorem can only hold for a fixed initial state and also that the correlation–exchange (XC) potential depends on that state.

$$H(t)\Phi(\mathbf{r}_1, \ldots \mathbf{r}_N; t) = i\hbar \frac{\partial}{\partial t} \Phi(\mathbf{r}_1, \ldots \mathbf{r}_N; t) \tag{5.10.4}$$

$$\text{where,} \quad \widehat{H}(t) = \widehat{T} + \widehat{V}_{ee} + \widehat{V}_{ext}(t) \tag{5.10.5}$$

$$\widehat{T} = -\frac{1}{2} \sum_{i=1}^{N} \nabla_i^2; \; \widehat{V}_{ee} = \frac{1}{2} \sum_{i \neq j}^{N} \frac{1}{|r_i - r_j|} \tag{5.10.6}$$

$$\text{and} \quad \widehat{V}_{ext}(t) = \sum_{i=1}^{N} v_{ext}(\mathbf{r}_i, t) \tag{5.10.7}$$

For example, $V_{ext}(\mathbf{r}_i, t)$ can represent the Coulomb interaction of the electrons with a set of nuclei or the interaction with external fields, as for a system illuminated by a laser beam.

As mentioned earlier, the DFT solves a time-independent second-order differential equation

$$H\Phi(\mathbf{r}_1, \ldots, \mathbf{r}_N) = E\Phi(\mathbf{r}_1, \ldots, \mathbf{r}_N) \tag{5.10.8}$$

and so it deals with a boundary value problem. This is contrast to the first-order time-dependent equation, as in TDDFT, which solves an initial value problem. In both the cases, however, the density of a system of interacting particles can be calculated as the density of an auxiliary system of noninteracting particles.

Though we shall not attempt to rigorously prove the Runge–Gross theorem, we may mention a few difficulties that one encounters while trying to prove it. There are several differences between a time-dependent and a static quantum mechanical problem that one should keep in mind while trying to prove the Runge–Gross theorem, the most important of which is that, unlike the time-independent systems, in the time-dependent systems, there is no variational principle for total energy as it is not a conserved quantity. There exists, however, a quantity analogous to the energy known as the quantum mechanical action.

$$A[\Phi] = \int_{t_0}^{t_1} dt \left\langle \Phi(t) \middle| i\frac{\partial}{\partial t} - H(t) \middle| \Phi(t) \right\rangle \tag{5.10.9}$$

where Φ is an N-body function defined in some convenient space. Equating the derivative of $A[\Phi]$ in terms of $\Phi(t)$ to zero, we arrive at the time-dependent Schrödinger equation. The time-dependent equation can therefore be solved by calculating the stationary point of the functional $A[\Phi]$. The function $\Psi(t)$ that makes $A[\Phi]$ stationary will be the solution of the Schrödinger equation. Thus, in time-dependent case, we have only a "stationary principle" and not an energy "minimum principle." The action is always zero at the solution point, that is $A[\Psi] = 0$. This is shown in Figure 5.4(b).

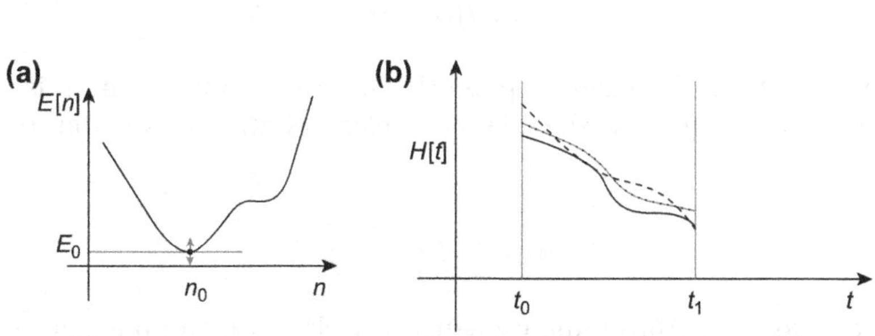

FIGURE 5.4 (a) In DFT, the ground state energy E_0 corresponding to the ground state density n_0. The total energy functional has a minimum, (b) In time-dependent Schrödinger equation, the initial condition ($\Phi(t=0) = \Phi_0$) corresponds to a stationary point of the Hamiltonian action.

Runge–Gross theorem has enormous advantages. If we know only the time-dependent density of a system, evolving from a given initial state, then it is possible to identify the external potential that produced this density. The external potential completely identifies the Hamiltonian as the other terms in Eq. (5.10.5) can be easily determined. The time-dependent Schrödinger equation can then be solved, in principle, and all properties of the system obtained. Thus for a given initial state, the electronic density, which is a function of just three spatial variables and time, determines all other properties of the interacting many-electron system.

5.10.2 Time-Dependent KS Equations

Just like DFT, instead of solving the interacting Schrödinger equation, the Kohn and Sham approach of utilizing noninteracting particles subject to an external local potential V_{KS} is adopted in TDDFT, as well. This potential $V_{KS}(n, t)$ of a noninteracting system is unique and is chosen such that the density of the noninteracting electrons is the same as the density $n(\mathbf{r}, t)$ of the original interacting system. In the time-dependent case, these KS electrons obey the time-dependent Schrödinger equation

$$i\frac{\partial}{\partial t}\Phi_i(\mathbf{r},t) = \left[-\frac{\nabla^2}{2} + V_{KS}(\mathbf{r},t)\right]\Phi_i(\mathbf{r},t). \tag{5.10.10}$$

The density of interacting system can be obtained from the time-dependent KS orbitals

$$n(\mathbf{r},t) = \sum_i^{occ} |\Phi_i(\mathbf{r},t)|^2 \tag{5.10.11}$$

Equation (5.10.10), having the form of a one-particle equation, is fairly easy to solve numerically. If we know the exact KS potential, V_{KS}, we can obtain from Eq. (5.10.11) the exact KS orbitals and from these the correct density of the system.

The KS potential is conventionally separated in the following way

$$V_{KS}(\mathbf{r},t) = V_{ext}(\mathbf{r},t) + V_H(\mathbf{r},t) + V_{XC}(\mathbf{r},t) \tag{5.10.12}$$

Here, the first term is the external potential and the second term V_H is the Hartree potential which accounts for the classical electrostatic interaction between the electrons

$$V_H(\mathbf{r},t) = \int d\mathbf{r}'\frac{n(\mathbf{r},t)}{|\mathbf{r}-\mathbf{r}'|} \tag{5.10.13}$$

The last term in Eq. (5.10.12), the exchange–correlation (XC) potential, comprises all the nontrivial many-body effects. While in DFT, V_{XC} is normally written as a functional derivative of the exchange–correlation energy, due to a problem related to causality this formulation cannot be directly extended to the time-dependent case. The time-dependent

XC potential can then be written as the functional derivative of the XC part of a new action functional $\tilde{A}$ introduced by Leeuwen [30],

$$V_{XC}([n],\boldsymbol{r},\tau) = \left.\frac{\delta \tilde{A}_{XC}[n]}{\delta n(\boldsymbol{r},\tau)}\right|_{n(r,\tau)} \tag{5.10.14}$$

where τ is the Keldish pseudotime.

The functional dependence of V_{ext} (the first term on the r.h.s. of Eq. (5.10.12)) is not important in practice since, for real calculations, the external potential is given by the problem at hand. In practice, only the XC potential needs to be approximated as a functional of the density, the true initial state, and the KS initial state. This functional given by Eq. (5.10.14) is a very complex one but by knowing this functional one can find solution of all time-dependent Coulomb interaction problems.

5.10.2.1 Steps in TDDFT Calculations

A TDDFT calculation proceeds as follows. An initial set of N orthonormal KS orbitals (Φ_i) is chosen, which must reproduce the exact density of the true initial state Ψ_0 of the problem at hand and its first time-derivative:

$$n(\mathbf{r},0) = \sum_{i=1}^{N} |\Phi_i(\mathbf{r},0)|^2 = N \sum_{\sigma,\sigma_2\ldots\sigma_N} \int d\mathbf{r}_2 \ldots \int d\mathbf{r}_N |\Psi_0(x,x_2\ldots x_N)|^2 \tag{5.10.15}$$

$$\begin{aligned} \dot{n}(\mathbf{r},0) &= -\nabla\cdot \mathrm{Im}\sum_{i=1}^{N}\sum_{\sigma}\Phi_i^*(\mathbf{r},0)\nabla\Phi_i(\mathbf{r},0) \\ &= -N\nabla\cdot\mathrm{Im}\sum_{\sigma,\sigma_2\ldots\sigma_N}\int d\mathbf{r}_2\ldots\int d\mathbf{r}_N \Psi_0^*(x,x_2\ldots x_N)\nabla\Psi_0(x,x_2\ldots x_N), \end{aligned} \tag{5.10.16}$$

where $x = (\mathbf{r}, \sigma)$.

The time-dependent KS Eq. (5.10.10) then propagates these initial orbitals, under the external potential (given by the problem at hand), the Hartree potential and the approximation for the XC potential in Eq. (5.10.12). The choice of the KS initial state, and the fact that the KS potential depends on this choice is a completely a new feature of TDDFT without a ground state analogue. Almost all applications of TDDFT make the adiabatic approximation which assumes that the XC potential reacts instantaneously and without memory to any change in the charge density with time.

Within TDDFT two cases can be considered:

1. If the external time-dependent potential is small, the complete numerical solution of the time-dependent KS equations can be avoided by the use of linear response theory. This is the case, for example, for the interaction of weak electromagnetic radiation with matter, i.e., the calculation of absorption spectra.
2. If the external potential is strong, a full solution of the time-dependent KS equations is required.

This situation is encountered, for instance, when matter interacts with intense laser fields.

5.10.3 Linear Response Theory

In circumstances where the external time-dependent potential is small, it may not be necessary to solve the full time-dependent KS equations. Instead, perturbation theory may prove sufficient to determine the behavior of the system. This is true for the calculation of excitation spectra.

As mentioned, according to the HK theorem for the ground state, all observables are functionals of the density and from a knowledge of the density it is possible to uniquely determine the external potential. Once we solve the many-body Schrödinger equation using this potential, we can find many-body energies and eigenfunctions and thus determine all observables. The energies of the excited states can also be viewed in the same manner, i.e., they are functionals of the ground state density. In general, we have little or no information on how to express an observable as a functional of the ground state density. This is also true for the excitation energies of the system. To circumvent this problem, several approaches, both *ad hoc* and those based on solid theoretical basis, have been proposed over the years. The degree of success varies considerably among the different techniques. Linear response time-dependent density functional theory (LR-TDDFT) is one such approach. This approach has the advantage that the variation of the system will depend only on the ground state wavefunction so that we can still use all the properties of DFT.

Suppose a system is initially in the ground state under the nuclear potential v_0 and electron density n_0. When it is subjected to a perturbation $v^{(1)}$, the total external potential will be $V_{ext}(t) = v_0 + v^{(1)}$ and the electron density of the system will change to $n(\mathbf{r}, t)$, such that

$$n(\mathbf{r}, t) = n_0(\mathbf{r}) + n^{(1)}(\mathbf{r}, t) + n^{(2)}(\mathbf{r}, t) + \ldots \tag{5.10.17}$$

where $n^{(1)}, n^{(2)}\ldots$ being the first-order, second-order, ... perturbation in the density. If the time-dependent external potential is small, we can restrict to the first-order term $n^{(1)}$ and write

$$n^{(1)}(\mathbf{r}, \omega) = \int \chi(\mathbf{r}, \mathbf{r}', \omega) v^{(1)}(\mathbf{r}', \omega) d\mathbf{r}', \tag{5.10.18}$$

where χ is called the linear density–density response function. Density–density response function is the basic quantity in the LR-TDDFT which relates the first-order density response to the applied perturbation

$$\chi(\mathbf{r}, \mathbf{r}', \omega) = \left. \frac{\delta n(\mathbf{r}, t)}{\delta V_{ext}(\mathbf{r}', \omega)} \right|_{v_0} \tag{5.10.19}$$

Since the density of the interacting system of electrons in the KS formalism is obtained from a fictitious system of noninteracting electrons, we can write Eq. (5.10.18) as

$$n^{(1)}(\mathbf{r}, \omega) = \int \chi_{KS}(\mathbf{r}, \mathbf{r}', \omega) v^{(1)}_{KS}(\mathbf{r}', \omega) \tag{5.10.20}$$

χ_{KS} is the density response function of the system of noninteracting particles and is therefore much more easier to calculate than χ. The potential $v_{KS}^{(1)}$ entering this equation is just the linear change of v_{KS} and following Eq. (5.10.12) can be given as

$$v_{KS}^{(1)}(\mathbf{r},t) = v^{(1)}(\mathbf{r},t) + v_{H}^{(1)}(\mathbf{r},t) + v_{XC}^{(1)}(\mathbf{r},t) \qquad (5.10.21)$$

where $v^{(1)}$ is the variation in the external potential, $v_{H}^{(1)}$ is the change in the Hartree potential given as

$$v_{H}^{(1)}(\mathbf{r},t) = \int \frac{n^{(1)}(\mathbf{r}',t)}{|\mathbf{r}-\mathbf{r}'|}d\mathbf{r}' \qquad (5.10.22)$$

and $v_{XC}^{(1)}$ is the linear part in $n^{(1)}$ of the functional $v_{XC}[\rho]$, given as

$$v_{XC}^{(1)}(\mathbf{r},t) = \int dt' \int \frac{\delta v_{XC}(\mathbf{r},t)}{\delta n(\mathbf{r}',t')} n^{(1)}(\mathbf{r}',t') \qquad (5.10.23)$$

In terms of unperturbed stationary KS orbitals the linear density–density response function χ_{KS} in Eq. (5.10.19) is given as

$$\chi_{KS}(\mathbf{r},\mathbf{r}',\omega) = \lim_{\gamma\to 0}\sum_{i,j}(f_j - f_i)\frac{\Phi_i(\mathbf{r})\Phi_i^*(\mathbf{r}')\Phi_j(\mathbf{r}')\Phi_j^*(\mathbf{r})}{\omega-(\varepsilon_i-\varepsilon_j)+i\gamma} \qquad (5.10.24)$$

where, f_i and f_j are the usual Fermi occupation numbers. This function has poles at the KS single-particle orbital energy differences which are actually not the poles of the true density-density response function $\chi(\mathbf{r},\mathbf{r}',\omega)$. Similarly, the strengths of the poles (the numerator in Eq.(5.10.24)), which is related to the optical strengths, are not those of the true system. The true density-response can be obtained from the KS system by using a Dyson-like equation

$$\chi(\mathbf{r},\mathbf{r}',\omega) = \chi_{KS}(\mathbf{r},\mathbf{r}',\omega) + \int d\boldsymbol{x}\int d\mathbf{r}'\chi_{KS}(\mathbf{r},\boldsymbol{x},\omega)\left[\frac{1}{|x-x'|}+f_{XC}(\boldsymbol{x},\boldsymbol{x}',\omega)\right]n^{(1)}(\mathbf{r}',\omega) \qquad (5.10.25a)$$

which, in the frequency space, gives

$$\chi(\omega) = \chi_{KS}(\omega) + \chi_{KS}(\omega) * f_{HXC}(\omega) * \chi(\omega) \qquad (5.10.25b)$$

where

$$f_{HXC}(\mathbf{r},\mathbf{r}',\omega) = \frac{1}{|\mathbf{r}-\mathbf{r}'|} + f_{XC}(\mathbf{r},\mathbf{r}',\omega) \qquad (5.10.25c)$$

is the Hartree XC kernel and

$$f_{XC} = \frac{\delta V_{XC}}{\delta n(\mathbf{r}',t')} \qquad (5.10.25d)$$

is the exchange-correlation kernel.

From, Eqs (5.10.25 (a–d)), it is possible to derive the excitation energies of the system as they are simply poles of the response function.

The LR-TDDFT theory is therefore based on the premise that the frequency-dependent linear response of a finite system with respect to time-dependent perturbation has discrete poles at the exact, correlated excitation energies of the unperturbed system.

Thus, the dynamic dipole polarizability is the response function that relates the external time-dependent electric field $\mathcal{E}(t)$ to the change in the molecular dipole

$$\delta\mu_x(\omega) = \alpha_{xz}(\omega)\varepsilon_z(\omega)$$

$$\text{where,}\quad \alpha_{xz}(\omega) = -\sum_{ij,kl} x_{ij}\chi_{ij,kl}(\omega)z_{kl} \quad \text{and} \quad x_{ij} = \langle\Phi_j|\hat{x}|\Phi_j\rangle$$

To be more specific, the frequency-dependent mean polarizability $\overline{\alpha}(\omega)$ describes the response of the dipole moment to a time-dependent electric field of frequency $\omega(t)$. It can be shown that $\overline{\alpha}(\omega)$ is related to the electronic excitation spectrum according to

$$\overline{\alpha}(\omega) = \sum_I \frac{f_I}{(\omega_I^2 - \omega^2)} \tag{5.10.26}$$

$$\text{where,}\quad \omega_I = E_I - E_0 \tag{5.10.27}$$

is the excitation energy and the summation is overall excited states of the system, and

$$f_I = \frac{2}{3}\omega_I\left(|\langle\Psi_0|\hat{x}|\Psi_I\rangle|^2 + |\langle\Psi_0|\hat{y}|\Psi_I\rangle|^2 + |\langle\Psi_0|\hat{z}|\Psi_I\rangle|^2\right) \tag{5.10.28}$$

Equation (5.10.26) is also called the sum-over-states (SOS) relation in spectroscopy. From this equation it follows that $\alpha(\omega)$ diverges at $\omega_I = \omega$, i.e., has a pole at the excitation energy ω_I. The zero order estimate of excitation energy is modified by the second term in Dyson relation given by Eq. (5.10.25a). The residues f_I are called the corresponding oscillator strengths.

It may also be noted from the above discussion that in the LR-TDDFT approach, the KS orbitals and their orbital energies obtained in the ground state calculations are involved. The KS eigenvalues do not have any strict physical meaning (except for the highest occupied one which represents the negative of the first IP), but in cases where no other calculations are available, they are often used to interpret experimental observations.

5.10.4 Few DFT Techniques to Calculate Excitation

Besides, LR-TDDFT, there are several other techniques to calculate excitations. Some of these are:

1. Restricting the variational principle to wavefunctions of a specified symmetry, thereby obtaining the energy of the lowest state for each symmetry class. The unrestricted variation will clearly yield the ground state. The states belonging to different symmetry classes will correspond to excited states. The excitations can then be calculated by simple total energy differences.
2. Generalized adiabatic connection formalism of Görling [31]. The so-called generalized adiabatic connection KS formalism is no longer based on the HK theorem but on generalized adiabatic connections associating a KS state with each state of the real system.
3. Searching for local extrema of the ground state energy functional. It was proved by Perdew and Levy [32] that every extremum density $n_i(\mathbf{r})$ of the ground state energy functional $E_v[n]$ yields the energy E_i of a stationary state of the system. However,

since not every excited state density, $n_i(\mathbf{r})$, corresponds to an extremum of $E_v[n]$, not all excitation energies can be obtained from this procedure.

4. Another technique, the so-called ensemble DFT, makes use of fractional occupation numbers. Ensemble DFT [33] is concerned with "ensembles" of the ground state with some excited states. In the simplest case, the ensemble is defined by the density matrix.

$$\widehat{D} = (1-\omega)|\Psi_1\rangle\langle\Psi_1| + \omega|\Psi_2\rangle\langle\Psi_2| \tag{5.10.29}$$

where Ψ_1 is the ground state and Ψ_2 the lowest excited state, and ω is some given weight between 0 and 1/2. Using the ensemble density it is then possible to construct a DFT and obtain the ensemble energy $E(\omega)$ from solving a set of Kohn–Sham-like equations. Knowing $E(\omega)$ and the ground state energy (which can be obtained from ordinary DFT) it is trivial to determine the energy of the first excited state. The procedure can obviously be continued to calculate all excited state energies. As in any other DFT method, the problem with the ensemble DFT is in the choice of good XC functionals.

A more promising approach, involving orbital functionals within an ensemble OEP method, has also been proposed recently [34,35]. This method involves the calculation of exact exchange (EXX) potential.

5.10.5 Matrix Formulation of TDDFT

The first matrix formulation of TDDFT linear response was developed in 1995 by Casida [36] by considering the response of the KS density matrix. It was shown that configuration interaction-like equations arise naturally out of TDDFT. Commonly known as "Casida's equations," these equations are similar in structure to time-dependent HF (TDHF) equations and are coded in most of the electronic structure softwares. Here only an outline of the TDDFT method is being given. (For details, see the review articles by Casida [37] and Marques and Gross [38].)

Properties of a stationary state may be defined without reference to the wavefunction by the dependence of the energy on an applied external perturbation. For example, the dipole moment may be defined as the first derivative of the energy (E) with respect to the components of the electric field ($\mathcal{E}_q$). It is a first-order property.

$$\mu_q = -\frac{\partial E}{\partial \mathcal{E}_q} \tag{5.10.30}$$

The dipole polarizability is a second-order property

$$\alpha_{q,q'} = -\frac{\partial^2 E}{\partial \mathcal{E}_q \partial \mathcal{E}_{q'}} \tag{5.10.31}$$

The excited state density can be defined as the functional derivative of the excited state energy with respect to an external perturbing potential at zero coupling. It is therefore sufficient to know the dependence of the excited state energy on the external potential to compute static excited state properties.

Casida considered an N-electron system, initially in its ground stationary state, exposed to a time-dependent perturbation $\delta\overrightarrow{P}(\omega)$ of frequency ω which was turned on adiabatically beginning at time $t = -\infty$ and deduced an equation for the dynamic response of the KS density matrix to the perturbation. The resultant dynamic equation is

$$\left\{\omega\begin{bmatrix}1 & 0\\ 0 & -1\end{bmatrix} - \begin{bmatrix}A & B\\ B^* & A^*\end{bmatrix}\right\}\begin{pmatrix}\delta\vec{P}(\omega)\\ \delta\vec{P}^*(\omega)\end{pmatrix} = \begin{pmatrix}\vec{V}_{\text{appl}}(\omega)\\ \vec{V}^*_{\text{appl}}(\omega)\end{pmatrix} \tag{5.10.32}$$

where $\delta\vec{P}(\omega)$ is the dynamic response of the density matrix, which may become infinite even for an arbitrarily small perturbation ($\vec{V}_{\text{appl}}(\omega)\rightarrow 0$) in case the exciting frequency matches a resonance frequency $\omega = \omega_I$ (Eq. (5.10.26)) capable of exciting an electron. At such frequencies, Eq. (5.10.32) takes the form $(x)(+\infty) = 0$, where x must be zero. The equation therefore takes the form

$$\begin{pmatrix}A & B\\ A^* & B^*\end{pmatrix}\begin{pmatrix}X_1\\ Y_1\end{pmatrix} = \omega_I\begin{pmatrix}1 & 0\\ 0 & -1\end{pmatrix}\begin{pmatrix}X_1\\ Y_1\end{pmatrix} \tag{5.10.33}$$

The matrix equations are often rewritten in the literature as

$$\begin{pmatrix}A & B\\ B^* & A^*\end{pmatrix}\begin{pmatrix}X\\ Y\end{pmatrix} = \omega\begin{pmatrix}-1 & 0\\ 0 & 1\end{pmatrix}\begin{pmatrix}X\\ Y\end{pmatrix} \tag{5.10.34}$$

$$\text{where,}\quad A_{ia,jb} = \delta_{ij}\delta_{ab}(\varepsilon_a - \varepsilon_i) + 2\int d\mathbf{r}\int d\mathbf{r}'\Phi_q^*(\mathbf{r})f_{\text{HXC}}(\mathbf{r},\mathbf{r}')\Phi_{q'}(\mathbf{r}') \tag{5.10.35}$$

$$B_{ia,jb} = 2\int d\mathbf{r}\int d\mathbf{r}'\Phi_q^*(\mathbf{r})f_{\text{HXC}}(\mathbf{r},\mathbf{r}')\Phi_{-q'}(\mathbf{r}') \tag{5.10.36}$$

where i and j represent the occupied and a, b the virtual KS orbitals. Here, $q = (i, a)$ is a double index with transition frequency $\omega_q = \varepsilon_a - \varepsilon_i$ and f_{HXC} is the Hartree–XC kernel

$$f_{\text{HXC}}(\mathbf{r},\mathbf{r}',\omega) = \frac{1}{|\mathbf{r}-\mathbf{r}'|} + f_{\text{XC}}(\mathbf{r},\mathbf{r}',\omega) \tag{5.10.37}$$

where $f_{\text{XC}}(\mathbf{r},\mathbf{r}',\omega)$ is the Fourier transform of the time-dependent kernel,

$$f_{\text{XC}}[n_0](\mathbf{r}t,\mathbf{r}'t') = \left.\frac{\delta v_{\text{XC}}[n](\mathbf{r},t)}{\delta n(\mathbf{r}',t')}\right|_{n=n_0} \tag{5.10.38}$$

Excitation energies and oscillator strengths can be extracted by rearranging this expression to take the form of the dynamic polarizability and using the SOS theorem of optical physics,

$$\alpha(\omega) = \sum_{I\neq 0}\frac{f_I}{\omega_I^2 - \omega^2} \tag{5.10.39}$$

$$\text{where,}\quad f_I = \frac{2}{3}\omega_I\sum_{r=xyz}|\langle\Psi_0|\mathbf{r}|\Psi_I\rangle|^2 \tag{5.10.40}$$

is the oscillator strength associated with the excitation energy,

$$\hbar\omega_I = E_I - E_0 \tag{5.10.41}$$

Ψ_0 and Ψ_I refer, respectively, to the ground and excited stationary states. Equation (5.10.40) is the same as Eq. (5.10.28). The KS orbitals are chosen to be real in this formulation. The matrix formulations are valid only for discrete spectra and hence are mostly used for finite systems.

Casida et al. [39] have suggested that TDDFT results are most reliable if the excitation energy is significantly smaller than the molecular IP and the promotion(s) does not take place into orbitals having positive KS eigenvalues. TDDFT is usually most successful for low-energy excitations because the KS orbital energies for orbitals that are high up in the virtual manifold are typically quite poor.

5.11 Approximate Exchange–Correlation Functionals for TDDFT

Approximate exchange–correlation functional can be constructed empirically as well as nonempirically. The difference between the two is that while empirical functionals contain a large number of parameters fitted to a reference set of accurate experimental or calculated data the nonempirical functionals contain few or no fitted parameters and are designed to satisfy known constraints. Also, the empirical functionals should be accurate for systems and properties contained in the training set, but they can fail for other systems. On the other hand the nonempirical functional usually exhibits a more uniform accuracy. The accuracy of approximate exchange–correlation functionals is limited by their form, i.e., there is a certain maximum accuracy that can be expected for local, semilocal, etc., functionals.

5.11.1 Local and Semilocal Functionals

Functionals such as LSDA (local spin density approximation), GGA (generalized gradient approximation), and meta-GGA have already been described in the context of ground state DFT. Semilocal functionals have the form

$$E^{\mathrm{XC}} = \int d\mathbf{r}\, f(\rho_\alpha(\mathbf{r}), \rho_\beta(\mathbf{r}), \nabla\rho_\alpha(\mathbf{r}), \nabla\rho_\beta(\mathbf{r}), \dots) \qquad (5.11.1)$$

In the LSDA, f depends on the spin densities at $\mathbf{r}$ only. In GGA, f also depends on the gradient of the spin densities. In meta-GGA functionals, f depends on additional local information such as the kinetic energy density or the Laplacian of the density.

An exchange–correlation functional, BmLBLYP has been proposed and assessed by Imamura and Nakai [40], to describe both core- and valence-excited states with high accuracy in the TDDFT calculations. The BmLBLYP functional is designed to adopt the modified van Leeuwen–Baerends (mLB) and the Becke 88 (B88) exchange functionals, combined with the LYP correlation functional. The combination of BmLBLYP is based on the analysis that the LB94 functional behaves better for the core excitations than the B88 functional, while the opposite is true for the valence excitations. Numerical assessment confirms the high accuracy and wide applicability of the BmLBLYP functional.

5.11.2 Hybrid Functionals

Popular hybrid functionals such as B3LYP, BH&HLYP, PBE0, B97-1, B97-2, M05 and M06 functionals have been described in Section 5.7.5. Hybrid functionals interpolate between HF theory and semilocal functionals and the fraction of HF exchange is controlled by the exchange-mixing parameter. This interpolation leads to an error compensation for many properties.

5.11.3 Asymptotic Corrections

The exchange–correlation potentials of semilocal functionals decay too fast in the asymptotic region outside a molecule. In most cases, the decay is exponential, instead of the correct $-1/r$ decay. As a result, diffuse excited states are often predicted too low in energy and higher Rydberg excitations may be absent from the bound spectrum. Various correction schemes have been suggested to remedy this problem.

5.11.4 Optimized Effective Potential (OEP)-Based Functionals

Orbital functionals within an ensemble OEP (optimized effective potential) method proposed by van Faassen and de Boeij [34] and Lanczos [35] involve the calculation of Exact Exchange potential. Full OEP calculations of the frequency-dependent exchange kernel have been reported for solids [41–42]. In most TDDFT applications, KS orbitals and orbital energies from an OEP calculation are combined with adiabatic LSDA or GGA exchange–correlation kernels.

5.11.5 Current-Dependent Functionals

Some deficiencies of semilocal functionals can be cured by using the current density j instead of the density. Vignale and Kohn [43,44] have shown that the time-dependent exchange correlation vector potential of weakly inhomogeneous systems possesses a gradient expansion as a functional of j but not of ρ. Current-dependent functionals can account for macroscopic polarization effects in solids which are ultra-nonlocal in the density.

5.12 Advantages of TDDFT

Some of the advantages of TDDFT are that

1. It is formally exact and further improvement of the XC functionals is still possible. On the other hand, improvement of wavefunction-based methods can only be done only at huge cost in computational time.
2. It is still computationally more efficient and scales better than *ab initio* methods.
3. It can be used for large systems (up to 1000 atoms).
4. It can easily be combined with MD though the total number of atoms that it can handle is limited to a few hundred.

References

[1] M. Gell-Mann, K.A. Brueckner, Phys. Rev. 106 (1957) 367.
[2] D.M. Caperley, B.J. Alder, Phys. Rev. Lett. 45 (1980) 566.
[3] O. Gunnarsson, B.I. Lundquist, Phys. Rev. B13 (1976) 4274.
[4] J.P. Perdew, A. Zunger, Phys. Rev. B23 (1981) 504.
[5] J.P. Perdew, Y. Wang, Phys. Rev. 845 (1992) 13244.
[6] S.J. Vosko, L. Wilk, M. Nusair, Can. J. Phys. 58 (1980) 1200.
[7] A.D. Becke, Phys. Rev. A38 (1988) 3098.
[8] J. Tao, J.P. Perdew, V.N. Staroverov, G.E. Scuseria, Phys. Rev. Lett. 91 (2003) 146401.
[9] A.D. Becke, J. Chem. Phys. 98 (1993) 5648.
[10] P.J. Stephens, F.J. Devlin, C.F. Chabalowski, H.J. Frisch, J. Phys. Chem. 98 (1994) 11623.
[11] A.D. Becke, J. Chem. Phys. 98 (1993) 1372.
[12] A.D. Becke, J. Chem. Phys. 107 (1997) 8554.
[13] F.A. Hamprecht, A.J. Cohen, D.J. Tozer, N.C. Handy, J. Chem. Phys. 109 (1998) 6264.
[14] Y. Zhao, N.E. Schultz, D.G. Truhlar, J. Chem. Theory Comput. 2 (2006) 364.
[15] B. Santra, Density-Functional Theory Exchange–Correlation Functionals for Hydrogen Bonds in Water (M.Sc. thesis), Berlin, 2010. www.chem.ucl.ac.uk/ice/docs/santra_biswajit.pdf.
[16] A.J. Cohen, P. Mori-Sanchez, W. Yang, Chem. Rev. 112 (2012) 289–320.
[17] A.D. Boese, J.M.L. Martin, N.C. Handy, J. Chem. Phys. 119 (2003) 3005.
[18] Y. Zhao, N.E. Schultz, D.G. Truhlar, J. Chem. Phys. 123 (2005) 161103.
[19] V.P. Gupta, Spectrochim. Acta A67 (2007) 870.
[20] A.P. Cox, I.F. Thomas, J. Sheridan, Spectrochim. Acta 15 (1959) 542.
[21] A.L.L. East, W.D. Allen, S.J. Klippenstein, J. Chem. Phys. 102 (1995) 8506.
[22] K. Burke, J. Chem. Phys. 136 (2012) 150801.
[23] Y. Andersson, D.C. Langreth, B.L. Lundquist, Phys. Rev. Lett. 76 (1996) 102.
[24] S. Grimme, J. Comput. Chem. 27 (2006) 1787.
[25] A. Tkatchenko, M. Scheffler, Phys. Rev. Lett. 102 (2009) 073005.
[26] W.T. Yang, P.W. Ayers, Q. Wu, Phys. Rev. Lett. 92 (2004) 146404.
[27] R.A. Donnelly, R.G. Parr, J. Chem. Phys. 69 (1978) 4431.
[28] S. Grimme, Chem. Phys. Lett. 259 (1996) 128.
[29] E. Runge, E.K.U. Gross, Phys. Rev. Lett. 52 (12) (1984) 997.
[30] R. van Leeuwen, Phys. Rev. Lett. 80 (1998) 1280.
[31] A. Görling, Phys. Rev. A 47 (1999) 3359.
[32] J.P. Perdew, M. Levy, Phys. Rev. B 31 (1985) 6264.
[33] A. Theophilou, J. Phys. C 12 (1979) 5419.
[34] M. van Faassen, P.L. de Boeij, J. Chem. Phys. 120 (2004) 8353.
[35] C. Lanczos, J. Res. Nat. Bur. Stand. 45 (1950) 255.
[36] M.E. Casida, J. Mol. Struct. Theochem 914 (2009) 3.

[37] M.E. Casida, Time-dependent density-functional response theory for molecules, in: D.P. Chong (Ed.), Recent Advances in Density Functional Methods Part I, World Scientific, Singapore, 1995, p. 155.

[38] M.A. Marques, E.K.U. Gross, Annu. Rev. Phys. Chem. 55 (2004) 427.

[39] M.E. Casida, K.C. Casida, D.R. Salahub, Int. J. Quantum Chem. 70 (1998) 933.

[40] Y. Imamura, H. Nakai, Chem. Phys. Lett. 419 (2006) 297.

[41] Y.H. Kim, A. Görling, Phys. Rev. Lett. 89 (2002) 096402.

[42] T. Grabo, T. Kreibich, S. Kurth, E.K.U. Gross, in: V.I. Anisimov (Ed.), Strong Coulomb Correlations in Electronic Structure Calculations: Beyond the Local Density Approximation, Gordon and Breach, Tokyo, 2000, p. 203.

[43] G. Vignale, W. Kohn, Phys. Rev. Lett. 77 (1996) 2037.

[44] G. Vignale, C.A. Ullrich, S. Conti, Phys. Rev. Lett. 79 (1997) 4878.

Further Reading

[1] G.C. Schatz, M.A. Ratner, Quantum Mechanics in Chemistry, Dover, 2002.

[2] D. Sholl, J.A. Stecke, Density Functional Theory: A Practical Introduction, Wiley, 2009.

[3] D.A. McQuarrie, Quantum Chemistry, Viva Books, New Delhi, 2003.

[4] R.G. Parr (1921), Density-functional Theory of Atoms and Molecules, Oxford University Press, USA, 1989.

[5] A. Leach, Molecular Modelling: Principles and Applications Pearson Education EMA, 2001.

[6] T. Veszprémi, M. Fehér, Quantum Chemistry: Fundamentals to Applications, Springer, India, 2008.

6 Electron Density Analysis and Electrostatic Potential

CHAPTER OUTLINE

6.1 Electron Density Distribution

In Chapter 3, we discussed the molecular orbital (MO) theory for polyatomic systems and introduced Born's probabilistic interpretation of the wavefunction according to which $\Psi^*(1, 2, 3,\ldots N)\ \Psi(1, 2, 3,\ldots N)d\tau$ gives the probability of finding the first particle between r_1 and $r_1 + dr_1$, second between r_2 and $r_2 + dr_2$, and the nth between r_n and $r_n + dr_n$, where r_i stands for all the space and spin coordinates of particle i. In order to obtain the electron distribution, we need to express the probability of an electron being

Principles and Applications of Quantum Chemistry. http://dx.doi.org/10.1016/B978-0-12-803478-1.00006-6

present in a given infinitesimal volume in space. Thus, the probability of finding electron 1 in the volume of space between r_1 and $r_1 + dr_1$ can be written as

$$p_1 = \int_{-\infty}^{+\infty}\int_{-\infty}^{+\infty}\cdots\int_{-\infty}^{+\infty} \Psi^*(1,2,\ldots N)\Psi(1,2,\ldots N)dr_2dr_3\ldots dr_N \quad (6.1.1)$$

In Eq. (6.1.1), the integration is over all other electrons except electron 1, because for all the others, the integration over all space will give unit. Similarly, the probability of finding electron 2 in the volume element between r_2 and $r_2 + dr_2$ shall be obtained by integrating over all space for all other electrons except electron 2,

$$p_2 = \int_{-\infty}^{+\infty}\int_{-\infty}^{+\infty}\cdots\int_{-\infty}^{+\infty} \Psi^*(1,2,\ldots N)\Psi^*(1,2,\ldots N)dr_1dr_3\ldots dr_N \quad (6.1.2)$$

Since the electrons are indistinguishable and the probability of finding an electron in the same volume will be the same. We expect,

$$p_1 = p_2 = \ldots = p_N$$

The probability of finding a single electron out of N, in a given volume of space will therefore be the sum of the probabilities for the individual electrons,

$$\rho(\mathbf{r}) = N\int_{-\infty}^{+\infty}\int_{-\infty}^{+\infty}\cdots\int_{-\infty}^{+\infty} \Psi^*(1,2,\ldots N)\Psi(1,2,\ldots N)dr_2dr_3\ldots dr_N \quad (6.1.3)$$

Here, $\rho(\mathbf{r})$ tells us about the probabilistic distribution of electrons. $\rho(\mathbf{r})$ is known as electron probability distribution function or electron density function.

If we integrate the electron density function over all space, then

$$\int \rho(\mathbf{r})dr = N\int_{-\infty}^{+\infty}\int_{-\infty}^{+\infty}\cdots\int_{-\infty}^{+\infty} \Psi^*(1,2,\ldots N)\Psi(1,2,\ldots N)dr_1dr_2\ldots dr_n \quad (6.1.4)$$

and if the wavefunctions are normalized then,

$$\int \rho(\mathbf{r})dr = N \quad (6.1.5)$$

or the integration of the electron density function over all space equals the number of electrons N.

Electron density function perfectly describes the charge distribution in a molecule. However it is rather complicated and cumbersome to use. It will be simpler to work out a method that would divide the total charge between atoms. Assigning charges to atoms can be very useful as they can give a rough idea of the charge distribution in a molecule. This can be achieved if the total electron density is expanded in terms of the MOs and then each of them is expanded in terms of a set of basis sets (atomic orbitals). Thus,

$$\rho(\mathbf{r}) = \sum_i n_i\rho_i(\mathbf{r}) \quad (6.1.6)$$

where $\rho_i(\mathbf{r}) = \Psi_i^*(\mathbf{r})\Psi_i(\mathbf{r})$ and n_i accounts for the occupation of the ith orbital. For a closed system of N electrons occupying $N/2$ real orbitals, we can write

$$\rho(\mathbf{r}) = 2\sum_{i=1}^{N/2}\Psi_i^*(r)\Psi_i(r) \tag{6.1.7}$$

If we express the MO Ψ_i as a linear combination of the basis set of K function $\{\Phi_1,\Phi_2...\Phi_K\}$, that is,
$\Psi_i = \sum_{\mu=1}^{K} C_{\mu i}\Phi_\mu(\mathbf{r})$ then from Eq. (6.1.7)

$$\begin{aligned}\rho(r) &= 2\sum_{i=1}^{N/2}\left(\sum_{\mu=1}^{K} C_{\mu i}\Phi_\mu(\mathbf{r})\right)\left(\sum_{\nu=1}^{K} C_{\nu i}\Phi_\nu(\mathbf{r})\right) = 2\sum_{i=1}^{N/2}\sum_{\mu=1}^{K} C_{\mu i}C_{\mu i}\Phi_\mu(\mathbf{r})\Phi_\mu(\mathbf{r})\\ &+ 2\sum_{i=1}^{N/2}\sum_{\mu=1}^{K}\sum_{\nu=\mu+1}^{K} 2C_{\mu i}C_{\nu i}\Phi_\mu(\mathbf{r})\Phi_\nu(\mathbf{r})\end{aligned} \tag{6.1.8}$$

where the first term corresponds to $v = \mu$ and the second to $v > \mu$.

This equation can be written in a simpler form by introducing the so-called density matrix elements

$$P_{\mu\nu} = 2\sum_{i=1}^{N/2} C_{\mu i}C_{\nu i} \tag{6.1.9}$$

Then,

$$\rho(\mathbf{r}) = \sum_{\mu=1}^{K}\sum_{\nu=1}^{K} P_{\mu\nu}\Phi_\mu(\mathbf{r})\Phi_\nu(\mathbf{r}) = \sum_{\mu=1}^{K} P_{\mu\mu}\Phi_\mu(\mathbf{r})\Phi_\mu(\mathbf{r}) + 2\sum_{\mu=1}^{K}\sum_{\nu=\mu+1}^{K} P_{\mu\nu}\Phi_\mu(\mathbf{r})\Phi_\nu(\mathbf{r}) \tag{6.1.10}$$

Using Eq. (6.1.5), we get

$$\begin{aligned}N &= \int \rho(\mathbf{r})d\mathbf{r} = \sum_{\mu} P_{\mu\mu}\int \Phi_\mu(\mathbf{r})\Phi_\mu(\mathbf{r})d\mathbf{r} + 2\sum_{\mu=1}^{K}\sum_{\nu=\mu+1}^{K} P_{\mu\nu}\int \Phi_\mu(\mathbf{r})\Phi_\nu(\mathbf{r})d\mathbf{r}\\ &= \sum_{\mu=1}^{K} P_{\mu\mu}S_{\mu\mu} + 2\sum_{\mu=1}^{K}\sum_{\nu=\mu+1}^{K} P_{\mu\nu}S_{\mu\nu}\end{aligned} \tag{6.1.11}$$

where

$$S_{\mu\nu} = \int \Phi_\mu(\mathbf{r})\Phi_\nu(\mathbf{r})d\tau \tag{6.1.12}$$

is the overlap matrix. If the basis functions are taken to be normalized, then $S_{\mu\mu} = 1$. Therefore,

$$N = \sum_{\mu=1}^{K} P_{\mu\mu} + 2\sum_{\mu=1}^{K}\sum_{\nu=\mu+1}^{K} P_{\mu\nu}S_{\mu\nu} \tag{6.1.13}$$

From Eq. (6.1.13) it follows that it is possible to equate the total number of electrons in a molecule to a sum of products of density matrix and overlap matrix elements. It is reasonable (but not necessarily correct) to assign electrons associated with a particular diagonal elements, $P_{\mu\mu}$, to that atom on which the basis function Φ_μ is located. It is also reasonable to assign electrons associated with off-diagonal elements, $P_{\mu\upsilon}$, where both Φ_μ and Φ_υ reside on the same atom, to that atom. However, it is not apparent how to partition electrons from density matrix element, $P_{\mu\upsilon}$, when Φ_μ and Φ_υ reside on different atoms. This leads to the concept of population analysis.

6.2 Population Analysis

The distribution of charges within molecules is conducted through population analysis. The intention is to work out the electron charge density on each atom and in the region between the atoms. Partial atomic charges, that is, charge on each atom, are not the observable characteristics of molecules and the idea of modeling electron population is not unique. Several different methods have been suggested to divide $P_{\mu\upsilon}S_{\mu\upsilon}$ in Eq. (6.1.13) to obtain numbers that tell us where the electron density in a system resides. This procedure is called population analysis. Population methods also assign each atom a partial charge. Each atomic center is a positively charged core (Atomic number, Z_A) surrounded by a shielding electron cloud. So, once we know the electron density on each atom, we can determine the atomic partial charge (Q_A),

$$Q_A = Z_A - \int \rho_A(\mathbf{r})d\mathbf{r} \tag{6.2.1}$$

There is no observable property associated with the partial charge and so there is no way to evaluate its accuracy. The charge can be distributed in qualitatively different ways by different methods. It is therefore very important not to over interpret data provided by population analysis methods. Some of the commonly used population analysis methods are the Mulliken method [1], the Löwdin [2] method, the natural bonding orbitals analysis, and methods based on electrostatic potential such as CHelpG. Partial charges are also calculated by other methods based on only the charge density such as by Bader analysis and Hirshfeld analysis. These will be taken up in Chapter 11.

6.2.1 Mulliken Population Analysis

Mulliken analysis is the most commonly used population analysis method and is popular because of its simplicity. The starting point is Eq. (6.1.13) which relates total number of electrons to density matrix and to overlap integrals. In this method, all of the electron density ($P_{\mu\mu}$) in an orbital is assigned to the atom on which Φ_μ is located. The remaining electron density is associated with the overlap population $\Phi_\mu\Phi_\upsilon$.

Mulliken arbitrarily assigned half of the electron density of the overlap matrix to each atom, regardless of the properties like electronegativity. Thus, half of the density is assigned to atom on which Φ_μ is located and half to the atom on which Φ_υ is located. The gross electron population q_μ for the basis function Φ_μ is therefore given as

$$q_\mu = P_{\mu\mu} + \sum_{\substack{\mu=1 \\ \mu\in A}}^{K} \sum_{\substack{\nu=1 \\ \nu\neq\mu}}^{K} P_{\mu\nu}S_{\mu\nu} \tag{6.2.2}$$

where $\mu\in$A stands for μ on atom A.

The atomic electron population q_A on atom A is therefore given as

$$q_A = \sum_{\mu \in A}^{K} q_\mu \tag{6.2.3}$$

and atomic charge as

$$Q_A = Z_A - q_A \tag{6.2.4}$$

The total overlap population between atoms A and B is given as

$$p_{AB} = \sum_{\mu \in A} \sum_{\nu \in B} P_{\mu\nu} S_{\mu\nu} \tag{6.2.5}$$

p_{AB} gives quantitative information on binding between two atoms. A large positive value indicates a large electron population and hence strong binding. On the other hand, a negative value shows that the electrons are displaced from the bond and hence its antibonding character.

The Mulliken population analysis method is computationally cheap and is available in most software programs for molecular modeling but it suffers from the major disadvantage that the populations are strongly basis dependent and have different values for the same system when different basis sets are used. Computations using different basis sets cannot, therefore, be compared. For example, the change of basis set from 6-31G* to 6-311++G** can cause the charge on the central carbon atom in isobutane to change from +0.1 to +1.0. There is no way to account for differences in electronegativities of atoms within the molecule because this method always equally distributes shared electrons between two atoms. This equal distribution of shared electrons can lead to quite unrealistic values for the net atomic charges. In some extreme cases, some orbitals may contain a negative number of electrons and in others, more than two electrons. Atomic charges on different atoms of 2-iminomalononitrile obtained by Mulliken population analysis are given in Table 6.1 and compared with values obtained by different methods.

6.2.2 Löwdin Population Analysis

The instability of predicted charges with increasing basis sets, which is the main drawback of the Mulliken theory, was corrected by Löwdin by transforming the atomic orbital basis functions into an orthonormal set of basis functions before conducting population analysis. The transformed orbitals Φ'_μ in the orthogonal set are given by

$$\Phi'_\mu = \sum_{\nu=1}^{K} \left(S^{-1/2}\right)_{\nu\mu} \Phi_\nu \tag{6.2.6}$$

and the atomic electron population becomes,

$$q_A = \sum_{\mu=1, \mu \in A}^{K} \left(S^{1/2} P S^{1/2}\right)_{\mu\mu} \tag{6.2.7}$$

Löwdin method avoids the problem of negative populations or populations greater than 2 and gives charges close to chemically intuitive values that are less sensitive to basis set. However, it still does not account for electronegativity of different atoms.

6.2.3 Natural Bonding Orbitals (NBO) and Natural Population Analysis (NPA)

The MOs that result from a quantum chemical calculation do not appear to be based on the so-called directed bond interpretation of chemical structure. Such orbitals are called delocalized MOs. The theories that give rise to the delocalized interpretation of bonding are well grounded and mathematical and cannot be ignored. The delocalized MOs reproduce the total molecular density exactly but they are not the only orbitals that can produce the same density. The delocalized MOs can be localized or transformed to conform to the empirical ideas of directional bonds. We can write the new MOs as a linear combination of the old MOs.

$$\Psi_{\text{new}} = c_1\Psi_1 + c_2\Psi_2 + \ldots + c_n\Psi_n$$

Localization refers to how we determine the new coefficients, the *c*'s in the above equation. There are several localization procedures each of which generates its own values of *c*. Thus, for example, Boys' localization method [3] depends on minimizing the spatial extent of the orbitals, as much as possible. Some other localization methods are due to Edmiston and Ruedenberg [4] and von Niessen [5]. The most general and nonarbitrary method [4] is based on the minimization of interorbital interactions. One localization and population analysis method that is very popular is the NBO analysis. This method can determine natural atomic orbitals (NAOs) that are the effective orbitals of an atom in the particular molecular environment (instead of the isolated and the gas phase environment). These are also the maximum occupancy orbitals. The NBOs are localized few center MOs that reflect Lewis-like bonding structure.

Natural bond analysis classifies and localizes orbitals into three distinct groups; nonbonding NAOs, orbitals involved in bonding and antibonding (NBOs), and Rydberg-type orbitals. In an initial step, orbitals that are associated almost entirely with a single atom, e.g., core orbitals and have lone pairs are localized and form the so-called NAOs. Next, orbitals involving bonding (or antibonding) between pairs of atoms are localized by using only the basis set AOs of those atoms. Finally, the Rydberg-like orbitals are identified and all orbitals are made orthogonal to one another. The result is that, except for very small contributions from other atomic orbitals to ensure orthogonality, all NAOs and Rydberg orbitals are described using the basis set AOs of a single atom and all the NBOs are described using the basis set AOs of two atoms. This is similar to our notion of core electrons, lone pair of electrons, and valence electrons, and works under the assumption that only the bonding orbitals should be made by combinations of two atoms basis sets. However, in those cases where resonance or other delocalization effects require orbitals delocalized over more than two atoms, additional efforts are required. Working on this model of electron partitioning, natural population analysis (NPA) then treats the NBOs just as Mulliken method treats all the orbitals. Results of NPA for 2-iminomalononitrile are given in Table 6.1.

Table 6.1 Comparison of Mulliken, NBO, and CHELPG charges of 2-iminomalononitrile by different DFT and RHF methods

	Mulliken		NBO									
	B3LYP/6-31G*	RHF/6-31G*	B3LYP/6-31G*					RHF/6-31G*				
Atom no.	Atomic charge	Atomic charge	Atomic charge	Core	Valence	Rydberg	Total	Atomic charge	Core	Valence	Rydberg	Total
1C	0.2532	0.2971	0.0750	1.9989	3.9075	0.0186	5.9250	0.1118	1.9989	3.8670	0.0223	5.8882
2C	0.3516	0.2964	0.2201	1.9993	3.7499	0.0307	5.7799	0.2468	1.9993	3.7226	0.0314	5.7533
3C	0.3847	0.3388	0.2489	1.9993	3.7205	0.0314	5.7512	0.2719	1.9993	3.6960	0.0328	5.7281
4 N	−0.4259	−0.3969	−0.2227	1.9997	5.2017	0.0213	7.2227	−0.2610	1.9996	5.2343	0.0270	7.2610
5 N	−0.4257	−0.4021	−0.2232	1.9997	5.2027	0.0209	7.2232	−0.2664	1.9996	5.2401	0.0267	7.2664
6 N	−0.4740	−0.5196	−0.4813	1.9995	5.4611	0.0208	7.4813	−0.4972	1.9995	5.4729	0.0248	7.4972
7H	0.3360	0.3863	0.3833	0.0000	0.6156	0.0012	0.6167	0.3940	0.0000	0.6052	0.0009	0.6060
Sum of charges	0.0000	0.0000	0.0000	11.9963	27.8590	0.1447		0.0000	11.9962	27.8380	0.1658	

	CHELPG	
	B3LYP/6-31G*	RHF/6-31G*
Atom no.	Atomic charge	Charge
1C	0.3838	0.4514
2C	0.2108	0.2146
3C	0.2802	0.2889
4 N	−0.3274	−0.3548
5 N	−0.3454	−0.3805
6 N	−0.5701	−0.6201
7H	0.3682	0.4004
Sum of charges	0.0000	0.0000

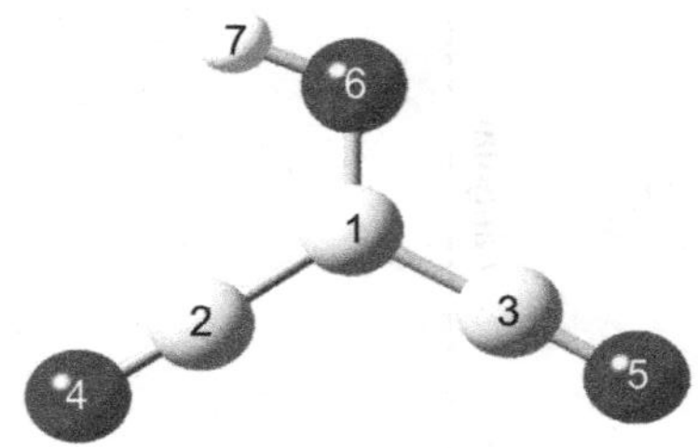

Table 6.2(a) Natural population analysis–natural atomic orbital occupancies in methyleneimine at B3LYP/6-31G* level

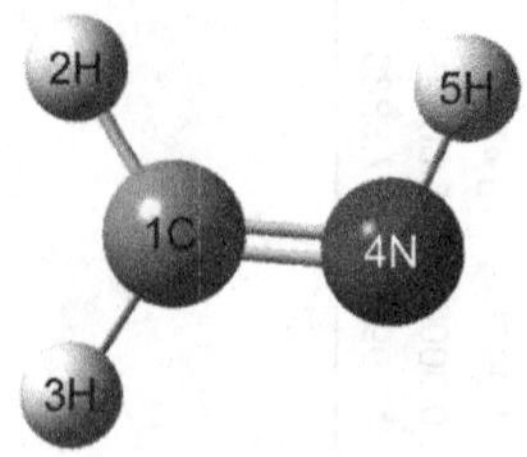

NAO	Atom	No	*l* angular	Type(AO)	Occupancy	Energy
1	C	1	*s*	Cor(1*s*)	1.99954	−10.11330
2	C	1	*s*	Val(2*s*)	1.04994	−0.24691
3	C	1	*s*	Ryd(3*s*)	0.00478	0.94859
4	C	1	*s*	Ryd(4*s*)	0.00004	3.84744
5	C	1	p_x	Val(2*p*)	0.98275	−0.03027
6	C	1	p_x	Ryd(3*p*)	0.00746	0.51603
7	C	1	p_y	Val(2*p*)	1.17929	−0.05873
8	C	1	p_y	Ryd(3*p*)	0.00202	0.72145
9	C	1	p_z	Val(2*p*)	0.86214	−0.12845
10	C	1	p_z	Ryd(3*p*)	0.00147	0.53607
11	C	1	d_{xy}	Ryd(3*d*)	0.00229	2.40018
12	C	1	d_{xz}	Ryd(3*d*)	0.00079	2.12413
13	C	1	d_{yz}	Ryd(3*d*)	0.00001	1.68354
14	C	1	$d_{x^2-y^2}$	Ryd(3*d*)	0.00072	2.32862
15	C	1	d_{z^2}	Ryd(3*d*)	0.00047	2.14021
16	H	2	*s*	Val(1*s*)	0.83019	0.03833
17	H	2	*s*	Ryd(2*s*)	0.00251	0.59695
18	H	3	*s*	Val(1*s*)	0.80889	0.05856
19	H	3	*s*	Ryd(2*s*)	0.00159	0.60449
20	N	4	*s*	Cor(1*s*)	1.99954	−14.14946
21	N	4	*s*	Val(2*s*)	1.50897	−0.57487
22	N	4	*s*	Ryd(3*s*)	0.00200	1.26228
23	N	4	*s*	Ryd(4*s*)	0.00003	3.67125
24	N	4	p_x	Val(2*p*)	1.36475	−0.17959
25	N	4	p_x	Ryd(3*p*)	0.00388	0.79086
26	N	4	p_y	Val(2*p*)	1.58559	−0.18455
27	N	4	p_y	Ryd(3*p*)	0.00512	0.92586
28	N	4	p_z	Val(2*p*)	1.13224	−0.17776
29	N	4	p_z	Ryd(3*p*)	0.00000	0.74739
30	N	4	d_{xy}	Ryd(3*d*)	0.00258	2.02823
31	N	4	d_{xz}	Ryd(3*d*)	0.00234	1.77263
32	N	4	d_{yz}	Ryd(3*d*)	0.00102	1.70736
33	N	4	$d_{x^2-y^2}$	Ryd(3*d*)	0.00384	2.26621
34	N	4	d_{z^2}	Ryd(3*d*)	0.00242	2.04408
35	H	5	*s*	Val(1*s*)	0.64675	0.10360
36	H	5	*s*	Ryd(2*s*)	0.00204	0.63006

Table 6.2(b) Summary of natural population analysis

Atom	No.	Natural Charge	Natural Population			
			Core	Valence	Rydberg	Total
C	1	−0.09371	1.99954	4.07412	0.02005	6.09371
H	2	0.16729	0.00000	0.83019	0.00251	0.83271
H	3	0.18952	0.00000	0.80889	0.00159	0.81048
N	4	−0.61431	1.99954	5.59154	0.02323	7.61431
H	5	0.35121	0.00000	0.64675	0.00204	0.64879
Total		0.00000	3.99908	11.95149	0.04943	16.00000

Table 6.2(c) Contribution of atomic orbitals to bond formation based on NPA

S. No.	Occupancy	Bond	Atom	Contribution (%)	Orbital/Coefficients/Hybrids (%)
1	1.99624	BD (1) C_1–H_2	C_1	59.42	*s* 0.77 (33.56) *p* 1.98 (66.37) *d* 0.00 (0.07)
			H_2	40.58	*s* 0.64 (100.00)
2	1.98571	BD (1) C_1–H_3	C_1	60.09	*s* 0.77 (31.43) *p* 2.18 (68.50) *d* 0.00 (0.07)
			H_3	39.91	*s* 0.63 (100.00)
3	2.00000	BD (1) C_1–N_4	C_1	43.22	*s* 0.66 (0.00) *p* 1.00 (99.1) *d* 0.00 (0.09)
			N_4	56.78	*s* 0.75 (0.00) *p* 1.00 (99.70) *d* 0.00 (0.30)
4	1.99895	BD (2) C_1–N_4	C_1	40.92	*s* 0.64 (35.19) *p* 1.84 (64.75) *d* 0.00 (0.06)
			N_4	59.08	*s* 0.77 (40.46) *p* 1.46 (59.26) *d* 0.01 (0.28)
5	1.97720	BD (1) N_4–H_5	N_4	67.68	*s* 0.83 (21.05) *p* 3.74 (78.73) *d* 0.01 (0.22)
			H_5	32.32	*s* 0.57 (100.00)

NPA differentiates between the orbitals that will overlap to form a bond and those that are too near the core of an atom to be involved in bonding. This results in the stabilization of the atomic partial charge as the basis set size is increased. The NBO localization scheme permits the assignment of hybridization both to the atomic lone pairs and to each atom's contribution to its bond orbitals. With NBO analysis, the percentage of *s, p, d, f,* etc., contribution becomes immediately evident from the coefficients of the AO basis functions from which the NAO or NBO is formed. This will be evident from Table 6.2 for methyleneimine which is based on DFT calculations using G09W software at the B3LYP/6-31G* level. This method is, however, computationally more expensive than the Mulliken or Löwdin methods. As with any population analysis method, NPA is best used for comparing differences rather than in determining absolute atomic charges. Thus, it is favored for understanding the trend of charges as a function of some variables.

6.3 Electrostatic Potential

The electrostatic potential at a point **r** is defined as the work done in bringing a unit positive charge from infinity to that point. The molecular electrostatic potential (ESP), also abbreviated as MEP or MESP and used interchangeably, is defined as the energy of interaction of a positive point charge with a molecule. ESP has contributions from both the nuclei and electrons and is given by the formula;

$$V_{\mathrm{ESP}} = \sum_A \frac{Z_A}{|R_A - \mathbf{r}|} - \int \frac{\rho(\mathbf{r}')d\mathbf{r}'}{|\mathbf{r}' - \mathbf{r}|}, \tag{6.3.1}$$

where Z_A is the nuclear charge of atom A located at position R_A, **r** is the position of the positive charge, and $\rho(\mathbf{r}')$ is the electron density at point $\mathbf{r}'$. Potential due to electron is obtained by appropriate integration of electron density. A positive value of V_{ESP} specifies repulsion ($V_{\mathrm{ESP}} > 0$) and a negative value specifies attraction. If we insert Eq. (6.1.10) for electron density in Eq. (6.3.1), we get an expression in terms of atomic orbitals that form the basis set

$$V_{\mathrm{ESP}} = \sum_A \frac{Z_A}{|R_A - \mathbf{r}|} - \sum_{\mu}^{AO} \sum_{\nu}^{AO} P_{\mu\nu} \int \frac{\Phi_\mu(\mathbf{r}')\Phi_\nu(\mathbf{r}')}{|\mathbf{r}' - \mathbf{r}|} d\mathbf{r}', \tag{6.3.2}$$

where $P_{\mu v}$ are the elements of the density matrix. The second summation is over all orbitals of the system. The integral in the second term represents Coulomb interaction between the probe positive charge and the electrons.

The electrostatic potential may be calculated uniquely from the electronic wavefunctions. Since ESP is related to electron density, it is a very useful descriptor in understanding electrophilic and nucleophilic attacks in chemical reaction. Thus, if we imagine a positive charge approaching a molecule, it will try to avoid regions where the potential is positive and favor negative, more electron-rich parts. This is a simplified model of electrophilic attack because of two reasons: (1) ESP correlates with dipole moment, partial charges, and site of chemical reactivity and provides a visual method to understand the relative polarity of a molecule, (2) It helps identifying regions of local negative and positive potential in a molecule. Since electrostatic forces are primarily responsible for long range interactions between molecules, ESP can rationalize intermolecular interactions between polar species. Noncovalent intramolecular interactions which are known to occur at distances equal to the sum of van der Waals radii are also reflected in ESP surfaces and so examination of ESP can be used to understand such interactions (Figure 6.1) [6].

6.3.1 Electron Density and ESP Isosurfaces

A combined plot of the electron density isosurface and ESP isosurface is an important visual method for indicating the reactive sites of a molecule. By definition, electron density isosurface is a three-dimensional surface at which molecule's electron density has a particular value and that encloses a specified fraction of its electron probability density. In order to construct an electron density isosurface, the electron density is

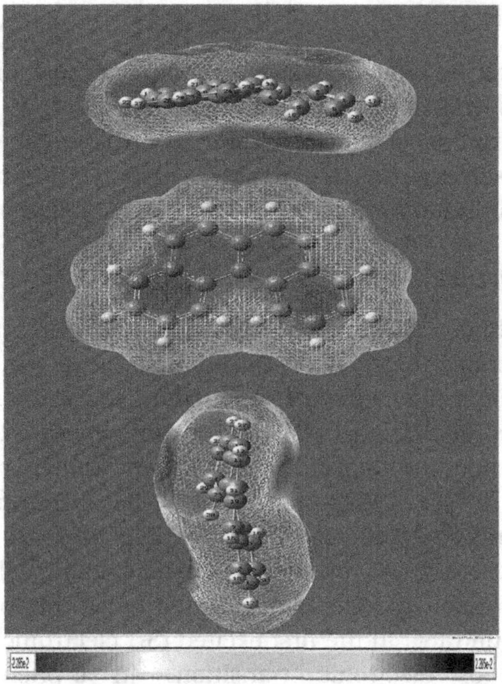

FIGURE 6.1 Molecular electrostatic potential mapped on the $\rho(r) = 0.0004$ au isodensity surface in the range from −2.285e−2 (red (gray in print versions)) to +2.285e−2 (blue (black in print versions)) for benzo[c]phenanthrene in three different orientations calculated at the B3LYP/6-31G** level of theory.

calculated at a large number of grid points within a three-dimensional cube; the points within the cube having the same given value of the electron density are then connected with lines to give the isosurface. The same is done in the case of potential isosurface except that now we consider ESP instead of electron density. The electrostatic potential at different points on the electron density isosurface is shown by coloring the isosurface with contours. Red represents regions of most negative electrostatic potential, blue represents regions of most positive electrostatic potential, and green represents region of zero potential. Potential increases in the order red < orange < yellow < green < blue. When the molecule is more polar, larger red/blue differences appear. In case of nonpolar molecules, the surface has lighter shades. Figure 6.1 shows for benzo[c]phenanthrene the electron density isosurface on which the electrostatic potential surface has been mapped. Such surfaces depict the size, shape, charge density, and site of chemical reactivity of molecules. Three projections of surfaces—one in the molecular plane and two in perpendicular planes in case of benzo[c]phenanthrene are included in Figure 6.1. These projections clearly show nucleophilic region (blue) sandwiching the π system leaving a more electrophilic region (red) in the plane of the hydrogen atoms. It is also seen that the maps show a region of positive electrostatic potential in the Fjord region of the molecule which includes the intramolecular H−H bond.

6.4 Analysis of Bonding and Interactions in Molecules

A theoretical insight into the interactions and bonding in molecules can be obtained by carrying out four types of analysis. These are: (1) molecular orbital analysis (2) population analysis (3) electron density analysis, and (4) energy analysis.

We shall restrict ourselves to only the first three analyses although the energy analysis has been included in computational packages like ADF (Amsterdam Density Functional).

6.4.1 Molecular Orbital Analysis

The MOs which go to build up the total wavefunction for the system provide valuable insight into the interactions that give rise to bonding. MOs give us more specific and detailed information about the electronic interactions occurring in a molecule and the changes in electronic configuration due to various influences. We cannot see the MOs directly, but we can look at the total molecular density which is just the square of the wavefunction. All the MOs together contribute to the electron density. Changes in molecular conformations and the nature of bonding in 2-oxo-2H-pyran derivatives following electronic excitation was reported by Gupta et al. [7], who used MOs to derive additional information about the electronic charge redistribution in the molecule on excitation from the ground (S_0) to the first excited (S_1) electronic state.

A detailed study of electronic transitions in 6-phenyl-4-methylsulfanyl-2-oxo-2H-pyran (Figure 6.2(a)) shows that its first excited state S_1 is a $^1(\pi\text{–}\pi^*)$ state. Plot of the highest occupied molecular orbital (HOMO) (Figure 6.2(b)) and the lowest unoccupied molecular orbital (LUMO) (Figure 6.2(c)) shows that the S_1 state arises out of π-electron transfer from the region of bond $C_3–C_4$ to $C_6–C_{15}$ followed by redistribution of electron charge. This may explain an increase in the $C_3–C_4$ bond length and a decrease in the $C_6–C_{15}$ bond length on electronic excitation. A study of the optimized structures in the two states shows that the length of bonds $C_1–C_2$, $C_4–C_5$, $C_6–C_{15}$, $C_{16}–C_{17}$, and $C_{19}–C_{20}$ in the S_1 state are shorter by as much as 0.064 Å than the corresponding bonds in the S_0 state. This follows an increase in the lengths of $C_3–C_4$, $C_5–C_6$,

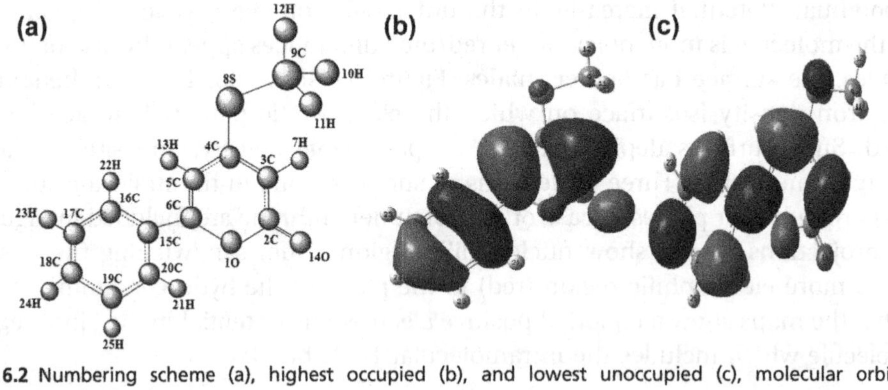

FIGURE 6.2 Numbering scheme (a), highest occupied (b), and lowest unoccupied (c), molecular orbitals of 6-phenyl-4-methylsulfanyl-2-oxo-2H-pyran.

C_{15}–C_{16}, and C_{15}–C_{20} bonds. It is also noted that electronic excitation causes a reduction in the C=O bond length from 1.204 Å (S_0) to 1.184 Å (S_1), possibly due to reduced conjugation of the C=O bond with its neighboring bonds in the S_1 state. This also follows from the shape of MOs in the highest occupied and lowest unoccupied states; the size of the π-electron lobe on the oxygen atom of the carbonyl group in LUMO gets localized and significantly reduced in size as compared to the HOMO where it is broad and close to the π-electron lobe of the conjugated atoms. The dihedral angles $C_1C_6C_{15}C_{20}$ and $C_5C_6C_{15}C_{16}$ between the benzene and pyran rings change from 12.95° and 13.83° in the S_0 state to −0.22° and −0.04°, respectively, in the S_1 state making the two rings almost coplanar in the first excited electronic state. Thus, the molecule is found to be nonplanar in the S_0 state with mean interplanar angle of about 15° but planar in the S_1 state.

6.4.2 Electron Density Analysis

Visualization of electron density gives a better indication of molecular size. For example, a solid state view of 2-iminomalononitrile (Figure 6.3(a)) gives information about the size of the molecule. Plot of electron density contour indicates electron density distribution around each atom of the molecule (Figure 6.3(b)). However, the total electron density does not give details about the rearrangement of electron density that occurs in bond formation. Hunt et al. [8] provided a method for getting this kind of information by subtracting the atomic density of the ground state of contributing atoms from the molecular density for 1-butyl-3-methylimidazolium chloride cation ($bmim^+$) (Figure 6.4). The white sections in the center show regions where the very dense contours have been removed. The accumulation of charge is represented by solid lines and the depletion by dotted lines. It is concluded that as compared to the total electron density, only a very small fraction is involved in bond formation. So chemistry is driven by these very small changes.

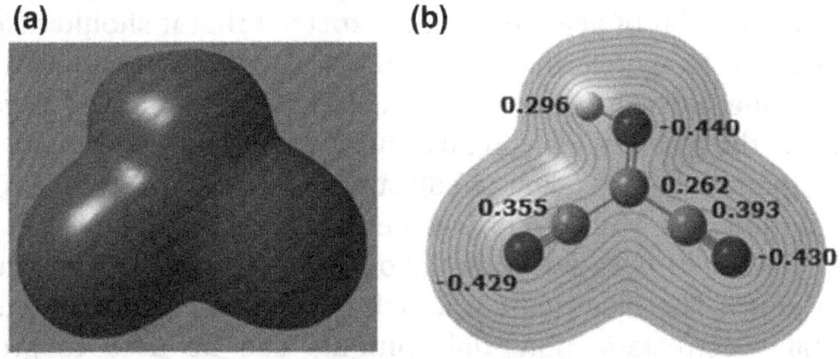

FIGURE 6.3 (a) Solid state view (b) electron density contour of 2-iminomalononitrile.

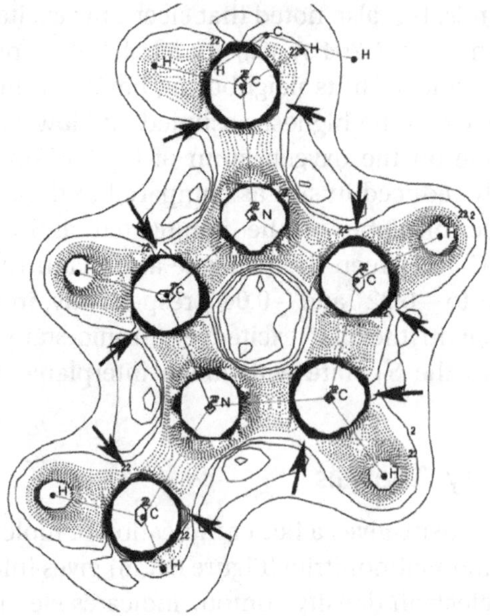

FIGURE 6.4 Electron density difference for 1-butyl-3-methylimidazolium chloride cation (bmim$^+$). *Reprinted with permission from Ref. [8], Copyright @ 2006, John Wiley and Sons Publishing.*

6.4.3 Population Analysis

This mathematical method of partitioning a wavefunction or electron density into charges on each atom has already been described in Section 6.2. The method involves condensing nuclear charge and electron density into an atomic partial charge which can help to understand reactivity, bond orders, etc. The total overlap population between two atoms gives information about bonding; large positive value reflects strong bonding and large negative value reflects antibonding. The information based only on population analysis can sometimes be misleading. Thus, for example, if one can considers protonation reaction of acetamide, it is expected that it should take place at an atom having the highest negative charge (Figure 6.5(a)). Thus, we should have expected it to take place at the nitrogen atom ($q = -0.605$) having the largest negative charge. This is not true and the reaction instead takes place at the oxygen atom ($q = -0.496$). In order to get true picture, help is taken of ESP surfaces. In Figure 6.5(b), an ESP contour map has been plotted over Mulliken charge density for acetamide. It can be seen that there is a major volume of negative potential over the oxygen atom with a smaller one at the nitrogen atom. The electrophilic attack is therefore primarily on the oxygen atom. In general, electrostatic potential contours can be used to propose where

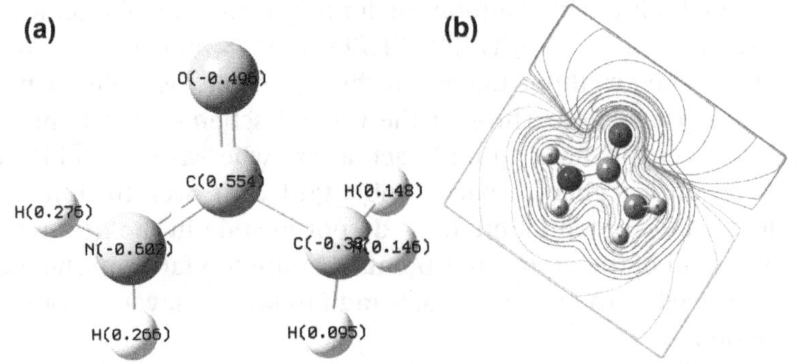

FIGURE 6.5 (a) Charge density distribution in acetamide based on Mulliken analysis, and (b) Mulliken charge density + ESP contour analysis at B3LYP/6-31G level.

electrophilic attack might occur; the electrophiles are often attracted to regions where the electrostatic potential is most negative. The prediction of nucleophilic activity is, however, trickier though ESP still has some limited applicability for predicting the direction of nucleophilic reactions at larger distances from the nuclei as in the case of van der Waals interactions.

6.5 Electrostatic Potential-Derived Charges

Potential-derived charges approach is based on the idea of having a suitable model for reproducing ESP. The principle of this approach is to fit atomic monopoles so that they are able to reproduce ESP defined by Eqs (6.3.1) and (6.3.2). All ESP charge-fitting schemes involve determining atomic partial charges q_k which, when used as a monopole expansion according to the relation,

$$V_{\mathrm{ESP}}(r) = \sum_{k}^{\mathrm{nuclei}} \frac{q_k}{|r - r_k|}, \tag{6.5.1}$$

minimize the difference between V_{ESP} and the correct V_{ESP} Eq. (6.3.2). In practice, a large number of points spaced evenly on a cubic grid surrounding the van der Waals surface of the molecule are selected. In order to ensure rotational invariance, a reasonable density of points is required.

After the calculation of the wavefunction using an appropriate quantum chemical technique, the grid points around the molecule are set and at each of these grid points ESP is evaluated. By using the method of least square minimization the atomic charges are fitted to each atom to reproduce the ESP to a good degree.

Four different methods are in common use for the calculation of atomic charges from electrostatic potential. These are CHELP, CHELPG, MKS, and restrained ESP (RESP) Schemes. All these schemes differ mainly in the choice of the points where the electrostatic potential is calculated. Thus, in the CHELP scheme, the points are selected symmetrically on spherical shells around each atom, whereas in CHELPG method the points are selected on a regularly spaced cubic grid with over 10 times higher point density. The electrostatic potential methods do not include in the fit the points within the van der Waals radii of the atoms, nor points that are too far from the molecule. The exclusion limits, as well as the van der Waals radii themselves, vary appreciably among the different methods.

The charges calculated from the potential-derived methods are less sensitive to the choice of basis set in comparison to the *ab initio* methods. Another positive aspect of ESP-derived charges is that they are able to represent electrostatic moment of the molecule much more reliably than *ab initio* methods. They are also able to describe the nonbonding interactions fairly well. However, these charges are not accurate for larger systems as ESP, is a characteristic of the surface.

6.5.1 CHELPG

CHELPG, the acronym of CHarges from ELectrostatic Potentials using a Grid-based method, is a charge calculation scheme developed by Breneman et al. [9]. Under this scheme the calculations are conducted in two steps. First, electrostatic potential is evaluated at each of the grid points which are 3 pm apart and distributed regularly in a large cube with molecule at the center. In the second step, the least squares fitting algorithm is applied and the charges are fitted to the potentials. The only constraint that is applied in the least square fit to the potential is that the total charge (sum of the obtained charges) has to be equal to the overall charge of the system. Also, there is a restraint on the way the grid is constructed because all the grid points that appear inside the molecule are discarded and only those outside the molecules are taken into account. CHELPG charges are often considered to be superior to Mulliken charges because they are not much dependent on the selection of basis set. The CHELPG scheme however fails in the treatment of larger systems where the internal atoms are far away from the grid points.

6.5.2 Merz–Kollman–Singh (MKS)

In this scheme the whole procedure is again divided into two steps. Firstly, the electrostatic potential (ESP) is calculated at a number of points located on several layers around the molecule. Usually ESP is calculated along four layers encompassing the molecule; each layer is a scaling factor larger than the van der Waals surface. For example, if layer 1 is 1.4 times, then layer 2 is 1.6 times, layer 3 is 1.8 times, and layer 4 is

2.0 times the van der Waals radii. In the second step, atomic charges are derived to reproduce the electrostatic potential as closely as possible taking into account the only constraint of this approach namely, the sum of the calculated atomic charges has to be equal to the total charge of the system. The number of layers can be increased for greater accuracy.

This method not only provides charges that are able to reproduce electrostatic potential to a high accuracy but also gives precise interaction and conformational energies. It is however unable to provide accurate dipole moments. It also suffers from the problem of not fully accounting for the buried atoms which lie inside the molecule and are hard to fit to electrostatic potential. Also, in this method it is difficult to decide about the conformational dependence of the partial charges. Thus, for example, in the case of the methylene group, for any particular conformation, not more than two electrons can be symmetrically related. Hence there will always be at least two different partial charges for the hydrogen atoms. However, in the case of a methyl group which is freely rotating in a nonconjugated system, it shall be unreasonable to distinguish between its three hydrogen atoms. In order to take into account the conformational dependence of partial charges, Cornell et al. [10] introduced a modification of ESP which is now known as the RESP approach. Another method to resolve such conformation-related conflicts is to undertake a more accurate simulation. However, this approach is not only expensive but it also complicates force-field energy derivatives. Another possible approach suggested by Basma et al. [11] is a compromise between the above two approaches. It takes a weighted average of the fixed partial charges to account for the possible conformations.

6.5.3 Restrained ESP

Bayly et al. [12] developed this technique to solve the problem of ESP of buried atoms. In this approach the ESP charges are adjusted to meet chemical intuition by adding hyperbolic restraint to the nonhydrogen atoms to keep their values near zero. The buried atoms are therefore rescaled without affecting the overall fit to the electrostatic potential.

As can be seen from Table 6.3, the charges obtained with the various methods vary appreciably (by up to 1.24e). In general, the charge on carbon atoms varies the most, but charges on other buried atom such as copper and nitrogen also vary appreciably. Interestingly, this variation is most pronounced in the molecules involving sulfur atoms. It has been noted that the Atoms in Molecules (AIM) charges and also the NBA charges differ appreciably from the other charges; they are often larger in magnitude. Among all the other charges, the RESP method gives charges with the lowest magnitude whereas the MKS method gives charges with the largest magnitudes. This is probably an effect of the fact that either fewer or none of the electrostatic potentials near the molecule are used in the fit of the MKS method.

Several computer softwares such as Gaussian, MOLPRO, GAMESS have provision for the calculation of ESP-based atomic charges using CHELPG and MKS schemes.

Table 6.3 Charges for molecules calculated by various methods

		Mulliken	NPA	AIM	CHELP	CHELPG	RESP	MK
CH_4	C	−0.63	−0.92	−0.07	0.00	−0.37	−0.23	−0.50
	H	0.16	0.23	0.02	0.00	0.09	0.06	0.12
C_2H_4	C	−0.29	−0.43	−0.28	−0.23	−0.25	−0.08	−0.30
	H	0.14	0.21	0.14	0.12	0.13	0.04	0.15
Pentane	C1	−0.44	−0.67		−0.13	−0.18	0.01	−0.20
	H1	0.14	0.22		0.02	0.04	0.00	0.05
	C2	−0.25	−0.45		0.13	0.14	−0.01	0.10
	H2	0.13	0.23		−0.03	−0.03	0.00	−0.01
	C3	−0.24	−0.46		−0.02	−0.04	−0.01	−0.09
	H3	0.13	0.22		−0.01	0.00	0.01	0.03
	C4	−0.25	−0.45		0.13	0.14	0.00	0.11
	H4	0.13	0.23		−0.03	−0.03	0.00	−0.01
	C5	−0.44	−0.67		−0.13	−0.18	−0.04	−0.22
	H5	0.14	0.22–0.23		0.02	0.04	0.01	0.05
CH_3OH	C	−0.21	−0.31	0.49	0.26	0.18	0.04	0.07
	H	0.13–0.16	0.18–0.21	0.00–0.03	−0.02	0.01	0.04	0.04
	O	−0.61	−0.74	−1.07	−0.55	−0.55	−0.52	−0.53
	H	0.39	0.47	0.54	0.33	0.35	0.35	0.35
H_2CO	H	0.12	0.13	0.04	−0.04	−0.01	0.02	0.00
	C	0.08	0.23	1.04	0.47	0.42	0.33	0.38
	O	−0.32	−0.49	−1.11	−0.39	−0.39	−0.36	−0.38
$(CH_3)_2O$	C	−0.19	−0.31		−0.08	0.01	−0.04	−0.14
	H	0.13–0.16	0.19–0.22		0.07	0.05	0.06	0.09
	O	−0.44	−0.56		−0.26	−0.30	−0.28	−0.24
CH_3CN	C	−0.52	−0.78	0.05	−0.37	−0.25	−0.07	−0.45
	H	0.21	0.28	0.08	0.14	0.11	0.07	0.17
	C	0.35	0.28	0.99	0.42	0.38	0.28	0.42
	N	−0.46	−0.33	−1.27	−0.48	−0.46	−0.43	−0.47
H_2NCHO	N	−0.70	−0.87	−1.24	−0.85	−0.86	−0.65	−0.87
	H	0.34	0.41	0.42–0.43	0.38	0.39	0.33	0.40
	C	0.36	0.51	1.55	0.62	0.62	0.40	0.58
	O	−0.45	−0.60	−1.20	−0.49	−0.51	−0.44	−0.49
	H	0.10	0.14	0.03	−0.04	−0.03	0.03	−0.02
Imidazole	C	0.02	−0.11	0.34	−0.29	−0.25	−0.24	−0.32
	H	0.15	0.23	0.08	0.15	0.16	0.16	0.19
	N	−0.55	−0.57	−1.28	−0.22	−0.21	−0.10	−0.16
	C	0.21	0.17	1.05	0.22	0.20	0.06	0.15
	H	0.15	0.22	0.08	0.04	0.07	0.12	0.10
	C	−0.03	−0.10	0.42	0.31	0.17	0.05	0.15
	H	0.14	0.23	0.06	−0.02	0.06	0.11	0.09
	N	−0.42	−0.49	−1.18	−0.49	−0.49	−0.42	−0.47
	H	0.33	0.43	0.43	0.29	0.29	0.26	0.29
CH_3SH	C	−0.58	−0.81	0.11	0.19	−0.06	−0.13	−0.32
	H	0.19	0.24–0.25	0.04–0.05	−0.02	0.07	0.09	0.14

Table 6.3 Charges for molecules calculated by various methods—cont'd

		Mulliken	NPA	AIM	CHELP	CHELPG	RESP	MK
	S	−0.08	−0.06	−0.05	−0.28	−0.30	−0.30	−0.28
	H	0.10	0.13	0.02	0.13	0.16	0.16	0.17
$(CH_3)_2SO$	C	−0.63	−0.90	−0.17	0.29	−0.25	−0.18	−0.49
	H	0.17–0.20	0.24–0.26	0.04–0.06	−0.05	0.11	0.09	0.18
	S	0.75	1.26	1.27	0.08	0.22	0.17	0.28
	O	−0.62	−0.96	−1.25	−0.39	−0.40	−0.39	−0.39
H_2O	O	−0.77	−0.93	−1.09	−0.75	−0.75	−0.74	−0.75
	H	0.39	0.47	0.55	0.38	0.37	0.37	0.37
$(H_2O)_2$	O_{donor}	−0.83	−0.97	−1.15	−0.79	−0.81	−0.80	−0.81
	H	0.40	0.48	0.58	0.35	0.37	0.36	0.37
	$H_{H\text{-}bond}$	0.38	0.46	0.54	0.39	0.38	0.38	0.38
	$O_{acceptor}$	−0.75	−0.94	−1.10	−0.76	−0.74	−0.73	−0.74
	H	0.41	0.49	0.57	0.41	0.40	0.40	0.40
NH_3	N	−0.89	−1.11	−1.04	−0.98	−1.01	−1.01	−1.02
	H	0.30	0.37	0.35	0.33	0.34	0.34	0.34
PH_3	P	−0.09	0.01	1.32	−0.19	−0.23	−0.25	−0.26
	H	0.03	0.00	−0.44	0.06	0.08	0.08	0.09
HCl	Cl	−0.23	−0.28	−0.26	−0.27	−0.26	−0.26	−0.26
	H	0.23	0.28	0.26	0.27	0.26	0.26	0.26

References

[1] R.S. Mulliken, J. Chem. Phys. 23 (1955) 1833.

[2] P.O. Löwdin, Adv. Phys. 5 (1956) 1.

[3] S.F. Boys, Rev. Mod. Phys. 32 (1960) 296.

[4] C. Edmiston, K. Ruedenberg, Rev. Mod. Phys. 35 (1963) 457.

[5] W. Von Niessen, J. Chem. Phys. 56 (1972) 4290.

[6] P. Thul, V.P. Gupta, V.J. Ram, P. Tandon, Spectrochim. Acta Part A 72 (2009) 82.

[7] P. Thul, V.P. Gupta, V.J. Ram, P. Tandon, Spectrochim. Acta Part A 75 (2010) 251.

[8] P. Hunt, B. kirchner, T. Welton, Chem. Eur. J. 12 (26) (2006) 6762.

[9] C.M. Breneman, K.B. Wiberg, J. Comp. Chem. 11 (1990) 361.

[10] W.C. Cornell, P. Cieplak, C.I. Bayly, P.A. Kollman, J. Am. Chem. Soc. 115 (1993) 9620.

[11] M. Basma, S. Sundara, D. Calgan, T. Vernali, R.J. Woods, J. Comput. Chem. 22 (2001) 1125.

[12] C. Bayly, J. Phys. Chem. 97 (1993) 10269.

Further Reading

[1] A. Hinchliffe, Computational Quantum Chemistry, John Wiley & Sons, 1989.

[2] J.J.W. McDouall, Computational Quantum Chemistry: Molecular Structure and Properties in Silico, by Joseph J.W. McDouall, RSC Publication, 2013.

[3] J.S. Murray, K. Sen (Eds.), Molecular Electrostatic Potentials: Concepts and Applications, Elsevier Science, Amsterdam, 1996.

[4] J.S. Murray, P. Politzer, Wiley Interdiscip. Review; Comput. Mol. Sci. 1 (2011) 153.

[5] R.S. Mulliken, W.C. Ermler, Diatomic Molecules, Acad. Press, New York, 1977.

[6] R.S. Mulliken, W.C. Ermler, Polyatomic Molecules, Acad. Press, New York, 1981.

7

Molecular Geometry Predictions

CHAPTER OUTLINE

Principles and Applications of Quantum Chemistry. http://dx.doi.org/10.1016/B978-0-12-803478-1.00007-8

7.1 Introduction

One of the fundamental issues that chemistry addresses is molecular structure, that is, how the atoms of the molecule are linked together by bonds and what the interatomic distances, angles, and dihedral angles are. A knowledge of structure is important because it has a very important role in determining the kinds of reactions a molecule will undergo, what kind of radiation it will absorb or emit, and to which active sites in neighboring molecules or materials it will bind. Various experimental techniques such as X-ray crystallography, electron diffraction measurements, and microwave and spectral techniques have been used for the determination of molecular geometry with varying degrees of accuracy depending upon the technique used and the physical state of the system. Quantum chemical techniques have provided alternative accurate methods of geometry determination through geometry optimization methods which are applicable not only to the stable molecules but also to transition states (TSs) and other meta-stable states where the experimental measurements are difficult to carry out. The quantum mechanical procedures lead to a priori prediction of structure without making use of the experimental data other than using the values of the fundamental constants. Geometry optimization is a key step in all theoretical studies that are concerned with the molecular structure and reactivity. A brief overview of the fundamental concepts related to the methods of geometry optimization of equilibrium structures, search for TSs and conical intersections, and basics of their theory and applications shall be discussed in this chapter.

7.2 Potential Energy Surface

The starting point of any geometry optimization process is the construction of the potential energy surface (PES) of the given molecular system. The way the energy of a molecular system varies with small changes in its structure is specified by its PES. The PES for a molecule thus describes the energy of the molecule as a function of the positions of the atoms or nuclei. In the case of a diatomic molecule, it is a two-dimensional curve with internuclear separation along x-axis and the total energy along the y-axis. In the case of a polyatomic molecule, it is a $(3N-6)$ dimensional surface; $(3N-5)$ in the case of a linear molecule. A more detailed description of the determination of PESs by molecular mechanics methods has been given in Chapter 12. Here, we shall be concerned with the basic outlines that can be used for geometry optimization purposes. Since the equilibrium configuration corresponds to the minimum of the PES, methods have been devised to calculate them from the *ab initio* and density functional theory (DFT) methods. In addition, the second derivatives of energy, the Hessians, are also calculated by using these theoretical approaches.

As an example of a two-dimensional potential energy curve we may consider the case of rotation of an amino group about a C–N bond (bond C1–N3 in Figure 7.1) in diaminofumaronitrile (DAFN). Figure 7.2 gives a plot between the dihedral angle ϕ(C2C1N3H5) and energy for this molecule which has been discussed in details Gupta et al. [1].

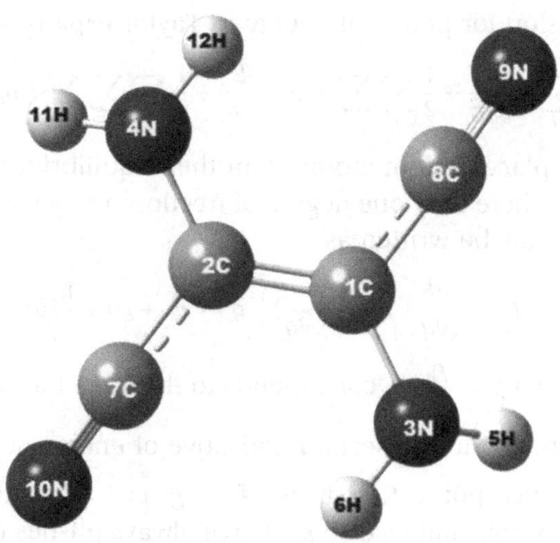

FIGURE 7.1 Geometry of diaminofumaronitrile.

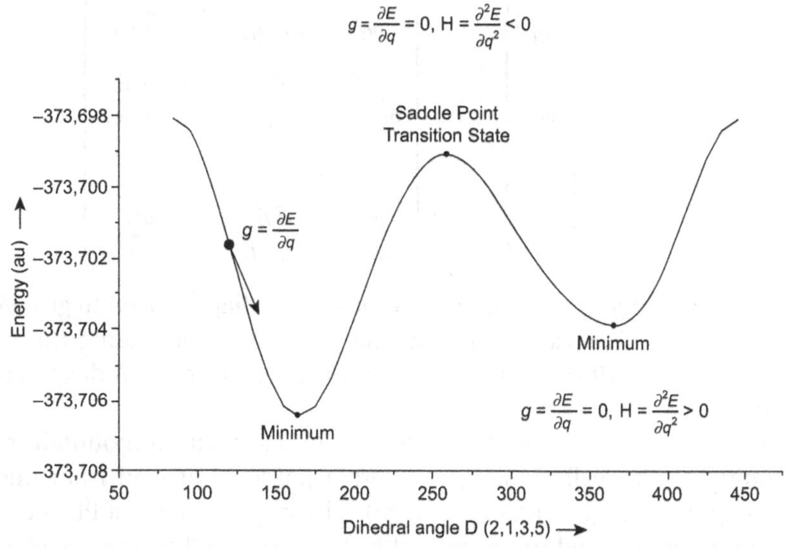

FIGURE 7.2 Potential energy curve of diaminofumaronitrile (DAFN) for rotation about C1–N3 bond.

The general expression for potential energy in Taylor expansion is

$$E = E_0 + \sum_i q_i \frac{\partial E}{\partial q_i} + \frac{1}{2}\sum_i q_i^2 \frac{\partial^2 E}{\partial q_i^2} + \frac{1}{2}\sum_i \sum_{j \neq i} q_i q_j \frac{\partial^2 E}{\partial q_i \partial q_j} + \frac{1}{6}\sum_i \sum_j \sum_k q_i q_j q_k \frac{\partial^3 E}{\partial q_i \partial q_j \partial q_k} + \ldots \quad (7.2.1)$$

where q represents displacement of atoms from their equilibrium positions.

In the present case where only one degree of freedom namely the change in dihedral angle is involved, this may be written as

$$E(q) = E_0 + \left(\frac{\partial E}{\partial q}\right)_0 q + \frac{1}{2}\left(\frac{\partial^2 E}{\partial q^2}\right)_0 q^2 = E_0 + gq + \frac{1}{2}Hq^2 + \ldots \quad (7.2.2)$$

where the energy gradient $g = \left(\frac{\partial E}{\partial q}\right)_0$ corresponds to the force **f** at the reference geometry and $H = \left(\frac{\partial^2 E}{\partial q^2}\right)_0$ corresponds to the second derivative of energy or curvature, also called "Hessian" at the reference point. Classically, $\mathbf{f} = -g$, i.e., when the energy gradient is negative, the force is positive and vice versa. Force always pushes the system toward the nearest minimum. The two prominent points of the curve shown in Figure 7.2 correspond to:

Minima—At this point the energy is minimum, $g = 0$ and $H > 0$.

Saddle point—It is an energy maximum, $g = 0$, $H < 0$. Physically, the saddle point corresponds to the TS.

We can generalize these ideas to the case of more than one degree of freedom for some set of coordinates $\{R_1, R_2, \ldots R_n\}$ on the potential energy curve, by writing

$$g(R_1 \ldots R_N) = \begin{pmatrix} \frac{\partial E}{\partial q_1} \\ \frac{\partial E}{\partial q_2} \\ \vdots \\ \frac{\partial E}{\partial q_n} \end{pmatrix}, \quad H = \begin{pmatrix} \frac{\partial^2 E}{\partial q_1^2} & \frac{\partial^2 E}{\partial q_1 \partial q_2} & \cdots & \frac{\partial^2 E}{\partial q_1 \partial q_n} \\ \frac{\partial^2 E}{\partial q_1 \partial q_2} & \frac{\partial^2 E}{\partial q_2^2} & \cdots & \frac{\partial^2 E}{\partial q_2 \partial q_n} \\ \vdots & \vdots & \ddots & \vdots \\ \frac{\partial^2 E}{\partial q_1 \partial q_n} & \frac{\partial^2 E}{\partial q_2 \partial q_n} & \cdots & \frac{\partial^2 E}{\partial q_n^2} \end{pmatrix} \quad (7.2.3)$$

The Hessian matrix being real and symmetric can be diagonalized to give eigenvalues and eigenvectors. The eigenvalues are the "natural" directions along the PES (which correspond to harmonic vibrational modes) and the eigenvectors indicate curvature in that direction.

In a multidimensional case, the PES can be compared with a mountain range complete with multiple hills, valleys along mountain passes that lead from one valley to another (cf. Chapter 12, Figure 12.3(a) and (b)). The importance of a PES can be judged by the fact that it can be used to describe the structural- and reactivity-related parameters of a molecule. Thus, the equilibrium geometry corresponds to a minimum on the PES. At this point $g = 0$, and all Hessian eigenvalues are positive. The position of the minimum gives information about the bond length and bond angles of the molecules; often a molecule may have more than one stable structure (isomer) for a given electronic state. In this case, several minima may exist on the PESs. The molecular geometries pertaining to the stable structures of the excited states may be different from those obtained for the ground state due to orbital occupancy and the nature of bonding. The ground and excited state potential energy curves are therefore different. Uncertainty in the prediction of molecular geometries may however arise when the potential energy minima is very broad or shallow.

While considering molecular reactions, we may divide the PES into reactant, intermediate, and product valleys and the reaction may be seen as movement from the reactant to the product valley. The highest point along the lowest energy reaction path connecting the two valleys is the TS which is the first-order saddle point corresponding to $g = 0$ with one and only one negative Hessian eigenvalue. It is also possible to have a higher order saddle point. A second-order saddle point is characterized by two negative eigenvalues of the Hessian or two negative frequencies. This point corresponds to a local maximum in two coordinates of the PES. The steepest descent path in mass-weighted coordinates from the TS to the nearby minimum is the minimum energy path or the intrinsic reaction coordinates (IRCs). Thus in order to get information about structure and reactivity, it is necessary to locate the minima, TS, and the reaction paths. We shall here be concerned with the first two, while the study of reaction paths has been left for Chapter 12.

A comparison of the structural features of the reactants and the TSs tells us about the structural changes that occur during the transition. The geometry of the TS will tell what product we can obtain. If two parallel reactions are possible, we can find out the difference between the geometries and energies of the two TSs. While the shape of the potential energy curve at the minima, that is, the second derivative of the PES $\left(\frac{\partial^2 E}{\partial q_i \partial q_j}\right)$ helps to calculate the vibrational frequencies and the zero-point energies, the shape of the PES at the TS is needed for the estimation of reaction rates [2]. Vibrational motion of the molecule about the equilibrium geometries of the reactants and products are used to compute zero-point energies and thermal corrections needed to calculate enthalpy and Gibbs free energy.

It may here be noted that the energy derivatives are useful for determining other molecular properties as well. While, the first and second derivatives of energy give the force and the Hessian, respectively, as discussed above, the third and higher derivatives such as the cubic, quartic, etc., give the cubic, quartic, etc., force constants which determine the anharmonic corrections to the vibrational frequencies. The derivatives of energy with respect to the electric field $\mathcal{E}$ give information about the components of the dipole moment and polarizability. Thus, $\frac{\partial E}{\partial \mathcal{E}_x}$ gives the component of electric dipole moment along the x-coordinate, $\frac{\partial^2 E}{\partial \mathcal{E}_x \partial \mathcal{E}_y}$ gives the xy-component of the polarizability, and $\frac{\partial^3 E}{\partial \mathcal{E}_x \partial \mathcal{E}_y \partial \mathcal{E}_z}$ gives the xyz-component of the first hyperpolarizability. A mixed derivative with respect to a nuclear coordinate and an electric field component such as $\frac{\partial^2 E}{\partial x \partial \mathcal{E}_x}$ gives the dipole moment derivative that determines the intensity of infrared bands in the harmonic approximation.

7.3 Conical Intersections and Avoided Crossings

Analysis of chemical reactions is usually based on the concept of PESs, which are derived from the Born–Oppenheimer approximation. Reactions usually start in the ground electronic state, where the reactant is found to be at a local minimum. In thermal reactions where the energy source is heat, the reaction proceeds along a trajectory that leads adiabatically to a TS, and subsequently the system goes down to the product. The entire route occurs on one potential surface (PES) only, along a single coordinate; this is thus a one-dimensional process. In photochemical reactions, on the other hand, the light energy required to initiate the reaction promotes the molecule to an electronic excited state; thus, the reaction is nonadiabatic. An allowed electronic transition leads to an excited state which may revert back to the ground state either by radiative transitions (lifetime $\approx 10^{-9}$ s) or by fast ($\approx 10^{-12}$ s) or ultrafast ($\approx 10^{-15}$ s) nonradiative transitions. It turns out that many interesting photochemical reactions proceed at an ultrafast speed, for instance, in biological systems.

In Figure 7.3, we have shown the path for standard photochemical reaction (a), for an ultrafast photochemical reaction (b) and ultrafast internal conversion (c). The occurrence of these ultrafast reactions can be explained by the fact that the excited paths are barrier less or have a negligible energy barrier. As such, they can move toward the conical intersection without any restraint. The process of ultrafast internal conversion also happens due to negligible or very small barriers. It has been demonstrated by experiments that a nonradiative transition between two states becomes faster as they approach each other; the probability of transition between two electronic states increases as the energy gap between their potential surfaces decreases. In some regions, ground and excited state potential surfaces may come closer together, touch or cross (Figure 7.4(a–c)). The adiabatic surfaces which follow the Born–Oppenheimer approximation but have the same symmetry avoid crossing, except where the coupling matrix element is zero. On the other hand, they will cross if the two states have different spatial or spin symmetry.

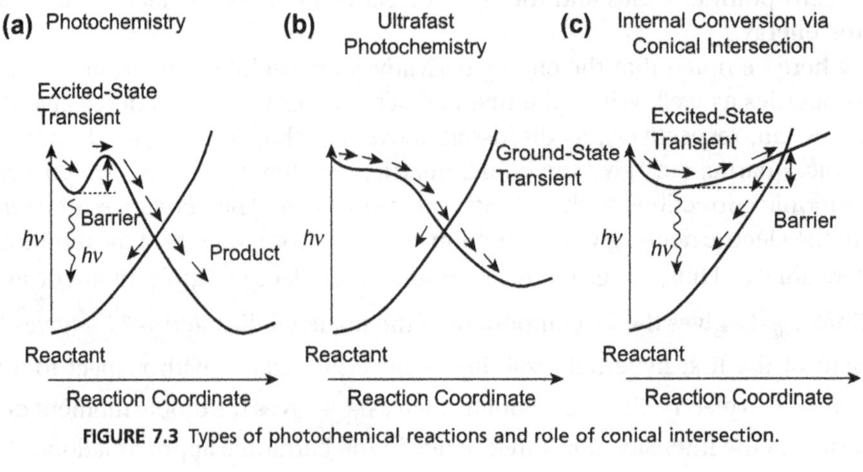

FIGURE 7.3 Types of photochemical reactions and role of conical intersection.

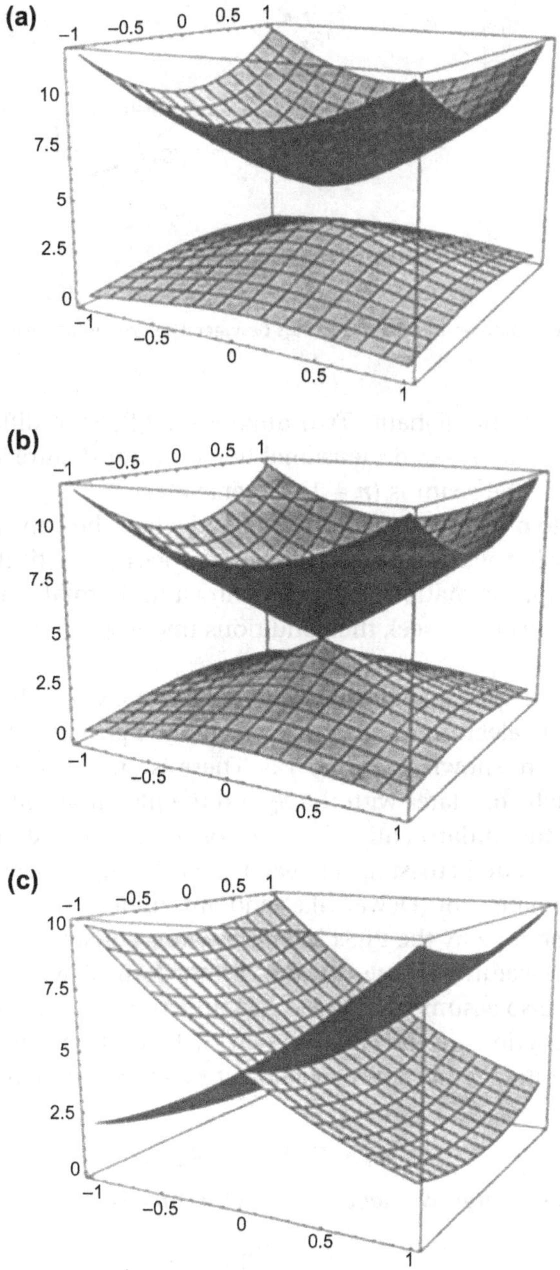

FIGURE 7.4 Interaction between two model potential energy surfaces showing (a) weakly avoided intersection (b) conical intersection and (c) a seam. *Reprinted with permission from Ref. [23], Copyright @ 1995, World Scientific Publishing.*

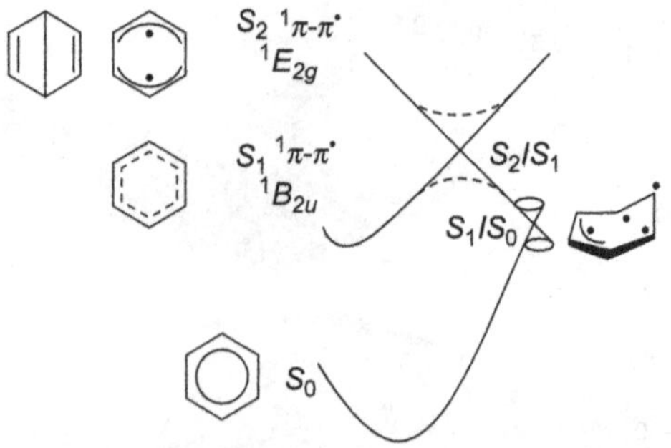

FIGURE 7.5 Conical intersection and avoided crossings between the ground and excited states of benzene.

Such surfaces are said to be diabatic. Two-dimensional PESs of different spatial or spin symmetry can intersect in a one-dimensional line or seam (Figure 7.4(c)). In case of an n-dimensional surface, the seam is $(n-1)$ dimensional.

The maximum rate of energy transfer is attained when the gap vanishes and the two PESs cross. This crossing, which is a nonadiabatic process, constitutes a violation of the Born–Oppenheimer approximation. Thus, in contrast to thermal reactions, the theory of photochemical reactions must seek the conditions under which nonadiabatic processes become efficient.

As an illustration, we may take the example of benzene which has been discussed by Palmer et al. [3]. Three electronic states of benzene, the ground state S_0, and the states $S_1(^1B_{2u})$ and $S_2(^1E_{2g})$ are shown in Figure 7.5. There is an S_1/S_0 conical intersection of the C_s symmetry which correlates with the S_2 and the planar ground S_0 states. It follows from the figure that the antiaromatic character of the S_1 surface changes along the S_1 reaction path due to avoided crossing between the S_2 and S_1 surfaces. The S_2 state can be described by a combination of (Dewar-like) and antiquinoid spin couplings.

In order to understand why the PESs intersect, let us consider two surfaces $V_1(q)$ and $V_2(q)$ described by wavefunctions $\Phi_1(q)$ and $\Phi_2(q)$, both of which depend on $3N-6$ coordinates q. Let us also assume that these wavefunctions are only approximations and not the exact eigenfunctions of the Hamiltonian H. To find improved functions Ψ_1 and Ψ_2 which more accurately represent the eigenstates, let us form a linear combination of Φ_1 and Φ_2.

$$\Psi_k = C_{K1}\Phi_1 + C_{K2}\Phi_2 \tag{7.3.1}$$

In order to find the energy we need to solve the matrix

$$\begin{vmatrix} H_{11}-E & H_{12} \\ H_{21} & H_{22}-E \end{vmatrix} = 0 \tag{7.3.2}$$

which gives

$$2E_\pm = (H_{11}+H_{22}) \pm \sqrt{(H_{11}-H_{22})^2 + 4H_{12}^2} \tag{7.3.3}$$

Two possible cases may arise:

1. $E_+ \neq E_-$. In this case the two energy surfaces in the $(3N-6)$ dimensions space will not cross. This is a more general situation.
2. $E_+ = E_-$. The two energy states will have the same energy provided the second term in Eq. (7.3.3) is zero. This is possible only if

$$H_{11} = H_{22} \tag{7.3.4}$$

$$\text{and,}\ \ H_{12} = 0 \tag{7.3.5}$$

Since these two conditions must simultaneously be obeyed, the two states may have equal energies in a lower dimensional space $(3N-8)$. It is therefore said that the PESs in $(3N-6)$ dimensions may undergo intersections in spaces of $(3N-8)$ dimension. The two PESs cross at the apex and the wavefunction at that place is degenerate. The energy surface near the crossing is called conical intersection which is in fact a seam and not a point. Conical intersections are not isolated points but are connected along a $(3N-8)$ dimensional hyperline.

If the two energy states have different symmetry, the nondiagonal element H_{12} vanishes automatically. In this case, only one condition ($H_{11} = H_{22}$) is needed for crossing. So, the two states of different symmetry may cross in a space of $(3N-7)$ dimension. Conical intersection geometries play a central role in our understanding of photochemical reactions [4–7].

7.4 Evaluation of Energy Gradients

7.4.1 Energy Gradients for Hartree–Fock SCF Theory

As a first step toward geometry optimization, it is necessary to calculate energy gradients with respect to the nuclear coordinates. In order to do this within the Hartree–Fock (HF) formalism, it required to calculate derivatives of one- and two-electron integrals over the basis functions. Since the basis functions are centered on atomic nuclei, we also need to calculate their derivatives with respect to the nuclear coordinates. The derivatives of the expansion coefficients may also have to be computed depending on whether or not they were determined variationally. While for fuller details of these calculations the reader may consult the article by Pulay et al. [8], an outline of the method for the evaluation of the energy gradients from HF self-consistent field (SCF) theory is given here.

Suppose a given electronic state Ψ is represented by a normalized linear combination of orthogonal configurations

$$\Psi = \sum_k C_k \Phi_k \tag{7.4.1}$$

$$\text{with}\ \langle \Phi_k | \Phi_l \rangle = \delta_{kl}\ \text{and}\ \sum_k C_k^2 = 1. \tag{7.4.2}$$

Φ_k, for example, could be a Slater determinant or a fixed linear combination of Slater determinants, each representing a different electronic state. Each of the Slater determinants may be constructed from n orthogonal orbitals, Φ_i which are linear combinations of m basis functions χ_r.

$$\Phi_i = \sum_r^m a_{ri}\chi_r \tag{7.4.3}$$

The expectation value of the energy for the wavefunction Eq. (7.4.1) can therefore be written as

$$E = \langle \Psi | H | \Psi \rangle \tag{7.4.4}$$

The expectation value of the energy for the wavefunction Ψ given by Eq. (7.4.4) depends on several variables such as nuclear coordinates R_α, the expansion coefficients C_k, the expansion coefficients a_{ri}, and a set of nonlinear parameters p_t, the most important of which are the positions of the orbital centers χ_r, and the orbital exponents ζ_r. It is also assumed that p_t has a functional dependence on the nuclear coordinates R_α and the basis functions explicitly depend on R_α through the parameters p_t. So, while carrying out energy optimization, we may have to consider two types of variables: those which may be allowed to vary such as nuclear positions R_α, expansion coefficients a_{ri}, nonlinear parameters defining positions of orbital centers p_t, etc., and those which have to be kept fixed during the calculations such as orbital exponents of the basis functions. If we wish to show the dependence of the energy expectation value on the various parameters, we may write Eq. (7.4.4) in the form

$$E(R_\alpha) = \langle \Psi(C_k(R_\alpha), a_{ri}(R_\alpha), p_t(R_\alpha)) | H(R_\alpha) | \Psi(C_k(R_\alpha), a_{ri}(R_\alpha), p_t(R_\alpha)) \rangle \tag{7.4.5}$$

When we take the derivative of $E(R_\alpha)$ with respect to R_α, we shall get a sum of four terms in the expression corresponding to direct dependence of E on R_α and indirect dependence on C_k, a_{ri}, and p_t. The derivative of energy with respect to these variables therefore need to be evaluated.

The derivative of the energy with respect to the nuclear coordinate is termed as *Hellmann–Feyman force*

$$-f_\alpha^{H-F} = \left\langle \Psi \middle| \frac{\partial H}{\partial R_\alpha} \middle| \Psi \right\rangle,$$

the derivative with respect to a_{ri} as the *density force* (because of its connection with the density matrix), and the negative of the derivative with respect to the nonlinear parameter p_t as the *integral force*. The derivative with respect to C_k vanishes. The sum of these three nonvanishing forces is also called exact force because its value should be equal to negative of the force from point-wise SCF calculation [8].

The total energy of a closed-shell polyatomic molecule in HF SCF theory (Eq. (3.8.6)) is given by the expression

$$E = 2\sum\sum P_{rs}\langle r|H^{core}|s\rangle + \sum\sum P_{rs}P_{tu}\left[2\left\langle rs\middle|\frac{1}{r_{12}}\middle|tu\right\rangle - \left\langle rt\middle|\frac{1}{r_{12}}\middle|su\right\rangle\right] \tag{7.4.6}$$

$$\text{where, } P_{rs} = 2\sum_{i}^{occ\ MO} C_{ri}C_{si} \tag{7.4.7}$$

is the density matrix

$$\langle r|H^{core}|s\rangle = \langle \chi_r|H^{core}|\chi_s\rangle \tag{7.4.8}$$

and H^{core} is the one-electron integral.

$$\left\langle rs\left|\frac{1}{r_{12}}\right|tu\right\rangle = \left\langle \chi_r(1)\chi_s(1)\left|\frac{1}{r_{12}}\right|\chi_t(2)\chi_u(2)\right\rangle \tag{7.4.9}$$

$$\left\langle rt\left|\frac{1}{r_{12}}\right|su\right\rangle = \left\langle \chi_r(1)\chi_t(1)\left|\frac{1}{r_{12}}\right|\chi_s(2)\chi_u(2)\right\rangle \tag{7.4.10}$$

and the last term V_{NN} is the repulsion energy between the nuclei α and β

$$V_{NN} = \sum_{\alpha>\beta}\frac{Z_\alpha Z_\beta}{R_{\alpha\beta}} \tag{7.4.11}$$

To find the energy gradient we need to differentiate Eq. (7.4.6). As a final result, we get

$$\begin{aligned}\frac{\partial E}{\partial C_k} &= 2\sum P_{rs}\left\langle r\left|\frac{\partial H^{core}}{\partial C_k}\right|s\right\rangle + 4\sum P_{rs}\left\langle\frac{\partial \chi_r}{\partial C_k}|H^{core}|\chi_s\right\rangle - \sum\sum\frac{\partial S_{ij}}{\partial C_k}X_{ij} \\ &+2\sum\sum P_{rs}P_{tu}\left[4\left\langle\frac{\partial\chi_r}{\partial C_k}\chi_s\left|\frac{1}{r_{12}}\right|\chi_t\chi_u\right\rangle - 2\left\langle\frac{\partial\chi_r}{\partial C_k}\chi_s\left|\frac{1}{r_{12}}\right|\chi_u\chi_t\right\rangle\right] + \frac{\partial V_{NN}}{\partial C_k}\end{aligned} \tag{7.4.12}$$

Using abbreviations $\frac{\partial\chi}{\partial C_k} = \dot{r}_k$, $\frac{\partial S_{ij}}{\partial C_k} = \dot{S}_{ij}$ and replacing $\chi_r, \chi_s, \ldots$ by letter r, s,..., this equation may be written in a simpler form as

$$\begin{aligned}\frac{\partial E}{\partial C_k} &= 2\sum P_{rs}\left\langle r\left|\frac{\partial H^{core}}{\partial C_k}\right|s\right\rangle + 4\sum P_{rs}\langle \dot{r}_k|H^{core}|s\rangle - \sum\sum \dot{S}_{ij}X_{ij} + 2\sum\sum P_{rs}P_{tu}\left[4\left\langle \dot{r}_k s\left|\frac{1}{r_{12}}\right|tu\right\rangle\right. \\ &\left. - 2\left\langle \dot{r}_k s\left|\frac{1}{r_{12}}\right|ut\right\rangle\right] + \frac{\partial V_{NN}}{\partial C_k}\end{aligned} \tag{7.4.13}$$

X and $\dot{S}$ matrices are defined in terms of the linear combination of atomic orbital coefficients a_{ri}, orbital energies, and the derivatives of the overlap integrals $s_{pq} = \langle\chi_p|\chi_q\rangle$.

In this equation second and fourth terms contain derivatives of the basis functions and hence contain nonlinear parameters p_t. They require most of the numerical work in calculating the gradient. Also, because the derivatives of a large number of one- and two-electron integrals have to be calculated, the geometry optimization takes a considerable amount of computer time.

This equation was derived by Bratoz [9] and implemented by Bratoz and Allavena [10] for one-center basis sets and forms the basis for the analytical second derivative formula.

It also follows from Eqs (7.4.5–7.4.13) that if the wavefunction Ψ depends on a single parameter λ, i.e., $\Psi = \Psi(\lambda)$ and the basis set does not depend on the parameters C_k and consequently on the nuclear coordinate, i.e., $\dot{r}_k = 0$, then Eq. (7.4.13) will reduce to the Hellmann–Feynman formula

$$\frac{\partial E}{\partial \lambda} = \int \Psi^* \frac{\partial H}{\partial \lambda} \Psi d\tau = \left\langle \frac{\partial H}{\partial \lambda} \right\rangle \tag{7.4.14}$$

The second derivative of energy can be obtained from the first derivative by numerical differentiation. Thus, in harmonic approximation, the force constant

$$f_{ij} = \frac{\partial^2 E}{\partial q_i \partial q_j} \tag{7.4.15}$$

can be obtained as

$$f_{ij} = -\frac{\Delta f_i}{\Delta q_j} = \frac{f_i(0) - f_i(q_j + \Delta q_j)}{\Delta q_j} \tag{7.4.16}$$

In this equation, the first term corresponds to the force acting on atom i at the reference geometry and the second corresponds to the force when the atom j is displaced by Δq_j. So, the second-order derivative is calculated as the difference of two first-order derivatives divided by the step size. This treatment, however, does not exclude the effect of anharmonicity. In order to partially eliminate the effect of cubic anharmonicity, instead of using a single displacement of Δq_j, two points at $q_j(0) = \pm\frac{1}{2}\Delta q_j$ located symmetrically around the reference geometry are chosen. Thus, from two evaluations of the gradient, a complete row of harmonic force constants f_{ij}, $j = 1, \ldots M$, can be determined. A total of $2M$ gradient evaluations are, therefore, needed to calculate all the harmonic force constants or the Hessian. Pulay [11] has shown that with small modification in this procedure, it is possible to evaluate all cubic diagonal F_{iii} and semidiagonal F_{iij} force constants.

7.4.2 Energy Gradients for DFT

The gradient and Hessian of the SCF energy are usually evaluated by solving coupled perturbed HF (CPHF) equations, These are also formulated for various other approaches such as multiconfiguration SCF, coupled cluster, Møller–Plesset 2 (MP2), and DFT [12,13]. DFT, although similar in implementation to standard SCF theory, differs from the latter in that it introduces an exchange–correlation term which is density dependent. The presence of such a quantity introduces additional derivative terms which are not present in standard approaches of electronic structure theory. Analytical first derivatives are available for almost all density functionals and as a result geometry optimization can be carried out with ease.

7.5 Optimization Methods and Algorithms

The quantum chemical calculations for equilibrium structures are carried out under Born–Oppenheimer approximation which helps to separate out the electronic and nuclear motions. Except when PESs for different states get too close to each other or cross, the Born–Oppenheimer approximation is usually quite good.

Under the Born–Oppenheimer approximation, the energy of a molecule is obtained by solving the electronic structure problem using Schrödinger or Kohn–Sham equations for a set of fixed nuclear positions. Thus, the energy E of a molecule can be described as a parametric function of the position of the nuclei $\{q_a\}$ to yield a PES. The technique of geometry optimization then relates to determination of minima on the position of PESs, i.e., the position at which the energy gradient $\frac{\partial E}{\partial q} = 0$ and $\left(\frac{\partial^2 E}{\partial q_a \partial q_b}\right) > 0$. By finding the local minima, one is able to determine a stable structure of the molecule.

The procedure of finding the energy minima involves calculation of energy derivatives and testing whether the energy decreases or increases as each geometrical coordinate is varied by a small amount. These derivatives can be calculated either by point-wise calculation of energy or by gradient methods. In the method of point-wise calculation, which may be useful for small molecules (mostly diatomic) and simplest possible cases (such as C=C in ethene) only, energies are calculated for different values of geometrical parameters followed by a fitting procedure. Thus, for example, in a diatomic molecule the theoretical bond length can be obtained by simply calculating the total energy as a function of internuclear separation, and then finding the minimum. This method, however, has some serious drawbacks both in accuracy and numerical efficiency [14,15]. The second method, that is the gradient method, involves direct analytical calculation of energy derivatives from the wavefunction. The first derivatives of the PES are known as the gradients.

If the gradients of energy (usually represented as $g(q)$) are known at a particular geometry, one can perform searches along the negative of the gradient $g(q)$ to move toward a lower energy geometry. Since the forces on the atoms or nuclei in a molecule are equal to the negative of the gradient $\left(f_i = -\frac{\partial E}{\partial q_i}\right)$, a search along a negative gradient also means a search toward a position where the forces get minimized. At the minima of the PESs, TSs and higher order saddle points, the gradients, or the forces exerted on a nucleus by electrons and other nucleus are zero. These equilibrium positions are therefore also called stationary points. While a zero gradient can characterize a stable point, it cannot differentiate between minima, maxima, and saddle points. To distinguish between the types of stationary points and to make more intelligent search toward a geometry of lower energy, one calculates the second derivatives of energy $\left(\frac{\partial^2 E}{\partial q_i \partial q_j}\right)$. With the possible exception of optimization of diatomic molecules, the gradient- or derivative-based geometry optimization methods are significantly more efficient than the energy only-based methods.

7.5.1 Basics of Gradient Methods

By definition, gradient of a scalar field, say $f(\mathbf{r})$, is a vector field at any point of which the vector has a magnitude equal to the most rapid rate of increase of the scalar $f(\mathbf{r})$ and points in the direction of this rapid increase. This is the intrinsic property of a scalar field quite independent of the particular system coordinate axes. The gradient is therefore invariant.

$$grad\, f(\mathbf{r}) = \frac{\partial f}{\partial n}\mathbf{n}, \tag{7.5.1}$$

where **n** is a unit vector in the direction of the fastest increase.

$$\text{So,}\ \ grad\, f \cdot d\mathbf{r} = \frac{\partial f}{\partial n}\mathbf{n} \cdot d\mathbf{r} = \frac{\partial f}{\partial n} d\mathbf{r} \cos\theta = \frac{\partial f}{\partial n} \cdot d\mathbf{n} = df$$

where θ is the angle between **n** and $d\mathbf{r}$ and df is the total increase in the value of $f(\mathbf{r})$.

From elementary calculus,

$$df(\mathbf{r}) = \frac{\partial f}{\partial x}dx + \frac{\partial f}{\partial y}dy + \frac{\partial f}{\partial z}dz = \left(\mathbf{i}\frac{\partial f}{\partial x} + j\frac{\partial f}{\partial y} + \mathbf{k}\frac{\partial f}{\partial z}\right) \cdot (idx + \mathbf{j}dy + \mathbf{k}dz) \quad \text{or} \quad df(\mathbf{r}) = \nabla f \cdot d\mathbf{r} \tag{7.5.2}$$

where **i**, **j**, and **k** are unit vectors along the x, y, z axes in a Cartesian coordinate system in which $\mathbf{r} = (x, y, z)$ is represented.

The vector ∇f is called the gradient of the scalar function f and ∇ is the gradient operator

$$\nabla = \mathbf{i}\frac{\partial}{\partial x} + \mathbf{j}\frac{\partial}{\partial y} + \mathbf{k}\frac{\partial}{\partial z} \tag{7.5.3}$$

In the case of geometry optimization we may replace the scalar function f by energy E. Since as per definition ∇f always points toward larger values of f, in the case of geometry optimization, where we are looking for the minimum value of the energy E, we are interested in $-\nabla E$.

For geometry optimization we need to calculate a gradient vector $\mathbf{g}(q)$ defined as

$$\mathbf{g}(q) = \begin{vmatrix} \frac{\partial E}{\partial q_1} \\ \frac{\partial E}{\partial q_2} \\ \vdots \\ \frac{\partial E}{\partial q_n} \end{vmatrix} \tag{7.5.4}$$

where q is an n-dimensional coordinate vector.

Looking back to Eq. (7.4.1), the expansion coefficients $C_1, C_2, \ldots, C_k$ are variables which determine the energy E of the state Ψ. We may think of variables $C_1, C_2, \ldots, C_k$ as a vector **C** on which the energy E depends, $E = E(\mathbf{C})$. There can be a very large number of variables: those which are allowed to vary such as nuclear positions, expansion coefficients, etc., or those which may be kept fixed during the calculations, such as the orbital exponents of the basis functions. During the calculations these variables are allowed to change in all possible ways so that the energy E gets optimized and reaches its minimum value. The normalization condition (Eq. (7.4.2)) of the molecular orbitals Φ represented by the Slater determinant is taken into account by using Lagrange's undetermined multipliers. Thus, following Eq. (7.5.3), we may introduce a gradient operator

$$\nabla = \widehat{C}_1 \frac{\partial}{\partial C_1} + \widehat{C}_2 \frac{\partial}{\partial C_2} + \cdots + \widehat{C}_k \frac{\partial}{\partial C_k} \tag{7.5.5}$$

$$\text{and so,} \quad \nabla E = \widehat{C}_1 \frac{\partial E}{\partial C_1} + \widehat{C}_2 \frac{\partial E}{\partial C_2} + \cdots + \widehat{C}_k \frac{\partial E}{\partial C_k} \tag{7.5.6}$$

where $\widehat{C}_1, \widehat{C}_2, \ldots, \widehat{C}_k$ are unit vectors defined in terms of C, which constitute a k-dimensional space.

The optimum value of E is therefore determined by pursuing $(-\nabla E)$ to its minimum value. This forms the basis of the so-called gradient technique for geometry optimization. In Eq. (7.5.6), the partial derivatives may be obtained by finite difference method.

$$\frac{\partial E}{\partial C_k} \approx \frac{E(C_1, C_2, \ldots, C_k + \Delta) - E(C_1, C_2, \ldots, C_k)}{\Delta} \tag{7.5.7}$$

The energy gradients can also be obtained analytically. In fact vast resources of analytical gradients are available in the literature for different levels of theory.

In case the potential energy curve has more than one minima, the choice of the starting point for optimization is important because the minimization procedure shall lead to the closest minimum, also called the local minimum, unless the step size is large enough to skip it.

The analytical gradient methods have a few advantages over the point-wise calculations. Some of these advantages are:

1. They are computationally faster and much more information can be obtained about the energy surface from a single SCF calculation. This is due to the fact that a single energy value can give information about $3N - 6$ independent forces for an N-atom molecule and hence all the first derivatives can be determined at the same time. Thus, a single gradient calculation is equivalent to $3N - 6$ energy calculations.
2. Methods which determine energy derivatives analytically are more accurate than the numerical differentiation methods. Equilibrium geometries can be more sharply determined by the condition of vanishing forces than by that of minimum energy.
3. The gradient methods are computationally much cheaper, particularly for large molecules. It is so because the time taken for SCF procedures increases much faster than the time required for gradient calculations.

Pulay [11,16] developed the first computer program for determining first derivatives of SCF energies analytically and used it to calculate the equilibrium geometry and force constants for water. During the last four decades rapid progress has been made and milestones have been set in the development of the analytic gradients of the potential energy and the methods based on them. Several surveys and reviews of geometry optimization are available [17–23]. Some of the more prominent analytical gradient methods are the Newton and quasi-Newton methods and their modifications, conjugate gradient methods, DIIS (direct inversion of iterative subspace), and its modification, Geometric Direct Inversion in the Iterative Subspace (GDIIS) [24,25].

Hessian update techniques allowed information to be collected for the PES and accelerate the optimization process. The quadratic line search (QLS) [26–30], rational functional optimization (RFO) [31], trust radius method (TRM) [21,23,26–30], and trust radius update [21,23,26–30] which control the step size, makes such techniques more reliable.

7.5.2 Algorithms for Finding Potential Energy Minima

The overall efficiency of optimization method depends upon the choice of the algorithm. Several different algorithms are now available for obtaining optimized geometries of molecules. These can broadly be classified into three categories:

1. Those using only the energy-such algorithms search for minima without any gradient/force information. Inference about gradients and Hessian are made from lots of displacements. These tend to be very slow to converge and are only used in specialized situations;
2. Those using both the energy and its analytical first derivatives; and
3. Those using energy together with the analytical first and second derivatives. These are also called the Hessian-based algorithms.

These algorithms are progressively more efficient and accurate and have been discussed in several review articles [26,28,29,32,33]. Most methods of geometry optimization rely on the use of first and second derivatives for an efficient performance. These derivatives are calculated analytically though the calculation of the second derivatives requires a much higher cost than the first derivatives. The steps followed in a geometry optimization program can be described by the flow chart given in Figure 7.6.

Without going into the various algorithms used for finding the energy minimum in details, we shall briefly consider a few of them such as steepest descent, the conjugate gradient, and the quasi-Newton–Raphson algorithms; the last being one of the best approaches for geometry optimization and can also be used effectively to find the TSs. They can greatly benefit from using redundant internal coordinates for the reason that a good approximation of the PES can be used for a wider range than in Cartesian coordinates.

7.5.2.1 *Method of Steepest Descent*

This is based on the method of line search. If we know the energy E and the energy gradients $\mathbf{g}$ on the potential energy curve for some set of coordinates $\mathbf{R}$, then the straightforward approach will be to move the atoms in the direction of $-\mathbf{g}$, and find λ such that the coordinates $\mathbf{R}' = \mathbf{R} - \lambda\mathbf{g}$ have minimum energy. The process is repeated

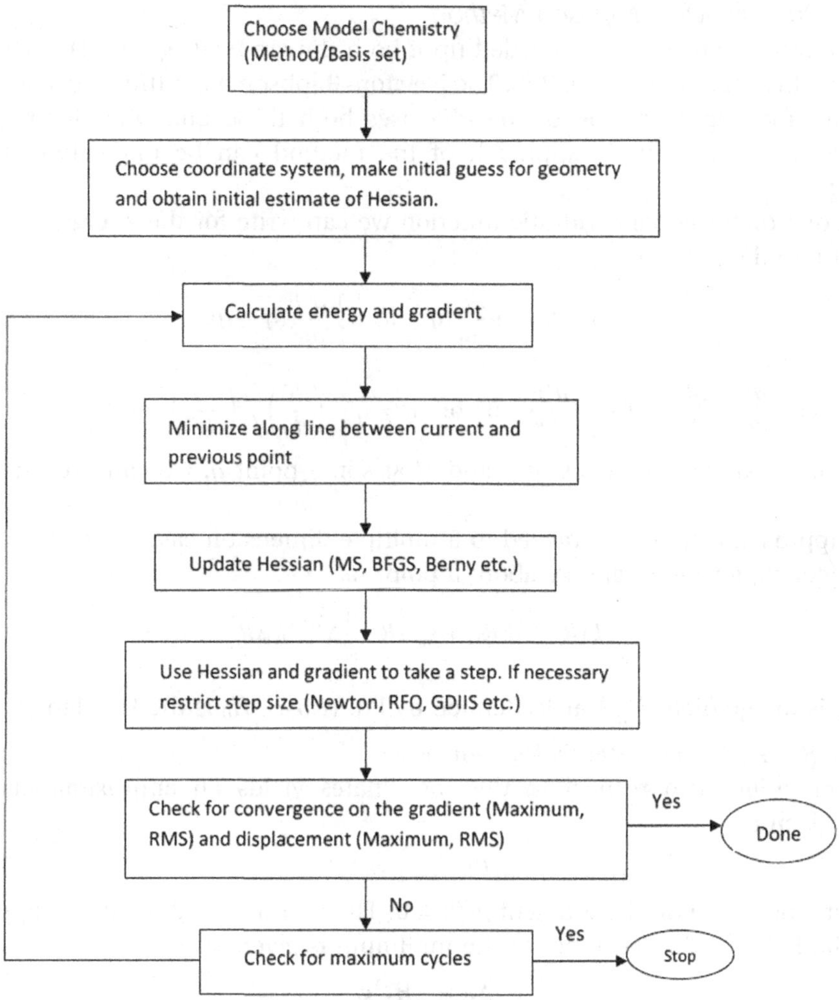

FIGURE 7.6 Flow chart for geometry optimization.

from the updated values $\mathbf{R}'$ until $\mathbf{g}$ or change in energy ΔE becomes very small. The drawback of this method is that while the convergence is very fast in the region away from the energy minima, it is very slow near a minimum.

7.5.2.2 Method of Conjugate Gradient

The difficulty in the method of steepest descent is that it is difficult to remember information about the earlier steps. In the method of conjugate gradient this difficulty is removed by taking each step orthogonal to some number of previous steps. It generally performs better than the method of steepest descent in regions far away from equilibrium geometry.

7.5.2.3 Quasi-Newton–Raphson Method

The optimization process gets speeded up if both the gradient (**g**) and Hessian (*H*) are known at different points on the PES. The Newton–Raphson algorithm which is based on local quadratic approximation of the PES uses both these quantities for finding the stationary points. The basic approach of the method can be understood from the following.

For a one-dimensional quadratic function we can write for the energy at a point q' quite close to the point q

$$E(q') = E(q) + \frac{dE}{dq}(q' - q) + \frac{1}{2}\frac{d^2E}{dq^2}(q' - q)^2 \tag{7.5.8}$$

$$\text{Hence,}\quad \frac{dE}{dq'} = \frac{dE}{dq} + (q' - q)\frac{d^2E}{dq^2} = 0 \quad \text{or} \quad q' = q - \left(\frac{dE}{dq}\right)\bigg/\left(\frac{d^2E}{dq^2}\right) = q - \frac{\mathbf{g}(q)}{H(q)} \tag{7.5.9}$$

Thus, if we know the values of g and H at some point q, we can predict the next point q'.

This approach can be generalized to a multiple dimension case by taking a second-order Taylor expansion of energy about a point R_0.

$$E(R) = E(R_0) + g_0^T \Delta R + \frac{1}{2}\Delta R^T H_0 \Delta R \tag{7.5.10}$$

where g_0 is the gradient $\left(\frac{dE}{dR}\right)$ at R_0 defined by Eq. (7.5.4), H_0 is the Hessian $\left(\frac{d^2E}{dR^2}\right)$ at R_0, and $\Delta R = R - R_0$. Letter T stands for transpose.

Differentiation with respect to the coordinates yields an approximation for the gradient given by:

$$g(R) = g_0 + H_0 \Delta R \tag{7.5.11}$$

At a stationary point, the gradient $g(R) = 0$. Thus, in the local quadratic approximation to the PES, the displacement to the minimum is given by:

$$\Delta R = -H_0^{-1} g_0 \tag{7.5.12}$$

Using the steps given by this equation, one can successfully find points of lower energy and ultimately the energy minima. The disadvantage of this method, however, is that at each step we should know the Hessian which is expensive to calculate. However, there are methods which may upgrade the Hessian given its value at one step. Large varieties of methods for Hessian updating such as Murtagh–Sargent (MS), Broyden–Fletcher–Goldfarb–Shano (BFGS), GDIIS etc., are presently used for the purpose. Thus, one initially starts with an estimated Hessian obtained by molecular mechanics or a lower level SCF calculation and the same is then updated at each step.

The BFGS method is one of the best Hessian-updating techniques and is included in most electronic structure softwares. It is based on the relation

$$H_{\text{New}} = H_{\text{Old}} + \Delta H_{\text{BFGS}} \tag{7.5.13}$$

$$\Delta H_{\text{BFGS}} = \frac{\Delta g \Delta g^T}{\Delta g^T \Delta R} - H_{\text{Old}}\,\Delta R \frac{\Delta R^T H_{\text{Old}}}{\Delta R^T H_{\text{Old}} \Delta R} \tag{7.5.14}$$

It has also been pointed out by Schlegel [19] that a modification of the Bofill's update [34] for the TSs is very useful for locating energy minima.

7.5.2.4 GDIIS Method

Another optimization method that is very efficient in the vicinity of the minimum is the GDIIS approach [35]. In this method the goal is to construct a new geometry as a linear combination of previous geometries so as to minimize the size of the Newton step. In the Newton–Raphson method, the step size $\Delta R(= H_0^{-1} g_0)$ gives some estimate of the error in the determination of the current position. If we have a series of such steps, we can predict the next step by requiring that it minimizes the error from the previous step. In other words, the length of an error vector can be minimized through a linear interpolation/extrapolation of the available structures. So,

$$\text{Error} = \sum_i c_i \Delta R = \sum_i c_i H^{-1} g, \ \sum_i c_i = 1 \tag{7.5.15}$$

The minimization of $|\Delta R|^2$ leads to a least squares problem that can be solved for the coefficients c_i.

The next point in optimization is then given by,

$$R_j = \sum_{i<j} c_i \left[R_i - H^{-1}(R_i) g \right] \tag{7.5.16}$$

While GDIIS method is very efficient near minima, it may misbehave farther from the minima, and may even converge to a nearby saddle point of higher order, or oscillate about an inflection point. A number of improvements have been suggested to overcome these difficulties. Thus, for example, the problem of higher order saddle point can be controlled by comparing the GDIIS step with a reference step such as the quasi-Newton step using RFO or trust radius method (TRM).

7.5.3 Transition State Structures

Finding TSs for chemical reactions is much more difficult task than finding equilibrium geometries or minima for the reason that while for minima it is easy to choose a good starting geometry and a suitable coordinate system, in the case of TSs there is only a vague idea that the saddle point should lie somewhere between the reactants and the products. As mentioned, a TS is a stationary point on a PES which has a local maximum in one and only one direction and a minimum in all other perpendicular directions. This unique direction is termed the transition vector and is the eigenvector associated with the negative eigenvalue of the Hessian. In general the orientation of the transition vector is not known a priori, and hence must be determined in the transition structure optimization. A transition structure optimization must, therefore, be found by moving uphill in one direction and downhill in all other orthogonal directions. Many algorithms have been developed to search for the TSs and have been reviewed in the literature [18,21,23,33]. These methods can be broadly divided into single-ended and double-ended methods [18]. The single-ended methods start with an initial structure and displace it toward the transition structure. Newton and quasi-Newton algorithms are the most efficient single-ended methods for optimizing transition structures. However, they require a starting geometry near the quadratic region of the transition structure. Double-ended methods start from the reactants and products and work from both sides to find the transition structure and the reaction path. Methods such as the nudged elastic band (NEB) method, string method (SM), and the growing SM belong to this category. (For greater details of these methods and associated references, see Ref. [19].)

7.5.3.1 Quasi-Newton–Raphson Methods for Transition Structures

One approach that has been used for optimizing TSs is to turn the saddle point search into a search for minimization. This is done by optimizing the gradient norm $|\mathbf{g}| = \sqrt{\mathbf{g}^T\mathbf{g}}$. Since $\mathbf{g} = 0$, both at the saddle point as well as the minima, if the starting structure is within the quadratic region to the saddle point, the minimization of the gradient norm shall directly lead to the saddle point. This procedure works for some TSs but the radius of convergence in this case is even smaller than in the conventional quasi-Newton–Raphson methods modified for TS search.

Newton and quasi-Newton algorithms are the most efficient single-ended methods for optimizing transition structures when the starting geometry is within the quadratic region of the transition structure. Even those cases where the starting geometry is outside the quadratic region of the transition structure can be handled by this method by using techniques such as TRM, RFO, etc., which control the optimization steps. Controlling the step size and direction are keys to the success for the quasi-Newton methods for the transition structures, especially when the particular structure is outside the quadratic region. Some of the important differences in the quasi-Newton methods for searching the minima and TSs are that in the latter case the Hessian update must allow for negative eigenvalues. The BFGS method is therefore no more appropriate as it yields positive definite updates. Also the line searches may, in general, no more be possible because the step toward the TS may be either uphill or downhill.

Sometimes, while using the quasi-Newton methods, it is necessary to gather information about the quadratic region of the TS. If the geometries of reactants and products are known, then a rough approximation to the reaction path is a linear interpolation between these two structures. The maxima on the reaction path are then a crude approximation to the transition structure. In such cases, the synchronous transit methods are used to find the quadratic region of the TS. The synchronous transit methods start from the reactant and product and find maximum along a linear or a quadratic path across the surface. When a linear path is chosen the method is called the linear synchronous transit (LST) path method. Thus, suppose **X**, **R**, and **P** are the coordinates of the current point, the reactants, and the products, respectively, then in the LST approach, **X** is on a path that is a linear interpolation between **R** and **P**. It gives an upper bound to the energy barrier to the reaction. However, it frequently yields a structure with two or more negative eigenvalues. An improvement over the LST approach is achieved by using quadratic synchronous transit (QST) method which searches for a maximum along a curved path (parabola) connecting reactants (**R**) and products (**P**) and for a minimum in all directions perpendicular to the curved path. By parabolic interpolation, this method gives lower bound of the energy barrier. The maximum along the QST path is a much better estimate of the transition structure. Quite often one can find a structure quite close to the quadratic region of the TS from where the quasi-Newton methods may take over.

When a transition structure for an analogous reaction is not available a QST2 approach [36] is used to generate an initial guess of the transition structure for optimization. In this approach, a structure on the reactant side and one on the product side are used to provide a crude estimate of the TS geometry and to approximate the direction of the reaction path. Another method known as the QST3 approach uses three structures to approximate a transition structure. A series of steps is taken along the path connecting these two or three structures until a maximum is reached. Quasi-Newton methods can then take over from this point to give the optimized transition structure. The algorithm that combines the synchronous transit and quasi-Newton methods to find transition structures are available in all computational softwares meant for electronic structure studies.

An appropriate choice of coordinates is very important for the TS geometry optimization. The interpolation by LST or QST methods can be done in Cartesian coordinates, internal coordinates, or distance matrix coordinates (i.e., $N(N-1)/2$ interatomic distances in an N-atom molecule). Often a combination of redundant internal coordinates of the reactants and products provides a good coordinate set for the transition structure. Sometimes, extra dummy atoms are included to avoid problems with the coordinate system.

Quasi-Newton methods start with an initial geometry that is somewhere near the quadratic region of the TS and an initial estimate of the Hessian that has a negative eigenvalue and the corresponding eigenvector which is roughly parallel to the required reaction path. The optimization process starts with an analytical or numerical calculation of the full Hessian. Updating the Hessian while searching uphill along the reaction path ensures that the Hessian has a negative eigenvalue. For QSTn methods, a suitable negative eigenvalue and the corresponding eigenvector are obtained by updating the Hessian during the initial maximization steps along the reaction path.

7.5.4 Algorithms for Conical Intersections

Different approaches have been proposed to optimize the structures where two electronic states become degenerate. Since the conical intersection in itself is a multidimensional surface, so the programs actually find the minimum energy crossing point on the seam. The algorithm incorporates the constraint that the states involved have the same energy. A conical intersection optimization requires a balanced description of the electronic structure of all the states involved. The algorithms currently available are based either upon Lagrange–Newton methods [37–39] or projection methods [40–44]. The algorithms for the Lagrange–Newton methods have the common feature of using variations of the classical Lagrangian multipliers method [34,45] which constrains the energy of the two electronic states to be equal. In order to find the lower-dimensional space in which the two surfaces cross, it is necessary to have information about the gradients and Hessians of functions V_1 and V_2 representing the two surfaces. This information is then used to locate a geometry at which the difference function $f = (V_1 - V_2)^2$ passes through zero. Conventional methods of finding the root which are designed to locate the points where $f = 0$ are used for the purpose. Once such a geometry has been found, the seam along which the function f remains zero is followed.

The projection methods have a dual approach. They reach the energy degeneracy by means of a displacement within the branching space, and optimize simultaneously the energy of the excited state within the intersection space [46]. *Ab initio* multireference methods such as complete active space SCF or CASPT2 are also widely used, but these are computationally expensive. Semiempirical configuration interaction methods provide a lower cost alternative.

7.6 Practical Aspects of Optimization

The application of computational methods and algorithms to molecular structure problems raises many practical problems which need to be solved for getting improved results. These are associated with the choice of the coordinate system and use of molecular symmetry, choice of the starting geometries and Hessians, choice of the quantum chemical method and basis sets, choice of the convergence limits in geometry optimization procedures, and finally testing the character of the stationary point. Some of these aspects shall be discussed here.

7.6.1 Choice of Coordinates

While, in principle, any complete set of coordinates can be used to represent a molecule and its PES, an appropriate choice of the coordinate system is also very important for an efficient geometry optimization. The choice of Cartesian coordinates is perhaps the most universal. While it has the advantage that most energy and derivative calculations are carried out in Cartesian coordinates but, as they do not reflect the chemical structure and bonding of a molecule, they are not well suited for geometry optimization. The internal coordinates such as bond lengths, valence angles, and torsional angles which describe the molecular structures are better suited for geometry optimization than the Cartesian coordinates. The internal valence coordinates also meet two practical requirements: (1) They facilitate transfer and comparison of force constants between related molecules. (2) They allow a simple representation of the anharmonic terms, i.e., cubic, quartic, and higher force constant terms in the Taylor expansion of potential energy.

A few variants of the internal coordinates are in vogue. These are:

1. Nonredundant internal coordinates or the *Z*-matrix. The *Z*-matrix approach is a convenient way to specify the geometry of a molecule. It uses a list of bond distances, bond angles, and dihedral angles. It has better performance than the Cartesian. This type internal coordinates are implemented in most molecular packages.
2. "Natural" internal coordinates—These coordinates introduced by Fogarasi et al. [47] are linear combinations of internal coordinates that match more closely to the normal modes of the molecules. These coordinates are hard to generate automatically although they can work very efficiently particularly for ring systems.
3. Redundant internal coordinates—The set of all bonds, valence angles, and dihedral angles and, if necessary, the out-of-plane bends and linear bends constitute the redundant system of coordinates. These are the overdetermined set of internal coordinates: in the case of acyclic molecules having N atoms one can easily select $3N - 6$ (or $3N - 5$) internal coordinates, but in the case of cyclic molecules their number can be more than $3N - 6$. The transformation from the redundant to Cartesian coordinates is a difficult task and can be obtained iteratively. These coordinates are however quite efficient for molecules.
4. Symmetry coordinates—These are the symmetrized internal coordinates corresponding to the molecular symmetry. They help to speed up the calculations but may create problems in those cases where the molecular symmetry changes during the calculations, for example during the study of internal rotation.

It is now generally agreed that in quantum chemical applications one should use redundant internal coordinates, Hessian updates, and GDIIS, line-search and RFO for efficient geometry optimization.

7.6.2 Use of Molecular Symmetry

Use of molecular symmetry in geometry optimization has its good and bad points. The use of symmetry reduces the number of coordinates that must be optimized. The number of coordinates to be optimized can be determined by looking at the sets of symmetry equivalent atoms and the symmetry elements. The problem with the use of molecular symmetry is that if we start an optimization in a certain point group, the molecule will be forced to stay in that point group. This may not be desirable in several applications, particularly those related to constrained optimization. Sometimes, one may also get significant imaginary frequencies indicative that the molecule wants to go to a lower symmetry. Sherrill [48] suggests that in this situation, it may be necessary to distort the molecule in the direction of the imaginary normal mode to achieve geometry optimization.

7.6.3 Choice of the Starting Geometries and Hessians

The rate of optimization is accelerated by making a proper choice for the starting geometry and initial Hessian. A suitable estimate of the geometry can be made either by using databases containing theoretical and experimental structures or by conducting semiempirical or lower level calculations using HF or DFT methods. An initial choice of the Hessian can be made on the basis of semiempirical and low level calculations though some precautions may be needed as the PESs may change significantly at the higher level calculations. As mentioned earlier, for optimizing the saddle point, the Hessian should have one negative eigenvalue and the corresponding eigenvector should suitably approximate the transition vector. Sometimes, it may also be necessary to calculate the full Hessian at the same level of theory at which the optimization is done.

7.6.4 Choice of the Quantum Chemical Method and Basis Sets

For a typical equilibrium structure, the HF level of theory, due to its overemphasis on the occupied orbitals, predicts bond lengths that are usually short. However, inclusion of configuration interaction with the excited states corrects this error to some extent as the orbitals into which excitation occurs typically have some antibonding character. For geometry optimization, the MP2 level is considered as an excellent choice as significant improvements in geometry at the minima can be obtained at low cost. Theories like G2, G2 (MP2), G3, and G4 for accurate energies also optimize geometries at the HF/6-31 G(d) level. If, ongoing from HF to MP2, a large change in some geometric properties is observed, it is worthwhile to try theories like MP3, MP4, etc. It is found that MP2 level theory with 6-31 G(d, p) basis sets gives an average error of 0.015 Å in bond length against 0.021 Å at the HF level. Bond angles at both the levels are equally correct. Improvement of the same order has been reported for smaller molecules by using aug-cc-pVnZ basis sets. The HF theories, however, do not give as accurate results for the TSs where the correlation effects may be large. In these cases, going beyond MP2 may become necessary.

In contrast, for minimum energy structure, the DFT gives very good results. Bond lengths at the local density approximation (LDA) level for molecules composed of first and second row of atoms are as good as those predicted by MP2 level. The use of generalized gradient approximation functionals does not lead to much improvement over the LDA functionals but they systematically overestimate the bond lengths. Since HF systematically underestimates and the LDA functionals overestimate the bond lengths, hybrid functionals which are mixture of the above two methods were developed. The hybrid functionals give an improvement of 0.005 Å in bond lengths with the result that accuracy of the order of 0.004 Å can be obtained by functionals such as BLYP and B3LYP. DFT methods also give bond angles accurate to within 1° and dihedral angles within 3°. Comparative performance of some *ab initio* and DFT methods is given in Table 7.1.

Table 7.1 Mean absolute errors in bond lengths for commonly used methods over test set of molecules including first and second row atoms[a]

Level of theory	Test sets
HF methods	
HF	0.021
MP2	0.014
LSDA functionals	0.016
SVWN	0.013[c]
GGA and MGGA functionals	
BLYP	0.021
	0.019[b]
BP86	0.022[c]
	0.018[c]
BPW91	0.017
	0.017[c]
PBE	0.016[c]
Hybrid functional	
B97-1	0.008[b]
PBE/PBE	0.010[c]
B3LYP	0.008[b]
	0.010[c]

LSDA, local spin density approximation.
[a]Scheiner et al. [49].
[b]Hamprecht et al. [50].
[c]Staroverov et al. [51].

As for TSs, both the HF and DFT methods have been used to calculate their structure. The performance of DFT methods for TSs is in general better than these of correlated HF methods. DFT methods, however, show weakness in the calculation of the following:

1. van der Waals complexes, where the intrafragment distances are too large. This is due to the fact that the dispersion-induced attraction is not properly included. Functionals have been developed to overcome this problem to a large extent (cf. Chapter 5).
2. Hydrogen bonds are generally too short.
3. Charge-transfer complexes where their polarities are overestimated and so they are too tightly bound.

The choice of basis sets is one of the most important factors in *ab initio* calculations, since it ultimately determines the accuracy of the calculation. A wrong choice may give meaningless results. In selecting a basis set for systematic calculations, there are two conflicting requirements: the basis set should be sufficiently complete to permit a good description of the wavefunction, and at the same time it should be small enough so that the calculations can be extended to larger systems without much cost. In Chapter 3 we have already discussed in details about the various Slater-type and Gaussian-type basis sets used in *ab initio* calculations and their relative merits and demerits. For minimum energy geometries, both the HF and DFT methods perform very well with basis sets of even very modest size and give bond lengths accurate within 0.01 Å, and bond angles within 1.0°. Expansion of basis sets to the double-zeta or split-valence level leads to an improvement in bond length predictions. The addition of polarization functions to the split valence basis sets also improves the prediction of the bond angles and dihedral angles.

Sherrill [48] has given a few suggestions for obtaining good results for geometry optimization. These are found to be quite useful and may be tried:

1. 6-31G* or 6-31G** give reasonable results and are better than cc-pVDZ.
2. For higher accuracy cc-pVTZ may be tried, which is better than 6-311G**.
3. For anions and excited states diffuse functions (aug – or +) may be tried. They are also good for dispersion-bonded complexes.
4. Beyond Ne, one may try cc-pV(X+d)Z basis sets.

Some other suggestions that may be useful while trying to calculate optimized geometry and correct energy values are:

1. In cases of very large molecules, such as biomolecules or polymers, when even B3LYP/6-31G(d, p) is too expensive for geometry optimization and frequency calculations, often HF/3-21G(d, p) can be used for geometries and B3LYP with zero-point energy correction for the final energy calculation.
2. In cases, where only the Austin Model-1 level geometries can be calculated, the single point energy if calculated at the B3LYP level can still significantly improve the accuracy of the final energy.

7.6.5 Choice of the Convergence Limits

Four parameters, namely, maximum force, root-mean-square (RMS) force, displacement for the next step, and RMS displacement are usually chosen to check if the molecular geometry has been optimized. The process of checking is repeated after each optimization cycle. This prevents premature identification of the minima. The calculated values of these four parameters should be below a threshold value which is usually taken as 0.00045 for maximum force, 0.0003 for RMS force, 0.0018 for displacement, and 0.0012 for RMS displacement as in Gaussian suite of programs. In practical situations when dealing with large molecules, sometimes the displacement remains larger than the cutoff value. In such cases, the geometry is treated as optimized if the forces are two orders of magnitude smaller than the cutoff value.

7.6.6 Testing the Character of the Stationary Point

After optimizing a transition structure it is necessary to confirm that it is indeed a transition structure. This is done by calculating the Hessian at the same level of theory that was used for optimization. It is not sufficient to examine the updated Hessian used in the quasi-Newton optimization process because it may not be sufficiently reliable to confirm the nature of the stationary point. As a side product, calculated Hessian also yields the vibrational frequencies and zero-point energies. As discussed, a minimum must have no negative eigenvalue of Hessian and so no imaginary frequency. The TS which is a first-order saddle point has one negative eigenvalue and hence only one imaginary frequency. More than one negative eigenvalue may indicate the existence of a lower energy stationary point. This needs to be tested by reoptimization. In the case of a TS, the transition vector, i.e., the vibrational normal mode associated with the imaginary frequency must be inspected by using graphical software to ensure that the motion corresponds to the desired reaction. While studying chemical reactions, it may be necessary to follow the reaction path for some distance down from the TS in both the forward and reverse directions to determine whether the TS connects the correct reactants and products and also to detect the existence of any intermediates along the reaction paths.

7.7 Illustrative Examples

7.7.1 Geometry Optimization of Cyanocarbene

Gupta et al. [52] have shown that the reaction leading to the formation of adenine in the interstellar space involves several simple neutral molecules and radicals such as HCN, cyanocarbene (HCCN), cynamide (NH_2CN), and CN. Cyanocarbene (HCCN) is found in high concentrations in space and has been subjected to various experimental and theoretical studies. HCCN molecule can have two possible geometries; a linear equilibrium geometry for the triplet electronic ground state and a bent structure consistent with the accepted name cyanocarbene. While initial studies on this molecule believed that the linear structure is the most plausible [53], theoretical studies by Kim et al. [54] have suggested that a bent geometry is more stable than the linear geometry. In order to confirm which conformation of HCCN takes part in the ring closure leading to the five-membered adenine ring, MP2 and DFT calculations were conducted for the singlet and triplet electronic states of the molecule using different basis sets. The results are given in Table 7.2. It is seen from the table that the molecule has a bent structure in both the electronic states but the energy of the triplet state is lower than the singlet by about 16.4 kcal/mol. Thus, only the bent conformation of the HCCN molecule in the triplet state can participate in the formation of the adenine ring.

7.7.2 Transition State in Isomerization of Carbonyl Cyanide

Carbonyl cyanide $CO(CN)_2$ is of great astrophysical significance due to the prevalence of carbonyl and cyano compounds in the interstellar medium [55]. It decomposes

Table 7.2 Optimized geometry of cyanocarbene in singlet and triplet states. Bond lengths in Å, angles in degrees

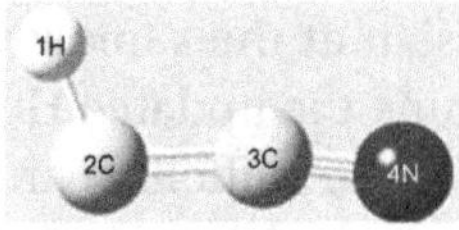

Singlet

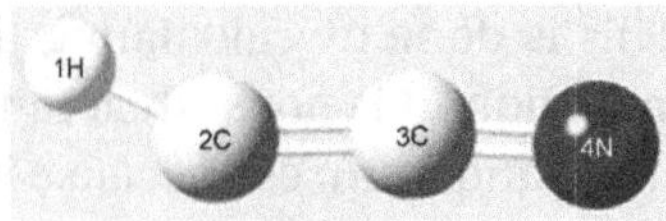

Triplet

	MP2/6-31G*	B3LYP/6-31G*	MP2/6-31G*	B3LYP/6-31G*	B3LYP/6-311+G**
	Singlet	Singlet	Triplet	Triplet	Triplet
R(1,2)	1.0978	1.101	1.0729	1.0709	1.0672
R(2,3)	1.4073	1.376	1.3785	1.3066	1.2955
R(3,4)	1.1886	1.1847	1.1517	1.2068	1.2021
A(1,2,3)	108.3292	111.3041	134.6223	153.1604	159.4287
L(2,3,4,1,−1)	171.3522	171.7131	182.0864	175.8955	183.4151
L(2,3,4,1,−2)	180	180	180	180	180
Energy (au)	−131.003923	−131.3969734	−131.011067	−131.4230634	−131.4600205

into OCCN radical which further dissociates into CO and CN radicals [56]. Recently, Gupta et al. [57] reported the possibility of its isomerization to carbonyl isocyanide and studied the process of interconversion by HF and DFT theories in a manner similar to the isomerization studies on HCN and CH_3CN [58,59], using the software Gaussian 03.

A search for TS structures in carbonyl cyanide by using STQN(QST2) procedure in MP2/6-31G* and B3LYP/6-311+G* calculations shows that, besides the molecular dissociation process, there is also the possibility of a photoisomerization process through a nonplanar TS. IRC calculations starting from the above nonplanar TS lead to a planar and stable isomeric structure CNC(O)CN having total energy of −187,140.4 kcal/mol in MP2/6-31G* and −187,698.7 kcal/mol in B3LYP/6-311+G*. The isomerization process is shown in Figure 7.7. Geometries of the reactant, product, and the TS structure are given in Table 7.3. These are compared with the corresponding geometric parameters for the TS of CH_3CN [59].

The computed TS geometry for carbonyl cyanide shows almost the same behavior as reported in methyl cyanide TS structure. The multiple CN bond and the C–C bond distances are 1.187 and 1.813 Å at B3LYP level and 1.197 and 1.777 Å at the MP2 level, respectively. The CNC bond angles are about 70°. The results at the two levels of theory differ by about 0.01 Å in bond length and by about 1° in bond angle. These studies confirm the experimental findings regarding the dissociation and isomerization processes.

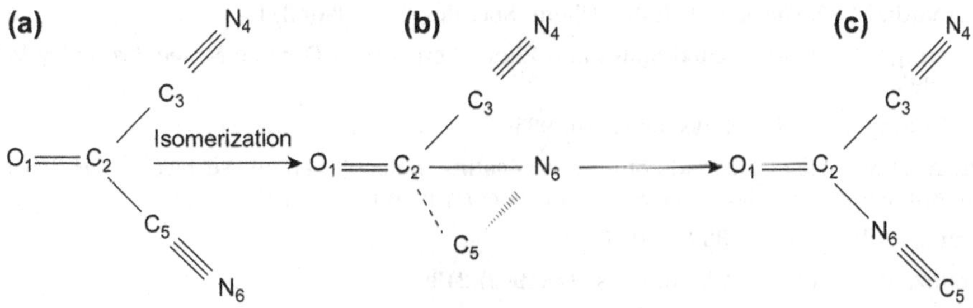

FIGURE 7.7 Molecular geometries of (a) reactant $CO(CN)_2$ (b) transition state and (c) isomerization product (CN–CO–CN).

Table 7.3 Molecular geometries of carbonyl cyanide, transition state (TS), and isocarbonyl cyanide in isomerization process

	$CO(CN)_2$		COCN–CN		CNC(O)CN	
	Reactant		TS		Product	
Coordinate	**MP2/ 6-31G***	**B3LYP/ 6-311+G***	**MP2/ 6-31G***	**B3LYP/ 6-311+G***	**MP2/ 6-31G***	**B3LYP/ 6-311+G***
R(O1C2)	1.2234	1.2016	1.1949	1.1735	1.2120	1.1944
R(C2C3)	1.4639	1.4634	1.4484	1.4410	1.4614	1.4609
R(C2C5)	1.4639	1.4634	1.7570 (1.777)[a]	1.8132 (1.805)[b]	2.5974	2.5755
R(C2N6)	–	–	1.7778	1.8041	1.4003	1.3950
R(C3N4)	1.1840	1.1533	1.1838	1.1531	1.1832	1.1528
R(C5N6)	1.1840	1.1533	1.2071 (1.197)[a]	1.1875 (1.187)[b]	1.1972	1.1805
A(O1C2C3)	122.51	122.70	127.39	128.15	123.69	123.28
A(C2C3N4)	180.00	179.36	178.26	178.94	177.41	177.83
A(C2C5N6)	180.00	179.36	70.96	70.42	–	–
A(C2N6C5)	–	–	69.11 (75.5)[a]	71.25 (73.1)[b]	178.84	178.82
A(O1C2C5)	122.51	122.70	120.80	120.31	–	–
A(O1C2N6)					123.39	123.63
D(O1C2C3N4)	0.0	0.0	2.84	11.92	0.0	0.0
D(O1C2C5N6)	0.0	0.0	99.79	99.55	–	–
D(O1C2N6C5)	–	–	−104.36	−103.25	0.0	0.0

[a]Values in TS of CH_3CN [59] at MP2/6-311+G(2d,2p) level.
[b]Values in TS of CH_3CN [59] at B3LYP/6-311+G(2d,2p) level.

References

[1] V.P. Gupta, P. Tandon, Spectrochim. Acta A89 (2012) 55.

[2] N. Koga, K. Morokuma, Chem. Phys. Lett. 119 (1985) 371.

[3] I.J. Palmer, I.N. Ragazos, F. Bernardi, M. Olivucci, M.A. Robb, J. Am. Chem. Soc. 115 (1993) 673.

[4] W. Domcke, D.R. Yarkony, H. Koppel, in: Conical Intersections: Electronic Structure, Dynamics & Spectroscopy, vol. 15, World Scientific Publishing Co, Singapore, 2004.

[5] F. Bernardi, M. Olivucci, M.A. Robb, Chem. Soc. Rev. 25 (1996) 321.

[6] M. Kessinger, J. Michl, Excited States and Photochemistry of Organic Molecules, Wiley-VCH, New York, 1995.

[7] D.R. Yarkony, Rev. Mod. Phys. 68 (1996) 985.

[8] P. Pulay, Direct use of the gradient for investigating molecular energy surfaces, in: H.F. Schaefer III (Ed.), Applications of Electronic Structural Theory, Plenun Press, 1977.

[9] S. Bratoz, Colloq. Int. CNRS 82 (1958) 287.

[10] S. Bratoz, M. Allavena, J. Chem. Phys. 37 (1962) 2138.

[11] P. Pulay, Mol. Phys. 17 (1969) 197.

[12] P. Deglmann, F. Furche, R. Ahlrichs, Chem. Phys. Lett. 362 (2002) 511.

[13] A. Komornicki, G. Fitzgerald, J. Chem. Phys. 98 (1993) 1398.

[14] D.R. Hartree, Numerical Analysis, Oxford University Press, Oxford, 1968.

[15] J. Gerratt, I.M. Mills, J. Chem. Phys. 49 (1968) 1719.

[16] P. Pulay, Mol. Phys. 18 (1970) 473.

[17] H.P. Hratchian, H.B. Schlegel, Finding minima, transition states, and following reaction pathways on *ab initio* potential energy surfaces, in: C.E. Dykstra, G. Frenking, K.S. Kim, G.E. Scuseria (Eds.), Theory and Applications of Computational Chemistry: The First Forty Years, Elsevier, New York, 2005, p. 195.

[18] D.J. Wales, Energy Landscapes, Cambridge University Press, Cambridge, 2003.

[19] H.B. Schlegel, J. Comput. Chem. 24 (2003) 1514.

[20] P. Pulay, J. Baker, Optimization and reaction path algorithms, in: N.D. Spencer, J.H. Moore (Eds.), Encyclopedia of Chemical Physics and Physical Chemistry, Institute of Physics, Bristol, 2001, p. 2061.

[21] H.B. Schlegel, Geometry optimization, in: P.v.R. Schleyer, N.L. Allinger, P.A. Kollman, T. Clark, H.F. Schaefer III, J. Gasteiger, P.R. Schreiner (Eds.), Encyclopedia of Computational Chemistry, vol. 2, John Wiley & Sons, Chichester, 1998, p. 1136.

[22] H.B. Schlegel, Reaction path following, in: PvR. Schleyer, N.L. Allinger, P.A. Kollman, T. Clark, H.F. Schaefer III, J. Gasteiger, P.R. Schreiner (Eds.), Encyclopedia of Computational Chemistry, vol. 4, JohnWiley & Sons, Chichester, 1998, p. 2432.

[23] H.B. Schlegel, Geometry optimization on potential energy surfaces, in: D.R. Yarkony (Ed.), Modern Electronic Structure Theory, World Scientific Publishing, Singapore, 1995, p. 459.

[24] P. Csaszar, P. Pulay, J. Mol. Struct. 114 (1984) 31.

[25] X. Li, M.J. Frisch, J. Chem. Theory Comput. 2 (2006) 835.

[26] R. Fletcher, Practical Methods of Optimization, Wiley, Chichester, 1987.

[27] P.E. Gill, W. Murray, M.H. Wright, Practical Optimization, Academic Press, London, 1981.

[28] L.E. Scales, Introduction to Non-linear Optimization, Springer, New York, 1985.

[29] J.E. Dennis, R.B. Schnabel, Numerical Methods for Unconstrained Optimization and Nonlinear Equations, Prentice-Hall, Englewood Cliffs, NJ, 1983.

[30] M.J.D. Powell (Ed.), Nonlinear Optimization, Academic Press, New York, 1982.

[31] A. Banerjee, N. Adams, J. Simons, R. Shepard, J. Phys. Chem. 89 (1985) 52.

[32] R. Fletcher, Practical Methods of Optimization, John Wiley, Chichester, 1981.

[33] F. Jensen, Transition structure optimization techniques, in: PvR. Schleyer, N.L. Allinger, P.A. Kollman, T. Clark, H.F. Schaefer III, J. Gasteiger, P.R. Schreiner (Eds.), Encyclopedia of Computational Chemistry, vol. 5, John Wiley & Sons, Chichester, 1998, p. 3114.

[34] J.M. Bofill, J. Comput. Chem. 15 (1994) 1.

[35] O. Farkas, H.B. Schlegel, Chem. Phys. 4 (2002) 11.

[36] C.Y. Peng, H.B. Schlegel, Isr. J. Chem. 33 (1993) 449.

[37] D.R. Yarkony, J. Chem. Phys. 92 (1990) 2457.

[38] M. Dallos, H. Lischka, R. Shepard, D.R. Yarkony, P.G. Szalay, J. Chem. Phys. 120 (2004) 7330.

[39] I.N. Ragazos, M.A. Robb, F. Bernardi, M. Olivucci, Chem. Phys. Lett. 197 (1992) 217.

[40] M.J. Bearpark, M.A. Robb, H.B. Schlegel, Chem. Phys. Lett. 223 (1994) 269.

[41] T. Chachiyo, J.H. Rodriguez, J. Chem. Phys. 123 (2005) 094711.

[42] S. Yamazaki, S. Kato, J. Chem. Phys. 123 (2005) 114510.

[43] R. Izzo, M. Klessinger, J. Comp. Chem. 21 (2000) 52.

[44] M. Olivncci, J. Comp. Chem. 24 (2003) 298.

[45] P.E. Gill, W. Murray, Numerical Methods for Constrained Optimization, Academic Press, London, 1974.

[46] G.J. Achity, S.S. Xantheas, K. Ruedenberg, J. Chem. Phys. 95 (1991) 1862.

[47] P. Pulay, G. Fogarasi, F. Pang, J.E. Boggs, J. Am. Chem. Soc. 101 (1979) 10.

[48] C. David Sherrill, http://vergil.chemistry.gatech.edu/courses/chem6485/pdf/quantum-practical.pdf.

[49] A.C. Scheiner, J. Baker, J.W. Andzelm, J. Comput. Chem. 18 (1997) 775.

[50] F.A. Hamprecht, A.J. Cohen, D.J. Tozer, N.C. Handy, J. Chem. Phys. 109 (1998) 6264.

[51] V.N. Staroverov, G.E. Scuseria, J. Tao, J.P. Perdew, J. Chem. Phys. 119 (2003) 12129.

[52] V.P. Gupta, P. Tandon, P. Rawat, R.N. Singh, A. Singh, Astron. Astrophys. A129 (2011) 528.

[53] S. Saito, Y. Endo, E. Hirota, J. Chem. Phys. 80 (1984) 1427.

[54] K.S. Kim, H.F. Schaefer, J.A. Pople, J.S. Binkley, J. Am. Chem. Soc. 105 (1983) 4148.

[55] J.J. Clouthier, D.C. Moule, J. Am. Chem. Soc. 109 (1987) 6259.

[56] A. Furlan, H.A. Scheld, J.R. Huber, Chem. Phys. Lett. 1 (1998) 282.

[57] V.P. Gupta, A. Sharma, Pramana J. Phys. 67 (2006) 487.

[58] L. Fan, T. Zigler, J. Am. Chem. Soc. 114 (1992) 10890.

[59] B.S. Jursic, Computing transition state structures with density functional theory methods, in: Recent Developments and Applications of Modern Density Functional theory, Theoretical and Computational Chemistry, vol. 4, Elsevier Science B.V., 1996.

Further Reading

[1] A. Hinchliffe, *Ab Initio* Determination of Molecular Properties, Adam, Hilger, Bristol, 1987.

[2] A. Miglani, M. Olivucci, Conical interaction and organic reaction mechanism, in: W. Domcke, D.R. Yarkony, H. Koppel (Eds.), Conical Intersections: Electronic Structure, Dynamics and Spectroscopy, World Scientific Publications, 2004.

[3] D. Young, Computational Chemistry: A Pratical Guide for Applying Techniques to Real World Problems, Wiley-Interscience, 2001.

[4] J.B. Foresman, E. Frisch, Exploring Chemistry with Electronic Structure Methods, Gaussian, Pittsburgh, 1996.

[5] W.J. Hehre, Practical Strategies for Electronic Structure Calculations, Wavefunction, Irvine, 1995.

[6] C.J. Cramers, Essentials of Computational Chemistry: Theories and Models, John Wiley & Sons, Ltd, 2004.

[7] F. Jensen, Introduction to Computational Chemistry, John Wiley & Sons, Ltd, 2007.

[8] H.B. Schlegel, in: D.R. Yarkony (Ed.), Modern Electronic Structure Theory, World Scientific, Singapore, 1995.

[9] P. Pulay, in: H.F. Schaefer III (Ed.), Applications of Electronic Structure Theory, Plenum Press, 1977.

8

Vibrational Frequencies and Intensities

CHAPTER OUTLINE

Principles and Applications of Quantum Chemistry. http://dx.doi.org/10.1016/B978-0-12-803478-1.00008-X

8.1 Introduction

Vibrational spectroscopy is a valuable tool for the elucidation of molecular structure. The transitions between vibrational states of a molecule are observed experimentally via infrared and Raman spectroscopy. These techniques can be used to determine a molecule's structure and environment since these factors affect the vibrational frequencies. Vibrational spectroscopy provides important information about the nature of chemical bond, intramolecular forces acting between the atoms in a molecule, and intermolecular forces in condensed phase. In order to gain such useful information, it is necessary to assign vibrational motion corresponding to each peak in the spectrum. This assignment can be quite difficult due to the large number of closely spaced peaks even in fairly simple molecules. In order to aid in this assignment, computer simulations are used to calculate the vibrational frequencies of molecules. Ever since Pulay and collaborators popularized the calculation of molecular force fields in the 1970s, there has been considerable interest among theoretical chemists to determine ever more accurate force fields for the prediction of rotational–vibrational spectra, the evaluation of vibrationally averaged properties and thermodynamic functions, the determination of many-body potentials, and for other purposes. As for the prediction of vibrational spectra of large molecules, the use of HF *ab initio* or DFT methods for determining whole or partial force fields has also benefitted greatly from the advances made in quantum chemical methodology. The rapidly growing impact that computational studies are having in this important area of research is amply demonstrated by the different types of systems under investigation, ranging from small molecules to polymeric systems and liquid crystals.

8.2 Quantum Mechanical Model for Diatomic Vibrator–Rotator

We start with the Schrödinger equation for two particles of masses m_1 and m_2 interacting via a potential $U(r)$ which depends upon the interatomic distance r only. After separating out the center of mass motion, we are left with the Schrödinger equation in the relative coordinates,

$$\frac{1}{r^2}\frac{\partial}{\partial r}\left(r^2\frac{\partial\Psi}{\partial r}\right)+\frac{1}{r^2\sin\theta}\frac{\partial}{\partial\theta}\left(\sin\theta\frac{\partial\Psi}{\partial\theta}\right)+\frac{1}{r^2\sin^2\theta}\frac{\partial^2\Psi}{\partial\phi^2}+\frac{2}{\hbar^2}\mu_r[E-U(r)]\Psi=0 \tag{8.2.1}$$

where $\mu_r=\frac{m_1m_2}{m_1+m_2}$ is the reduced mass.

Since $U(r)$ is not dependent upon the angular coordinates θ or ϕ, the above equation is again separable into the angular and the radial equations which are

$$\frac{1}{\sin\theta}\frac{\partial}{\partial\theta}\left(\sin\theta\frac{\partial Y}{\partial\theta}\right)+\frac{1}{\sin^2\theta}\frac{\partial^2 Y}{\partial\phi^2}=-J(J+1)Y \tag{8.2.2}$$

and

$$\frac{d}{dr}\left(r^2\frac{dR}{dr}\right)+\left[\frac{2\mu_r r^2}{\hbar^2}(E-U(r))-J(J+1)\right]R=0, \tag{8.2.3}$$

respectively, such that $\Psi(r,\theta,\phi)=R(r)Y(\theta,\phi)$

We know that the solutions of the angular equation $Y_{JM}(\theta, \phi)$ are the spherical harmonics such that,

$$\Psi_{nJm}(r,\theta,\phi) = R_{nJ}(r)Y_{JM}(\theta,\phi) = R(r)\,\Theta_{Jm}(\theta)\Phi_m(\phi)$$

As in the case of the hydrogen atom, the solutions of the angular equation $Y_{JM}(\theta, \phi)$ are the spherical harmonics.

$$\Theta_{Jm}(\theta) = \left\{\frac{(2J+1)(J-|m|)!}{2(J+|m|)!}\right\}^{1/2} P_J^{|m|}(\cos\theta) \tag{8.2.4}$$

and

$$\Phi_m(\phi) = \frac{1}{\sqrt{2\pi}} e^{\pm im\phi} \tag{8.2.5}$$

where $J = 0, 1, 2, 3, \ldots$ are the rotational quantum numbers and $m_J = 0, \pm1, \pm2, \ldots \pm J$ are the magnetic quantum numbers.

To solve the radial equation we need to know $U(r)$. In general, $U(r)$ looks like Figure 8.1; that is, it represents an attractive potential for $r > r_e$, but constant (zero force) for $r = r_e$, and a repulsive potential for $r < r_e$, where r_e is the equilibrium distance. The analytical expression for the potential function is given in Eq. (8.2.6).

$$U(q) = \frac{1}{2!}\left(\frac{\partial^2 U}{\partial q^2}\right)_{q=0} q^2 + \frac{1}{3!}\left(\frac{\partial^3 U}{\partial q^3}\right)_{q=0} q^3 + \ldots = \frac{1}{2}kq^2 + aq^3 + bq^4 + \ldots \tag{8.2.6}$$

where $q = r - r_e$ is the distortion of the bond from its equilibrium length. k is called the spring constant or the harmonic force constant, while a, b,… are called anharmonic force constants.

If we are interested in internuclear distance to have only small variations from r_e then, in this region

$$U(r) = \frac{1}{2}k(r - r_e)^2$$

The potential in this case shall be of the simple harmonic type represented by a parabola (Figure 8.2).

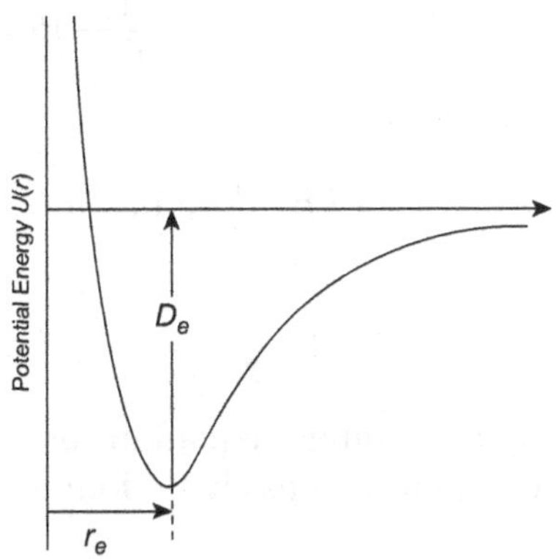

FIGURE 8.1 Potential energy curve of a diatomic molecule, the minimum energy point corresponds to equilibrium bond length of the molecule.

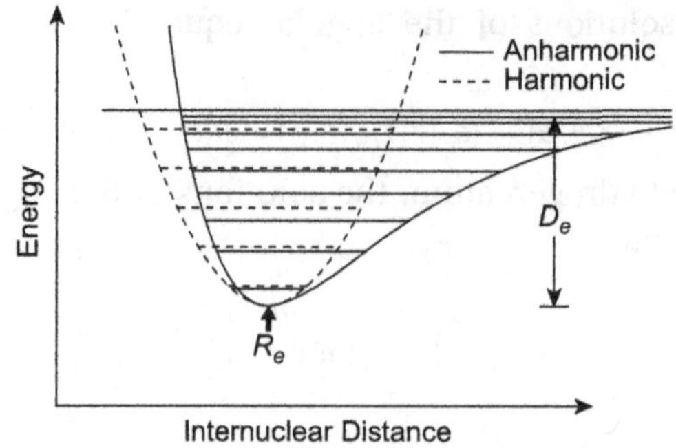

FIGURE 8.2 Harmonic oscillator (dotted line) and anharmonic oscillator (solid line) potential energy curves.

Inserting the expression for $U(r)$ into the radial Eq. (8.2.3) we get,

$$\frac{1}{r^2}\frac{d}{dr}\left(r^2\frac{dR}{dr}\right)+\left[\frac{2\mu_r}{\hbar^2}\left\{E-\frac{1}{2}k(r-r_e)^2\right\}-\frac{J(J+1)}{r^2}\right]R=0 \quad (8.2.7)$$

Substituting $S(r) = r\,R(r)$, this equation reduces to

$$\frac{d^2S}{dr^2}+\left[\frac{2\mu_r}{\hbar^2}\left\{E-\frac{1}{2}k(r-r_e)^2\right\}-\frac{J(J+1)}{r^2}\right]S(r)=0 \quad (8.2.8)$$

If we introduce a new variable $\rho = r - r_e$, then, $\frac{d^2S}{dr^2}=\frac{d^2S}{d\rho^2}$ and the above equation becomes

$$\frac{d^2S}{d\rho^2}+\frac{2\mu_r}{\hbar^2}\left(E-\frac{1}{2}k\rho^2-\frac{\hbar^2}{2\mu}\frac{J(J+1)}{(r_e+\rho)^2}\right)S=0 \quad (8.2.9)$$

We expect the potential $U(r)$ and hence the above equation to be valid only for $r \cong r_e$, that is, for small values of $\rho = r - r_e$. Hence we make a Taylor expansion of $\frac{1}{(r_e+\rho)^2}$

$$\left(\frac{1}{r_e+\rho}\right)^2=\frac{1}{r_e^2}\left(1-\frac{2\rho}{r_e}+\frac{3\rho^2}{r_e^2}-\ldots\right)$$

In the first approximation, if we ignore all powers of $\frac{\rho}{r_e}$, this gives

$$\frac{d^2S}{d\rho^2}+\frac{2\mu_r}{\hbar^2}\left(E-\frac{1}{2}k\rho^2-\frac{\hbar^2}{2\mu}\frac{J(J+1)}{r_e^2}\right)S=0$$

This equation takes a familiar form

$$\frac{d^2S}{d\rho^2}+\frac{2\mu_r}{\hbar^2}\left(E'-\frac{1}{2}k\rho^2\right)S=0 \quad (8.2.10)$$

$$\text{with } E'=E-\frac{\hbar^2}{2\mu_r}\frac{J(J+1)}{r_e^2} \quad (8.2.11)$$

Equation (8.2.10) is just the Schrödinger equation for a one-dimensional harmonic oscillator. A solution of this equation can easily be found if it is reduced to a Hermite equation of the type

$$\frac{d^2H}{d\eta^2} - 2\eta\left(\frac{dH}{d\eta}\right) + \left(\frac{\lambda}{\alpha} - 1\right)H = 0 \tag{8.2.12}$$

$$\text{by putting } \lambda = \frac{2\mu_r E'}{\hbar^2}, \quad \alpha = \frac{\sqrt{\mu_r k}}{\hbar} \text{ and } \eta = \sqrt{\alpha}\rho \tag{8.2.13}$$

The solution of Hermite equation is in the form of polynomials of degree n, called Hermite polynomials, defined as

$$H_n(\eta) = (-1)^n e^{\eta^2} \frac{d^n e^{-\eta^2}}{d\eta^n} \tag{8.2.14}$$

This polynomial terminates and gives finite solutions for

$$\lambda = (2n+1)\alpha, \tag{8.2.15}$$

with n having integer values 0, 1, 2, ….

Functions of the type

$$\Psi_n(\rho) = N_n e^{-\frac{\eta^2}{2}} H_n(\eta), \tag{8.2.16}$$

are called Hermite orthogonal functions. They have the property that

$$\begin{aligned} \int_{-\infty}^{+\infty} \Psi_n(\rho)\Psi_m(\rho)d\rho &= 0, \quad \text{if } n \neq m \\ &= 1, \quad \text{if } n = m \text{ for } N_n = \left\{\sqrt{\frac{\alpha}{\pi}} \frac{1}{2^n n!}\right\}^{\frac{1}{2}} \end{aligned} \tag{8.2.17}$$

where N_n is called normalization constant.

Hermite polynomials follow the recursion formula

$$2\eta\, H(\eta) = H_{n+1}(\eta) + 2nH_{n-1}(\eta) \tag{8.2.18}$$

Using Eqs (8.2.13)–(8.2.16) the solution of Eq. (8.2.10) gives the eigenvalues and eigenfunctions of the harmonic oscillator. These are,

$$E'_n = \hbar\omega\left(n + \frac{1}{2}\right) \tag{8.2.19}$$

$$\text{and } S_n(\rho) = H_n(\eta) e^{-\frac{1}{2}\eta^2}$$

$$\text{or, } S_n(\rho) = H_n\left(\sqrt{\frac{\mu_r \omega}{\hbar}}\rho\right) e^{-\frac{1}{2}\frac{\mu_r \omega}{\hbar}\rho^2} \tag{8.2.20}$$

where $\omega = \sqrt{\frac{k}{\mu_r}}$ is the angular frequency and $n = 0, 1, 2, 3, \ldots$ are called the vibrational quantum numbers.

Thus, a harmonic oscillator has equally spaced quantized energy levels E'_n.

For $n = 0$ the oscillator has energy $E'_0 = \frac{1}{2}\hbar\omega$, which is known as the zero-point energy.

Using Eq. (8.2.19), we may write Eq. (8.2.11) as,

$$E = E' + \frac{\hbar^2}{2\mu_r}\frac{J(J+1)}{r_e^2} = \hbar\omega\left(n + \frac{1}{2}\right) + \frac{\hbar^2}{2\mu_r r_e^2} J(J+1) \tag{8.2.21}$$

In the simplest approximation where we have assumed that Ψ vanishes for $\rho \neq 0$ or $r \neq r_e$, we see that the molecule behaves as a rigid rotator with moment of inertia μr_e^2 and a harmonic oscillator of angular frequency ω. The first term in this equation represents the energy of the harmonic oscillator Eq. (8.2.19) and the second term represents the energy of the rigid rotator

$$E_{\mathrm{rot}} = \frac{\hbar^2}{2\mu_r r_e^2} J(J+1) \tag{8.2.22}$$

where $J = 0, 1, 2\ldots$, are called the rotational quantum numbers.

In this approximation the wavefunction for relative motion is given by

$$\Psi_{n,J,M}(r,\theta,\phi) = S(r)Y_{JM}(\theta,\phi) = e^{-\frac{1}{2}\frac{\mu_r\omega}{\hbar}\rho^2} H_n\left(\sqrt{\frac{\mu_r\omega}{\hbar}}\rho\right) Y_{JM}(\theta,\phi) \tag{8.2.23}$$

that is, the angular degrees of freedom all go into the rotational motion and the radial degree of freedom into a one-dimensional harmonic oscillator. The spherical harmonics representing the rotational wavefunctions have the same form as for hydrogen atom except that l is now replaced by J.

Now let us see some of the effects of nonrigidity, that is, the interatomic distance is no more fixed but the variation from equilibrium is still small and the motion is harmonic. In this case, in the Taylor expansion of $\left(\frac{1}{r_e+\rho}\right)^2$ we keep up to the third term $\frac{3\rho^2}{r_e^2}$ and ignore all higher powers of $\frac{\rho}{r_e}$. The radial Eq. (8.2.8) now takes the form.

The radial Eq. (8.2.8) now takes the form

$$\frac{d^2S}{d\rho^2} + \frac{2\mu_r}{\hbar^2}\left[E - J(J+1)\sigma + 2J(J+1)\frac{\sigma}{r_e}\rho - 3J(J+1)\frac{\sigma}{r_e^2}\rho^2 - \frac{1}{2}k\rho^2\right]S = 0 \tag{8.2.24}$$

where $\sigma = \frac{\hbar^2}{2\mu_r r_e^2}$.

In the above equation we can eliminate the term linear in ρ by a transformation $\rho = \xi + a$, where a is a constant

$$a = \frac{J(J+1)\sigma r_e}{3J(J+1)\sigma + \frac{1}{2}kr_e^2}$$

Hence, $\frac{d^2S}{d\rho^2} = \frac{d^2S}{d\xi^2}$ and the radial equation becomes

$$\frac{d^2S}{d\xi^2} + \frac{2\mu_r}{\hbar^2}(E'' - k'\xi^2)S = 0 \tag{8.2.25}$$

$$\text{where, } E'' = E - \frac{J(J+1)\hbar^2}{2\mu_r r_e^2} + \frac{[J(J+1)]^2\sigma^2}{3J(J+1)\sigma + \frac{1}{2}kr_e^2} \tag{8.2.26}$$

$$\text{and } k' = k + \frac{3}{2}\frac{J(J+1)\sigma}{r_e^2} \tag{8.2.27}$$

Equation (8.2.26) is once again a harmonic oscillator-type equation but we see the effect of rotation on the vibration. The energy has been further modified by a $[J(J+1)]^2$

dependence and now the spring constant k has also been modified to k'. Solution of Eq. (8.2.25) gives

$$E'' = \hbar\omega'\left(n+\frac{1}{2}\right);\ \text{where,}\ \omega' = \sqrt{\frac{k'}{\mu_r}} \tag{8.2.28}$$

Substituting for E'' and σ into Eq. (8.2.26), we get

$$E = \hbar\omega'\left(n+\frac{1}{2}\right) + \frac{\hbar^2}{2\mu_r r_e^2}J(J+1) - \left(\frac{\hbar^2}{2\mu_r r_e^2}\right)^2 \frac{[J(J+1)]^2}{3J(J+1)\sigma + \frac{1}{2}kr_e^2} \tag{8.2.29}$$

Now we can make Taylor expansion of the denominator of the last term, that is

$$\frac{1}{3J(J+1)\sigma + \frac{1}{2}kr_e^2} = \frac{1}{\frac{1}{2}kr_e^2}\left[1 - \frac{6J(J+1)\sigma}{kr_e^2} + \ldots\right]$$

which gives, $$E'' = \hbar\omega'\left(n+\frac{1}{2}\right) + \frac{\hbar^2}{2\mu_r r_e^2}J(J+1) - \frac{[J(J+1)]^2\sigma^2}{\frac{1}{2}kr_e^2} + \ldots \tag{8.2.30}$$

From Eq. (8.2.30) it follows that the nonrigidity of a rotator gives the energy as a series expansion in powers of $J(J+1)$.

A more accurate treatment is still necessary because even though a nonrigid rotator modifies the spring constant k to k', we still get equally spaced vibrational levels. The reason is that we have assumed the interatomic potential to be that of a harmonic oscillator whereas in reality the potential is anharmonic.

8.2.1 Diatomic Anharmonic Oscillator

When the amplitude of vibration is very large, that is, $(r - r_e)$ is comparable to r_e, the motion ceases to be harmonic and Eq. (8.2.6) of the potential energy function containing cubic, quartic, etc., terms has to be used. Solution of the Schrödinger equation in this case gives the energy levels for the anharmonic oscillator and also accounts for the vibration–rotation interaction. An exact solution of the Schrödinger equation by using potential energy function in the form of Eq. (8.2.6) is not possible even for simple systems and hence, an approximate solution is obtained by using the perturbation or variation technique. The energy levels for the anharmonic oscillator may be given by Eq. (8.2.31).

$$E = h\nu_e\left(n+\frac{1}{2}\right) - h\nu_e x_e\left(n+\frac{1}{2}\right)^2 + hB_e J(J+1) - h\alpha_e\left(n+\frac{1}{2}\right)J(J+1) - hD_e J^2(J+1)^2 \tag{8.2.31}$$

where $B_e \equiv \frac{h}{8\pi^2 I_e}$, $I_e \equiv \mu_r R_e^2$, $\nu_e x_e$ and α_e are the anharmonicity constants, D_e is the dissociation energy given as $D_e \equiv \frac{4B_e^3}{\nu_e^2}$, and $J = 0, 1, 2, \ldots$ are the rotational quantum numbers of the molecule.

The use of anharmonic potential function gives rise to unequally spaced vibrational levels as shown in Figure 8.2, besides the inclusion of vibrational–rotational interactions.

8.2.2 Selection Rules for Harmonic and Anharmonic Oscillators

From quantum mechanical considerations it shall be shown in Chapter 9 that when radiation interacts with a molecule having dipole moment μ transition between two states m and n shall be allowed only if the transition moment

$$\mu_{mn} = \int \Psi_m \mu \Psi_n d\tau \neq 0.$$

The intensity of the transition is proportional to the square of the transition moment.

Thus, in the case of a harmonic oscillator if Ψ_m and Ψ_n, the wavefunctions of the two states m and n, are given by Eq. (8.2.20) and $\mu = q\rho$ is the dipole moment, where q is the charge on the oscillator and $\rho = r - r_e$ is the atomic displacement from the equilibrium position, then, the transition moment

$$\mu_{mn} = \int_{-\infty}^{+\infty} \Psi_m^*(\rho)\mu\Psi_n(\rho)d\rho = q\int_{-\infty}^{+\infty} \Psi_m^*(\rho)\rho\Psi_n(\rho)d\rho$$

Since $\eta = \sqrt{\alpha}\rho$, we get

$$\mu_{mn} = \frac{1}{\alpha}N_m N_n \int_{-\infty}^{+\infty} e^{-\frac{1}{2}\eta^2} H_m(\eta)\eta H_n(\eta) e^{-\frac{1}{2}\eta^2} d\eta \tag{8.2.32}$$

Using the recursion formula (Eq. (8.2.18)) for the Hermite polynomial, this may be written as

$$\mu_{mn} = \frac{1}{\alpha}N_m N_n \left[\frac{1}{2}\int_{-\infty}^{+\infty} H_m(\eta)e^{-\frac{1}{2}\eta^2} H_{n+1}(\eta)e^{-\frac{1}{2}\eta^2} d\eta + \int_{-\infty}^{+\infty} nH_m(\eta)e^{-\frac{1}{2}\eta^2} H_{n-1}(\eta)e^{-\frac{1}{2}\eta^2} d\eta\right] \tag{8.2.33}$$

Using the orthogonality condition for the Hermite polynomials (Eq. (8.2.17)), the transition moment μ_{mn} shall be different from zero only when

$$m = n + 1 \quad \text{or,} \quad m = n - 1$$

This gives the selection rule,

$$\Delta(m - n) = \pm 1 \tag{8.2.34}$$

Thus, the allowed electric dipole transitions for the harmonic oscillator involve a change of ± 1 in the vibrational quantum number.

In the case of an anharmonic oscillator, the selection rule is

$$\Delta(m - n) = \pm 1, \pm 2, \pm 3, \ldots \tag{8.2.35}$$

8.3 Vibrations of Polyatomic Molecules

The difficulty in the solution of vibrational problem of a polyatomic molecule arises from the fact that there is a coupling between the various vibrational modes. This coupling can be removed by the introduction of normal coordinates, each of which represents a particular vibrational mode.

The Schrödinger equation for polyatomic molecules can be solved in terms of normal coordinates which are linear combination of Cartesian or mass-weighted Cartesian coordinates. A normal coordinate can be used to represent a particular vibrational mode such that for a molecule of N atoms having $3N-6$ ($3N-5$ in the case of linear molecules) vibrational modes, a corresponding number of normal coordinates are used. The kinetic energy and potential energy of a polyatomic molecule can be represented by diagonal matrix in normal coordinate formalism. Before considering quantum mechanical formalism of molecular vibrations, we shall first consider how normal coordinates can be determined using classical mechanics.

8.3.1 Classical Formulation of Molecular Vibrations—Coupled Oscillators

Consider a polyatomic molecule having N atoms. In Cartesian coordinate system, each of the atom may be represented by three coordinates and so the molecule shall be represented by $3N$ coordinates $\xi_1, \xi_2, \ldots \xi_{3N}$. Its kinetic and potential energies may be written as

$$T = \frac{1}{2}\sum_i m_i \dot{\xi}_i^2 \tag{8.3.1}$$

and,

$$V = V_0 + \sum_i \left(\frac{\partial V}{\partial \xi_i}\right)_0 \xi_i + \frac{1}{2}\sum_{i,j}\left(\frac{\partial^2 V}{\partial \xi_i \partial \xi_j}\right)_0 \xi_i \xi_j + \frac{1}{6}\sum_{i,j}\left(\frac{\partial^2 V}{\partial \xi_i \partial \xi_j \partial \xi_k}\right)_0 \xi_i \xi_j \xi_k + \ldots \tag{8.3.2}$$

If $V_0 = 0$ at equilibrium and $\left(\frac{\partial V}{\partial \xi_i}\right) = 0$, being the condition of minima, then restricting to the second-order term corresponding to harmonic oscillator approximation, we get

$$V = \frac{1}{2}\sum_{i,j}\left(\frac{\partial^2 V}{\partial \xi_i \partial \xi_j}\right)_0 \xi_i \xi_j = \frac{1}{2}\sum_{i,j} f'_{ij}\, \xi_i \xi_j \tag{8.3.3}$$

If we use mass-weighted Cartesian coordinates η_i in place of ξ_i, such that

$$\eta_i = \sqrt{m_i}\xi_i \tag{8.3.4}$$

then from Eqs (8.3.1) and (8.3.3), we get

$$T = \sum_i \eta_i^2 \tag{8.3.5}$$

and

$$V = \frac{1}{2}\sum_{i,j} f_{ij}\eta_i \eta_j \tag{8.3.6}$$

where $f_{ij} = \sqrt{m_i m_j}\, f'_{ij}$ represents force constant.

Using Eqs (8.3.5) and (8.3.6), we may now solve the Lagrange's equation of motion

$$\frac{d}{dt}\left(\frac{\partial L}{\partial \dot{\eta}_i}\right) - \frac{\partial L}{\partial \eta_i} = 0 \tag{8.3.7}$$

where $L = T - V$, to give

$$\ddot{\eta}_i + \sum_j f_{ij}\eta_j = 0 \tag{8.3.8}$$

A solution to Eq. (8.3.8) may be found in the form

$$\eta_i = \eta_i^0 \sin\left(t\sqrt{\lambda} + \delta\right), \quad i = 1, 2, \ldots 3N \tag{8.3.9}$$

Substitution of Eq. (8.3.9) in Eq. (8.3.8) gives a set of equations

$$\sum_j f_{ij}\eta_j^0 - \lambda\eta_i^0 = 0$$

$$\text{or } (f_{ii} - \lambda)\eta_i^0 + \sum_{i \neq j} f_{ij}\eta_j^0 = 0 \tag{8.3.10}$$

which may be written in the matrix form as

$$|F - \lambda I|\eta^0 = 0 \tag{8.3.11}$$

where F is a square matrix representing the force constants (also called Hessians in quantum mechanics).

$$F = \begin{pmatrix} f_{11} & f_{12} & \cdots & f_{13N} \\ f_{21} & f_{22} & \cdots & f_{23N} \\ f_{3N1} & f_{3N2} & \cdots & f_{3N,3N} \end{pmatrix} \tag{8.3.12}$$

and η^0 is a column matrix

$$\eta^0 = \begin{pmatrix} \eta_1^0 \\ \eta_2^0 \\ \vdots \\ \eta_{3N}^0 \end{pmatrix} \tag{8.3.13}$$

The set of equations given by Eq. (8.3.11) shall have nontrivial solutions if the determinant

$$|F - \lambda I| = 0 \tag{8.3.14}$$

A solution of these equations shall have $3N$ roots of λ. Each one of them can be substituted into Eq. (8.3.10) to determine the eigenvectors corresponding to the given root λ_i. This will show how coordinates η_i vary with time. Each positive root represents the mode of simple harmonic oscillation of all the atoms at the same frequency, where $\lambda = 4\pi^2\nu^2$ and ν is the normal frequency. Six roots of the equation in case of a nonlinear and five for a linear molecule are zero. $\lambda < 0$ does not occur for molecules with stable equilibrium configuration.

Thus, for $\lambda = \lambda_k$, from Eq. (8.3.10) we can obtain the ratio $\eta^0_{1k} : \eta^0_{2k} : \ldots : \eta^0_{3Nk}$, which may be normalized to give the coefficients $l_{1k}, l_{2k}, \ldots l_{3Nk}$ such that

$$\sum l^2_{ik} = 1 \tag{8.3.15}$$

A coefficient of proportionality K_k relates the corresponding members of the two sets

$$\eta^0_{ik} = K_k l_{ik} \; (i = 1, 2, \ldots 3N) \tag{8.3.16}$$

Thus, we can write each set in the form of a column vector so that altogether the sets are

$$\left|\begin{matrix} l_{11} \\ l_{21} \\ \vdots \\ l_{3N1} \end{matrix}\right|, \left|\begin{matrix} l_{12} \\ l_{22} \\ \vdots \\ l_{3N2} \end{matrix}\right| \cdots \left|\begin{matrix} l_{13N} \\ l_{23N} \\ \vdots \\ l_{3N3N} \end{matrix}\right| \tag{8.3.17}$$

Each of these sets may be represented by a column matrix, say, l_k.

Thus, we get information about the displacement of the nuclei and the relative amplitudes of displacement. The normal modes of vibration for two molecules H_2O and HCN are shown in Figure 8.3.

In the matrix form Eq. (8.3.15) can also be written as

$$\tilde{l}_k l_k = I \tag{8.3.18}$$

where I is a unit matrix. When F is a symmetric matrix ($F = \tilde{F}$), it can be shown that the eigenvectors l_k are orthogonal, that is, for any pair

$$\tilde{l}_j l_k = 0, \quad \text{if } j \neq k \tag{8.3.19}$$

The elements of eigenvectors in Eq. (8.3.13) can be written side by side to form a matrix L of $3N$ dimension.

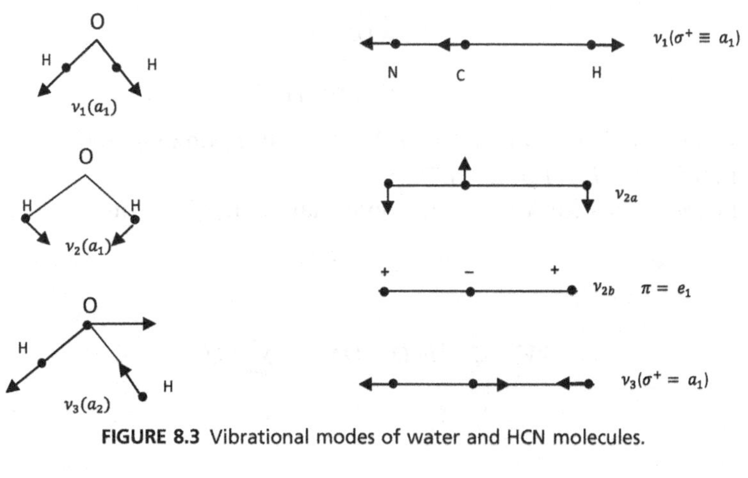

FIGURE 8.3 Vibrational modes of water and HCN molecules.

$$L = \begin{pmatrix} l_{11} & l_{12} & l_{13N} \\ l_{21} & l_{22} & l_{23N} \\ \vdots & \vdots & \vdots \\ l_{3N1} & l_{3N2} & l_{3N3N} \end{pmatrix} \tag{8.3.20}$$

It may be seen that matrix L is also orthogonal and has the property $L^{-1} = \tilde{L}$ or $L\tilde{L} = I$.

8.3.2 Motion in Normal Coordinates

If we formulate the problem in matrix representation, the kinetic energy (Eq. (8.3.5)) and potential energy (Eq. (8.3.6)) may be written as

$$2T = \tilde{\dot{\eta}}\dot{\eta} \tag{8.3.21}$$

$$2V = \tilde{\eta}F\eta \tag{8.3.22}$$

where η is a column matrix of the $3N$ mass-weighted Cartesian displacement coordinates η_i and F is the square matrix representing force constants given by Eq. (8.3.12). Thus, in mass-weighted Cartesian coordinate, the kinetic energy matrix takes up a diagonal form but the potential energy matrix is nondiagonal. This complicates solving of the determinantal Eq. (8.3.14).

Both the kinetic energy and potential energy matrices can be reduced to a diagonal form in terms of what are known as normal coordinates Q_1, Q_2,...Q_{3N} which can be obtained by a linear transformation from another set of displacement coordinates. The transformation from mass-weighted Cartesian coordinates to normal coordinates is easiest to carry out because in both the systems the kinetic energy matrix is diagonal, that is, a sum of squared terms only and the transformation is orthogonal.

$$\eta = LQ \tag{8.3.23}$$

where L is the matrix defined in Eq. (8.3.20).

Equations (8.3.21) and (8.3.22) may then be written in terms of the normal coordinates Q as

$$2T = \tilde{\dot{Q}}\tilde{L}L\dot{Q} \tag{8.3.24}$$

$$\text{and,}\quad 2V = \tilde{Q}\tilde{L}FLQ \tag{8.3.25}$$

In order for the kinetic energy matrix in Eq. (8.3.20) to be diagonal, the L matrix must be a unitary matrix, or $\tilde{L}L = \mathrm{I}$ or $\tilde{L} = L^{-1}$.

Hence the expressions for kinetic and potential energies become

$$2T = \tilde{\dot{Q}}\dot{Q} = \sum_i \dot{Q}_i^2 \tag{8.3.26}$$

$$2V = \tilde{Q}L^{-1}FLQ = \tilde{Q}\Lambda Q = \sum_i \lambda_i \dot{Q}_i^2 \tag{8.3.27}$$

$$\text{where,}\quad \Lambda = L^{-1}FL = \begin{pmatrix} \lambda_1 & 0 & \dots & 0 \\ 0 & \lambda_2 & \dots & 0 \\ \dots & \dots & \dots & \dots \\ 0 & \dots & \dots & \lambda_{3N} \end{pmatrix} \tag{8.3.28}$$

There are no mixed terms in the expression for potential energy because of the similarity transformation. The determinant F therefore has a diagonal form and the secular Eq. (8.3.14) reduces to

$$|\Lambda - \lambda I| = 0 \tag{8.3.29}$$

which has $3N$ roots.

In terms of normal coordinates, the Lagrange's equation becomes

$$\ddot{Q} + \lambda Q = 0$$

$$\text{or } \ddot{Q}_i + \lambda_i Q_i = 0, \quad (i = 1, 2, \dots 3N) \tag{8.3.30}$$

and gives the solution

$$Q_i = Q_i^0 \sin\left(\sqrt{\lambda_i} t + \delta_i\right) \tag{8.3.31}$$

These solutions are orthogonal. Those with $\lambda = 0$ correspond to translations and rotations.

Each normal coordinate represents one normal mode of vibration in which each atom moves about its equilibrium position in the same phase with the same frequency. From Eq. (8.3.19), the displacement η_i in mass-weighted Cartesian coordinates is

$$\eta_i = \sum_k l_{ik} Q_k, \quad i = 1, \dots 3N \tag{8.3.32}$$

Hence, using Eq. (8.3.31)

$$\eta_i = \sum_k l_{ik} Q_k^0 \sin\left(\sqrt{\lambda_i} t + \delta_i\right) = \sum_k A_{ik} \sin\left(\sqrt{\lambda_i} t + \delta_i\right) \tag{8.3.33}$$

$$\text{where,}\quad A_{ik} = l_{ik} Q_k^0 \tag{8.3.34}$$

Let us look at the physical nature of Eq. (8.3.33). Suppose, in a special case all $Q_k^0 = 0$ except for Q_m^0, i.e., $Q_m^0 \neq 0$. In this case Eq. (8.3.28) becomes

$$\eta_i = A_{im} \sin\left(\sqrt{\lambda_m} t + \delta_m\right), \quad i = 1, \dots 3N \tag{8.3.35}$$

A vibration represented by Eq. (8.3.30) is called a normal mode of vibration. In this case the coordinates of each atom vibrate in phase with one another with the same frequency ν_m; obviously from Eq. (8.3.35) $2\pi\nu_m = \sqrt{\lambda_m}$ or $\lambda_m = 4\pi^2\nu_m^2$. For each normal mode, the vibrational amplitude A_{im} of each atomic coordinate is constant, but the amplitudes of different coordinates are, in general, different. The nature of normal modes depends on molecular geometry, nuclear masses, and the value of force constants f_{ik}.

8.3.3 Solution of Vibrational Problem in Internal Coordinates—Wilson GF-Matrix Method

Traditionally the vibrational problem is solved in terms of internal coordinates instead of Cartesian coordinates though in some cases the latter may be more convenient and computationally economical to handle. The internal coordinates may be classified into six different categories namely, bond stretching, angle bending, rocking, wagging, twisting, and out-of- plane deformation. There are $3N-6$ internal coordinates in nonlinear molecules and $3N-5$ in linear molecules. The internal coordinates (S) may be constructed from Cartesian coordinates matrix (X) using a transformation matrix B.

$$S = BX \tag{8.3.36}$$

Since there are $3N-6$ ($3N-5$ for a linear molecule) internal coordinates S and $3N$ Cartesian coordinates X, the transformation matrix B is a $(3N-6 \times 3N)$ rectangular matrix.

Also,

$$X = B^{-1}S \tag{8.3.37}$$

Since B is a rectangular matrix, the inverse can be found by adding zeros to make B a square matrix.

In matrix representation, the kinetic energy is

$$2T = \tilde{\dot{X}}M\dot{X} \tag{8.3.38}$$

where M is a diagonal matrix of order $3N$ whose elements are atomic masses, each being present three times. Then,

$$2T = \tilde{\dot{S}}\tilde{B}^{-1}MB^{-1}\dot{S} \tag{8.3.39}$$

Likewise, the potential energy in the matrix form is

$$2V = \tilde{X}fX = \tilde{S}\tilde{B}^{-1}fB^{-1}S \tag{8.3.40}$$

Let

$$\tilde{B}^{-1}MB^{-1} = G^{-1} \tag{8.3.41}$$

$$\text{and,}\quad \tilde{B}^{-1}fB^{-1} = F \tag{8.3.42}$$

Then, we can write Eqs (8.3.39) and (8.3.40) as

$$2T = \tilde{\dot{S}}G^{-1}\dot{S} \tag{8.3.43}$$

$$\text{and,}\quad 2V = \tilde{S}FS \tag{8.3.44}$$

Equations (8.3.43) and (8.3.44) represent the kinetic and potential energies in internal coordinates. F is the force constant matrix in internal coordinates.

The internal coordinate S may be transformed into normal coordinates Q by using the transformation matrix L

$$S = LQ \quad \text{or} \quad Q = L^{-1}S \tag{8.3.45}$$

We get from Eqs (8.3.44) and (8.3.45)

$$2T = \tilde{\dot{Q}}\tilde{L}G^{-1}L\dot{Q} \tag{8.3.46}$$

$$2V = \tilde{Q}\tilde{L}FLQ \tag{8.3.47}$$

We have seen from Eqs (8.3.26) and (8.3.27) that the kinetic and potential energies take diagonal form in normal coordinates, hence comparing Eq. (8.3.46) with Eq. (8.3.26) and Eq. (8.3.47) with Eq. (8.3.27), we get

$$\tilde{L}G^{-1}L = I$$

$$\text{or,} \quad G = L\tilde{L} \tag{8.3.48}$$

$$\text{and} \quad \tilde{L}FL = \Lambda \tag{8.3.49}$$

On multiplying Eq. (8.3.49) by L and using Eq. (8.3.48), we get

$$L\tilde{L}FL = L\Lambda$$

$$GFL = L\Lambda$$

$$\text{or,} \ (GF - \Lambda I)L = 0 \tag{8.3.50}$$

Finally, from Eq. (8.3.41),

$$G = BM^{-1}\tilde{B} \tag{8.3.51}$$

In Wilson representation, G is the kinetic energy matrix which depends upon the relationship between the internal coordinates and the Cartesian coordinates as expressed through matrix B. M^{-1} is a diagonal matrix of order $3N$ whose elements are the inverses of atomic masses. The potential energy matrix F is obtained from Eq. (8.3.44). In the classical approach of normal coordinate analysis, the choice of force constants is the most critical step. This results from the fact that some force constants must either be assigned arbitrary values, or be constrained by an arbitrary choice of force field to bear certain relationships to other force constants. The principle of transferability of force constants is also applied and the force constants for larger molecules may be given the same initial values as to similar functional groups in smaller molecules. Equation (8.3.50) is then solved, which will give nontrivial solutions if

$$|GF - \Lambda I| = 0 \tag{8.3.52}$$

$3N-6$ values of λ (where $\lambda = 4\pi^2\nu^2$) are obtained to give $(3N-6)$ vibrational frequencies. The force constants matrix F is refined to give a correct agreement between the calculated and the experimental vibrational frequencies obtained through infrared or Raman Spectroscopy measurements. The solution of Eq. (8.3.50) also provides L matrix and hence the atomic displacements for each vibrational modes. These help in the analysis of the vibrational spectral bands in terms of the frequencies and forms of vibrations.

8.3.4 Quantum Mechanics of Molecular Vibrations

In terms of normal coordinates, the total classical vibrational energy of a molecule is the sum of kinetic and potential energies (Eqs (8.3.26) and (8.3.27)) is given as

$$E_v = \frac{1}{2}\sum_i \dot{Q}_i^2 + \frac{1}{2}\sum_i \lambda_i Q_i^2 \tag{8.3.53}$$

where summation is over $(3N-6)$ normal coordinates that represent genuine vibrations.

Generalized momentum in classical mechanics is given by the relation $P_j = \frac{\partial L}{\partial \dot{q}_j}$.

$$\text{In our case,}\quad 2L = 2T - 2V = \sum_i \dot{Q}_i^2 - \sum_i \lambda_i Q_i^2$$

and so,

$$P_{Q_i} = \frac{\partial L}{\partial \dot{Q}_i} = \dot{Q}_i \tag{8.3.54}$$

$\dot{Q}_i$ may, therefore, be taken as the momentum conjugate to Q_i and replaced by the operator $\left(-i\hbar\frac{\partial}{\partial Q_i}\right)$. The quantum mechanical vibrational Hamiltonian operator will therefore be

$$H_v = -\frac{\hbar^2}{2}\sum_i \frac{\partial^2}{\partial Q_i^2} + \frac{1}{2}\sum_i \lambda_i Q_i^2 \tag{8.3.55}$$

and the Schrödinger equation is

$$\Psi_v H_v = E_v \Psi_v \tag{8.3.56}$$

Since the normal coordinates $Q_1, Q_2, \ldots, Q_{3N-6}$ are independent of each other, the total wavefunction can be written as product of wavefunctions,

$$\Psi_v = \Psi(Q_1)\Psi(Q_2)\ldots\Psi(Q_{3N-6}) = \prod_{k=1}^{3N-6}\Psi(Q_K) \tag{8.3.57}$$

The wave Eq. (8.3.56) will then separate out into $3N-6$ equations

$$-\frac{\hbar^2}{2}\frac{\partial^2\Psi(Q_i)}{\partial Q_i^2} + \frac{1}{2}\lambda_i Q_i^2\Psi(Q_i) = E_i\Psi(Q_i) \quad i = 1, 2, \ldots 3N-6 \tag{8.3.58}$$

with eigenvalues E_i, such that the total energy is

$$E_v = \sum_i E_i \tag{8.3.59}$$

Equation (8.3.58) can be rewritten as,

$$\frac{\partial^2\Psi(Q_i)}{\partial Q_i^2} + \frac{2}{\hbar^2}\left(E_i - \frac{1}{2}\lambda_i Q_i^2\right)\Psi(Q_i) = 0 \tag{8.3.60}$$

which is the same equation as that for the harmonic oscillator. So, in terms of normal coordinates the wave equation for molecular vibrations splits up into $3N-6$ separate

Harmonic oscillator equations and each of these can be solved separately to obtain eigenvalues and eigenfunctions.

The eigenvalues of the ith equation are

$$E_i = h\nu_i\left(v_i + \frac{1}{2}\right) \tag{8.3.61}$$

where ν_i is the classical vibrational frequency for normal vibrational mode associated with normal coordinate Q_i, with $\lambda_i = 4\pi^2\nu_i^2$; λ_i is the root of the secular equation. $v_i = 0, 1, 2, \ldots \infty$ are the vibrational quantum numbers of the ith normal mode.

The vibrational eigenfunctions are

$$\Psi_{v_i}(Q_i) = N_{v_i}H_{v_i}(\zeta_i)e^{\frac{\zeta_i^2}{2}} \tag{8.3.62}$$

where H_{v_i} is the Hermitian polynomial of degree v_i and

$$\zeta_i = \frac{Q_i}{\hbar}\sqrt{\lambda_i^{1/2}} \tag{8.3.63}$$

Equations (8.3.59) and (8.3.61) give total energy

$$E_v = h\nu_1\left(v_1 + \frac{1}{2}\right) + h\nu_2\left(v_2 + \frac{1}{2}\right) + \ldots + h\nu_{3N-6}\left(v_{3N-6} + \frac{1}{2}\right) = \sum_{i=1}^{3N-6} h\nu_i\left(v_i + \frac{1}{2}\right) \tag{8.3.64}$$

For each normal mode the vibrational quantum number v_i can independently be zero or can take positive integral values, $v_i = 0, 1, 2, 3\ldots.$

The total eigenfunctions from Eqs (8.3.57) and (8.3.62) would be

$$\Psi_v = NH_{v_1}(\zeta_1)H_{v_2}(\zeta_2)\ldots\exp\left[-\frac{1}{2}\left(\zeta_1^2 + \zeta_2^2 + \ldots + \zeta_{3N-6}^2\right)\right] \tag{8.3.65}$$

where N is a normalization constant for the wavefunction.

If a doubly degenerate vibration occurs, then two of the frequencies in Eq. (8.3.64) are equal. If these are called ν_m and ν_n, then the corresponding energies are

$$h\nu_m\left(v_m + \frac{1}{2}\right) + h\nu_n\left(v_n + \frac{1}{2}\right) = h\nu_m(v_m + v_n + 1) \tag{8.3.66}$$

In general the energy expression can be written as

$$E_v = \sum_{i=1}^{3N-6} h\nu_i\left(v_i + \frac{d_i}{2}\right) \tag{8.3.67}$$

where d_i represents the degree of degeneracy.

The energy of the lowest vibrational state corresponding to $v_i = 0$ is known as the zero-point energy. Thus, from Eq. (8.3.64), the zero-point energy E_0 is

$$E_0 = \frac{1}{2}\sum_{i=1}^{3N-6} h\nu_i \tag{8.3.68}$$

Because of large numbers of vibrational modes in a polyatomic molecule, the zero-point energy can be substantial.

8.3.5 Selection Rules for Vibrational Transitions in Polyatomic Molecules

We now consider the electric dipole selection rules for radiative transitions between the vibrational levels of the same electronic state of a polyatomic molecule. A transition between the vibrational levels v' and v'' will be allowed if the transition moment $\boldsymbol{\mu}_{v'v''}$ has a nonzero value.

$$\boldsymbol{\mu}_{v'v''} = \int \Psi^*_{v'} \boldsymbol{\mu} \Psi_{v''} d\tau \neq 0 \tag{8.3.69}$$

where $\boldsymbol{\mu}$ is the permanent electric dipole moment.

$$\boldsymbol{\mu} = \mathbf{i}\mu_x + \mathbf{j}\mu_y + \mathbf{k}\mu_z \tag{8.3.70}$$

where x, y, z are the molecule-fixed principal axes of inertia and **i**, **j**, **k** are unit vectors along these axes.

We can expand μ_x, μ_y, and μ_z in Taylor series about the equilibrium nuclear configuration for which all $Q_k = 0$.

$$\mu_x(Q_1, Q_2, \ldots Q_{3N-6}) = \mu_{x_0} + \sum_{k=1}^{3N-6} \left(\frac{\partial \mu_x}{\partial Q_k}\right)_0 Q_k + \frac{1}{2} \sum_{j=1}^{3N-6} \sum_{k=1}^{3N-6} \left(\frac{\partial^2 \mu_x}{\partial Q_j \partial Q_k}\right)_0 Q_j Q_k + \ldots \tag{8.3.71}$$

where μ_{x_0} is the x component of the dipole moment for the equilibrium nuclear configuration. The quadratic and higher terms in Eq. (8.3.71) are believed to be very small and so we may neglect them. Hence, from Eqs (8.3.69) and (8.3.71) we get,

$$\int \Psi^*_{v'} \mu_x \Psi_{v''} d\tau = \mu_{x_0} \int \Psi^*_{v'} \Psi_{v''} d\tau + \sum_{k=1}^{3N-6} \left(\frac{\partial \mu_x}{\partial Q_k}\right)_0 \int \Psi^*_{v'} Q_k \Psi_{v''} d\tau \tag{8.3.72}$$

The vibrational wavefunction is a product of harmonic oscillator functions, one for each normal coordinate. Hence

$$\begin{aligned} \int \Psi^*_{v'} \Psi_{v''} d\tau &= \prod_{k=1}^{3N-6} \int \Psi^*_{v'_k}(Q_k) \Psi_{v''_k}(Q_k) dQ_k \\ &= \int \Psi_{v'_k}(Q_k) \Psi_{v''_k}(Q_k) dQ_k \prod_{j \neq k} \Psi_{v'_j}(Q_j) \Psi_{v''_j}(Q_j) dQ_j \end{aligned} \tag{8.3.73}$$

and

$$\int \Psi^*_{v'} Q_k \Psi_{v''} d\tau = \int \Psi^*_{v'_k}(Q_k) Q_k \Psi_{v''_k}(Q_k) dQ_k \prod_{j \neq k} \Psi_{v'_j}(Q_j) \Psi_{v''_j}(Q_j) dQ_j \tag{8.3.74}$$

Since the wavefunctions $\Psi_{v'_k}$ given by Eq. (8.3.62) for the harmonic oscillator form an orthonormal set, the integral Eq. (8.3.73) will vanish unless

$$v'_j = v''_k, \quad k = 1, 2, \ldots 3N-6 \tag{8.3.75}$$

So, the first term on the right of Eq. (8.3.72) shall correspond to a transition with no change in the vibrational quantum number and hence to a pure rotational transition. Also, if $\mu_{x_0} = 0$, the first term in Eq. (8.3.72) shall be zero. Hence, it is an essential condition that at equilibrium the molecule must have a nonzero permanent dipole moment to exhibit pure rotational transitions.

Similarly, since the wavefunctions of the harmonic oscillator are expressed in terms of Hermite polynomials, the second term in Eq. (8.3.72) would be different from zero only if

$$\Delta v_k = v'_k - v''_k = \pm 1, \quad k = 1, 2, \ldots 3N - 6$$
$$\text{and } \Delta v_j = v'_j - v''_j = 0 \quad \text{for } j \neq k \qquad (8.3.76)$$

Thus, the second term in Eq. (8.3.72) gives rise to vibrational transitions in which one vibrational quantum number changes by unity while all other vibrational quantum numbers are unchanged. Equation (8.3.76) gives the selection rules for a harmonic oscillator.

8.3.6 Fundamental Bands, Overtones, and Combination Tones

Vibrational energy levels for a molecule with three normal modes are shown in Figure 8.4. The vibrational quantum numbers of each mode are given in parenthesis like $(v_1, v_2, \ldots v_{3N-6})$. The levels with one $v_i = 1$ and all vibrational quantum numbers equal to zero are called fundamental levels. For example, levels (1,0,0), (0,1,0), (0,0,1) in Figure 8.4 are called fundamental levels. Levels with one $v_i > 1$ and all others zero are called overtone levels. Levels with more than one nonzero v_i are called combination levels. While in the case of diatomic molecules the vibrational frequencies are high enough to make the population of excited vibrational levels negligible at room temperatures, it is not so in polyatomic molecules. Molecules having five or more atoms generally have one or more

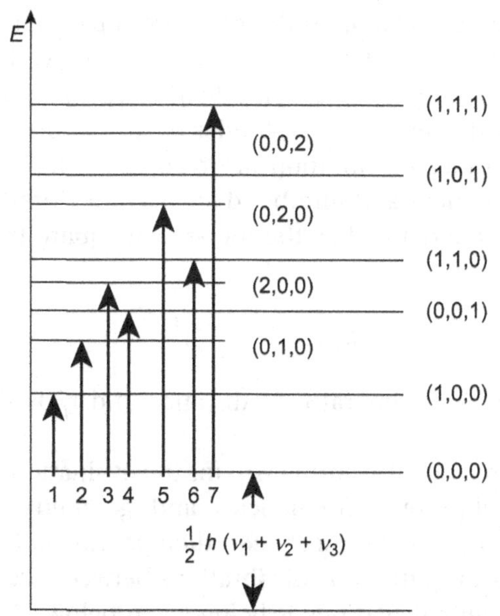

FIGURE 8.4 Vibrational energy levels, overtones and combination tones for a molecule with three normal modes.

low-lying vibrational levels which are appreciably populated at room temperature. The transition energies between the vibrational levels may be given as

$$h\nu_v = E_{v'} - E_{v''} = h\sum_k \nu_k(v'_k - v''_k)$$

where $E_{v'}$ and $E_{v''}$ are energies of the upper and lower vibrational states. Thus, a transition from the zero-point level (0,0,0) to (1,0,0) level shall lead to a fundamental band of frequency ν_1. A transition from the (0,0,0) level to one of the overtone or combination levels shall give rise to overtones and combination tone frequencies. Thus, a transition from (0,0,0) to (2,0,0) shall lead to a band of frequency $2\nu_1$ (the first overtone) and from (0,0,0) to (1,1,0) to a band of frequency $(\nu_1 + \nu_2)$, the first combination tone. It is also possible to have a band of frequency $(\nu_1 - \nu_2)$, if the transition occurs between levels (1,0,0) and (0,1,0).

As for the degeneracies of the vibrational levels in the harmonic oscillator approximation, the ground vibrational level can have only one set of quantum numbers (0,0,..., 0) and hence it is always nondegenerate. If none of the normal modes are degenerate, then each different set of vibrational quantum numbers gives a different value to E_v (Eq. (8.3.64)), and each vibrational level is nondegenerate. If at least one vibrational mode is degenerate, then there is degeneracy in the energy levels. For example, in HCN (Figure 8.3) (or any linear triatomic molecule), the modes ν_{2a} and ν_{2b} are doubly degenerate. These may be represented as $(v_1, 01, v_3)$ and $(v_1, 10, v_3)$ where the representation is $(v_1, v_{2a}, v_{2b}, v_3)$.

8.3.7 Mean Amplitude of Vibration

It is well known that vibrational amplitudes of an atom pair in a molecule are indicators of the electron overlap between atoms and hence the energy of the system during their vibration. In general terms the mean amplitude of vibration characterizes the looseness or rigidity of a molecule. Large mean amplitudes occur for loosely bonded atoms or vice versa. The mean amplitudes of vibration are recognized as important parameters, in addition to the bond distances and interbond angles of a vibrating molecule. The mean amplitudes of vibration are defined as the root-mean-square (rms) values according to the relation

$$\langle l_{ij}\rangle = \left\langle \left(r_{ij} - r_{ij}^e\right)^2 \right\rangle^{1/2}$$

where r_{ij} is the instantaneous interatomic distance and r_{ij}^e is the corresponding equilibrium distance.

Cyvin [1,2] developed a detailed formalism for the evaluation of these quantities from a knowledge of the vibrational frequencies and geometry of the molecule. This formalism involves the use of symmetry coordinates. An outline of the principles for calculating mean square amplitudes l_{ij}^2 of vibrations between atoms i and j in a form that is convenient to computer calculations is being given below.

Consider an arbitrary displacement d_{ij} in the interatomic distance between two atoms i and j (bonded or nonbonded) for which we wish to calculate the mean amplitude.

In terms of Cartesian coordinates $(x_1, y_1, z_1, \ldots x_n, y_n, z_n)$, we may write

$$d_{ij} = l_{ij}^x(x_i - x_j) + l_{ij}^y(y_i - y_j) + l_{ij}^z(z_i - z_j) \tag{8.3.77}$$

where $l_{ij}^x, l_{ij}^y, l_{ij}^z$ are the components of a unit vector from atom j to i, given by

$$l_{ij}^x = \frac{\left(x_i^e - x_j^e\right)}{R_{ij}^e}, \ldots \tag{8.3.78}$$

x_i^e, y_i^e, z_i^e are the Cartesian coordinates of the equilibrium positions of atoms i and j, and R_{ij}^e is the interatomic distance,

$$R_{ij}^e = \left[\left(x_i^e - x_j^e\right)^2 + \left(y_i^e - y_j^e\right)^2 + \left(z_i^e - z_j^e\right)^2\right]^{1/2} \tag{8.3.79}$$

Equation (8.3.77) can be written in matrix representation as

$$d = lx \tag{8.3.80}$$

where x is a column matrix of Cartesian displacement coordinates of all atoms in the molecule and rows of l contain the appropriate elements l_{ij} from Eq. (8.3.77).

In terms of normal coordinates Q_k, we may write d_{ij} as,

$$d_{ij} = \sum_k \left(\gamma_{ij}\right)_k Q_k \tag{8.3.81}$$

where γ_{ij} are the elements of the transformation matrix between atomic displacements d_{ij} and the normal coordinates. In matrix representation, this equation may be written as

$$d = \gamma Q \tag{8.3.82}$$

where γ matrix has as its elements γ_{ij}. The mean square amplitude of vibration $\langle d_{ij}^2 \rangle$ may then be given as

$$\left\langle l_{ij}^2 \right\rangle = \left\langle d_{ij}^2 \right\rangle = \sum_k \left(\gamma_{ij}\right)_k^2 \langle Q_k^2 \rangle \tag{8.3.83}$$

where,

$$\langle Q_k^2 \rangle = \frac{h}{8\pi^2 c\omega_k} \coth\left(\frac{hc\omega_k}{2kT}\right) \tag{8.3.84}$$

is the well-known temperature-dependent frequency parameter (δ_k).

If we are working in internal coordinates (valence or symmetry) designated by column matrix S, using the standard relations, we can write

$$X = AS, \quad A = M^{-1}\tilde{B}G^{-1} \tag{8.3.85}$$

Thus, using Eq. (8.3.80), we get

$$d = lM^{-1}\tilde{B}G^{-1}S, \quad \text{where, } S = LQ \tag{8.3.86}$$

Consequently

$$d = lM^{-1}\tilde{B}G^{-1}LQ \tag{8.3.87}$$

Thus, the coefficients γ_{ij} of Eq. (8.3.81) are the elements in the row of the matrix $(lM^{-1}\tilde{B}G^{-1}L)$.

Defining Γ as a matrix formed from the squares of the γ matrix, one has the mean square amplitudes in the form of a column matrix

$$|l^2| = \Gamma|\delta_k| \tag{8.3.88}$$

where $|\delta_k|$ is a column matrix of the frequency parameters defined by Eq. (8.3.84).

8.3.7.1 Practical Applications of Mean Amplitudes of Vibration

Use of calculated mean amplitudes of vibration in the interpretation of gas phase electron diffraction measurements is still the most common application of mean amplitudes. Attempts have also been made to obtain mean amplitudes from electron diffraction measurements and then use these values as constraints in normal coordinate analysis of vibrational motion. However, these cannot be used to calculate force constants with sufficient accuracy. The situation is much more favorable for Coriolis coupling constants, centrifugal distortion constants, and isotopic frequency shifts. The calculated mean amplitudes may be decisive, however, when choosing between substantially different sets of force constants or gross trends in frequency assignments.

It should however be recognized that the use of mean amplitudes of vibration is not a very sensitive method in the structural analysis of simple molecules, but the procedure introduces a new constraint namely, minimization of amplitudes to fix the geometry.

8.3.8 Potential Energy Distribution

Potential energy distribution is defined as the way in which each coordinate contributes to the potential energy of each normal vibration. Potential energy of the kth normal mode is given from Eq. (8.3.27) as $\lambda_k Q_k^2$. The potential energy for a unit displacement $Q_k = 1$ is thus equal to λ_k, which is given as

$$\lambda_k = \sum_m \sum_n l_{mk} l_{nk} f_{mn}$$

$$= \sum_m l_{mk}^2 f_{mm} + \sum_{m \neq n} l_{mk} l_{nk} f_{mn} \tag{8.3.89}$$

$$\text{or} \sum_m \frac{l_{mk}^2 f_{mm}}{\lambda_k} + \sum_{m \neq n} \frac{l_{mk} l_{nk}}{\lambda_k} f_{mn} = 1 \tag{8.3.90}$$

Here, l_{mk} and l_{nk} are the elements of the transformation matrix L connecting the internal coordinates R and the normal coordinates Q,

$$R_m = \sum_k l_{mk} Q_k$$

$$\text{or } R = LQ \tag{8.3.91}$$

The first term in Eq. (8.3.90) corresponds to the potential energy contribution of the principal force constant f_{mm} to the normal mode λ_k and the second represents the contribution of the interaction force constant f_{mn}. For the purpose of potential energy distribution, the term in Eq. (8.3.89) involving the off-diagonal force constants f_{mn} is usually ignored so that potential energy distribution for unit displacement Q_k is given as

$$l_{1k}^2 f_{11} + l_{2k}^2 f_{22} + \ldots l_{nk}^2 f_{nn} \tag{8.3.92}$$

Similar equations can be obtained for all the normal modes and if each term is constituted as a separate element, there will result a matrix V whose elements are given as $V_{ij} = l_{ji}^2 F_{jj}$. Each element V_{ij} gives the potential energy contribution from the internal coordinate R_j to the normal coordinate Q_i. Thus, each row corresponds to one normal coordinate and the elements of that row give the relative contribution from each internal coordinate to the total potential energy associated with that normal coordinate.

In general $\frac{L_{ms}^2}{\lambda_s}$ terms are large and provide a reasonable measure of the contribution of the internal coordinates R_m to the normal coordinates Q_s.

8.3.9 Intensity of Infrared Vibrational Bands

The intensities of absorption bands in infrared spectra help to a great extent in the study of molecular structures. It is possible to characterize the properties of electron clouds which are reflected in the intensities and polarization of spectral bands in terms of dipole moments of the molecular groups. The intensity of absorption bands is proportional to $\left(\frac{\partial \mu}{\partial Q_i}\right)^2$, where μ is the dipole moment and Q_i is the normal coordinate corresponding to the given normal mode of vibration. Infrared intensities are evaluated by taking the first derivative of the dipole moment. These calculations typically use numerical differentiation.

The dipole moment relative to the normal coordinate is written as

$$\frac{\partial \mu}{\partial Q_i} = \sum_k L_{ik} \frac{\partial \mu}{\partial R_k} \tag{8.3.93}$$

where the internal coordinates and normal coordinates are related through Eq. (8.3.91). In the numerical differentiation technique the dipole moments are calculated by quantum chemical methods while the dipole moment derivatives $\frac{\partial \mu}{\partial R_k}$ are evaluated by numerical differentiation. In general, the derivatives of dipole moments and polarizabilities (in the case of Raman spectra) are calculated for several different distorted structures to determine the numerical derivatives according to a central difference formula. If a 3-point central difference formula is to be used, the data must be calculated for $6N$ different structures of the molecule because every atomic Cartesian coordinate has to be distorted in positive and negative directions. Usually, the distortion in bond length is taken to be ±0.001 Å and in bond angles to be ±0.025 radians. The computational effort is reduced by taking advantage of molecular symmetry.

The use of numerical derivatives is time-consuming and may affect precision. Yamaguchi et al. [3] have reported expression for the analytical evaluation of the dipole moment derivatives which are derived by solving coupled perturbed Hartree–Fock (CPHF) equations. In this formulation, the dipole moment is derived as derivative of energy E_{SCF} with respect to an external field $\mathcal{E}$ and the derivative of the dipole moment with respect to a nuclear coordinate "a" is then used to obtain the intensity of vibrational bands. The analytical expression of Yamaguchi et al. [3] for closed-shell self-consistent field (SCF) wavefunctions is

$$\frac{\partial^2 E_{SCF}}{\partial \mathcal{E} \partial a} = 2 \sum_{i}^{\text{occ MO}} \frac{\partial^2}{\partial \mathcal{E} \partial a}(h_{ii}) + 4 \sum_{i}^{\text{occ MO}} \sum_{j}^{\text{all}} \frac{\partial (U_{ji})}{\partial a} \frac{\partial}{\partial \mathcal{E}}(h_{ij}) \tag{8.3.94}$$

where

$$E_{SCF} = 2 \sum_{i}^{\text{occ MO}} h_{ii} + \sum_{i,j}^{\text{occ MO}} \{2(ii|jj) - (ij|ji)\}$$

h_{ii} and h_{ij} are matrix elements entering the Fock-operator (Eq. (3.8.23) and $\frac{\partial U_{ji}}{\partial a}$ is defined by

$$\frac{\partial C_{\mu i}}{\partial a} = \sum_{m}^{\text{all}} \frac{\partial (U_{mi})}{\partial a} C_{\mu m}^{0} \tag{8.3.95}$$

where $C_{\mu i}$ is the μth coefficient of the ith molecular orbital. The superscript "'0"' designates an unperturbed value and "all" implies a sum over all occupied and virtual molecular orbitals.

While using this analytical expression, the diagonalization of the force constant matrix in mass-weighted Cartesian coordinates provides the normalized modes and vibrational frequencies. Transformation of the dipole moment derivatives with respect to nuclear coordinates (Eq. (8.3.94)) to normal coordinates by using Eq. (8.3.93) will then provide the infrared intensities.

8.4 Quantum Chemical Determination of Force Field

Vibrational analysis in the quantum chemical method is performed in a manner identical to the normal coordinate analysis by the GF matrix method (Eq. (8.3.45)) proposed by Wilson, Decius, and Cross [4]. In the quantum chemical methods, the force constants f_{ij} or the Hessian are obtained as the second derivative of the energy, that is,

$$f_{ij} = H_{ij} = \frac{\partial^2 E}{\partial x_i \partial x_j} \tag{8.4.1}$$

These force constants constitute the Hessian matrix H. Thus, the Hessian matrix is the matrix of second derivatives of the energy with respect to geometry. The Hessian is based on Cartesian rather than internal coordinates. In the calculation of Hessian matrix a harmonic oscillator approximation is assumed for making the calculations more tractable.

Although the first-order derivatives of energy are relatively easy to calculate, the second-order derivatives are not. The simplest way to calculate second-order derivatives is to calculate first-order derivatives for a given geometry, then perturb the geometry, do an SCF calculation on the new geometry, and then recalculate the derivatives. The second-order derivatives can then be calculated from the difference of two first derivatives divided by the step size. Thus, the first step is to calculate a gradient vector **g** defined in the last chapter (Eq. (7.2.3)),

$$\mathbf{g}(q) = \begin{vmatrix} \dfrac{\partial E}{\partial q_1} \\ \dfrac{\partial E}{\partial q_2} \\ \vdots \\ \dfrac{\partial E}{\partial q_n} \end{vmatrix} \tag{8.4.2}$$

If we are working in the internal coordinates, $n = 3N - 6$, where N is the number of atoms, but if we are working in Cartesian coordinates $n = 3N$. If we cannot compute the partial derivatives that make up the gradient vector **g** analytically, we can do so numerically. The Hessian H_{ij} can be written as

$$H_{ij} = \frac{g_i^{+\delta_j} - g_i^{-\delta_j}}{2\delta} \tag{8.4.3}$$

The first term in the numerator represents gradient of energy at a point incremented by $+\delta_j$ and the second term represents the gradient at a second point decremented by the same amount from the initial position, say, x_i. Thus, for $\delta_j = 0.5\Delta x_j$, we get,

$$H_{ij} = \frac{\left(\frac{\delta E}{\delta x_i}\right)_{+0.5\Delta x_j} - \left(\frac{\delta E}{\delta x_i}\right)_{-0.5\Delta x_j}}{\Delta x_j} \tag{8.4.4}$$

This is done for all $3N$ Cartesian coordinates. Since the Hessian is symmetric, that is $H_{ij} = H_{ji}$, the random errors that occur in the gradient calculation can be reduced by redefining the Hessian as

$$H_{ij} = \frac{1}{2}\left[\frac{\left(\frac{\delta E}{\delta x_i}\right)_{+0.5\Delta x_j} - \left(\frac{\delta E}{\delta x_i}\right)_{-0.5\Delta x_j}}{\Delta x_j} + \frac{\left(\frac{\delta E}{\delta x_j}\right)_{+0.5\Delta x_i} - \left(\frac{\delta E}{\delta x_j}\right)_{-0.5\Delta x_i}}{\Delta x_i}\right] \tag{8.4.4}$$

It must be noted that the vibrational analysis using the Hessian matrix is valid only when the first derivatives of the energy with respect to displacement of the atoms are zero. This implies that the geometry used for vibrational analysis must be optimized at the same level of theory and with the same basis set using which the second derivatives were generated.

The Hessians are easily derived if analytic first derivatives are available, otherwise it requires a lot of computational time to construct the Hessian. The quality of Hessian matrix depends upon the method used. Improved values of Hessian result in improved vibrational frequencies.

8.5 Scaling Procedures

The vibrational frequencies calculated by *ab initio* and density functional theory (DFT) methods are consistently higher than the experimental fundamental frequencies by about 10% because of the neglect of electron correlation ad anharmonicity effects, except for the very low frequencies (below about 200 cm^{-1}) which are often quite far from the experimental values. The Hartree–Fock (HF) theory overemphasizes bonding resulting in the force constants being systematically too large and so are all the frequencies. Two approaches namely the scaled quantum mechanical (SQM) method [5] and the direct scaling method [6] have been suggested for obtaining good agreement between the calculated and experimental frequencies. In SQM, Pulay et al. [7–9] suggested refining of the force constants by using scaling factors and then conducting normal coordinate analysis by Wilson's method. In the direct scaling approach, the calculated frequencies are directly scaled using appropriate scaling factors to match the experimental frequencies. Many studies have been reported for the calculation of vibrational frequencies by using *ab initio* methods and multiplying the resulting frequencies by about 0.9 to get a good estimate of the experimental results. Several scaling factors for HF and DFT calculations for different basis sets have been suggested and the errors in calculation have been determined statistically.

The basic features of SQM suggested by Pulay [5] are as follows:

1. The experimental geometry instead of the optimized geometry from the quantum chemical methods is used as the reference geometry. In practice, when a reliable equilibrium structure is not available, the theoretical geometry is corrected on the basis of experience on small molecules.
2. The force constants of similar chemical fragments share the same scaling factor.
3. The diagonal internal force constants are scaled using different scale factors. These scale factors are then used to scale the nondiagonal force constants by using an expression.

$$f'_{ij} = \sqrt{s_i s_j} f_{ij}$$

where s_i and s_j denote the scale factors for the diagonal force constants f_{ii} and f_{jj}.

The advantage of the Pulay's method is that it makes it possible to calculate scale factors that are transferable between similar molecules if suitable internal coordinates are chosen. In a systematic study, Gupta et al. [10] have reported and checked the transferability of force constants and the scaling factors based on SQM calculations on the *cis* and *gauche* conformers and isotopomers of propanal for which complete vibrational spectra in nitrogen and argon matrices were recorded. The scaling factors for the *cis* and the *gauche* conformers were found to be identical indicating transferability of scaling factors in molecules having similar structures.

A group of scientists lead by Radom [11,13] and Wong [12] adopted the second approach, namely directly scaling the calculated frequencies. Such scale factors depend upon the level of electron structure theory and the one-electron basis sets. The frequencies may be scaled in different ways. For example, one may scale them for reproducing the true vibrational harmonic frequencies, the observed vibrational fundamental frequencies, or the vibrational zero-point energy. Separate sets of scaling factors have also been suggested for the low- and high-vibrational frequencies. Some of the suggested scaling factors for different levels of theory and the commonly used basis sets as provided by Radom et al. [13] are listed in Tables 8.1–8.3.

Table 8.1 Frequency scale factors suitable for fundamental vibrations

	6-31G(d)	6-31+G(d,p)	6-31G(2df,p)	6-311+G(d,p)	6-311+G(2df,p)
Method	**Scale factor**	**Scale factor**	**Scale factor**	**Scale factor**	**Scale factor**
HF	0.8953	0.9007	0.9035	0.9059	0.9073
MP2	0.9441	0.9418	0.9462	0.9523	0.9569
QCISD	0.9536	0.9454	–	0.9560	–
QCISD(*T*)	0.9611	0.9529	–	0.9647	–
CCSD	0.9516	0.9436	–	0.9542	–
CCSD(*T*)	0.9603	0.9522	–	0.9639	–
B-B95	0.9919	0.9944	0.9943	–	–
B-LYP	0.9940	0.9969	0.9964	1.0001	0.9994
B-P86	0.9914	0.9939	0.9943	0.9978	0.9976
G96-LYP	0.9923	0.9944	0.9948	0.9977	0.9970
HCTH147	0.9748	0.9771	0.9781	0.9812	0.9820
HCTH407	0.9698	0.9728	0.9735	0.9765	0.9778
HCTH93	0.9734	0.9760	0.9768	0.9799	0.9810
O-LYP	0.9775	0.9802	0.9811	0.9842	0.9854
PBE	0.9875	0.9904	0.9907	0.9944	0.9948
TPSS	0.9741	0.9767	0.9787	0.9818	0.9821
VSXC	0.9681	0.9700	0.9700	0.9738	0.9746
B1-B95	0.9501	0.9535	0.9544	–	–
B1-LYP	0.9561	0.9599	0.9603	0.9640	0.9639
B3-LYP	0.9613	0.9648	0.9652	0.9688	0.9686
B3-P86	0.9557	0.9588	0.9600	0.9632	0.9635
B3-PW91	0.9571	0.9602	0.9615	0.9648	0.9652
B971	0.9609	0.9644	0.9651	0.9684	0.9685
B972	0.9512	0.9543	0.9558	0.9587	0.9596
B98	0.9602	0.9635	0.9642	0.9676	0.9675
BB1K	0.9322	0.9359	–	–	–
BMK	0.9475	0.9533	0.9520	0.9566	0.9551
EDF1	0.9798	0.9820	–	0.9858	–
EDF2	0.9595	0.9627	–	0.9668	–
M05	0.9492	0.9542	0.9582	0.9601	0.9614
M05-2X	0.9373	0.9419	0.9450	0.9446	0.9444
MPW1K	0.9278	0.9315	0.9336	0.9365	0.9375
mPW1PW91	0.9499	0.9532	0.9546	0.9580	0.9587
MPWB1K	0.9295	0.9335	0.9348	–	–
O3-LYP	0.9617	0.9648	0.9658	0.9690	0.9701
PBE0	0.9512	0.9547	0.9561	0.9594	0.9602

Table 8.2 Frequency scale factors suitable for low-frequency vibrations

	6-31G(d)	6-31+G(d,p)	6-31G(2df,p)	6-311+G(d,p)	6-311+G(2df,p)
Method	**Scale factor**	**Scale factor**	**Scale factor**	**Scale factor**	**Scale factor**
HF	0.9062	0.9146	0.9048	0.9146	0.9075
MP2	1.0139	1.0333	0.9833	1.0157	0.9917
QCISD	1.0014	1.0203	–	1.0086	–
QCISD(*T*)	1.0375	1.0591	–	1.0429	–
CCSD	0.9971	1.0153	–	1.0034	–
CCSD(*T*)	1.0347	1.0567	–	1.0399	–
B-B95	1.0489	1.0617	1.0428	–	–
B-LYP	1.0627	1.0738	1.0540	1.0915	1.0615
B-P86	1.0497	1.0608	1.0480	1.0848	1.0498
G96-LYP	1.0605	1.0670	1.0552	1.0859	1.0574
HCTH147	1.0304	1.0421	1.0248	1.0531	1.0300
HCTH407	1.0239	1.0374	1.0205	1.0457	1.0275
HCTH93	1.0301	1.0424	1.0253	1.0519	1.0308
O-LYP	1.0355	1.0500	1.0302	1.0609	1.0378
PBE	1.0411	1.0539	1.0373	1.0714	1.0437
TPSS	1.0352	1.0459	1.0315	1.0568	1.0345
VSXC	1.0273	1.0353	1.0109	1.0380	1.0100
B1-B95	0.9804	0.9897	0.9838	–	–
B1-LYP	0.9930	1.0028	0.9915	1.0101	0.9971
B3-LYP	1.0007	1.0117	0.9998	1.0189	1.0053
B3-P86	0.9921	1.0012	0.9929	1.0081	0.9962
B3-PW91	0.9937	1.0032	0.9949	1.0086	0.9984
B971	0.9991	1.0083	0.9983	1.0162	1.0034
B972	0.9879	0.9976	0.9879	1.0027	0.9920
B98	0.9977	1.0069	0.9973	1.0132	1.0016
BB1K	0.9570	0.9654	–	–	–
BH and H	0.9370	0.9474	0.9389	0.9495	0.9426
BH and HLYP	0.9484	0.9581	0.9482	0.9608	0.9528
BMK	0.9776	0.9844	0.9771	0.9879	0.9800
EDF1	1.0397	1.0522	–	1.0646	–
EDF2	0.9958	1.0067	–	1.0197	–
M05	0.9691	0.9786	0.9603	0.9757	0.9566
M05-2X	0.9364	0.9386	0.9194	0.9336	0.9168
MPW1K	0.9525	0.9621	0.9543	0.9648	0.9584
mPW1PW91	0.9828	0.9932	0.9848	0.9984	0.9888
MPWB1K	0.9536	0.9627	0.9610	–	–
O3-LYP	1.0042	1.0156	1.0043	1.0230	1.0099
PBE0	0.9821	0.9916	0.9834	0.9964	0.9873

Table 8.3 Frequency scale factors derived from a least-squares fit of zero-point vibrational energy (ZPVE)

	6-31G(d)	6-31+G(d,p)	6-31G(2df,p)	6-311+G(d,p)	6-311+G(2df,p)
Method	**Scale factor**	**Scale factor**	**Scale factor**	**Scale factor**	**Scale factor**
HF	0.9135	0.9200	0.9222	0.9255	0.9268
MP2	0.9670	0.9657	0.9678	0.9768	0.9777
QCISD	0.9777	0.9703	0.9707	0.9812	0.9803
QCISD(*T*)	0.9859	0.9786	–	0.9907	–
CCSD	0.9758	0.9686	0.9691	0.9795	0.9786
CCSD(*T*)	0.9851	0.9779	–	0.9897	–
B-B95	1.0129	1.0162	1.0152	–	–
B-LYP	1.0135	1.0169	1.0158	1.0189	1.0186
B-P86	1.0121	1.0155	1.0150	1.0183	1.0185
G96-LYP	1.0121	1.0148	1.0143	1.0168	1.0166
HCTH147	0.9964	0.9999	0.9998	1.0028	1.0037
HCTH407	0.9911	0.9950	0.9951	0.9981	0.9993
HCTH93	0.9957	0.9991	0.9992	1.0022	1.0033
O-LYP	0.9985	1.0022	1.0024	1.0056	1.0070
PBE	1.0085	1.0123	1.0120	1.0154	1.0161
TPSS	0.9925	0.9957	0.9971	0.9999	1.0007
VSXC	0.9877	0.9904	0.9896	0.9937	0.9947
B1-B95	0.9716	0.9760	0.9762	–	–
B1-LYP	0.9760	0.9805	0.9802	0.9838	0.9840
B3-LYP	0.9813	0.9857	0.9853	0.9887	0.9889
B3-P86	0.9768	0.9809	0.9814	0.9845	0.9852
B3-PW91	0.9780	0.9819	0.9825	0.9858	0.9865
B971	0.9817	0.9859	0.9859	0.9893	0.9899
B972	0.9719	0.9760	0.9768	0.9799	0.9809
B98	0.9809	0.9850	0.9849	0.9884	0.9886
BB1K	0.9539	0.9587	0.9595	–	–
BH and H	0.9500	0.9562	0.9568	0.9607	0.9620
BH and HLYP	0.9446	0.9498	0.9506	0.9540	0.9547
BMK	0.9709	0.9773	0.9752	0.9794	0.9787
EDF1	1.0006	1.0037	–	1.0066	–
EDF2	0.9805	0.9847	–	0.9879	–
M05	0.9736	0.9787	0.9809	0.9841	0.9851
M05-2X	0.9580	0.9631	0.9657	0.9658	0.9663
MPW1K	0.9489	0.9537	0.9552	0.9584	0.9596
mPW1PW91	0.9708	0.9751	0.9759	0.9793	0.9804
MPWB1K	0.9513	0.9563	0.9569	–	–
O3-LYP	0.9826	0.9867	0.9872	0.9904	0.9918
PBE0	0.9726	0.9771	0.9779	0.9812	0.9824

Adapted with permission from Ref. [13]. Copyright @ 2007, American Chemical Society.

DFT accounts for correlation effects to a large extent and hence the calculated frequencies from DFT are much closer to the experimental values. Different approaches have been adopted for scaling the vibrational frequencies based on DFT calculations. These include using a single scaling factor for all the frequencies, two sets of scaling factors (one for the higher and other for the lower frequencies), and a linear scaling method, where a linear relationship is provided between the scaling factor and vibrational wave numbers. Yoshida et al. [14,15] suggested a relationship between the observed and calculated frequencies given by the relation.

$$\nu_{obs} = (1.0087 - 0.0000163\ \nu_{calc})\nu_{calc}$$

This method has been widely used by Gupta et al. [16–20] for scaling DFT-based vibrational frequencies of polyatomic organic molecules. The calculated values have been found to be in agreement with the experimental values within 10–20 cm^{-1}.

The vibrational scaling factors for vibrational frequencies and zero-point energy calculations are being maintained by NIST-USA for different levels of theory and basis sets. These are available at www.cccbdb.nist.gov/vibscalcjust.asp.

8.6 Vibrational Analysis and Thermodynamic Parameters

The spectroscopic measured vibrational frequencies and moments of inertia are important variables in determining the thermodynamic functions of a molecule. If the vibrational frequency of a molecule can be calculated, it is possible to find out the contribution of the vibrational energy to the total energy of the molecule. Similarly, if the moment of inertia of a molecule is known, the rotational contribution to the total energy can also be calculated.

The key feature in statistical mechanics is the *partition function* as it allows calculation of all thermodynamic functions of the molecule. Thermodynamic parameters can be calculated using the concept of partition function [21] given by

$$Q = \sum g_i e^{-\frac{\varepsilon_i}{kT}} \tag{8.6.1}$$

where g_i is the degeneracy factor that indicates the number of states with the same energy ε_i. The thermodynamic functions are calculated by taking the molecule as an ideal gas in its standard state. Following expressions are used for the calculation of various thermodynamic functions in terms of partition function:

1. The total internal energy E^0 of 1 mol of an ideal gas, if no reference is made to the zero-point energy, is given as

$$E^0 = RT^2 \frac{dlnQ}{dT} \tag{8.6.2}$$

if no reference is made to the zero-point energy.

2. Enthalpy function

$$\frac{H^0 - E^0}{T} = RT\frac{dlnQ}{dT} + R \tag{8.6.3}$$

where H^0 and E^0 represent the enthalpy and the total internal energy of an ideal gas, respectively.

3. Heat capacity

$$C_V^0 = \frac{R}{T^2}\left[\frac{d^2Q/d(1/T)^2}{Q} - \left(\frac{dQ/d(\frac{1}{T})}{Q}\right)^2\right] \tag{8.6.4}$$

4. Entropy

$$S^0 = RT\frac{dlnQ}{dT} + RlnQ - RlnN + R \tag{8.6.5}$$

5. Free energy function

The free energy of an ideal gas is given as

$$G^0 = H^0 - TS^0 \tag{8.6.6}$$

In terms of partition function, it is given as

$$\frac{G^0 - E_0^0}{T} = -R\,ln\,\frac{Q}{N} \tag{8.6.7}$$

where E_0^0 is the energy at absolute zero.

8.6.1 Vibrational Partition Function and Vibrational Energy

Vibrational partition function is written as

$$Q_{vib} = \sum g_i e^{-\varepsilon_i/kT} = \sum_v e^{-(v+1/2)h\nu'/kT} \tag{8.6.8}$$

where $g_i = 1$, and v is the vibrational quantum number. On simplification, this may be written as

$$Q_{vib} = e^{-x/2}\left(\frac{1}{1 - e^x}\right) \tag{8.6.9}$$

where $x = h\nu'/kT$ or $h\nu c/kT$ and the constant term has been removed.

The vibrational energy contribution toward the internal energy may be given as

$$E_{vib}^0 = RT^2\frac{dlnQ_{vib}}{dT} = RT\left(\frac{xe^{-x}}{1 - e^{-x}}\right) + \frac{R}{k}\frac{h\nu c}{2} \tag{8.6.10}$$

where k is the Boltzmann constant.

Since the second term in this equation is the zero-point energy per mole (E_0^0),

$$\left(E^0 - E_0^0\right)_{\text{vib}} = RT\frac{xe^{-x}}{1 - e^{-x}} \tag{8.6.11}$$

which is the internal energy after accounting for the zero-point contribution. Expressions for other thermodynamical functions may, therefore, be written as:
Enthalpy function:

$$\left(\frac{H^0 - E_0^0}{T}\right)_{\text{vib}} = \frac{Rx}{e^x - 1} \tag{8.6.12}$$

Heat capacity:

$$\left(C_v^0\right)_{\text{vib}} = R\left(\frac{h\nu c}{kT}\right)^2 \frac{e^x}{\left(e^x - 1\right)^2} = \frac{Rx^2 e^x}{\left(e^x - 1\right)^2} \tag{8.6.13}$$

Free energy function:

$$\left(\frac{G^0 - E_0^0}{T}\right)_{\text{vib}} = -Rln\left(\frac{1}{1 - e^{-x}}\right) = Rln(1 - e^{-x}) \tag{8.6.14}$$

8.6.2 Illustration—Thermodynamic Parameter for 2-Butanone

Using the above sets of relations, thermodynamic parameters for the *trans* and *gauche* conformers of 2-butanone (Figure 8.5) have been calculated by Gupta et al. [22] at different levels of theory. These are given in Table 8.4 and compared with the experimental values obtained through calorimetric measurements [23], chemical equilibrium studies [24], and spectroscopic measurements [25]. The calculated thermodynamic parameters for both the *trans* and *gauche* conformers were used to verify the correctness of vibrational assignments for these conformers of 2-butanone. It follows from Table 8.4 that the values of entropy, enthalpy, and Gibbs free energy for these two conformers are very close to each other and are in agreement with the values reported by Chao et al. [25]. It may be noted that the results from the DFT for the trans conformer are in better agreement with the experimental values than the results of RHF calculations using the same basis set (6-311G**).

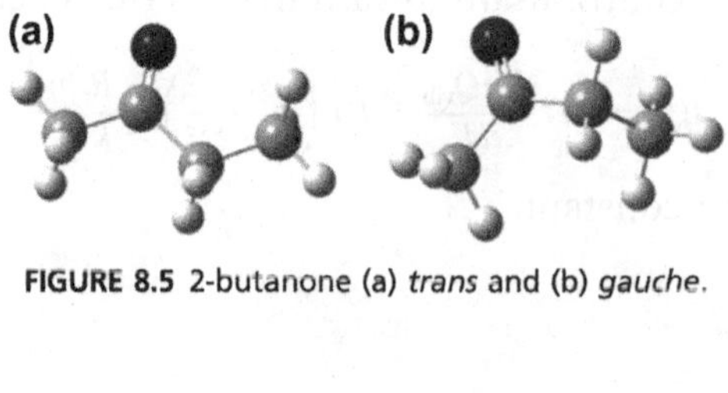

FIGURE 8.5 2-butanone (a) *trans* and (b) *gauche*.

Table 8.4 Thermodynamic properties of *trans* and *cis* conformers of 2-butanone

		RHF/4-31G		RHF/6-31G**		RHF/6-311G**		DFT/6-311G**		Exptl. value
S.No.	**Parameter**	***trans***	***gauche***	***trans***	***gauche***	***trans***	***gauche***	***trans***	***gauche***	***trans***
1	$H-H^0$ (kcal mol^{-1})	4.531	4.488	4.581	4.524	4.041	4.512	4.775	4.148	5.01
2	$G-H^0$ (kcal mol^{-1})	18.951	18.878	19.164	19.025	17.978	18.851	19.377	18.003	19.174
3	S (cal/mol K)	78.757	78.373	79.640	78.985	73.841	78.364	80.994	74.285	81.111
4	C_v (cal/mol K)	19.994	19.884	20.321	20.161	18.409	20.217	21.699	19.526	-
5	C_p (cal/mol K)	21.981	21.871	22.308	22.148	20.396	22.204	23.686	21.513	24.681

8.7 Anharmonic Polyatomic Oscillator—Anharmonicity and Vibrational Parameter

In the case of polyatomic molecules, anharmonic motion significantly affects the vibrational frequencies and results in considerable vibration–rotation interaction. The inclusion of anharmonicity of molecular vibrations is therefore an important aspect of the goal of making highly accurate theoretical predictions of spectroscopic properties of molecule. While in the case of small molecules converged rovibrational levels can be obtained by fully variational methods, in large molecules some approximation becomes necessary both for the form of the potential and the rovibrational treatment. Three approaches are commonly in use for the anharmonic vibrational treatments. These are based on truncated two- or three- mode potentials, self-consistent vibrational treatment (VSCF) [26–29], and the second-order perturbative vibrational treatment [30–33]. Very accurate (in terms of line shapes etc.) vibrational spectra can be calculated using quantum molecular dynamics (MD) and Fourier transformation of the trajectories. Classical MD can also be used for these calculations.

Recently developed analytic third derivative methods for SCF wavefunctions have made it possible to determine the complete cubic and quartic force fields of polyatomic molecules, thus allowing the treatment of anharmonic effect. A computational code for the treatment of anharmonicity by using the perturbative approach has been developed by Barone [34] and adopted in the Gaussian suite of programs [35]. A set of equations used for calculating the various anharmonicity-related constants are given below.

The vibration–rotation term values of a polyatomic molecules may be expressed empirically as the sum of a vibrational term which is independent of the rotational quantum numbers and a rotational term which is largely independent of the vibrational quantum numbers.

$$T(v,J) = G(v) + F_v(J) \tag{8.7.1}$$

Instead of the normal coordinate Q, the anharmonic oscillator problem is solved by using reduced normal coordinates q, as suggested by Neilsen [36–38].

$$q_i = \gamma_i^{1/2} Q_i \tag{8.7.2}$$

where $\gamma_i = \frac{\lambda_i^{1/2}}{\hbar} = \frac{2\pi c\omega_i}{\hbar}$.

The vibrational potential energy is then expanded in terms of these reduced normal coordinates as

$$\frac{V}{hc}=\frac{1}{2}\sum_{r}\omega_r q_r^2+\frac{1}{2}\sum_{rst}\Phi_{rst}q_r q_s q_t+\frac{1}{24}\sum_{rstu}\Phi_{rstu}q_r q_s q_t q_u+\ldots \tag{8.7.3}$$

$$\text{where,}\quad \Phi=\tilde{L}M^{-1/2}FM^{-1/2}L \tag{8.7.4}$$

is the second derivative matrix over normal coordinates, Φ_{rst} and Φ_{rstu} are the cubic and quartic force constants, respectively. L expresses the relationship between the normal coordinates Q and the Cartesian displacement coordinates x:

$$Q=\tilde{L}M^{1/2}x \tag{8.7.5}$$

The vibrational Hamiltonian (H_{vib}) can therefore be expressed in wave number units in the form

$$H_{vib}=H_{vib}^0+H_{vib}^1+H_{vib}^2=\frac{1}{2}\sum_{r}\omega_r\left(p_r^2+q_r^2\right)+\frac{1}{6}\sum_{rst}\phi_{rst}q_r q_s q_t+\frac{1}{24}\sum_{rstu}\phi_{rstu}q_r q_s q_t q_u+\sum_{\alpha}B_{\alpha}^{e}j_{\alpha}^{2} \tag{8.7.6}$$

where $j_\alpha=\sum_{i<j}\zeta_{ij}^{\alpha}(q_i p_j-q_j p_i)$ is the kinetic contribution arising from vibrational angular momentum and B_α^e is the corresponding equilibrium rotational constant. H^0 is the zeroth-order harmonic term and H_{vib}^1 is the first-order perturbation which contains contributions of the cubic, quartic, etc., components of the potential whereas the second-order term H_{vib}^2 also includes contribution from the vibrational angular momentum J_α. ζ_{ij}^{α} is the Coriolis coupling constant which couples the normal vibrational modes Q_i and Q_j through rotation about the axis α. It is expressed as

$$\zeta_{ij}^{\alpha}=\sum_{k}\left(L_{ik}^{\beta}L_{jk}^{\gamma}-L_{ik}^{\gamma}L_{jl}^{\beta}\right) \tag{8.7.7}$$

where L is the column-wise matrix of the eigenvectors of the mass-weighted Cartesian force matrix $M^{-1/2}FM^{-1/2}$.

The vibrational term values F(v) of a polyatomic molecule can be expressed as

$$G(v)=\sum_{i}\omega_i\left(v_i+\frac{1}{2}d_i\right)+\sum_{i\geq j}\xi_{ij}\left(v_i+\frac{1}{2}d_i\right)\left(v_j+\frac{1}{2}d_j\right)+\ldots \tag{8.7.8}$$

In Eq. (8.7.8), ω_i is the ith harmonic frequency, d_i is the degeneracy of the ith normal mode, and ξ_{ij} are the vibrational anharmonic constants. While terms like ξ_{ii} characterize anharmonicity of the given vibration, the terms ξ_{ij} characterize coupling between different normal modes resulting from anharmonicity and are determined from cubic and quartic force constants. The fundamental frequencies ν_i, overtones ($2\nu_i$), and combination tones ($\nu_i\,\nu_j$) are given by the relations

$$\nu_i=\omega_i+2\xi_{ii}+\frac{1}{2}\sum_{j\neq i}\xi_{ij}, \tag{8.7.9}$$

$$[2\nu_i] = 2\omega_i + 6\xi_{ii} + \sum_{j \neq i} \xi_{ij} = 2\nu_i + 2\xi_{ii}. \tag{8.7.10}$$

$$[\nu_i \nu_j] = \omega_i + \omega_j + 2\xi_{ii} + 2\xi_{jj} + 2\xi_{ij} + \frac{1}{2} \sum_{(l \neq i,j)} (\xi_{il} + \xi_{jl})$$

$$= \nu_i + \nu_j + \xi_{ij}, \tag{8.7.11}$$

Following Eq. (8.7.8), the zero-point energy is given as

$$ZPE = \frac{1}{2} \sum_i \left(\omega_i + \frac{1}{2} \xi_{ii} + \sum_{j>i} \frac{1}{2} \xi_{ij} \right) \tag{8.7.12}$$

The vibrational dependence of the effective rotational constant around axis α and vibrational state n for a polyatomic molecule is given as

$$B_\alpha^n = B_\alpha^e - \sum_i a_i^\alpha \left(v_i + \frac{1}{2} \right) \tag{8.7.13}$$

where B_α^e is the equilibrium rotational constant, a_i^α are the vibration–rotation interactions constants. The summation runs over all the normal modes i. The constant a_i^α and explicit expressions for all the other constants can be given in terms of quartic centrifugal distortion constants $\tau_{\alpha\beta\gamma\delta}$.

$$\tau_{\alpha\beta\gamma\delta} = - \frac{\hbar^4}{2hcI_\alpha I_\beta I_\gamma I_\delta} \sum_k \frac{I_{\alpha\beta}^k a_{\gamma\delta}^k}{\lambda_k}. \tag{8.7.14}$$

where $I_{\alpha\beta}^k$ is an inertial derivative.

$$I_{\alpha\beta}^k = \left(\frac{\partial I_{\alpha\beta}}{\partial Q_k} \right)_e \tag{8.7.15}$$

The quartic centrifugal distortion constants depend only on the quadratic (harmonic) part of the vibrational Hamiltonian and are expected to be about four orders of magnitude smaller than the corresponding rotational constants.

In the Kivelson–Wilson formalism the vibrationally averaged rotational constants for the ground vibrational state used to fit the observed energy levels are

$$B_\alpha'^0 = B_\alpha^0 + \frac{1}{4} (3\tau_{\beta\gamma\beta\gamma} - 2\tau_{\gamma\alpha\gamma\alpha} - 2\tau_{\alpha\beta\alpha\beta}) \tag{8.7.16}$$

$$B_\beta'^0 = B_\beta^0 + \frac{1}{4} (3\tau_{\gamma\alpha\gamma\alpha} - 2\tau_{\alpha\beta\alpha\beta} - 2\tau_{\beta\gamma\beta\gamma}) \tag{8.7.17}$$

$$B_\gamma'^0 = B_\gamma^0 + \frac{1}{4} (3\tau_{\alpha\beta\alpha\beta} - 2\tau_{\beta\gamma\beta\gamma} - 2\tau_{\gamma\alpha\gamma\alpha}) \tag{8.7.18}$$

and the asymmetry parameter σ is defined as

$$\sigma = \frac{2B_\alpha'^0 - B_\beta'^0 - B_\gamma'^0}{B_\beta'^0 - B_\gamma'^0} \tag{8.7.19}$$

The quartic centrifugal distortion constants appearing in the Kivelson–Wilson formalism for an asymmetric top are

$$D_J = -\frac{1}{32}[3\tau_{\beta\beta\beta\beta} + 3\tau_{\gamma\gamma\gamma\gamma} + 2(\tau_{\beta\beta\gamma\gamma} + 2\tau_{\beta\gamma\beta\gamma})] \tag{8.7.20}$$

$$D_K = D_J - \frac{1}{4}[\tau_{\alpha\alpha\alpha\alpha} - (\tau_{\alpha\alpha\beta\beta} + 2\tau_{\alpha\beta\alpha\beta}) - (\tau_{\gamma\gamma\alpha\alpha} + 2\tau_{\gamma\alpha\gamma\alpha})] \tag{8.7.21}$$

$$D_{JK} = -D_J - D_K - \frac{1}{4}\tau_{\alpha\alpha\alpha\alpha} \tag{8.7.22}$$

$$R_5 = -\frac{1}{32}[\tau_{\beta\beta\beta\beta} - (\tau_{\gamma\gamma\gamma\gamma} - 2(\tau_{\alpha\alpha\beta\beta}) + 2\tau_{\alpha\beta\alpha\beta}) + 2(\tau_{\gamma\gamma\alpha\alpha} + 2\tau_{\gamma\alpha\gamma\alpha})] \tag{8.7.23}$$

$$R_6 = \frac{1}{64}[\tau_{\beta\beta\beta\beta} + \tau_{\gamma\gamma\gamma\gamma} - 2(\tau_{\beta\beta\gamma\gamma} + 2\tau_{\beta\gamma\beta\gamma})] \tag{8.7.24}$$

$$\text{and } \delta_J = -\frac{1}{16}(\tau_{\beta\beta\beta\beta} - \tau_{\gamma\gamma\gamma\gamma}) \tag{8.7.25}$$

Using the perturbation theory, the vibrationally averaged properties $\langle P \rangle$ can be determined by using the relations

$$\langle P \rangle_n = P_e + \sum_i \alpha_i \langle Q_i \rangle_n + \sum_{ij} \beta_{ij} \langle Q_i Q_j \rangle_n, \tag{8.7.26}$$

$$\langle Q_i \rangle_n = -\frac{\hbar}{4\omega_i^2} \sum_j \frac{\Phi_{ijj}}{\omega_j}\left(n_j + \frac{1}{2}\right) \tag{8.7.27}$$

$$\langle Q_i Q_j \rangle_n = -\frac{\hbar}{2\omega_i}\delta_{ij}\left(n_j + \frac{1}{2}\right) \tag{8.7.28}$$

where,

$$\alpha_i = \left(\frac{\partial P}{\partial Q_i}\right)_e, \quad \beta_{ij} = \left(\frac{\partial^2 P}{\partial Q_i \partial Q_j}\right)_e, \tag{8.7.29}$$

and δ_{ij} is the Kronecker delta.

8.8 Illustration—Anharmonic Vibrational Analysis of Ketene

Gupta et al. [39] conducted anharmonic analysis of the vibrational spectrum of ketene by DFT using the second-order perturbative approach (PT2) described above. Harmonic and anharmonic vibrational frequencies, overtones and combination tones, the anharmonicity constants, rotational and rotation vibrational coupling constants, the Coriolis coupling constants, and the Nielsen centrifugal distortion constants were calculated for the molecule and compared with the experimental data as well as data from calculations based on different theoretical methods. These are given in Tables 8.5–8.10. The two different models—a homogeneous model using the same density functionals and basis

Table 8.5 Vibrational frequencies (cm^{-1}), intensities (km/mol), and assignments of ketene

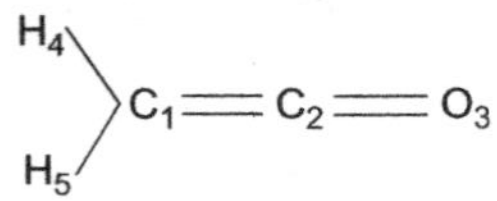

		B97-1/aug-cc-pvtz// B3LYP/6-31+G**			B97-1/aug-cc-pvtz// B97-1/aug-cc-pvtz			B97-1/6-311++G**// B97-1/6-311++G**			B3LYP/6-311++G**// B3LYP/6-311++G**			B3LYP/6-31+G**// B3LYP/6-31+G**			
		Harmonic		Anharmonic	Harmonic		Anharmonic	Harmonic		Anharmonic	Harmonic		Anharmonic	Harmonic		Anharmonic	
Freq	**Exptl.**	**ν calc**	**Int.**	**ν calc**	**ν calc**	**Int.**	**ν calc**	**ν calc**	**Int.**	**ν calc**	**ν calc**	**Int.**	**ν calc**	**ν calc**	**Int.**	**ν calc**	**Assignments**
Symmetry A1																	
ν1	3070	3220.4	29.0	3086.9	3174.1	27.6	3072.2	3175.5	30.5	3031.6	3177.5	29.9	3047.4	3197.8	28.9	3063.7	CH sym str.
ν2	2151	2279.9	673.8	2237.2	2213.6	670.3	2173.0	2226.8	682.6	2185.6	2223.4	693.1	2180.3	2213.0	660.6	2170.3	CCO asym str.
ν3	1388	1419.5	18.3	1392.9	1404.2	20.5	1375.8	1399.6	19.6	1372.6	1406.0	18.4	1378.7	1416.1	17.5	1388.6	CCH i.p.bend
ν4	1117.8	1187.2	6.3	1200.4	1166.9	4.2	1119.3	1166.7	5.8	1115.2	1170.8	6.6	1139.8	1170.8	6.7	1133.8	CCO sym str.
Symmetry B1																	
ν5	588	574.6	125.5	568.5	586.6	11.0	577.4	594.7	87.9	576.6	600.8	107.9	587.6	599.9	122.9	593.5	CCH o.p.bend
ν6	528	507.5	33.7	508.1	535.1	1.6	530.2	544.3	56.5	533.6	554.9	34.0	550.5	531.4	34.3	531.9	CCO o.p.bend
Symmetry B2																	
ν7	3166	3318.6	9.51	3169.3	3270.3	9.6	3164.7	3272.2	11.0	3111.8	3270.6	9.7	3124.8	3293.4	9.4	3142.3	CH asym. Str.
ν8	976.7	984.1	9.27	968.6	986.7	8.8	971.2	984.9	7.7	969.4	987.0	8.0	971.2	990.7	9.0	975.2	CCH i.p.bend
ν9	438	404.0	2.76	402.8	442.1	2.8	441.2	441.9	2.1	440.0	443.7	2.2	441.3	437.9	2.6	436.8	CCO i.p.bend

The two methods indicated in the table refer to harmonic frequencies and anharmonic corrections, respectively.

Table 8.6 Overtones and combination tones (cm^{-1}) of ketene and their assignments

	Calculated								
	B3LYP/6-31+G**// B3LYP/6-31+G**		B97-1/6-311++G**// B97-1/6-311++G**		B97-1/aug-cc-pVTZ// B97-1/aug-cc-pVTZ		B97-1/aug-cc-pVTZ// B3LYP/6-31+G**		
Experimental	**Harmonic**	**Anharmonic**	**Harmonic**	**Anharmonic**	**Harmonic**	**Anharmonic**	**Harmonic**	**Anharmonic**	**Assignments**
–	4426.0	4318.9	4452.9	4349.8	4427.3	4325.9	4559.1	4452.0	$2\nu_2$
1952.4[a],1947[f]	1981.3	1947.6	1972.3	1939.7	1973.5	1940.2	1968.8	1935.1	$2\nu_8$
–	1182.9	1191.1	1173.1	1184.5	1173.1	1184.4	1149.4	1110.2	$2\nu_5$
–	1062.8	1059.7	1088.4	1059.2	1070.2	1064.9	1015.9	1014.2	$2\nu_6$
6113.3[b]	6491.6	6085.3	6449.7	6015.9	6444.5	5990.8	6537.5	6134.6	$\nu_1 + \nu_7$
3513.0[f]	3629.1	3552.9	3625.2	3551.3	3617.8	3543.1	3699.3	3623.2	$\nu_2 + \nu_3$
3255[f],3266.0[d]	3383.6	3291.9	3392.4	3272.3	3380.6	3313.1	3467.1	3425.7	$\nu_2 + \nu_4$
3126.7[c]	3203.7	3139.5	3212.6	3147.9	3200.4	3136.7	3263.9	3197.5	$\nu_2 + \nu_8$
2513.6[b]	2586.6	2518.9	2564.6	2467.9	2751.1	2524.5	2607.3	2590.3	$\nu_3 + \nu_4$
1126.0[f]	1131.3	1127.0	1135.6	1097.8	1121.7	1105.3	1082.7	1081.1	$\nu_5 + \nu_6$
940–970[e]	969.3	976.8	985.8	976.8	977.2	973.8	912.6	920.2	$\nu_6 + \nu_9$

[a]Duncan et al. [43].
[b]Butler et al. [44].
[c]Duncan et al. [45].
[d]Arendale and Fletcher [46].
[e]Gruebele et al. [47].
[f]Moore and Pimentel [48].

Table 8.7 Anharmonicity constants ξ_{ij} (cm^{-1}) for ketene

		ξ_{ij}					ξ_{ij}		
I	**J**	**Method 1[a]**	**Method 2[b]**	**Ref. [41]**	**I**	**J**	**Method 1[a]**	**Method 2[b]**	**Ref. [41]**
1	1	−31.14	−27.79	−26.66	8	1	−11.38	−9.63	−9.94
2	1	−1.89	−1.76	−2.04	8	2	−7.51	−8.19	−3.28
2	2	−10.05	−10.58	−11.15	8	3	−9.94	−9.16	−9.06
3	1	−5.53	−2.53	−5.70	8	4	−1.23	−1.49	−1.25
3	2	−5.69	−6.75	−4.76	8	5	1.34	4.84	0.45
3	3	−6.34	−6.54	−6.29	8	6	12.67	8.52	8.02
4	1	−4.67	−5.16	−3.85	8	7	−9.59	−7.04	−12.03
4	2	−11.68	−11.82	−13.97	8	8	−1.12	−1.31	−1.83
4	3	−3.03	−3.59	−2.88	9	1	−4.52	−2.01	−1.22
4	4	−1.06	−0.85	−1.15	9	2	−5.22	−6.00	−5.96
5	1	−9.03	−11.91	−5.87	9	3	−2.30	−1.19	−0.86
5	2	−3.90	−2.61	−4.25	9	4	3.92	2.84	4.06
5	3	8.43	9.27	11.12	9	5	10.05	4.03	10.89
5	4	−4.04	51.70	−3.78	9	6	2.31	8.25	3.76
5	5	−1.41	−13.32	0.28	9	7	−5.16	−2.93	−1.89
6	1	−10.39	−3.10	−8.81	9	8	−0.93	−3.64	−1.45
6	2	−9.29	−11.11	−9.83	9	9	0.04	−0.43	0.68
6	3	4.59	1.85	6.78					
6	4	−1.03	2.65	−5.53					
6	5	−2.33	4.05	14.92					
6	6	2.23	−1.45	14.83					
7	1	−132.32	−119.88	−112.25					
7	2	4.12	−5.11	−0.14					
7	3	−17.98	−14.99	−18.42					
7	4	−4.35	−5.24	−3.41					
7	5	−13.14	−18.60	−10.42					
7	6	−15.25	−4.10	−16.05					
7	7	−36.18	−32.79	−31.17					

[a]Homogeneous model B97-1/aug-cc-pVTZ//B-97-1/aug-cc-pVTZ.[b]Hybrid model B97-1/aug-cc-pVTZ//B3LYP/6-31+G**.

sets for the harmonic calculations and anharmonic corrections and a hybrid model in which the two parts of the calculation are conducted using different density functionals and basis sets were employed in these calculations. The results show that the former is superior to the latter in several respects. The anharmonic frequencies obtained by the second-order perturbative approach are close to the experimental frequencies and do not need any *ad hoc* scaling that is required in the case of harmonic frequencies. The absolute average error of the fundamental bands is usually below 10 cm^{-1} and never exceeds 35 cm^{-1} in the homogeneous model. However, in the hybrid model, the two figures could be as large as 30 and 90 cm^{-1}, respectively. These inadequacies in the fundamental modes of ketene in the hybrid model are also reflected in the overtones and combination bands where large deviations of up to 180 cm^{-1} from the experimental

Table 8.8 Rotational constants (cm^{-1}) including terms due to quartic centrifugal distortion constants and rotational–vibrational coupling constants (cm^{-1}) of ketene

	Rotational constants					
	B3LYP/6-31+G//B3LYP/6-31+G****	**B3LYP/6-311++G**//B3LYP/6-311++G****	**B97-1/6-311++G**//B97-1/6-311 ++G****	**B97-1/aug-cc-pVTZ//B97-1/aug-cc-pVTZ**	**B97-1/aug-cc-pVTZ//B3LYP/6-31+G****	**Ref. [41]**
A_e	9.450	9.494	9.435	9.452	9.452	–
B_e	0.340	0.344	0.343	0.344	0.344	–
C_e	0.328	0.332	0.331	0.332	0.332	–
A_0	9.450[a]	9.494	9.435	9.452	9.452	9.410
B_0	0.340	0.344	0.343	0.344	0.344	0.343
C_0	0.328	0.332	0.331	0.332	0.332	0.331

[a]Experimental 9.36 ± 0.02 cm^{-1}.

Table 8.9 Rotation–Vibration coupling constants (10^{-3} cm^{-1})

	B3LYP/6-31+G//B3LYP/6-31+G****			**B97-1/aug-cc-pVTZ//B97-1/aug-cc-pVTZ**			**B97-1/aug-cc-pVTZ//B3LYP/6-31+G****			**Ref. [41]**		
	a	b	c	a	b	c	a	b	c	a	b	c
α_1	159.1	0.27	0.44	156.2	0.27	0.44	159.4	0.27	0.45	160.7	0.27	0.44
α_2	11.8	23.6	2.20	11.7	2.37	2.21	11.3	2.31	2.15	9.4	2.51	2.35
α_3	−125.4	−0.06	0.52	−124.8	0.07	0.54	−120.2	0.02	0.58	−128.8	−0.10	0.51
α_4	−0.92	−0.52	0.49	−0.73	0.53	0.51	−6.37	0.41	0.44	1.90	0.71	0.62
α_5	−382.9	0.11	−0.62	−1219.7	−0.11	−0.76	−327.6	0.10	−0.67	−1218.8	−0.10	−0.74
α_6	−1336.6	−0.21	−0.72	−136.9	0.10	−0.57	−1237.8	−0.26	−0.80	−442.1	0.14	−0.55
α_7	114.2	0.36	0.42	109.4	0.37	0.41	115.4	0.37	0.43	117.3	0.36	0.42
α_8	−557.3	−0.42	0.15	−567.8	−0.54	0.16	−542.7	−0.56	0.13	−514.7	−0.51	0.23
α_9	2208.2	−1.45	−0.51	1856.8	−1.46	−0.50	2036.6	−1.62	−0.60	2125.4	−1.45	−0.49

values are observed. In contrast, the homogeneous model gives values of the overtones and combination bands within a maximum of 35 cm^{-1} of the experimental values.

The rotational constants A_0, B_0, C_0 which contain effects due to zero-point vibrations and centrifugal distortions are close to the experimental values reported by Cox et al. [40]. Also the level of calculation does not seem to have much influence on the calculated values. The rotational–vibrational coupling constants based on the second-order perturbative vibrational treatment are also in good agreement with the results obtained by East et al. [41] from QZ (2d, 2p) SQM (CCSD) + MP2//EXPT anharmonic force field calculations. Also the values obtained from homogeneous model are in close agreement with the experimental values of Cox et al. [40]. The calculated values of Coriolis constant

Table 8.10(a) Coriolis coupling constants Z(I,J) and Nielsen's centrifugal distortion constants for ketene

Modes		Coriolis constants (cm^{-1})		
		|Z(I,J)|		
I	**J**	**Method 1[a]**	**Method 2[b]**	**Ref. [41]**
X-components				
7	1	0.049	0.051	0.054
7	2	0.055	0.058	0.062
7	3	0.934	0.912	0.949
7	4	0.349	0.403	0.305
8	1	0.773	0.786	0.792
8	2	0.585	0.566	0.560
8	3	0.016	0.028	0.003
8	4	0.244	0.245	0.243
9	1	0.602	0.585	0.577
9	2	0.798	0.811	0.817
9	3	0.017	0.016	0.022
9	4	0.004	0.005	0.003
Y-components				
5	1	0.393	0.502	0.382
5	2	0.602	0.222	0.614
5	3	0.644	0.785	0.653
5	4	0.259	0.288	0.226
6	1	0.348	0.156	–
6	2	0.784	0.962	–
6	3	0.514	0.204	–
6	4	0.012	0.087	–
Z-Components				
7	5	0.679	0.848	0.671
7	6	0.545	0.215	0.562
8	5	0.100	0.220	0.098
8	6	0.733	0.695	0.715
9	5	0.727	0.488	0.735
9	6	0.407	0.686	0.417

Table 8.10(b) Nielsen's centrifugal distortion constants (MHz)

	Method 1[a]	**Method 2[b]**	**Ref. [41]**
D_J	0.003186	0.003042	0.003253
D_{JK}	0.489299	0.547935	0.485300
D_K	21.554	21.154	21.071
R_5	−0.124918	−0.140382	–
$\|R_6\| \times 10^{-6}$	3.447	3.385	–
$\Delta J \times 10^{-4}$	1.242	1.201	–

[a]B97-1/aug-cc-pVTZ//B97-1/aug-cc-pVTZ.

[b]B97-1/aug-cc-pVTZ//B3LYP/6-31+G**.

of ketene are also in broad agreement with the experimental values of Nemes et al. [42] obtained by direct spectroscopic analysis. Nielsen's centrifugal distortion constants given in Table 8.6 were also compared with the values of East et al. [41]. As in the case of Coriolis coupling constants, the centrifugal distortion constants from homogeneous model are in satisfactory agreement with the experimental values [42].

References

[1] S.J. Cyvin, Molecular Vibrations and Mean Square Amplitudes, Elsevier, Amsterdam, 1968.

[2] S.J. Cyvin, Spectrochim. Acta 15 (1959) 828.

[3] Y. Yamaguchi, M. Frisch, J. Gaw, H.F. Schaefer III, J. Stephen Binkley, J. Chem. Phys. 84 (1986) 2262.

[4] E.B. Wilson, J.C. Decius, P.C. Cross, Dover Publications, New York, 1980.

[5] P. Pulay, G. Fograsi, J.E. Boggs, J. Chem. Phys. 74 (1981) 3999.

[6] A.P. Scott, L. Radom, J. Phys. Chem. 100 (1996) 16502.

[7] P. Pulay, G. Fograsi, J.E. Boggs, J. Chem. Phys. 74 (1981) 3999.

[8] P. Pulay, G. Fograsi, G. Ponger, J.E. Boggs, A.J. Vargha, J. Am. Chem. Soc. 105 (1983) 7037.

[9] G. Fogarasi, P. Pulay, Annu. Rev. Phys. Chem. 35 (1984) 191.

[10] V.P. Gupta, B. Ram, S. Vaish, Indian J. Pure Appl. Phys. 30 (1992) 452.

[11] A.P. Scott, L. Radom, J. Phys. Chem. 100 (1996) 16502.

[12] M.W. Wong, Chem. Phys. Lett. 256 (1996) 391.

[13] J.P. Merrick, D. Moran, L. Radom, J. Phys. Chem. A 111 (2007) 11683.

[14] H. Yoshida, K. Takeda, J. Okamura, A. Ehara, H. Matsuura, J. Phys. Chem. A 106 (2002) 35807.

[15] H. Yoshida, A. Ehara, H. Matsuura, Chem. Phys. Lett. 325 (2000) 477.

[16] V.P. Gupta, A. Sharma, A. Virdi, V. Ram, Spectrochim. Acta Part A 64 (1) (2006) 57.

[17] V.P. Gupta, P. Thul, S. Misra, R. Pratap, V.J. Ram, Spectrochim. Acta Part A Mol. Biomol. Spectrosc. 72 (1) (2009) 82.

[18] V.P. Gupta, P. Thul, D. Chaturvedi, P. Tandon, J. Struct. Chem. (Zh. Struktur. Chim.) 52 (2011) 259.

[19] V.P. Gupta, P. Thul, P. Tandon, Spectrochimica. Acta Part A 78 (2011) 1090.

[20] V.P. Gupta, P. Tandon, Spectrochimica. Acta Part A 89 (2012) 55.

[21] I.N. Levine, Physical Chemistry, McGraw-Hill, 1983. K. Lucas, Applied Statistical Thermodynamics, Springer-Verlag, 1991.

[22] A. Sharma, V.P. Gupta, A. Virdi, Indian J. Pure Appl. Phys. 40 (2002) 246.

[23] G.C. Sinke, F.L. Oetting, J. Phys. Chem. 68 (1964) 1354.

[24] E. Buckley, E.F.G. Herington, Trans. Faraday Soc. 61 (1965) 1618.

[25] J. Chao, B.J. Zwolinski, J. Phys. Chem. Ref. Data 5 (1976) 319.

[26] J.M. Bowman, Acc. Chem. Res. 19 (1986) 202.

[27] N.J. Wright, R.B. Gerber, D.J. Tozer, Chem. Phys. Lett. 324 (2000) 206.

[28] S.K. Gregurick, G.M. Chaban, R.B. Gerber, J. Phys. Chem. A 106 (2002) 8696.

[29] K. Yagi, K. Hirao, T. Taketsugu, M.W. Schmidt, M.S. Gordon, J. Chem. Phys. 121 (2004) 1383.

[30] D.A. Clabo, W.D. Allen, R.B. Remington, Y. Yamaguchi, H.F. Schaefer III, Chem. Phys. 123 (1988) 187.

[31] W. Schneider, W. Thiel, Chem. Phys. Lett. 157 (1989) 367.
[32] O. Christiansen, J. Chem. Phys. 119 (2003) 5773.
[33] T.A. Ruden, P.R. Taylor, T. Helgaker, J. Chem. Phys. 119 (2003) 1951.
[34] V. Barone, J. Chem. Phys. 122 (2005) 014108.
[35] Gaussian 09, Gaussian, Inc. 340 Quinnipiac St., Bldg.40 Wallingford, CT 06492 USA.
[36] H.H. Nielsen, Phys. Rev. 60 (1941) 794.
[37] H.H. Nielsen, Phys. Rev. 68 (1945) 181.
[38] H.H. Nielsen, Rev. Mod. Phys. 23 (1951) 90.
[39] V.P. Gupta, Spectrochimica. Acta Part A 67 (2007) 870.
[40] A.P. Cox, I.F. Thomas, J. Sheridan, Spectrochim. Acta 15 (1959) 542.
[41] A.L.L. East, W.D. Allen, S.J. Klippenstein, J. Chem. Phys. 102 (1995) 8506.
[42] J. Nemes, Mol. Spectrosc. 72 (1978) 102.
[43] J.L. Duncan, A.M. Ferguson, J. Harper, K.H. Tonge, F. Hegelund, J. Mol. Spectrosc. 122 (1987) 72.
[44] P.E.B. Butler, D.R. Eaton, H.W. Thompson, Spectrochim. Acta 13 (1958) 223.
[45] J.L. Duncan, A.M. Ferguson, Spectrochim. Acta A43 (1987) 1081.
[46] W.F. Arendale, W.H. Fletcher, J. Chem. Phys. 24 (1956) 581.
[47] M. Gruebele, J.W. Johns, L. Nemes, J. Mol. Spectrosc. 198 (1999) 376.
[48] C.B. Moore, G.C. Pimentel, J. Chem. Phys. 38 (1963) 2816.

Further Reading

[1] L. Pauling, E.B. Wilson, Introduction to Quantum Mechanics with Applications to Chemistry, McGraw Hill, 1935.
[2] E.B. Wilson, J.C. Decius, P.C. Cross, Dover Publications, New York, 1980.
[3] N.B. Colthup, L.H. Daly, S.E. Wiberley, Introduction to Infrared and Raman Spectroscopy, third ed., Academic Press, 1990.
[4] G. Herzberg, Molecular Spectra Vol. I, Van Nostrand, 1950.
[5] J.D. Graybeal, Molecular Spectroscopy, McGraw Hill Company, 1988.
[6] C.J. Cramers, Essentials of Computational Chemistry: Theories and Models, Wiley, 2004.

9

Interaction of Radiation and Matter and Electronic Spectra

CHAPTER OUTLINE

Principles and Applications of Quantum Chemistry. http://dx.doi.org/10.1016/B978-0-12-803478-1.00009-1

9.1 Introduction

Methods of experimental and theoretical spectroscopy and quantum chemical calculations in various approximations have played a vital role in understanding the structure of diatomic and polyatomic molecules, radicals, and ions and in interpreting their spectra. These techniques are useful in the study of structure and dynamical properties of molecules and to obtain various physicochemical and thermodynamic parameters. Several levels of sophistication starting from semiempirical to post-Hartree–Fock (post-HF) and density functional theory (DFT) are presently available for understanding structural, thermodynamic, and spectral features of molecules. We shall here be concerned with the evolution of wavefunction with time and formulation of spectroscopic response of molecules when they interact with electromagnetic radiation.

9.2 Time-Dependent Perturbation Theory

In order to understand the interaction of radiation with matter which results in the emission and absorption of radiation and the resulting electronic, vibrational, and rotational spectra we need to develop time-dependent perturbation theory. In Chapter 1, we have already considered the time-independent perturbation theory and its applications to stationary states. We shall here be concerned only with weak perturbations as caused by the interaction of weak radiation with matter.

We start with the time-dependent Schrödinger equation given by

$$H\Psi(q,t) = i\hbar\frac{\partial\Psi(q,t)}{\partial t} \tag{9.2.1}$$

where H is the Hamiltonian and Ψ is the wavefunction which may depend on all the electronic and nuclear coordinates and time. In the presence of the perturbation, the Hamiltonian can be written as

$$H = H^0 + H'$$

where H^0 is independent of time and H' is the time-dependent perturbation.

Let $\Psi_1^0, \Psi_2^0, \ldots, \Psi_n^0$ be an orthonormal set of functions which are the eigenfunctions of the unperturbed Hamiltonian H^0 corresponding to the eigenstates $E_1^0, E_2^0, \ldots, E_n^0$. They then satisfy equations

$$H^0\Psi^0(q,t) = i\hbar\frac{\partial\Psi^0(q,t)}{\partial t} \tag{9.2.2}$$

As usual, $\Psi^0(q,t)$ may be divided into the position- and time-dependent components and for a given state n may be written as

$$\Psi_n^0(q,t) = \Phi_n^0(q)e^{-\frac{i}{\hbar}E_nt} \tag{9.2.3}$$

In order to obtain a solution of Eq. (9.2.1), we expand the function $\Psi(q, t)$ in terms of the set of unperturbed eigenfunctions $\Psi_n^0(q,t)$ with coefficients that are functions of time.

$$\Psi(q,t) = \sum_n c_n(t)\Psi_n^0(q,t) \tag{9.2.4}$$

Substituting Eq. (9.2.4) into Eq. (9.2.1), we get

$$\sum_n c_n(t)H^0\Psi_n^0 + \sum_n c_n(t)H'\Psi_n^0 = i\hbar\sum_n\frac{\partial c_n(t)}{\partial t}\Psi_n^0 + i\hbar\sum_n c_n(t)\frac{\partial\Psi_n^0}{\partial t} \tag{9.2.5}$$

The first and fourth terms of this equation are equal. Hence,

$$\sum_n c_n(t)H'\Psi_n^0 = i\hbar\sum_n\frac{\partial c_n(t)}{\partial t}\Psi_n^0 \tag{9.2.6}$$

On multiplying the two sides by Ψ_m^{0*}, integrating and using orthogonality condition of the wavefunctions, we get

$$\frac{dc_m(t)}{dt} = -\frac{i}{\hbar}\sum_n c_n(t)\int\Psi_m^{0*}H'\Psi_n^0 d\tau \tag{9.2.7}$$

Equation (9.2.7) represents a set of simultaneous differential equations that may be solved to give explicit expressions for all the c_n coefficients. These may be used in Eq. (9.2.4) to understand the evolution of the wavefunction $\Psi(q,t)$ with time.

9.3 Interaction of Radiation with Matter—Semiclassical Theory

9.3.1 Hamiltonian of Charged Particles

In order to understand the interaction of radiation with matter we need to write the wave equation for a system of charged particles under the influence of an external electromagnetic field. An electromagnetic field of radiation is represented by an oscillating electrical field **E** and an oscillating magnetic field **H** which are perpendicular to each other and also to the direction of propagation of the electromagnetic waves. The interaction of the electromagnetic radiation with charged particles is mainly due to the electric field **E** rather than its magnetic field **H**. This is due to the fact that the force exerted by the electric field on charge q is $q\mathbf{E}$ whereas the force due to the magnetic field is $\frac{q\mathbf{v}\times\mathbf{H}}{c}$, where **v** is the velocity of the charge and c is the velocity of light. Since **v** is very much less than c for most of the systems, the contribution of the magnetic field to the interaction is very much smaller than the contribution of the electrical field. Hence while considering the interaction of electromagnetic field with radiation, usually, the magnetic contribution is ignored.

In order to derive an expression for the Hamiltonian of a system of charged particles one starts with Maxwell's equation

$$\mathbf{H} = \nabla \times \mathbf{A}$$

$$\text{and,}\quad \mathbf{E} = -\frac{1}{c}\frac{\partial \mathbf{A}}{\partial t} - \nabla\Phi \tag{9.3.1}$$

where **A** and Φ are, respectively, the vector and scalar potentials. For small intensity radiation fields **A** is very small. Also it can be shown that it is possible to choose gauge function so that $\nabla\cdot\mathbf{A} = 0$, and the scalar potential for the electromagnetic wave, $\Phi = 0$ for all **r** and t without loss of generality.

The classical expression for the Hamiltonian of a single charged particle in the presence of the electromagnetic field may be written as

$$H = \frac{1}{2m}\left[\left(p_x - \frac{q}{c}A_x\right)^2 + \left(p_y - \frac{q}{c}A_y\right)^2 + \left(p_z - \frac{q}{c}A_z\right)^2\right] + q\Phi \tag{9.3.2}$$

Replacing **p** by the quantum mechanical operator $-i\hbar\nabla$ we get the relationship

$$H = \frac{1}{2m}\left[-\hbar^2\nabla^2 + i\hbar\frac{q}{c}\nabla\cdot\mathbf{A} + 2iq\frac{\hbar}{c}\mathbf{A}\cdot\nabla + \frac{q^2}{c^2}A^2\right] + q\Phi \tag{9.3.3}$$

Using the relations $\nabla\cdot\mathbf{A} = 0$ and $\Phi = 0$ and neglecting the term $\frac{q^2}{c^2}A^2$ for weak interactions, the Hamiltonian H for a single particle can be written as

$$H = -\frac{\hbar^2}{2m}\nabla^2 + \frac{iq\hbar}{mc}\mathbf{A}\cdot\nabla \tag{9.3.4}$$

For a system of charged particles having an internal potential energy V' (independent of time), the Hamiltonian H would become

$$H = \sum_j -\frac{\hbar^2}{2m_j}\nabla_j^2 + \sum_j \frac{iq_j\hbar}{m_j c}\mathbf{A_j}\cdot\nabla_j + V \tag{9.3.5}$$

The right-hand side of this equation is made up of two parts:

(1) the time-independent unperturbed part

$$H^0 = -\sum_j \frac{\hbar^2}{2m_j}\nabla_j^2 + V \tag{9.3.6}$$

and (2) the time-dependent perturbation part due to the electromagnetic radiation field

$$H' = \frac{i\hbar}{c}\sum \frac{q_j}{m_j}\mathbf{A_j}\cdot\nabla_j = -\sum_j \frac{q_j}{m_j c}\mathbf{A_j}\cdot\mathbf{p_j} \tag{9.3.7}$$

9.3.2 Induced Emission and Absorption of Radiation—Periodic Perturbation Coupling Two Discrete States

9.3.2.1 Polarized Radiation

Let us now consider an atomic or molecular system subjected to perturbation H' of an electromagnetic field given by Eq. (9.3.7). For simplicity we first assume that the incident beam of light falling on the system is plane polarized, that is $A_x \neq 0$ but $A_y = A_z = 0$.

In order to see how the wavefunction develops with time, we need to calculate the matrix element H'_{mn} on the right-hand side of Eq. (9.2.7), by using Eq. (9.3.7)

$$H'_{mn} = \int \Psi_m^{0*} H' \Psi_n^0 d\tau = \int \Psi_m^{0*} \left(-\sum_j \frac{q_j}{m_j c} \mathbf{A_j} \cdot \mathbf{p_j} \right) \Psi_n^0 \, d\tau,$$

which for the x-component gives

$$\int \Psi_m^{0*} H' \Psi_n^0 d\tau = \int \Psi_m^{0*} \left(-\sum_j \frac{q_j}{m_j c} A_{xj} p_{xj} \right) \Psi_n^0 d\tau \tag{9.3.8}$$

Since the atomic and molecular dimensions are of the order of a few angstrom whereas the wavelength of light in the UV or visible regions are of the order of 1000 Å, a sufficiently good approximation in our case will be to treat **A** as constant over the molecule. Replacing q_j and m_j by q and m as all electrons have the same charge and mass, Eq. (9.3.8) may then be written as

$$H_{mn} = -\frac{q}{mc} A_x \sum_j \int \Psi_m^{0*} p_{xj} \Psi_n^0 d\tau = i\hbar \frac{q}{mc} A_x \exp\left(i \frac{E_m - E_n}{\hbar} t \right) \sum_j \int \Phi_m^{0*} \frac{\partial}{\partial x_j} \Phi_n^0 dx \tag{9.3.9}$$

where use has been made of Eq. (9.2.3).

The integral in the above equation can be evaluated by considering that Φ_m^{0*} and Φ_n^0 are the eigenfunctions of the time-independent Schrödinger equations

$$\frac{d^2 \Phi_m^{0*}}{dx^2} + \frac{2m}{\hbar^2} [E_m - V(x)] \Phi_m^{0*} = 0 \tag{9.3.10}$$

$$\frac{d^2 \Phi_n^0}{dx^2} + \frac{2m}{\hbar^2} [E_n - V(x)] \Phi_n^0 = 0 \tag{9.3.11}$$

Multiplying Eq. (9.3.10) by $x\Phi_n^0$ and Eq. (9.3.11) by $x\Phi_m^{0*}$, subtracting and integrating, we get

$$\int_{-\infty}^{\infty} \left(x\Phi_n^0 \frac{d^2 \Phi_m^{0*}}{dx^2} - x\Phi_m^{0*} \frac{d^2 \Phi_n^{0*}}{dx^2} \right) dx = \frac{2m}{\hbar^2} (E_n - E_m) \int_{-\infty}^{\infty} \Phi_m^{0*} x \Phi_n^0 dx \tag{9.3.12}$$

Integrating Eq. (9.3.12) by parts and considering that the wavefunctions vanish at $x \to \infty$, we get

$$\int_{-\infty}^{\infty} \left(\Phi_n^0 \frac{d\Phi_m^{0*}}{dx} - \Phi_m^{0*} \frac{d\Phi_n^0}{dx} \right) dx = \frac{2m}{\hbar^2} (E_m - E_n) \int_{-\infty}^{\infty} \Phi_m^{0*} x \Phi_n^0 dx \tag{9.3.13}$$

Integrating by parts the first term in the integral on the left-hand side of Eq. (9.3.13) shows that it is equal to the second integral and so we get

$$\int \Phi_m^{0*} \frac{d}{dx} \Phi_n^0 dx = -\frac{m}{\hbar^2}(E_m - E_n) \int \Phi_m^{0*} x \Phi_n^0 dx \tag{9.3.14}$$

Substituting Eq. (9.3.14) into Eq. (9.3.9), we get

$$\int \Psi_m^{0*} H' \Psi_n^0 d\tau = -\frac{1}{c} A_x \frac{i}{\hbar}(E_m - E_n)\mu_{xmn} \exp\left\{ i \frac{E_m - E_n}{\hbar} t \right\} \tag{9.3.15}$$

$$\text{where,} \quad \mu_{xmn} = \int \Phi_m^{0*} q \sum_j x_j \Phi_n^0 dx \tag{9.3.16}$$

is the matrix element for the x-component of the dipole moment $\mu_x = \Sigma_j q x_j$. $\boldsymbol{\mu}_{mn}$ is called the transition moment of the dipole transition as it determines whether or not a given transition between the states m and n is allowed, and μ_{xmn} as its x-component. It also determines the intensity of the transition.

Let us now introduce certain approximations. Since the perturbation H' is small, the rate of change of the coefficient c_m in Eq. (9.2.7) is also small. Suppose the system was initially in the state n such that at time $t = 0$, $c_n = 1$ and all the other c's are zero, then on substituting Eq. (9.3.10) into Eq. (9.2.7) we get,

$$\frac{dc_m}{dt} = -\frac{i}{\hbar} \int \Psi_m^{0*} H' \Psi_n^0 d\tau$$

$$\frac{dc_m}{dt} = -\frac{A_x}{c\hbar^2} \mu_{xmn}(E_m - E_n) e^{\frac{i}{\hbar}(E_m - E_n)t} \tag{9.3.17}$$

If the oscillating electric field associated with the electromagnetic radiation has a frequency ν, then the time dependence of A_x may be expressed as

$$A_x = A_x^0 \cos 2\pi\nu t = \frac{1}{2} A_x^0 \left(e^{2\pi i\nu t} + e^{-2\pi i\nu t}\right) = \frac{1}{2} A_x^0 \left(e^{\frac{i}{\hbar} h\nu t} + e^{-\frac{i}{\hbar} h\nu t}\right) \tag{9.3.18}$$

Hence,

$$\frac{dc_m}{dt} = -\frac{1}{2c\hbar^2} A_x^0 \mu_{xmn}(E_m - E_n) \left\{ e^{\frac{i(E_m - E_n + h\nu)t}{\hbar}} + e^{\frac{i(E_m - E_n - h\nu)t}{\hbar}} \right\} \tag{9.3.19}$$

On integrating this equation and using the condition that, at $t = 0$, $c_m = 0$ we get,

$$c_m = \frac{i}{2c\hbar} A_x^0 \mu_{xmn}(E_m - E_n) \left\{ \frac{e^{\frac{i}{\hbar}(E_m - E_n + h\nu)t} - 1}{(E_m - E_n + h\nu)} + \frac{e^{\frac{i}{\hbar}(E_m - E_n - h\nu)t} - 1}{(E_m - E_n - h\nu)} \right\} \tag{9.3.20}$$

Suppose we consider the case of energy absorption such that the system lying in the state n moves on interaction with the electromagnetic radiation to the higher energy state m, then $E_m > E_n$. In this case, c_m will be large only when $E_m - E_n \simeq h\nu$ and the first term in the brackets in Eq. (9.3.20) can be neglected. We get

$$c_m = \frac{i}{2c\hbar} A_x^0 \mu_{xmn}(E_m - E_n) \left\{ \frac{e^{\frac{i}{\hbar}(E_m - E_n - h\nu)t} - 1}{(E_m - E_n - h\nu)} \right\} \tag{9.3.21}$$

The probability that at time *t*, the system is in the state *m* will be given by the product

$$c_m^* c_m = \frac{1}{4c^2\hbar^4}\left|A_x^0\right|^2 |\mu_{xmn}|^2 \frac{(E_m - E_n)^2}{(E_m - E_n - h\nu)^2}\left\{2 - e^{\frac{i}{\hbar}(E_m - E_n - h\nu)t} - e^{-\frac{i}{\hbar}(E_m - E_n - h\nu)t}\right\}$$

$$c_m^* c_m = \frac{1}{4c^2\hbar^4}\left|A_x^0\right|^2 |\mu_{xmn}|^2 (E_m - E_n)^2 \frac{\sin^2\left\{\frac{E_m - E_n - h\nu}{2\hbar}t\right\}}{\left\{\frac{E_m - E_n - h\nu}{2\hbar}\right\}^2} \qquad (9.3.22)$$

$$\text{or,} \quad c_m^* c_m = |c_m|^2 = \frac{1}{4c^2\hbar^4}\left|A_x^0\right|^2 |\mu_{xmn}|^2 (E_m - E_n)^2 \frac{\sin^2(\alpha t)}{\alpha^2} \qquad (9.3.23)$$

$$\text{where,} \quad \alpha = \frac{E_m - E_n - h\nu}{2\hbar} = \frac{\omega_{mn} - \omega}{2} \quad \text{and} \quad \omega = 2\pi\nu \qquad (9.3.24)$$

The term $c_m^* c_m$, which may be taken as the probability of affecting transition to state *m* after time *t* depends upon the sine function of the form $\frac{\sin^2(\alpha t)}{\alpha^2}$.

The function $\frac{\sin^2(\alpha t)}{\alpha^2}$ is plotted against α in Figure 9.1.

It is seen that the maximum value of this term is obtained at $\alpha = 0$, i.e., $E_m - E_n \simeq h\nu$. Neglecting the higher powers of α,

$$f(\alpha) = \frac{\sin^2(\alpha t)}{\alpha^2} = \frac{1}{\alpha^2}\left[\alpha t - \frac{\alpha^3 t^3}{3!} + \ldots\right]^2 = t^2$$

This function $f(\alpha)$ will be zero, if $\alpha t = \pm n\pi$ or $\alpha = \pm\frac{n\pi}{t}$.

Thus it will have a zero value at $\alpha = \pm\frac{\pi}{t}, \pm\frac{2\pi}{t}, \ldots$ and more peak values at $\alpha = \frac{1}{t}\left(n + \frac{1}{2}\right)\pi$. These are bounded by the denominator α^2. The height of the main peak increases as t^2 while its width decreases inversely as *t*. It means that the area under the curve is proportional to *t*. Thus, at $t \to \infty$, $\frac{\sin^2(\alpha t)}{\alpha^2}$ is peaked at $\alpha = 0$ and behaves like

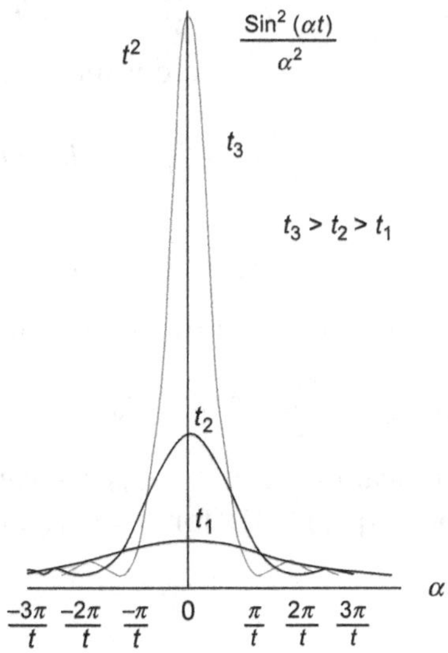

FIGURE 9.1 Plot of $\frac{\sin^2(\alpha t)}{\alpha^2}$ versus α for a fixed time *t*. At $t \to \infty$, the function asymptotes to a Dirac's δ function.

Dirac's delta function $\delta(\omega_{mn} - \omega)$. From Figure 9.1 it also follows that if only a short time has elapsed since the field started to interact with the microsystem, the transition probability is small even when the external frequency $\omega = \omega_{mn}$, i.e., $\alpha = 0$. If the interaction lasts longer, the transition probability at resonance increases steadily.

In a real system one employs a radiation field spread over a range of frequencies. So, in order to obtain the correct value of $c_m^* c_m$ one has to integrate over a range of frequencies. $c_m^* c_m$ shall, however, have an appreciable magnitude only for $E_m - E_n \simeq h\nu$ or $\nu_{mn} \simeq \nu$, that is, when the frequency of the incident radiation matches with the transition frequency between states m and n. We must therefore integrate $c_m^* c_m$ over the frequency range $-\infty$ to $+\infty$ and also treat A_x^0 as a constant and equal to $A_x^0(\nu_{mn})$. Hence, from Eq. (9.3.22) we get

$$|c_m|^2 = c_m^* c_m = \frac{1}{4c^2\hbar^4}|A_x^0|^2 |\mu_{xmn}|^2 (E_m - E_n)^2 \int_{-\infty}^{+\infty} \frac{\sin^2\left(\frac{E_m - E_n - h\nu}{2\hbar} t\right)}{\left(\frac{E_m - E_n - h\nu}{2\hbar}\right)^2} d\nu \tag{9.3.25}$$

$$\text{or,}\quad |c_m|^2 = c_m^* c_m = \frac{1}{4\pi c^2\hbar^4}|A_x^0|^2 |\mu_{xmn}|^2 (E_m - E_n)^2 \int_{-\infty}^{+\infty} \frac{\sin^2(\alpha t)}{\alpha^2} d\alpha \tag{9.3.26}$$

with normalization, $\int_{-\infty}^{+\infty} \frac{\sin^2(\alpha t)}{\alpha^2} d\alpha = \pi t$.

Hence

$$c_m^* c_m = \frac{\pi^2 \nu_{mn}^2}{c^2\hbar^2} |A_x^0(\nu_{mn})|^2 |\mu_{xmn}|^2 t \tag{9.3.27}$$

where $E_m - E_n$ has been replaced by $h\nu_{mn}$.

Equation (9.3.27) gives the probability that the system is in the state m at a time t. Since at time $t = 0$ this probability was zero and the system was only in the state n, hence the transition probability per unit time W_{mn} for transition $n \to m$ is given as

$$W_{n\to m} = \frac{c_m^*(t) c_m(t)}{t} = \frac{\pi^2 \nu_{mn}^2}{c^2\hbar^2} |A_x^0(\nu_{mn})|^2 |\mu_{xmn}|^2 \tag{9.3.28}$$

This equation can also be derived in a different manner from Eq. (9.3.23)

$$W_{n\to m} = \frac{c_m^*(t) c_m(t)}{t} = \frac{1}{4c^2\hbar^4 t} |A_x^0|^2 |\mu_{xmn}|^2 (E_m - E_n)^2 \frac{\sin^2(\alpha t)}{\alpha^2} \tag{9.3.29}$$

$$\text{We can replace } \lim_{t\to\infty} \frac{1}{t} \frac{\sin^2(\alpha t)}{\alpha^2} = 2\pi\delta(2\alpha) = 2\pi\delta(\omega_{mn} - \omega) = \delta(\nu_{mn} - \nu) \tag{9.3.30}$$

where use has been made of the property of the δ function, $\delta(ax) = a^{-1}\delta(x)$. This gives,

$$W_{n\to m} = \frac{\pi^2 \nu_{mn}^2}{c^2\hbar^2} |A_x^0|^2 |\mu_{xmn}|^2 \delta(\nu_{mn} - \nu) \tag{9.3.31}$$

Since $W_{n\to m}$ would be significant only at $\nu = \nu_{mn}$, we can replace $|A_x^0|^2$ by $|A_x^0(\nu_{mn})|^2$. Equation (9.3.28) follows from Eq. (9.3.31) for $\nu = \nu_{mn}$. Relation (9.3.31) is known as **Fermi's Golden Rule**.

From the expression for the Golden rule (Eq. (9.3.31)) it follows that, for transitions to occur and to satisfy energy conservation:

1. The final states must exist over a continuous energy range to match $\Delta E = \hbar\omega$ for fixed perturbation frequency ω, or
2. The perturbation must cover a sufficiently wide spectrum of frequency so that a discrete transition with a fixed $\Delta E = \hbar\omega$ is possible.

9.3.2.2 Unpolarized Isotropic Radiation

In the case of polarized radiation we assumed that A_y and A_z components of the electrical vector of the incident radiation were zero. Thus, we had neglected the contributions of these components in determining the total transition probabilities from energy state n to m. In the general case of an unpolarized radiation all the components A_x, A_y, and A_z are different from zero and so Eq. (9.3.27) shall have additional terms in A_y and A_z. In this case, this equation may be written as

$$c_m^* c_m = \frac{\pi^2 \nu_{mn}^2}{c^2 \hbar^2}\left[\left|A_x^0(\nu_{mn})\right|^2 |\mu_{xmn}|^2 + \left|A_y^0(\nu_{mn})\right|^2 |\mu_{ymn}|^2 + \left|A_z^0(\nu_{mm})\right|^2 |\mu_z(\nu_{mm})|^2\right] t \tag{9.3.32}$$

If the incident radiation is isotopic then all the three components A_x, A_y, and A_z are equal, then

$$\left|A_x^0(\nu_{mn})\right|^2 = \left|A_y^0(\nu_{mn})\right|^2 = \left|A_z^0(\nu_{mn})\right|^2 = \frac{1}{3}\left|A^0(\nu_{mn})\right|^2$$

$$\text{where,} \quad \left|A^0(\nu_{mn})\right|^2 = \left|A_x^0(\nu_{mn})\right|^2 + \left|A_y^0(\nu_{mn})\right|^2 + \left|A_z^0(\nu_{mn})\right|^2 \tag{9.3.33}$$

Using Eq. (9.3.33) into Eq. (9.3.32) we get,

$$c_m^* c_m = \frac{1}{3}\frac{\pi^2 \nu_{mn}^2}{c^2 \hbar^2}\left|A^0(\nu_{mn})\right|^2 t |\mu_{mn}|^2 \tag{9.3.34}$$

By analogy with Eq. (9.3.31), the transition probability per unit time W_{mn} for transition $n \to m$ is given as

$$W_{n\to m} = \frac{1}{3}\frac{\pi^2 \nu_{mn}^2}{c^2 \hbar^2}\left|A^0(\nu_{mn})\right|^2 |\mu_{mn}|^2 \delta(\nu_{mn} - \nu) \tag{9.3.35}$$

where $|\mu_{mn}|^2 = |\mu_{xmn}|^2 + \left|\mu_{ymn}\right|^2 + |\mu_{zmn}|^2$

From electromagnetic theory, it is possible to express $|A_0(\nu_{mn})|^2$ in terms of the radiation density $\rho(\nu_{mn})$.

$$E = -\frac{1}{c}\frac{\partial \mathbf{A}}{\partial t} - \nabla\Phi$$

Since $\nabla\Phi = 0$ and $\mathbf{A} = A^0 \cos 2\pi\nu t$, we get

$$E(\nu_{mn}) = -\frac{1}{c}\frac{\partial}{\partial t}\left[A^0(\nu_{mn})\cos 2\,\pi\nu t\right] = \frac{2\pi\nu_{mn}}{c}A^0(\nu_{mn})\sin 2\,\pi\nu t$$

$$\text{or,} \quad E^2(\nu_{mn}) = \frac{4\pi^2\nu_{mn}^2}{c^2}\left|A^0(\nu_{mn})\right|^2 \sin^2 2\pi\nu t \tag{9.3.36}$$

Since $\overline{\sin^2 2\pi\nu t} = \frac{1}{2}$, where the bar represents the average value, we get,

$$\overline{E^2}(\nu_{mn}) = \frac{2\pi^2\nu_{mn}^2}{c^2}\left|A^0(\nu_{mn})\right|^2 \tag{9.3.37}$$

Since radiation density is related to the electric field by the relation

$$\rho(\nu_{mn}) = \frac{\overline{E^2}(\nu_{mn})}{4\pi} \tag{9.3.38}$$

from Eqs (9.3.37) and (9.3.38), we get

$$\rho(\nu_{mn}) = \frac{\pi\nu_{mn}^2}{2c^2}\left|A^0(\nu_{mn})\right|^2$$

$$\text{or} \quad \left|A^0(\nu_{mn})\right|^2 = \frac{2c^2}{\pi\nu_{mn}^2}\rho(\nu_{mn}) \tag{9.3.39}$$

Equation (9.3.34) then gives,

$$c_m^* c_m = \frac{2}{3}\frac{\pi}{\hbar^2}|\mu_{mn}|^2\rho(\nu_{mn})t \tag{9.3.40}$$

Since at time $t = 0$ the probability for the system to exist in the state m was zero and its probability at time t is given by Eq. (9.3.40), the transition probability leading to the absorption of radiation would be

$$W_{n\to m} = \frac{c_m^* c_m}{t} = \frac{2}{3}\frac{\pi}{\hbar^2}|\mu_{mn}|^2\rho(\nu_{mn}) \tag{9.3.41}$$

$W_{n\to m}$ is also represented as $B_{n\to m}\rho(\nu_{mn})$. Hence,

$$B_{n\to m}\rho(\nu_{mn}) = \frac{2}{3}\frac{\pi}{\hbar^2}|\mu_{mn}|^2\rho(\nu_{mn}) \tag{9.3.42}$$

The probability of absorption of radiation is thus proportional to the density of radiation. It can similarly be shown that if the system was initially in the state m and then induced by some electromagnetic radiation perturbation to move to the lower energy state n resulting in the emission of radiation, the transition probability for this induced emission would also have the same value as in the case of absorption.

$$B_{m\to n}\rho(\nu_{mn}) = B_{n\to m}\rho(\nu_{mn}) \tag{9.3.43}$$

The coefficients $B_{m\to n}$ and $B_{n\to m}$ are known as the Einstein's transition probability coefficients for induced emission and absorption or Einstein's coefficients of emission and absorption, respectively.

9.3.3 Periodic Perturbation Coupling with a Continuum of Final States

In the derivation of Eq. (9.3.25) it was assumed that the transition from the state n takes place to a defined state m. In reality, however, one may observe transitions not to a definite state but to a range of states. Let us therefore consider the case when the final state is part of a continuum of states. In this case the probability of transition to a range of states will be the sum of the probabilities in Eq. (9.3.27). Thus the total probability of transition is

$$P(t) = \sum_m |c_m(t)|^2 \quad (9.3.44)$$

where the summation is over the range of states at m. Since, the final state m may occur as a narrow band of levels within a frequency range ν_m and $\nu_m + d\nu$, we may introduce here number density, i.e., the number of states in a unit frequency range of final states, $n(\nu_m)$. It is reasonable to assume that $n(\nu_m)$ does not change much in the narrow range of frequencies about the resonance frequency $\nu = \nu_{mn}$. Hence the total probability of transitions to a band of levels may be approximated as

$$P(t) = |c_m(t)|^2 n(\nu_m) = \frac{\pi^2 \nu_{mn}^2}{c^2\hbar^2} \left|A_x^0(\nu_{mn})\right|^2 |\mu_{xmn}|^2 n(\nu_m) t \quad (9.3.45)$$

Usually the density of states is expressed in terms of number of states in an energy range instead of the frequency range. If the density of states is written as $\rho(E)$ such that $\rho(E)d(E)$ is a number of continuum states in the range E and $E + dE$, then the total transition probability P(t) is given as

$$\mathrm{P}(t) = \int_{\text{range}} |c_m(t)|^2 \rho(E) dE \quad (9.3.46)$$

where the integration is over the range of frequencies around the resonance frequency $\nu = \nu_{mn}$. $|c_m(t)|^2$ is again given by Eq. (9.3.23) as before

$$\mathrm{P}(t) = \int_{\text{range}} \frac{1}{4c^2\hbar^4} |A_x^0|^2 |\mu_{xmn}|^2 (E_m - E_n)^2 \frac{\sin^2(\alpha t)}{\alpha^2} \rho(E) dE \quad (9.3.47)$$

where α is given by Eq. (9.3.24)

To evaluate this integral, the transition frequency ω_{mn} is written as $\omega_{mn} = \frac{E}{\hbar}$. A simplification is introduced by the fact that $\frac{\sin^2(\alpha t)}{\alpha^2}$ behaves like a δ function and is sharply peaked at $\alpha = 0$, i.e., $\omega = \omega_{mn}$. Hence, only at $\omega \approx \omega_{mn}$ one can expect maximum transition probability. We can therefore evaluate density of states $\rho(E)$ at $E = \hbar\omega_{mn}$ and treat it as a constant.

The integrand has a large value only in a narrow range around ω_{mn} and so the actual range of the integral can be expanded from $-\infty$ to $+\infty$. This approximation shall introduce no significant error because the integrand is very small outside the actual range. Equation (9.3.47) can therefore be written as

$$\mathrm{P}(t) = \frac{t^2}{4c^2\hbar^4} \left|A_x^0(\nu_{mn})\right|^2 |\mu_{xmn}|^2 (E_m - E_n)^2 \rho(E_{mn}) \int_{-\infty}^{+\infty} \frac{\sin^2\left[\frac{1}{2}\left(\frac{E}{\hbar} - \omega\right)t\right]}{\left[\frac{1}{2}\left(\frac{E}{\hbar} - \omega\right)t\right]^2} dE \quad (9.3.48)$$

Using standard integral of the type $\int_{-\infty}^{+\infty} \frac{\sin^2\beta}{\beta^2} d\beta = \pi$ and on simplification, this equation gives the transition probability of absorption $W_{n\to m}$ as,

$$W_{n\to m} = \frac{\mathrm{P}(t)}{t} = \frac{2\pi^3 \rho(E_{mn}) \nu_{mn}^2}{\hbar c^2} \left|A_x^0(\nu_{mn})\right|^2 |\mu_{xmn}|^2 \quad (9.3.49)$$

Equation (9.3.49) may also be derived from Eq. (9.3.31) after including the $\rho(E_{mn})$ term and by noting that

$$\delta(\nu_{mn} - \nu) = \delta\left[\frac{1}{h}(E_{mn} - E)\right] = h\delta(E_{mn} - E)$$

This gives,

$$W_{n\to m} = \frac{2\pi^3 \rho(E_{mn})\nu_{mn}^2}{\hbar c^2}\left|A_x^0(\nu_{mn})\right|^2 |\mu_{xmn}|^2 \delta(E_{mn} - E) \tag{9.3.50}$$

9.3.4 Einstein Relations

When a system is placed in an energy bath of radiation of density $\rho(\nu_{mn})$ it may either undergo induced absorption of radiation from the lower energy state to the higher energy state, or it may emit radiation due to stimulated emission or due to spontaneous emission (after a definite life time) from the higher to the lower state. The transition probabilities for the absorption ($W_{n\to m}$) and emission ($W_{m\to n}$) of radiation have been calculated earlier and are given by Eqs (9.3.41–9.3.43) in terms of the Einstein coefficients $B_{n\to m}$ and $B_{m\to n}$. In order to complete the theory of radiation, we need to calculate the transition probability coefficient $A_{m\to n}$ for spontaneous emission. A direct quantum mechanical calculation of this quantity is a difficult problem but Einstein determined its value from a consideration of equilibrium between two states of different energy. Einstein showed that the three processes of stimulated absorption, stimulated emission, and spontaneous emission are related mathematically through the requirement that for a system of atoms in thermal equilibrium with its own radiation, the rate of upward transitions from E_n to E_m must be equal to the rate of downward transitions (from E_m to E_n).

If there are N_n systems in the state n with energy E_n, then the upward transition rate is proportional to both N_n and the number of photons present with appropriate frequency ν. The energy density ρ_ν of such photons is simply, $\rho_\nu = Nh\nu$, where N is the number of photons per unit volume.

The upward transition rate can be written as:

$$\text{Stimulated absorption rate} = N_n B_{n\to m}\rho_\nu, \tag{9.3.51}$$

where $B_{n\to m}$ is a constant for a given pair of energy levels.

Similarly, if there are N_m systems in the state m with energy E_m, then we have

$$\text{Stimulated emission rate} = N_m B_{m\to n}\rho_\nu \tag{9.3.52}$$

where, once again, $B_{m\to n}$ is a constant for the pair of energy levels involved.

It may here be noted that since initially $N_n > N_m$, $B_{n\to m} = B_{m\to n}$ and ρ_ν is the same, it follows from Eqs (9.3.51) and (9.3.52) that the rate of stimulated absorption is larger than the rate of stimulated emission and so a situation may arise when the number of systems in the state m may become equal to the number in state n. This will be in contradiction to the equilibrium statistical mechanics which demonstrates that the number of molecules in the lower state must be larger than the number in the higher state. In order to resolve this paradox, Einstein proposed the existence of another process, the spontaneous emission. According to this proposition there exists a small probability for an excited molecule or atom to release a photon even in the absence of an electromagnetic field. The rate of spontaneous emission, which is denoted by $A_{m\to n}$ will be thus independent of the energy density of the radiation field.

The spontaneous transition rate depends upon the average life time τ_{mn} of the atoms in the excited state. The probability that a particular atom will undergo a spontaneous transition in a time dt is $A_{m\to n}dt = \frac{dt}{\tau_{mn}}$, where $A_{m\to n}$ is a constant.

Thus, if there are N_m atoms in the upper level, then

$$\text{Spontaneous emission rate} = N_m A_{m\to n} \tag{9.3.53}$$

Since the relationship between the Einstein coefficients $A_{m\to n}$, $B_{n\to m}$, and $B_{m\to n}$ is to be established by considering the condition for the assembly of systems to be in thermal equilibrium, we must have

$$N_n\rho_\nu B_{n\to m} = N_m\rho_\nu B_{m\to n} + N_m A_{m\to n}$$

or $\rho_\nu(N_n B_{n\to m} - N_m B_{m\to n}) = N_m A_{m\to n}$

$$\text{or,}\quad \rho_\nu = \frac{(A_{m\to n}/B_{m\to n})}{\left(\frac{N_n B_{n\to m}}{N_m B_{m\to n}} - 1\right)} \tag{9.3.54}$$

In the state of thermal equilibrium the number N_j of systems in the jth level (or the population of the state) is given by Maxwell–Boltzmann statistics

$$N_j = N_0 \frac{e^{-E_j/kT}}{\sum_i e^{-E_i/kT}} \tag{9.3.55}$$

where $N_0 = \Sigma_i N_i$ is the total number of systems and E_j is the energy of the jth level. The summation is overall the energy states i of the system. In this equation, we have assumed that all the available states or energy levels have the same probability of being occupied. In general this is not so, as each state may have different probability g_i of occupation (g is often called the degeneracy). Thus, in principle, Eqs (9.3.56–9.3.60) derived below by considering Maxwell-Boltzmann statistical distribution of states should be modified by including the degeneracies g_m and g_n. These parameters usually appear in the equation as ratio g_m/g_n, which is often of the order of unity and have, therefore, been omitted for simplicity.

From Eq. (9.3.55) it follows that

$$\frac{N_n}{N_m} = e^{(E_m - E_n)/kT} \tag{9.3.56}$$

Substituting Eq. (9.3.56) into Eq. (9.3.54), and remembering that $E_m - E_n = h\nu$, we get

$$\rho_\nu = \frac{A_{m\to n}/B_{m\to n}}{\left(\frac{B_{n\to m}}{B_{m\to n}}\, e^{h\nu/kT} - 1\right)} \tag{9.3.57}$$

Also, because the system is in equilibrium, the radiation within the assembly of atoms must be identical to black-body radiation which can be described by Planck's radiation formula

$$\rho_\nu = \frac{8\pi h\nu^3}{c^3}\left(\frac{1}{e^{h\nu/kT} - 1}\right) \tag{9.3.58}$$

Comparing Eqs (9.3.57) and (9.3.58), we see that

$$B_{n\to m} = B_{m\to n} \tag{9.3.59}$$

$$\text{and } A_{m\to n} = \frac{8\pi h}{c^3}\nu_{mn}^3 B_{m\to n} \tag{9.3.60}$$

Equations (9.3.42), (9.3.43), and (9.3.60) are known as Einstein relations which enable us to calculate the Einstein coefficients *A* and *B*. Equations (9.3.52) and (9.3.53) can be used to calculate the rates of spontaneous and stimulated emission for a given pair of energy levels in thermal equilibrium with the radiation.

9.4 Lasers

It is a common observation that under conditions of thermal equilibrium radiation is given out by the process of spontaneous emission. This is not difficult to understand if we use Eqs (9.3.52) and (9.3.53) to compare the rates of stimulated and spontaneous emission for a given pair of energy levels in thermal equilibrium with the radiation. If we work out the ratio of the rates of spontaneous and stimulated emission, *R*, then

$$R = \frac{N_m A_{m\to n}}{N_m B_{m\to n}\rho_\nu} = \frac{8\pi h\nu^3}{c^3 \rho_\nu} \tag{9.4.1}$$

On substituting the value of ρ_ν from Eq. (9.3.58), we get

$$R = \left[e^{h\nu/kT} - 1\right] \tag{9.4.2}$$

For example, if $\lambda = 6983$ Å (corresponding to He–Ne laser) and $T = 370$ K

$$R = \exp\left(\frac{h\nu}{kT}\right) - 1 = \exp\left(\frac{hc}{\lambda kT}\right) - 1 = e^{61.5} = 5 \times 10^{26} \tag{9.4.3}$$

Thus, under conditions of thermal equilibrium, spontaneous emission is 5×10^{26} times most likely to occur than the stimulated emission. We also note that the higher is the frequency (i.e., $e^{h\nu/kT}$), larger is *R* and so less likely is the process of stimulated emission. Thus, stimulated emission is more probable with smaller frequencies or larger wavelengths. It is therefore not surprising that the first light amplification by stimulated emission of radiation (abbreviated as LASERS) was witnessed with microwaves (MASERS).

The process of stimulated emission competes with the processes of spontaneous emission and stimulated absorption. Clearly, then, if we wish to amplify a beam of light by stimulated emission we must increase the rate of this process relative to the other two. It follows from Eqs (9.3.52) and (9.3.53) that in order to achieve a higher rate of stimulated emission, both the population density (N_m) of the upper level m and the radiation density ρ_ν should be increased while decreasing the population density (N_n) of the lower level n. Indeed, because of the Einstein relation $B_{n\to m} = B_{m\to n}$, we must ensure that $N_m > N_n$ (even though $E_m > E_n$), that is, we must create population inversion so that more photons are emitted than absorbed. If the emitted photons are in phase with the incident beam of the same frequency, amplification of light can result.

9.5 Magnetic Dipole and Electrical Quadrupole Transitions

In the derivation of transition probabilities $A_{m\to n}$, $B_{m\to n}$, and $B_{n\to m}$, in the last section, we had made a few approximations. Two of these are neglect of interaction of the magnetic field **H** of the radiation with the atomic and molecular electrons and nuclei and the neglect of spatial variation of radiation electric field. On account of these approximations, the transition coefficients depend only on the matrix element for the electric dipole moment (μ_{mn}) between the two states (the transition moment) and the transitions are called dipole transitions. A transition between the two states will be allowed only if $\mu_{mn} \neq 0$ and shall be disallowed or forbidden if $\mu_{mn} = 0$. This criterion for the transition between the two states is called the selection rule. However, it has been found quite often that the transition is not forbidden even when $\mu_{mn} = 0$, but only has a small probability. Corrections, therefore, need to be made in the expressions for transition probability so as to account for the above two approximations.

The application of an external magnetic field, **H**, influences the energies of atoms and molecules and can split their energy levels into states having different magnetic quantum numbers. This is called the Zeeman effect. The electronic orbital motion defined in terms of the orbital angular momentum **L** gives rise to an orbital magnetic dipole moment

$$\boldsymbol{\mu}_l = -\frac{e}{2mc}\mathbf{L} \tag{9.5.1}$$

and the spin motion defined in terms of spin angular momentum **S** gives rise to the spin magnetic dipole moment

$$\boldsymbol{\mu}_s = -\frac{e}{mc}\mathbf{S} \tag{9.5.2}$$

The total magnetic moment $\boldsymbol{\mu}_M$ is therefore,

$$\boldsymbol{\mu}_M = \boldsymbol{\mu}_l + \boldsymbol{\mu}_s \tag{9.5.3}$$

The interaction energy between the applied magnetic field **H** and a magnetic dipole $\boldsymbol{\mu}_M$ is therefore, from classical laws of electrostatics

$$E_M = -\boldsymbol{\mu}_M \cdot \mathbf{H} \tag{9.5.4}$$

and the corresponding Hamiltonian for Zeeman interaction is

$$\widehat{H}_M = -(\mu_l + \mu_s) \cdot \mathbf{H} = \beta_e \hbar \left(\widehat{L} + 2\widehat{S} \right) \cdot \mathbf{H} \tag{9.5.5}$$

where β_e, the Bohr magneton, is defined as $\beta_e = \frac{e\hbar}{2mc}$.

Thus, if we include the interaction of the magnetic field **H** of the electromagnetic radiation, the Hamiltonian of the molecular system will change by $-\widehat{\mu}_M \cdot \mathbf{H}$, where $\widehat{\mu}_M$ is the magnetic dipole moment operator. The perturbation in the Hamiltonian H' given by Eq. (9.3.7) shall therefore be modified to include this term. This will give additional terms in $c_m(t)$ (Eq. (9.3.21)) that are proportional to

$$\langle \Phi_m^0 | \widehat{\mu}_{M_x} | \Phi_n^0 \rangle, \left\langle \Phi_m^0 \middle| \widehat{\mu}_{M_y} \middle| \Phi_n^0 \right\rangle, \text{ and } \langle \Phi_m^0 | \widehat{\mu}_{M_z} | \Phi_n^0 \rangle \tag{9.5.6}$$

The magnetic dipole moment operators can be written in terms of spin and orbital angular momentum operators. As mentioned before, the interaction between the magnetic field **H** and the charges is about $\frac{1}{137}$ of the interaction between the electric field **E** and the charges. Since the transition probability is given by $\left| c_m^*(t) c_m(t) \right|^2$, the contribution of the magnetic dipole moment matrix element μ_{Mmn} is only a small fraction 10^{-4} or 10^{-5} of the electric dipole matrix element μ_{mn}. So, in those cases where $\mu_{mn} \neq 0$, we can neglect the contributions of transition moment μ_{Mmn}. If $\mu_{mn} = 0$, but $\mu_{Mmn} \neq 0$, we still have the possibility of a transition between the two states due to magnetic effects. Such transitions are called the magnetic dipole transitions and μ_{Mmn} are called magnetic dipole transition moments.

As mentioned, the second approximation was to consider that the electric field is uniform over the atoms and molecules by virtue of long wavelength of radiation. In reality the electric field may not be uniform over the system. If we take into account the spatial variation of the radiation electric field, we get additional contributions of the type

$$\left\langle \Phi_m^0 \sum_i q_i x_i^2 \middle| \Phi_n^0 \right\rangle, \left\langle \Phi_m^0 \middle| \sum_i q_i x_i y_i \middle| \Phi_n^0 \right\rangle, \left\langle \phi_m^0 \middle| \sum_i q_i x_i z_i \middle| \Phi_n^0 \right\rangle \ldots \tag{9.5.7}$$

to expression (9.3.32) for $c_m{}^*(t) c_m(t)$ involving matrix elements μ_{mn}. Transitions due to such matrix elements are called electric quadrupole transitions and the radiation emitted or absorbed in the process is called quadrupole radiation. Intensity of electric quadrupole radiation is about $\frac{1}{10^6}$ of the intensity of electric dipole radiation.

It may here be noted that while the dipole moment of a system of charges is a vector with three components given by $\mu_x = \Sigma_i q_i x_i$, $\mu_y = \Sigma_i q_i y_i$, and $\mu_z = \Sigma_i q_i z_i$, the quadrupole moment Q is a tensor of rank 2 (3×3 matrix). It has nine components (of which six are distinct) defined by

$$Q_{xx} = \sum_i q_i x_i^2, \; Q_{xy} = \sum_i q_i x_i y_i, \; Q_{xz} = \sum_i q_i x_i z_i,$$

$$Q_{yy} = \sum_i q_i y_i^2, \; Q_{yz} = \sum_i q_i y_i z_i, \text{ and } Q_{zz} = \sum_i q_i z_i^2 \tag{9.5.8}$$

Since the tensor is traceless, that is, $Q_{xx} + Q_{yy} + Q_{ZZ} = 0$, only five of the six components are independent. For a continuous charge distribution with density ρ, the components will be given as $Q_{xx} = \int x^2 d\tau$, etc., where $d\tau$ is a volume element ($dx\, dy\, dz$ or $r^2 \sin\theta\, dr\, d\theta\, d\phi$).

Based on the foregoing it follows that a complete expression for Einstein coefficient of spontaneous emission, Eq. (9.3.60) may be written as

$$A_{m\to n} = \frac{8\pi h\nu_{mn}^3}{c^3} B_{m\to n} = \frac{32\pi^3\nu_{mn}^3}{3\hbar c^3}|\mu_{mn}|^2$$

$$= \frac{32\pi^3\nu_{mn}^3}{3\hbar c^3}\left[|\langle m|q\mathbf{r}|n\rangle|^2 + |\langle m|\mu_M|n\rangle|^2 + |\langle m|q\mathbf{rr}|n\rangle|^2\right] \qquad (9.5.9)$$

The three terms on the right-hand side corresponds to electric dipole, magnetic dipole, and electric quadrupole radiation, respectively. The term **rr** represents a tensor of rank 2. For the case of zero spin ($S = 0$), the orbital angular momentum **L** alone may be incorporated in the calculation of μ_M so that

$$\mu_{Ml} = -\frac{e}{2mc}\mathbf{L} = -\frac{e}{2mc}\mathbf{r}\times\mathbf{p} \qquad (9.5.10)$$

Condon and Shortley [1] have shown that,

$$A_{m\to n} = \frac{32\pi^3\nu_{mn}^2}{3\hbar c^3}\left[|\langle m|q\mathbf{r}|n\rangle|^2 + \left|\left\langle m\left|\frac{e}{2mc}\mathbf{r}\times\mathbf{p}\right|n\right\rangle\right|^2 + \frac{3}{10}\pi^3\frac{\nu_{mn}^2}{c^2}|\langle m|q\mathbf{rr}|n\rangle|^2\right] \qquad (9.5.11)$$

In the case of an electron with charge $e = 4.8 \times 10^{-10}$ esu and mass $m_e = 9.035 \times 10^{-28}$ g the relative orders of magnitude of the three terms on the right-hand side, disregarding the constant term, are

first term $\sim (ea_o)^2 \sim 6.45 \times 10^{-36}$ cgs; a_o—radius of Bohr's first orbit (0.529 Å)

second term $\sim \left(\frac{e\hbar}{2m_e c}\right)^2 \sim 8.40 \times 10^{-41}$ cgs

third term $\sim 6.8 \times 10^{-43}$ cgs for $\lambda = 5000$ Å

It therefore follows that the transition probabilities of the magnetic dipole transitions and the electric quadrupole transitions are about 10^{-5} and 10^{-7} of the electric dipole transitions. The last two terms may play an important rule when the electric dipole transitions are forbidden either due to symmetry considerations (transitions between even–even or odd–odd states) or spin considerations (transitions between states of different spins, like singlet–triplet).

9.6 Selection Rules

We have seen that an induced transition is forbidden and its probability decreases drastically when the transition moment $\langle\Phi_m^0|\mu|\Phi_n^0\rangle$ vanishes. The same remarks apply for the probability of spontaneous emission since the same integral is involved in this case. The conditions on Φ_m^0 and Φ_n^0 for which the matrix element $\langle\Phi_m^0|\mu|\Phi_n^0\rangle$ is different from zero constitute the selection rule. Thus, a transition is allowed if the transition moment

$$\mu_{mn} = \int \Phi_m^{0*} \mu \Phi_n^0 d\tau \neq 0 \tag{9.6.1}$$

and forbidden if $\mu_{mn} = 0$.

On the basis of the selection rules, it is possible to make quantitative statements about which transitions can be induced by radiation and which transitions cannot be so induced.

9.6.1 Selection Rules for Surface Harmonic Wavefunctions

The selection rules can easily be formulated if the potential $V(\mathbf{r})$ that appears in the unperturbed Hamiltonian Eq. (9.3.6) is spherically symmetric. In this case, the potential function $V(\mathbf{r})$ is of the form $R_{nl}(r)\Theta_{lm}(\theta)\Phi_m(\phi)$, where $\Theta_{lm}(\theta)\Phi_n(\phi)$ are surface harmonics.

Thus, for example, the eigenfunctions of a hydrogen atom can be written in the form

$$\Psi_{nlm}(r,\theta,\phi) = \Psi_{nl}(r)\Psi_{lm}(\theta,\phi) = R_{nl}(r)P_l^{|m|}(\cos\theta)e^{im\phi} \tag{9.6.2}$$

The dipole moment of the hydrogen atom is $\mu(\mathbf{r}) = e\mathbf{r}$, where $\mathbf{r}$ is the distance between the two particles, the electron and proton, having charges $-e$ and $+e$.

The components of the electric dipole moment $\mu(\mathbf{r})$ in spherical polar coordinates can be written as

$$\mu_x = \mu(\mathbf{r})\sin\theta\cos\phi$$

$$\mu_y = \mu(\mathbf{r})\sin\theta\sin\phi \tag{9.6.3}$$

and $\mu_z = \mu(\mathbf{r})\cos\theta$.

If we consider a transition between the initial state i and final state f, we need to calculate the transition moment μ_{if}

$$\mu_{if} = \langle i|\mu|f\rangle = \langle i|\mu_x|f\rangle + \langle i|\mu_y|f\rangle + \langle i|\mu_z|f\rangle$$

So, we must evaluate integrals like

$$\int \Psi_{nlm}^* \mu_z \Psi_{n'l'm'} d\tau = \int \Psi_{nlm}^* \mu(r)\cos\theta\, \Psi_{n'l'm'} d\tau \tag{9.6.4}$$

for light polarized along z-axis and similar integrals for light polarized along the x and y axes.

Using Eq. (9.6.2), we may write Eq. (9.6.4) as

$$\int \Psi_{nlm}^* \mu_z \Psi_{n'l'm'} d\tau = \int R_{nl}^*(r)\Theta_{lm}^*(\theta)\Phi_m^*(\phi)\mu(r)\cos\theta R_{n'l'}(r)\Theta_{l'm'}(\theta)\Phi_{m'}(\phi)r^2\sin\theta drd\theta d\phi \tag{9.6.5}$$

On separating the variables and taking

$$\mu_{nln'l'} = \int_0^\infty R_{nl}^*(r)\mu(r)R_{n'l'}(r)r^2 dr \tag{9.6.6}$$

we can write Eq. (9.6.5) as

$$\Psi^*_{nlm}\mu_z\Psi_{n'l'm'} = \mu_{nln'l'}\int_0^{\pi}\Theta^*_{lm}(\theta)\cos\theta\,\Theta_{l'm'}(\theta)\sin\theta d\theta\int_0^{2\pi}\Phi_m(\phi)\Phi_{m'}(\phi)d\phi \tag{9.6.7}$$

Treating the integral in **r** nonvanishing, we may determine the integrals in θ and ϕ in a manner as was done for the case of hydrogen atom (Section 1.8)

$$\begin{aligned}&\int_0^{\pi}\Theta^*_{lm}(\theta)\cos\theta\Theta_{l'm'}(\theta)\sin\theta d\theta\int_0^{2\pi}\Phi^*_m(\phi)\Phi_{m'}(\phi)d\phi\\&=\int_0^{\pi}P_l^{|m|}(\cos\theta)\cos\theta P_{l'}^{|m'|}(\cos\theta)\sin\theta d\theta\int_0^{2\pi}e^{i(m-m')\phi}d\phi\end{aligned} \tag{9.6.8}$$

If follows from the recursion formula for the associated Legendre polynomial Eq. (1.8.20)

$$\cos\theta P_l^{|m|}(\cos\theta) = \frac{(l+|m|)}{(2l+1)}P_{l-1}^{|m|}(\cos\theta) + \frac{(l-|m|+1)}{(2l+1)}P_{l+1}^{|m|}(\cos\theta) \tag{9.6.9}$$

that the first integral in Eq. (9.6.8) will vanish except when $l' = l+1$ or $l-1$. The second integral in this equation be nonvanishing only for $m = m'$. Thus, the integral Eq. (9.6.8) and the *z*-component of the transition moment $\langle i|\mu_z|f\rangle$ will vanish except when $m = m'$ and $l' = l \pm 1$. In a similar treatment of the integrals for x and y it is found that they would be nonzero only when m changes by $+1$ or -1 and l changes by $+1$ or -1. Thus, we get the selection rule

$$\begin{aligned}\Delta m &= m - m' = 0, \pm 1 \text{ and}\\ \Delta l &= l - l' = \pm 1\end{aligned} \tag{9.6.10}$$

The evaluation of the integral $\mu_{nln'l'}$ is involved and it may simply be stated that the integral is nonzero for all combinations of the principal quantum n. We have

$$\Delta n = n - n' = 0, \pm 1, \pm 2, \pm 3, \ldots \tag{9.6.11}$$

Also the factors x, y, and z in Eq. (9.6.3) are a constant as far as the spin variable is concerned. The spin functions α and β are orthogonal. This makes all the integrals in Eq. (9.6.3) to vanish unless the spin magnetic quantum number $m_s = m_s'$. This gives the selection rule

$$\Delta m_s = m_s - m_s' = 0$$

So transitions between states with the same multiplicity alone can occur, both in absorption and emission of radiation.

Combining all the above results for the hydrogen atom, the electric dipole selection rules for isotropic radiation are:

$$\Delta n = 0, \pm 1, \pm 2, \ldots$$

$$\Delta l = \pm 1$$

$$\Delta m = 0, \pm 1$$

$$\text{and} \quad \Delta m_s = 0 \tag{9.6.12}$$

9.6.2 Selection Rules for Transitions in Diatomic Molecules

In order to get details of the electronic spectra one needs to know selection rules governing rotational, vibrational, and electronic transitions between the different electronic states. This is done by evaluating the corresponding dipole transition moments μ_{mn} by Eq. (9.6.1). Without going into intricate mathematical details one can also find out from simple symmetry considerations whether or not the transition between two states is allowed. Thus, for the transition moment to be different from zero, it is necessary that the integrand in Eq. (9.6.1) is totally symmetric under the symmetry operation on the system. In the case of diatomic molecules, the easiest operation to consider is inversion or reflection about the origin under which $(x_\alpha, y_\alpha, z_\alpha)$ change to $(-x_\alpha, -y_\alpha, -z_\alpha)$. Since the total molecular wavefunction $\Psi = \Psi_e \frac{1}{r}\Psi_v\Psi_r$, the total symmetry is determined by the symmetry of each of the components. However, as the vibrational contribution $\frac{1}{r}\Psi_v$ always remains unchanged by reflection at the origin. The contribution of the electronic and rotational wavefunctions are important in determining symmetry of the integral. Thus, for example, in the case of a homonuclear molecule, the components of the dipole moment

$$\mu_x = \sum_\alpha q_\alpha x_\alpha, \; \mu_y = \sum_\alpha q_\alpha y_\alpha, \; \text{and} \; \mu_z = \sum_\alpha q_\alpha z_\alpha,$$

where q_α is the charge and x_α, y_α, and z_α are the coordinates of the nuclei, remain unchanged if the two identical nuclei are exchanged. Therefore, if Φ_m refers to a symmetric and Φ_n to an antisymmetric state, the integral $\int \Phi_m^* \mu \Phi_n d\tau$ changes sign for an exchange of nuclei; μ and Φ_m would not change sign being symmetric but Φ_n would change sign being antisymmetric. As the value of the integral cannot depend on the labels assigned to the nuclei, it must be zero. Hence, there will be no transition between a symmetric and an antisymmetric state. The selection rule in this case of homonuclear molecule is

$$\text{sym} \leftrightarrow \text{sym}, \quad \text{antisym} \leftrightarrow \text{antisym}$$

Against this, in the case of a heteronuclear diatomic molecule where the transition is accompanied with a change in the dipole moment, μ may change sign on inversion and so a symmetric state can combine with an antisymmetric state

$$\text{sym} \leftrightarrow \text{antisym}$$

9.6.3 Selection Rules for a Rotator

The same selection rules apply in the case of a diatomic rigid rotator as in the case of the hydrogen atom as the two problems are identical. The selection rules for the rotational quantum number J are $\Delta J = \pm 1$ for simple rotator and $\Delta J = 0, \pm 1$ for a symmetric top with $\Lambda \neq 0$. Differences however arise when molecular symmetry is taken into account. A rotational level is called positive (+) or negative (−) depending on whether the total eigenfunction remains unchanged or changes sign for reflection at the origin. In the case of a rigid rotator, the reflection at the origin is obtained by replacing θ by $\pi - \theta$ and ϕ by $\pi + \phi$. Thus, it may be seen that the angular wavefunction $\Psi_{JM}(\theta, \phi)$ for such a reflection remains unchanged for even values of J but changes sign for odd values. This property, positive or negative, is also called "parity" and corresponds to the property even or odd of atomic energy levels.

For example, if the electronic orbital angular momentum $\Lambda = 0$ (corresponding to Σ state), then electronic wavefunction Ψ_e remains unchanged by reflection at the origin and the parity of the rotational level will be determined by Ψ_r only. In this case, the rotational levels will be positive or negative accordingly at J is even or odd. The selection rule for a homonuclear molecule is then

$$+ \to +, - \to -, + \nrightarrow -, \text{ and } - \nrightarrow +,$$

whereas for a heteronuclear molecule it is

$$+ \to -, \quad - \to +, \quad + \nrightarrow +, \text{ and } - \nrightarrow -$$

9.6.4 Selection Rules for Vibrational Transitions

These have been discussed in Section 7.2.2. While for a harmonic oscillator the selection rules are

$$\Delta v = \pm 1 \tag{9.6.13}$$

that is, the allowed electric dipole transitions for the harmonic oscillator involve a change of ± 1 in the vibrational quantum number; in the case of an anharmonic oscillator, the selection rule is

$$\Delta v = \pm 1, \pm 2, \pm 3, \ldots \tag{9.6.14}$$

When two different electronic states are involved in the emission or absorption of radiation, a vibrational fine structure may appear in the electronic band. The selection rule in this case is $\Delta v = 0, \pm 1, \pm 2, \pm 3, \ldots$ In absorption, transitions may occur between a given vibrational state (v'') of the lower electronic state and any of the vibrational states (v') of the upper electronic state giving rise to a so-called $v'-$ progression. In emission a given v' state may combine with any of the v'' states to give a $v''-$ progression.

9.6.5 Selection Rules for Electronic Transitions

As described earlier (Chapter 3), the electronic state configuration for molecules can be described in terms of primary quantum number n, angular momentum quantum number Λ, spin quantum number S and its projection along the internuclear axis

$\Sigma(M_s = S, S-1, \ldots, -S)$, and the projection of the total angular momentum along the molecular symmetry axis Ω. The term $\Omega = |\Lambda + \Sigma|$ and can have integer or half-integer values. Some of the selection rules involving the above mentioned quantum numbers are:

1. **Angular momentum rule**: According to this rule the total orbital angular momentum can change by 0, ±1, that is $\Delta\Lambda = 0, \pm 1$.
2. **Spin selection rule**: If the spin–orbit coupling is weak, the electron spin wavefunction can be separated from the electronic wavefunction. In this case, the selection rule is $\Delta S = 0$, $\Delta\Sigma = 0$.
   In this case, the selection rule is
   $$\Delta S = 0, \quad \Delta\Sigma = 0$$
   However, if the spin–orbit coupling is large, the selection rule is, $\Delta\Omega = 0, \pm 1$.
   If $\Omega = 0$ for both the electronic states, $J = 0$ is forbidden and only $J = \pm 1$ is allowed.
3. **Symmetry rule**: For a molecule with a center of inversion, allowed transitions must involve a change in parity,
   $$g \to u, \quad u \to g, \text{ but } g \nrightarrow g \text{ and } u \nrightarrow u$$
   In electron absorption spectroscopy this is called **Laporte rule** and applies to centrosymmetric molecules. Thus, if a molecule is centrosymmetric, transitions within a given set of *p* or *d* orbitals are forbidden. From the above considerations it follows that transitions such as $(^1\Sigma - {}^1\Sigma)$, $(^2\pi_{1/2} - {}^2\pi_{1/2})$, $(^2\Delta_{5/2} - {}^2\Delta_{5/2})$, $(^1\Sigma_g^+ \leftrightarrow {}^1\Sigma_u^+)$, $(^1\Sigma_g^- \leftrightarrow {}^1\Sigma_u^-)$, etc., will be allowed whereas $(^3\Sigma_u^+ - {}^1\Sigma_g^+)$ would be forbidden.
   The forbidden transitions are, however, allowed if the center of symmetry is disrupted, and, as such, apparently forbidden transitions are observed experimentally. Disruption of the center of symmetry occurs for various reasons, such as the Jahn–Teller effect and asymmetric vibrations. Transitions that occur as a result of an asymmetrical vibration of a molecule are called vibronic transitions, such as those caused by vibronic coupling. Such transitions are weakly allowed.

9.6.6 Selection Rules for Electronic Transitions in Polyatomic Molecules

As in the case of diatomic molecules, the general selection rule for polyatomic molecules is based on the Born–Oppenheimer approximation. In other words we consider an electronic transition between two states as allowed if it can occur for fixed nuclei and follow the criterion that the transition moment between the two states is different from zero.

$$\mu_{e'e''} = \int \Psi'_e \mu \Psi''_e d\tau_e \neq 0 \tag{9.6.15}$$

Here, μ is the dipole moment vector having components $\Sigma_i e_i x_i$, $\Sigma_i e_i y_i$, and $\Sigma_i e_i z_i$ along the space-fixed axes. If vibrational and rotational contributions to the electronic transitions are to be considered, then ψ_e is replaced by $\Psi_e\Psi_v\Psi_r$. Thus, x-component of the transition moment would be

$$\mu_{xe'e''} = \int \Psi^*_{e'v'r'}\mu_x\Psi_{e''v''r''}d\tau_e d\tau_v d\tau_r \qquad (9.6.16)$$

Several integrals in the expansions of the x, y, and z components of the transition moments will vanish unless certain relations between the vibrational and rotational quantum numbers of the two electronic states are fulfilled. They would determine the vibrational and rotational selection rules. In general, the rotational fine structure in the gas phase electronic transitions is identical to the vibrational–rotational spectra of the polyatomic molecules. Franck–Condon principle for vibrational and electronic interaction, i.e., vibronic coupling, which determines the vibrational band intensities in electronic transitions in diatomic molecules, is equally valid for polyatomic molecules. The presence of nontotally symmetric vibrational modes in polyatomic molecules can, however, allow electronic transitions which would otherwise be forbidden.

As long as spin–orbit interaction is small, the spin-selection rule $\Delta S = 0$ for diatomic molecules is equally applicable for polyatomic molecules. Thus only states of the same multiplicity combine with each other and so intercombination of states is forbidden. However, for higher atomic number of the constituent atoms, this rule is no longer strictly valid.

9.7 Electronic Spectra and Vibronic Transitions in Molecules

Molecular electronic transitions are generally accompanied by simultaneous changes in vibrational and rotational states. Molecular energy can be approximated as the sum of electronic vibrational and rotational energies and the total wavefunction can be approximated as the product of wavefunctions for these three kinds of motions

$$E = E_e + E_v + E_r \qquad (9.7.1)$$

$$\Psi(r, R) = \Psi_e(r, R)\Psi_v(R)\Psi_r(R) \qquad (9.7.2)$$

where E_e, E_v, and E_r are the energies for the electronic, vibrational, and rotational motions and Ψ_e, Ψ_v, and Ψ_r are the wavefunctions for these motions, respectively.

The Born–Oppenheimer approximation assumes that the electronic wavefunction Ψ_e is approximated by electronic coordinates (**r**) at the equilibrium nuclear distance (R_e). Since the mass of the electrons is much smaller than nuclear mass, the rotational wavefunction Ψ_r depends only on nuclear coordinates. The rotational wavefunction provides important information for rotational selection rules. In the following treatment, for the sake of simplicity, we shall not consider this function because most of the spectra are not rotationally resolved.

If we represent the electronic, vibrational, and rotational energies in wave numbers (cm^{-1}) as T_e, $G(v)$, and $F(J)$, and upper and lower energy states by single and double primes, respectively, then

$$T_e = \frac{E_e}{hc}, \quad G(v) = \frac{E_v}{hc}, \quad \text{and } F(J) = \frac{E_r}{hc} \tag{9.7.3}$$

The transition frequency in wavenumbes (cm^{-1}) will then be given as

$$\overline{v} = (T'_e - T''_e) + [G'(v') - G''(v'')] + [F'(J') - F''(J'')] \tag{9.7.4}$$

Thus, a transition between two given electronic states shows many bands corresponding to different combinations of initial and final states (v', J') and (v'', J''). Each electronic transitions is accompanied by a number of vibrational bands corresponding to different pairs of initial and final vibrational states, and each of these bands shows many closely spaced lines, each corresponding to a different pair of initial and final rotational states. In the case of molecules, therefore, an electronic spectrum is a band spectrum instead of a line spectrum in atoms. In order to completely understand an electronic spectrum, we need selection rules for electronic, vibrational, and rotational transitions. This is done by evaluating the transition moment μ_{mn} for the two combining states m and n defined as

$$\mu_{mn} = \int \Psi^*_m \mu \Psi_n d\tau \tag{9.7.5}$$

Since in the case of electronic transitions in molecules, both the electrons and the nuclei are involved, the total dipole moment μ can simply be defined as the sum of the electronic and nuclear dipole moments

$$\mu = \mu_e(\mathbf{r}) + \mu_N(\mathbf{R}) = -\sum_i \mathbf{r}_e + \sum_N Z_N \mathbf{R}_N \tag{9.7.6}$$

where $\mathbf{r}_e$ and $\mathbf{R}_N$ represent the electron and nuclear coordinates, respectively, in atomic units.

On using the Born–Oppenheimer approximation and ignoring the rotational function, the wavefunction of the molecule can be written as the product of electronic and nuclear terms

$$\Psi(r, R) = \Psi_e(r, R)\Psi_N(R) \tag{9.7.7}$$

It may be noted that while the nuclear part depends on the nuclear coordinates, the electronic part depends directly on the electronic coordinates and also parametrically and weakly on the nuclear coordinates.

We can therefore write for the transition moment

$$\begin{aligned} \langle \Psi' | \mu | \Psi'' \rangle &= \langle \Psi_{e'} \Psi_{v'} | \mu_e + \mu_N | \Psi_{e''} \Psi_{v''} \rangle \\ &= \langle \Psi_{e'} | \Psi_{e''} \rangle \langle \Psi'_{v'} | \mu_N | \Psi_{v''} + \langle \Psi_{e'} \Psi_{v'} | \mu_e | \Psi_{e''} \Psi_{v''} \rangle \end{aligned} \tag{9.7.8}$$

where the suffix e and v represent electronic and vibrational motions, respectively.

The electronic wavefunctions $\Psi_{e'}$ and $\Psi_{e''}$ are eigenfunctions of the Hermitian operator $\widehat{H}_e$ with different eigenvalues, hence they are orthogonal. The first term on the right-hand side of Eq. (9.7.8) therefore vanishes. This equation then reduces to

$$\langle\Psi'|\mu|\Psi''\rangle = \langle\Psi_{e'}\Psi_{v'}|\mu_e|\Psi_{e''}\Psi_{v''}\rangle \tag{9.7.9}$$

and so we are left only with the electronic part of the dipole matrix element. It may be noted here $\langle\Psi_{e'}\Psi_{v'}|\mu_e|\Psi_{e''}\Psi_{v''}\rangle$ cannot be split into $\langle\Psi_{v'}\Psi_{v''}\rangle\langle\Psi_{e'}|\mu_e|\Psi_{e''}\rangle$ for the reason that the electronic states $|\Psi_{e'}\rangle$ and $|\Psi_{e''}\rangle$ depend parametrically on **R** as well as on **r**. We can instead write this matrix element as

$$\langle\Psi'|\mu|\Psi''\rangle = \int \Psi_{v'}(R)\Psi_{v''}(R)\left[\int \Psi_{e'}(r,R)\mu_e\Psi_{e''}(r,R)dr_e\right]dR \tag{9.7.10}$$

It is convenient to define an electronic transition moment

$$M_e(R) = \int \Psi_{e'}(r,R)\mu_e\Psi_{e''}(r,R) \tag{9.7.11}$$

So, the overall electronic dipole matrix element can be simplified to

$$\langle\Psi'|\mu|\Psi''\rangle = \int \Psi_{v'}(R)M_e(R)\Psi_{v''}(R)dR \tag{9.7.12}$$

A further approximation which is generally made is that the electronic transition moment $M_e(R)$ does not vary with nuclear position or varies slowly over the range R for which $\Psi_{v'}(R)$ and $\Psi_{v''}(R)$ have substantial values. In this case, we can take

$$M_e(R) \approx M_e(R_e) \quad \text{or,} \quad M_e(R) = \overline{M_e}(R)$$

where $\overline{M_e}(R)$ is an averaged value of the electronic transition moment. This approximation known as the Condon approximation, is not quantitatively accurate because the electronic wavefunctions do vary with nuclear coordinates but is very useful for qualitative purposes. Equation (9.7.12) can therefore be written as

$$\langle\Psi'|\mu|\Psi''\rangle = \overline{M_e}(R)\int \Psi_{v'}(R)\Psi_{v''}(R)dR = \overline{M_e}(R)\langle\Psi_{v'}|\Psi_{v''}\rangle \tag{9.7.13}$$

The last term in this expression is the overlap integral between the vibrational levels of the upper and lower electronic states and is the basis of vibrational selection rules. $M_e(R)$, which is calculated at any convenient nuclear geometry (usually the minimum of the ground state electronic energy surface), defines the electronic selection rules. The probability of electronic transition is then given as

$$W_{i\to f} = \frac{1}{3}\frac{\pi^2\nu_{if}^2}{c^2\hbar^2}\left|A^0(\nu_{if})\right|^2\left|\overline{M_e}(\mathbf{R})\right|^2\langle\Psi_{v'}|\Psi_{v''}\rangle^2\delta(\nu_{if}-\nu) \tag{9.7.14}$$

where, $\langle\Psi_{v'}|\Psi_{v''}\rangle^2$, the squared overlap integral of the vibrational wavefunctions in the initial and final electronic states, is called the Franck-Condon factor for the $v' \to v''$ vibrational band of the electronic transition. Equation (9.7.14) is also known as the **Fermi's Golden rule** for electronic spectroscopy. It must be noted that since v' and v''

belong to two different electronic states, they shall no more be orthogonal and transitions between different vibrational states corresponding to $\Delta v = 0, \pm 1, \pm 2, \pm 3, \ldots$ are allowed. The electronic transitions between the two states are forbidden when the overlap integral $\langle \Psi_{v'} | \Psi_{v''} \rangle$ vanishes. Equation (9.7.14) also reflects the conservation of energy because the Dirac function $\delta(\nu_{if} - \nu)$ will be zero and hence the probability of transition shall be zero unless the energy absorbed or emitted by the molecule is almost equal to the energy difference between its two states. A detailed interpretation of this relation shall be taken up in the next section.

9.8 Franck–Condon Principle and Intensity Distribution in Electronic Bands

Franck–Condon principle relates to the interaction between the electronic and vibrational motions and like the Born–Oppenheimer approximation follows from the fact that the nuclear masses are much larger than the electronic mass. It states that the electronic transitions are so rapid and their timescale is so fast as compared with the nuclear motion that we may consider the nuclei to be fixed during the transition. Also, an electronic transition is most likely to occur when the nuclei are in their extreme positions on the potential energy curve. The charge redistribution in a molecule that follows electronic transition results in a change in Coulombic forces on the nuclei and causes changes in the vibrational state of the molecule. Simultaneous occurrence of electronic and vibrational transitions, called vibronic transitions, give rise to the vibrational structure of the electronic bands. Franck–Condon principle helps to analyze these vibronic transitions and explains the intensity distribution in the vibrational bands. A quantum mechanical explanation of the Franck–Condon principle follows from Eq. (9.7.14).

While analyzing the vibrational fine structure of electronic spectra, three types of intensity distributions have been observed in the spectra of molecules like O_2, CO, and I_2. In the case of O_2, the 0–0 band corresponding to a transition $v'' = 0 \rightarrow v' = 0$ is found to be most intense and the intensity gradually but sharply decreases on moving along the v'–progression ($v'' = 0$, $v' = 0, 1, 2, \ldots$). In the case of CO, the intensity of the bands first increases, reaches a maximum, and then decreases. Finally, in the case of I_2, the intensity of the bands in the v'–progression increases sharply and the most intense band may appear in the continuum with large value of v'. These different types of intensity distributions can be explained in terms of the Franck–Condon principle by considering the shapes of the potential energy curves of the two electronic states and the position of their potential energy minima. Three cases may be considered.

Case 1: The internuclear distance at the potential energy minima in the two electronic states are almost equal, i.e., $R_e'' \simeq R_e'$, and the two curves lie almost one above another. This is shown in Figure 9.2(a).

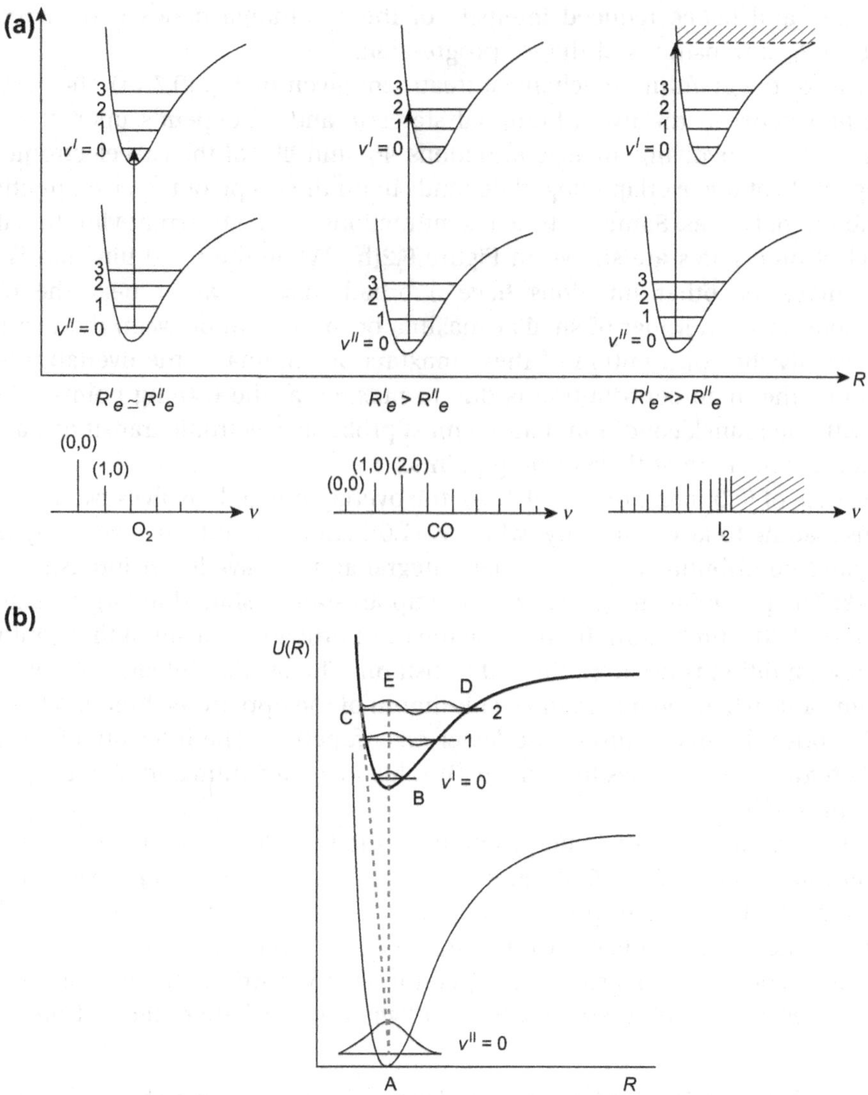

FIGURE 9.2 (a) Intensity distribution in vibronic spectra according to Franck–Condon principle, (b) Quantum mechanical explanation of Franck–Condon principle.

The most probable transition satisfying the Franck condition of no change in the position or velocity of nuclear motion, in this case is the vertical transition from A to B. The transition from A to C involves a change in the internuclear distance while the transition from A to E involves change in the kinetic energy. The (0, 0) band in this case is the most intense. The change in the position or velocity during the electronic transition increases as one moves toward higher values of v'. This results in decreasing probability of transition and hence reduced intensity of the vibrational bands corresponding to $0 \rightarrow 1, 0 \rightarrow 2, \ldots$ transitions of the v'−progression.

In terms of the quantum mechanical treatment given by Eq. (9.7.14), the probability of transition between the two vibrational states v' and v'' depends upon the overlap integral $\langle \Psi_{v'} | \Psi_{v''} \rangle$ involving the eigenfunctions $\Psi_{v'}$ and $\Psi_{v''}$ of the two electronic states. The magnitude of the overlap integral depends upon the shape of the eigenfunctions Ψ_v of the vibrational states. Some of these eigenfunctions for the different vibrational states in two electronic states are shown in Figure 9.2(b). While the $v = 0$ function is a bell-shaped curve, the other functions have a broad maxima at or near the classical turning point and a number of smaller maxima or minima in between the two turning points. Usually the contribution of these maxima or minima to the overlap integral is zero, and so the main contribution is due to maxima at the turning points. This is in accord with the Franck condition that the most probable electronic transitions are those at which the nuclei are at their turning points.

In the case of (0,0) band (Figure 9.2(b)), the overlap integral involves two bell-shaped functions leading to large intensity, whereas (1,0), (2,0),... bands involve many positive and negative contributions to the overlap integral and so have lesser intensity.

Case 2: The potential energy curve of the upper state is shifted to larger values of R, that is, $R'_e > R''_e$. (Figure 9.2(a)). In this case the most probable transition that conforms to the Franck condition is no more the 0–0 transition. The overlap integral $\langle \Psi_{v'} | \Psi_{v''} \rangle$ has a maximum value when the maximum or minimum of the upper wavefunction lies almost vertically above the maximum of the lower eigenfunction. The intensity of vibrational bands therefore first increases from the (0,0) value to a maximum and then decreases for higher values of v'.

Case 3: When the potential energy curves of the two electronic states are shifted by still larger values, that is, $R'_e \gg R''_e$ (Figure 9.2(a)). In this case the most probable transition will be from $v'' = 0$ to a state with larger values of v', which may even correspond to the continuum. The overlap integral will be large only for larger values of v'.

The above considerations can also explain intensity distributions in vibronic transitions for the v'' and v'−progressions, for case when v' or v'' are different from zero.

9.9 Oscillator Strength and Intensity of Absorption Bands

The absorption of radiation by a molecule in the visible and ultraviolet regions may cause electronic transitions from a vibrational state in the lower electronic state to the vibrational states in its upper electronic states. The intensity of these bands is related to the probability of the electronic transitions given by Einstein's coefficient of absorption (Eq. (9.3.42)) and are usually expressed in terms of oscillator strength.

Let us consider the case of a polyatomic molecule in which absorptive transition starts from the ground vibrational state $v'' = 0$ in the electronic state e'' and terminates at the vibrational state v' in the electronic state e'. From Eq. (9.3.42), the Einstein coefficient for this transition is

$$B_{e''v''\to e'v'} = \frac{2}{3}\frac{\pi}{\hbar^2}|\langle \Psi_{e''}\Psi_{v''}|\mu|\Psi_{e'}\Psi_{v'}\rangle|^2 \tag{9.9.1}$$

For the sake of simplicity the term in bracket is being abbreviated as $\langle e''v''|\mu|e'v'\rangle$ to give

$$B_{e''v''\to e'v'} = \frac{2}{3}\frac{\pi}{\hbar^2}|\langle e''v''|\mu|e'v'\rangle|^2 \tag{9.9.2}$$

Following Eq. (9.7.13), within the Born–Oppenheimer approximation, this may be written as

$$B_{e''v''\to e'v'} = \frac{2}{3}\frac{\pi}{\hbar^2}\langle e''|\mu_e|e'\rangle^2|\langle v''|v'\rangle|^2 = \frac{2}{3}\frac{\pi}{\hbar^2}|M_e|^2|\langle v''|v'\rangle|^2 \tag{9.9.3}$$

where $M_e = \langle e''|\mu_e|e'\rangle$ is the electronic moment and $\langle v''|v'\rangle^2$ is the Franck–Condon factor.

Summing Eq. (9.9.3) over all the vibrational states v' which are being reached from the vibrational state $v'' = 0$, we get

$$\sum_{v'} B_{e''v''\to e'v'} = \frac{2\pi}{3\hbar^2}|M_e|^2\sum_{v'}\langle v''|v'\rangle\langle v'|v''\rangle \tag{9.9.4}$$

Using the sum rule,

$$\sum_{v'}\langle v''|v'\rangle\langle v'|v''\rangle = \langle v''|v''\rangle = 1, \tag{9.9.5}$$

we get

$$\sum_{v'} B_{e''v'\to e'v'} = \frac{2\pi}{3\hbar^2}|M_e|^2 \tag{9.9.6}$$

which depends only upon the electronic transition moment M_e and is independent of the vibrational terms of the upper electronic state e''. This sum therefore measures the probability and hence the intensity of the electronic transition. In order to describe this quantity in dimensionless units a term oscillator strength is used. The oscillator strength of a given transition is defined in terms of the transition probability of a three-dimensional isotropic harmonic oscillator having mass and charge equal to the electronic mass and charge and vibrational frequency equal to the frequency of the electronic transition in question. Thus, the right-hand side of Eq. (9.9.6) is divided by a factor of $\frac{e^2}{2m_e\overline{\nu}\hbar}$ to get the oscillator strength

$$f_{e''e'} = \frac{4\pi m_e}{3\hbar e^2}\,\overline{\nu}|M_e|^2 \tag{9.9.7}$$

where m_e is the electron mass, e is the electron charge, $\overline{\nu}$ is mean frequency of the electronic transition. The summation $\Sigma_{e'} f_{e''e'}$ is obviously equal to 1 as it represents the sum of oscillator strengths of all the transitions starting from electronic state e'' to all other electronic states. A transition with $f_{e''e'} \simeq 1$ is called a strongly allowed transition of high intensity, as for example π–π^* transitions in aromatic hydrocarbons, and a transition with $f_{e''e'} \simeq 10^{-4}$ to 10^{-2} is called a forbidden transition such as n–π^* transitions in aliphatic aldehydes and ketones.

The calculated value of the oscillator strength is compared with the experimental value of intensity by the relation

$$f = 4.33 \times 10^{-9} \int \alpha(\nu) d\nu \tag{9.9.8}$$

where the integral term is the integrated absorption intensity.

9.10 Electronic Spectra of Polyatomic Molecules

Electronic spectra of polyatomic molecules are more complicated than those of the diatomic molecules for various reasons. The number of vibrational modes in a polyatomic molecule are $3N - 6$ ($3N - 5$ in a linear molecule), where N is the number of atoms and so larger the number of atoms in the molecule, greater the number of vibrational modes, and hence more congested the electronic band spectra due to many more vibrational transitions. The congestion further arises due to rotational fine structure. Polyatomic molecules have larger moment of inertia than the diatomic molecules and hence the rotational levels come so close together in medium- to large-sized molecules that it is not possible to resolve the rotational fine structure in the electronic transitions. Such is the case of naphthalene. Some more complexities arise due to different molecular structures in the ground and excited states and hence due to different vibrational motions in these electronic states. Thus, for example, formaldehyde and ethylene are planar in the ground state but nonplanar in one or more of their excited states. This may result in vibronic coupling between the electronic states of different symmetry and may be cause even the forbidden transitions to appear with appreciable intensity. One such example is the 2600 Å $S_0 \rightarrow S_1$ band system of benzene.

Some simplifications in the electronic spectra of polyatomic molecules, particularly in organic molecules, may however arise due to the fact that the energy of the electronically excited molecule is usually redistributed and can easily break a weak chemical bond in the molecule. The electronic transitions in polyatomic molecules are usually in the absorption spectra where transition occurs between the ground state and one of the excited states. In this case, only a few transitions due to electronic excitation are observed.

Since the electronic spectra of polyatomic molecules are usually observed in solutions, collision between the solvent and solute species suppresses the rotational fine structure. Solute–solvent interaction also affects the vibrational fine structure. While in nonpolar molecules, to some extent, the vibrational fine structure may still be present, as for example in saturated molecules, such as hydrocarbons, in polar solvents even that may be suppressed. The electron absorption bands usually appear as broad and relatively structure less. A particular simplification that helps in the study of the electronic spectra, especially of organic molecules, is that often the absorption in a particular region of the spectrum may be assigned to the transitions in a particular group of atoms in the molecule, called the chromophore. Historically, a chromophore is a molecular group which has the potentiality of color; for example, $-NO_2$, $-N{=}N-$ and $>C{=}O$ are typical chromophores, and the molecules containing these groups such as nitrobenzene, azobenzene, and phenolphthalein are chromogens. The more common chromophores are the carbonyl and nitro groups and the carbon–carbon double bond (C=C). While the electronic transition in small polyatomic molecule can be discussed in terms of the entire molecule, the spectra of larger molecules are often discussed in terms of their chromophores.

A study of the absorption bands due to electronic transitions, which usually appear in the visible and ultraviolet spectral regions, is an invaluable tool for understanding the structure of molecules. A number of important generalizations have been reported from experimental studies which correlate the change in the frequency and intensity of absorption bands with the change in molecular structure. Based on the semiempirical, *ab initio*, and DFT methods, it is possible to calculate the position and intensity of the electronic transitions and correlate them with the molecular structure. These studies are broadly divided into three categories: (1) study of valence states, (2) study of Rydberg states, and (3) study of core electrons. These will be taken up in the next few sections.

9.11 Electronic Transitions and Absorption Bands

A study of the electronic spectra in UV-VIS range has been an invaluable tool for understanding the nature of chemical bonds and molecular structure. In earlier chapters, the molecular orbital theories of bond formation were developed and the concepts of bonding and antibonding orbitals (σ, π, σ^*, π^*) introduced. Self-consistent field (SCF) molecular orbital theories both *ab initio* and semiempirical were used to calculate molecular energy levels and explain the origin and nature of electronic transitions. Besides the σ and π orbitals, another orbital of interest in the study of electronic transitions is an n-orbital which is associated with a "lone pair" or "nonbonding" electrons in atoms such as oxygen, nitrogen, sulfur, etc. Since these electrons do not form bonds, there is no antibonding orbitals associated with them. Since the absorption of light energy by organic compounds involves promotion of electrons in σ, π, and n-orbitals in the ground electronic state to the higher energy σ^* and π^* orbitals, we may have electronic transitions such as σ–σ^*, π–π^*, n–σ^*, π–σ^*, and n–π^* (Figure 9.3). All these transitions, with the exception of the σ–σ^* transitions, appear in the UV and visible regions. σ–σ^* transitions appear in the vacuum UV region.

The π–π^* transitions are allowed and are of high intensity ($f \simeq 1$) whereas the n–π^* are symmetry forbidden and are of rather low intensity ($f \sim 10^{-4}$ to 10^{-2}). Thus, for example, a π–π^* transition in a double bond, as in ethylene, has high intensity because in this case the symmetry of the initial and final states is the same and hence a large

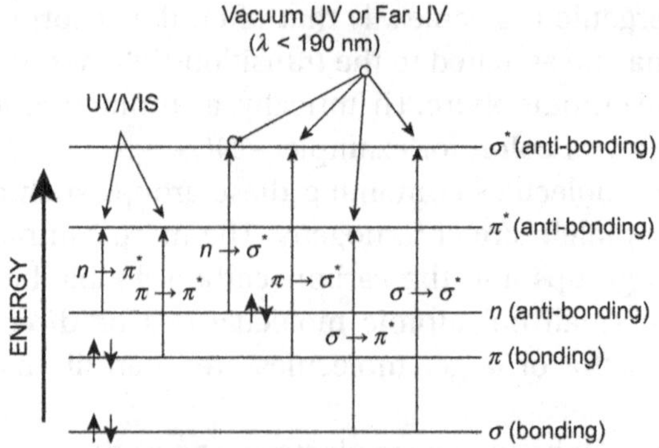

FIGURE 9.3 Electronic transitions in molecules.

transition moment occurs to give high intensity of the band. In the case of a n–π^* transition in saturated aldehydes and ketones, a carbonyl n-electron (on the oxygen atom) is promoted to an orthogonal π^* orbital and so there is no orbital overlap. This makes the transition forbidden and the resulting transition moment is zero. The weak intensity of n–π^* transition in some cases may be due to vibronic coupling. Sharp lines in ethylene spectrum forming a Rydberg progression arise from π–σ^* transition. The σ-type orbitals in this case are sufficiently large and resemble orbitals of one-electron atom.

9.12 Theoretical Studies on Valence States

In Chapters 3 and 4, we had developed the SCF molecular orbital theories both *ab initio* and semiempirical to calculate the energy levels of both saturated as well as unsaturated organic compounds and to explain the origin and nature of transitions involving the valence electrons. It may, however, be noted that when the optimized geometry of the ground state is not the same as that of the excited state, one must view the absorption of radiation as taking place at the ground state geometry. The energy involved in electronic excitation in this case is the vertical excitation energy. The use of configuration interaction with the ground state geometry only tends to include electron correlation and tends to improve the description of the ground state as in Section (3.13.1). The use of ground state orbitals is not appropriate for determining the excited state wavefunctions. If we use the orbitals of the ground state to describe the excited state, then in order to evaluate the energy of the excited state we need to evaluate the Hamiltonian for the Slater determinant formed after the promotion of the excited electron. This approach has the drawback that while these singly excited single-configuration wavefunctions are orthogonal to the ground state because of the Brillouin theorem, they are unlikely to be orthogonal to each other. This drawback has been removed in the configuration interaction-singles (CIS) method given below. Since the geometries of the ground and excited states are usually different, it is necessary to optimize the geometry of the excited state to get meaningful transition energies. The energy difference between the two states having their own optimal geometries is called "adiabatic" excitation energy. The same conceptual framework applies to the process of emission.

In order to predict the energy, properties, and reactivity of molecules in their excited states, several methods based on HF theory such as CIS (configuration interaction-singles or configuration interaction with all single-excited determinants) [2], MCSCF (multiconfigurational SCF) [3], and MR-CI (multireference CI) treatments [4,5] such as CASSCF (complete active space SCF)-SDCI or high level coupled-cluster methods and DFT have been described in the literature. A complete review of such methods has been given by Foresman et al. [6]. The time-dependent density functional theory (TDDFT) has also been described in Chapter 5. Here, we shall be restricting ourselves to the CIS method and application of some of the above methods for predicting the electronic transitions in molecular systems.

9.12.1 Molecular Orbital Treatment—CI-Singles

In Section 3.13.1 we discussed the configuration interaction techniques to improve the ground state energy and wavefunctions and introduced the concept of CIS. It was indicated that due to Brillouin theorem the interaction between the singly excited states and the ground state was zero due to the orthogonality of the wavefunctions of the excited states and the ground state. However the wavefunctions of the excited states may not be themselves orthogonal. Orthogonalization of the excited state wavefunctions is therefore the essence of the CIS technique as the configuration interaction matrix is formed by restricting to only singly excited states. These states are constructed by the orbitals of the ground state in which an electron jumps from an occupied molecular orbital to a vacant virtual orbital. Thus in CIS calculations the ground state is important only to the extent that it determines the orbitals and that configuration interaction is carried out to orthogonalize the singly excited states. This issue of orthogonality is an important one; every excited state should be orthogonal to the ground state as well as to all the other excited states. Any method that does not satisfy the requirement of orthogonality should be looked with caution [7] although, in reality, we are never working with an exact ground state function but only with its approximation.

9.12.1.1 Basics of CIS Method

In order to understand the basics of CIS theory, we start with a reference HF single-determinant closed-shell ground state wavefunction for an n-electron system

$$\Psi_{\mathrm{HF}} = \frac{1}{\sqrt{n!}}\det\{\chi_1\chi_2\cdots\chi_i\chi_j\cdots\chi_n\} \tag{9.12.1}$$

The spin orbital χ_p can be expressed in terms of a set of N atomic basis functions Φ_μ.

$$\chi_p = \sum_{\mu}^{N} C_{\mu p}\Phi_\mu \tag{9.12.2}$$

In the treatment given below we shall be using *i*, *j*, *k*, etc., for occupied orbitals, *a*, *b*, *c*, etc., for virtual orbitals, μ, ν, λ, σ, etc., for atomic basis functions, and *p*, *q*, *r*, etc., for general orbitals. The ground state wavefunction Ψ_{HF} can be determined by solving the time-independent Hartree–Fock equation which is given as

$$\sum_{\mu}(F_{\mu\nu} - \varepsilon_p S_{\mu\nu})C_{\mu p} = 0 \tag{9.12.3}$$

Here, $F_{\mu\nu}$ represents the Fock matrix element

$$F_{\mu\nu} = H_{\mu\nu} + \sum_{i}\sum_{\lambda\sigma} C_{\mu i}C_{\nu i}(\mu\lambda\|\nu\sigma) \tag{9.12.4}$$

$H_{\mu\nu}$ is the one-electron core Hamiltonian and $\langle\mu\lambda\|\nu\sigma\rangle$ is the antisymmetrized two-electron integral

$$\begin{aligned}\langle\mu\lambda\|\nu\sigma\rangle &= \langle\mu\lambda|\nu\sigma\rangle - \langle\mu\lambda|\sigma\nu\rangle \\ &= \iint \Phi_\mu(1)\Phi_\lambda(2)\frac{1}{r_{12}}[\Phi_\nu(1)\Phi_\sigma(2) - \Phi_\sigma(1)\Phi_\nu(2)]d\tau_1 d\tau_2\end{aligned} \tag{9.12.5}$$

$S_{\mu\nu}$ is the overlap integral matrix

$$S_{\mu\nu} = \int \Phi_\mu \Phi_\nu d\tau \tag{9.12.6}$$

and ε_p is the one-electron energy of orbital *p*. Solution of Eq. (9.12.3) gives the total energy of the ground state

$$E_{\text{HF}} = \sum_{\mu\nu} P_{\mu\nu}H_{\mu\nu} + \frac{1}{2}\sum_{\mu\nu\lambda\sigma} P_{\mu\nu}P_{\lambda\sigma}(\mu\lambda\|\nu\sigma) + V_{NN} \tag{9.12.7}$$

where,

$$P_{\mu\nu} = \sum_{i=1}^{n} C_{\mu i}C_{\nu i} \tag{9.12.8}$$

gives the electron density matrix as a sum over the occupied molecular orbitals and V_{NN} denotes the nuclear–nuclear repulsion. From this solution, we shall get *n*-occupied molecular orbitals, χ_i and $(N - n)$ virtual orbitals, χ_a. We can therefore get a total of $n(N - n)$ singly excited determinants of the type

$$\Psi_i^a = \frac{1}{\sqrt{n!}}\det|\chi_1\chi_2\cdots\chi_a\chi_j\cdots\chi_n| \tag{9.12.9}$$

by replacing occupied orbitals *i*, *j*, *k*, etc., of the ground state (Eq. 9.12.1) by the virtual orbitals *a*, *b*, *c*, etc. The energies of these orbitals are

$$E_i^a = E_{\text{HF}} + \varepsilon_a - \varepsilon_i - (ia\|ia) \tag{9.12.10}$$

where ε_a and ε_i are the orbital energies of the single electron orbitals χ_a and χ_i. These orbitals Ψ_i^a and their energies E_i^a can be used as a first approximation for the molecular excited states. To get better results configuration interaction is used.

In configuration interaction, the electronic wavefunctions is constructed as a linear combination of the ground state Slater determinant and all possible singly excited determinants. The CIS wavefunction can therefore be written as

$$\Psi_{\text{CIS}} = \sum_{ia} C_i^a \Psi_i^a \tag{9.12.11}$$

The summation runs over index pairs i and a with dimension $n(N-n)$.

The configuration interaction coefficients can be calculated as normalized eigenvectors of the Hamiltonian matrix

$$\hat{H}\Psi_{\text{CIS}} = E_{\text{CIS}}\Psi_{\text{CIS}} \tag{9.12.12}$$

Since we are dealing with a full configuration interaction in the space of single substitutions, we must take projections onto the space of singly excited determinants. We must, therefore, multiply Eq. (9.12.12) by $\langle\Psi_j^b|$ from the left and use Eq. (9.12.11)

$$\sum_{ia}\left\langle\Psi_j^b\middle|\hat{H}\middle|\Psi_i^a\right\rangle C_i^a = E_{\text{CIS}}\sum_{ia}\left\langle\Psi_j^b\middle|\Psi_i^a\right\rangle C_i^a = E_{\text{CIS}}\sum_{ia} C_i^a\delta_{ij}\delta_{ab} \tag{9.12.13}$$

After including configuration interaction, we also get a relationship analogous to Eq. (9.12.10)

$$\left\langle\Psi_j^b\middle|\hat{H}\middle|\Psi_i^a\right\rangle = (E_{\text{HF}} + \varepsilon_a - \varepsilon_i)\delta_{ij}\delta_{ab} + \langle ia\|jb\rangle \tag{9.12.14}$$

From Eqs (9.12.13) and (9.12.14), we get

$$(E_{\text{CIS}} - E_{\text{HF}})\sum_{ia} C_i^a\delta_{ij}\delta_{ab} = \sum_{ia}[(\varepsilon_a - \varepsilon_i)\delta_{ij}\delta_{ab} + \langle ia\|jb\rangle]C_i^a \tag{9.12.15}$$

where $\langle ia\|jb\rangle$ corresponds to antisymmetrized two-electron integrals defined as

$$\langle ia\|jb\rangle = \frac{\int[\chi_i(1)\chi_a(1)\chi_j(2)\chi_b(2) - \chi_i(1)\chi_j(1)\chi_a(2)\chi_b(2)]}{r_{12}}d\tau_1 d\tau_2 \tag{9.12.16}$$

The two-electron integrals in the molecular orbital basis may also be written as

$$\langle pq\|rs\rangle = \sum_{\mu\nu\lambda\sigma} C_{\mu p}C_{\nu q}C_{\chi r}C_{\sigma S}\langle\mu\nu\|\lambda\sigma\rangle \tag{9.12.17}$$

$(E_{\text{CIS}} - E_{\text{HF}})$ gives the excitation energy.

Both, the excitation energies as well as the corresponding CIS expansion coefficients (Eq. (9.12.11)) can be easily obtained from Eq. (9.12.15), which may be written in the form of an eigenvalue equation

$$H^s\mathbf{X} = \omega\mathbf{X} \tag{9.12.18}$$

where H^s is the matrix representation of the Hamiltonian in the space of singly excited determinants and $\omega = (E_{\text{CIS}} - E_{\text{HF}})$ is the diagonal matrix of the excitation energies. **X** is the matrix of CIS expansion coefficients. From Eq. (9.12.15), the matrix elements of H^s are

$$H^s_{ia,jb} = (\varepsilon_a - \varepsilon_i)\delta_{ij}\delta_{jb} + \langle ia\|jb\rangle \tag{9.12.19}$$

The diagonalization of matrix H^s can be carried out in a direct fashion to yield the excitation energies ω of the excited electronic states. The corresponding eigenvectors provide the expansion coefficients in Eq. (9.12.11) and hence the weightage of each singly excited state for a given electronic transition.

It may be mentioned that there are two different derivation routes for CIS: one via projection of the Hamiltonian operator $\hat{H}$ onto the space of singly excited determinants as discussed above and the other by time-dependent response theory discussed in Chapter 5. When the B matrix elements in the Casida matrix formulation of TDDFT (Eq. (5.10.32)) are set to zero, it reduces to CIS scheme. In nuclear physics, this approximation is well known as Tamm–Dancoff approximation. From the ongoing discussion it follows that since the expansion coefficients in CIS are determined by variational method, Ψ_{CIS} is a more relaxed wavefunction than Ψ_i^a and represents the excited states more accurately. The total energy in this case corresponds to the upper bound of the true ground state energy and, likewise, in the case of excited states the total energies are the upper bound of their respective exact values. Since HF method is size-consistent, so also is the CIS method. It is possible to obtain from CIS pure singlet and triplet states starting from a doubly occupied orbital by allowing positive and negative combinations of α and β excitations.

Finally, since CIS method provides well-defined wavefunctions and differentiable energy, it can be used to obtain the excited state-optimized geometries and other molecular properties with the help of analytical gradient techniques. An analytical expression for the total energy of excited states can be obtained from Eq. (9.12.15) by multiplying from the left with the corresponding CIS vector.

$$E_{\mathrm{CIS}} = E_{\mathrm{HF}} + \sum_{ia} \left(C_i^a\right)^2 (\varepsilon_a - \varepsilon_i) + \sum_{ia,jb} C_i^a C_j^b \langle ia \| jb \rangle \tag{9.12.20}$$

E_{CIS} is analytically differentiable and so Eq. (9.12.20) can be used to get the first and second derivatives of energy with respect to external parameters such as nuclear displacements and external fields and to calculate excited-state properties such as equilibrium geometries and vibrational frequencies.

Some of the major limitations of the CIS method are that the computed excitation energies are usually overestimated by about 0.5–2 eV compared with the experimental values. The neglect of correlation energies within the CIS method may cause error of the order of 1 eV. The transition moments and hence the intensities of electronic transitions are only qualitatively accurate. While CIS gives best results for lowest lying excited states it fails miserably for states dominated by double excitation. Several softwares such as Gaussian [8], GAMESS [9], Turbomole [10], CADPAC [11], etc., are available for CIS calculations.

9.12.2 Illustrative Example—Electronic Spectra of Conformers of Methyl Trans-Crotonate by CIS Method

In an attempt to interpret the experimental UV spectrum of methyl *trans*-crotonate and to understand the nature of its electronic transitions, Virdi et al. [12] conducted CIS

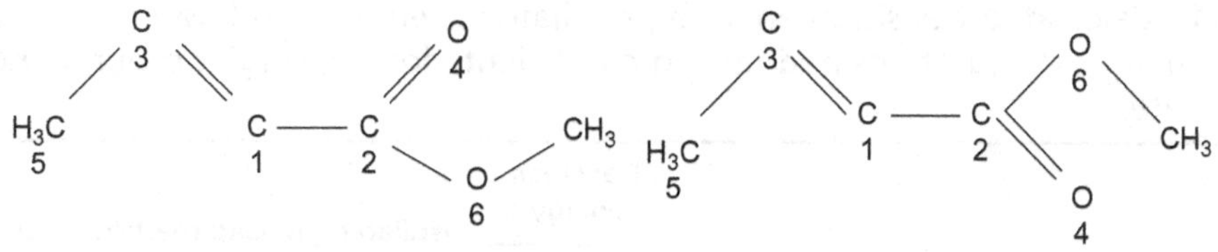

FIGURE 9.4 Geometry of *cis* and *trans* conformers of methyl *trans*-crotonate.

calculations using 6-31 G** basis set. As shown by detailed quantum chemical calculations by HF and DFT methods, the molecule has two stable conformers Cc and Tc. The former, in which the C=C and C=O bonds are in *cis* position, is more stable than the latter in which these bonds are in trans position. A hindered rotation of the methyl group about the C–O bond is also possible. Geometry of the Cc and Tc conformers is given in Figure 9.4. Based on CIS calculations on these two conformers of methyl *trans*-crotonate, transition energies, oscillator strengths, symmetry of each excited state, main configurations, and the mixing coefficients of the singlet ground and excited states were determined and assignments were given to the different transitions. The results are given in Table 9.1. The first electronic transition in Cc conformer has been predicted at 5.96 eV with oscillator strength (OS) of 0.6216, which is close to the experimental value of at 5.85 eV (212 nm) with extinction coefficient $\log \varepsilon = 4.16$ in methanol solution. On the basis of the mixing coefficients and molecular orbital coefficients of the involved states, this band may principally be assigned to $\pi_e - \pi_e^*$ transition in the ethylenic group slightly perturbed by the $\pi_e - \pi_{CO}^*$ of the carbonyl group. Another allowed transition $(^1A' \rightarrow {}^1A')$ with mixing coefficient 0.641 is predicated at 9.19 eV (OS = 0.1667) which may be assigned to $n\pi_{O6} - \pi_e^*$ involving the lone pair of the oxygen atom of the $COCH_3$ chromophoric group. In addition to the above, three forbidden bands in the 0–10 eV range corresponding to transition $(^1A' \rightarrow {}^1A'')$ have been predicted at 6.42 eV (OS = 0.0002), 7.91 eV (OS = 0.0012), and 8.99 eV (OS = 0.0009). These correspond to $n\sigma_{O4} - \pi_e^*$, $n\sigma_{O6} - \pi_e^*$, and $\sigma_{mix} \rightarrow \pi_e^*$ transitions as given in the Table 9.1. An electronic band has been reported near-UV spectrum of the molecule at 250 nm (4.96 eV) ($\log \varepsilon = 2.3$) in ethanol. While no such band has been predicted by CIS calculations, unrestricted HF calculations predict this band at 252.10 nm (4.92 eV) (OS = 0.138) corresponding to a triplet state.

9.12.3 Time-Dependent Density Function Theory

In Chapter 5, we have discussed a few other single-reference *ab initio* methods such as time-dependent Hartree-Fock (TDHF), linear response theory, and the TDDFT for the excited state calculations which are particularly useful for large molecules. The latter two methods have a conceptually different approach to include electron correlation and are built upon the electron density obtained from the ground state DFT. As mentioned, DFT and TDDFT methods are formally exact theories whose accuracy depends upon the

Table 9.1 Calculated transition energies, oscillator strengths, and assignments along with the main configurations and mixing coefficients for the singlet ground and excited states

	Symmetry	Assignments	Transition energy eV	Transition energy nm	Oscillator strength	Mixing coefficients and main configurations[a]
Cc						
Ground state[b] $E = -343.735558$ h	$^1A'$					1.000 [2222222/000000]
a	X^1A'	$\pi_e \rightarrow \pi_e^*$ $\pi_e \rightarrow \pi_{CO}^*$	5.96	207.87[c]	0.6216	0.685 [2222221/100000]
b	X^1A'	$n\pi_{O6} \rightarrow \pi_e^*$	9.19	134.93	0.1667	0.641 [2222122/100000] −0.196 [2222122/000010]
a	X^1A''	$n\sigma_{O4} \rightarrow \pi_e^*$	6.42	193.20	0.0002	0.616 [2222212/100000] −0.302 [2222212/000010]
b	X^1A''	$n\sigma_{O6} \rightarrow \pi_e^*$	7.91	156.81	0.0012	0.465 [2212222/100000] 0.374 [2221222/100000]
c	X^1A''	$\sigma_{mix} \rightarrow \pi_e^*$	8.99	137.88	0.0009	0.561 [1222222/100000] 0.292 [2212222/100000] −0.166 [2221000/100000]
Tc						
Ground state[b] $E = -343.734301$ h	$^1A'$					1.000 [2222222/100000]
a	X^1A'	$\pi_e \rightarrow \pi_e^*$ $\pi_e \rightarrow \pi_{CO}^*$	6.05	205.00	0.6956	0.683 [2222221/100000]
b	X^1A'	$n\pi_{O6} \rightarrow \pi_e^*$	9.14	135.65	0.0890	0.649 [2222122/100000] −0.193 [2222122/000010]
a	X^1A''	$n\sigma_{O4} \rightarrow \pi_e^*$	6.32	196.13	0.0002	0.618 [2222212/100000] −0.274 [2222212/000010]
b	X^1A''	$n\sigma_{O6} \rightarrow \pi_e^*$	8.08	153.39	0.0021	0.414 [2212222/100000] 0.373 [2221222/100000]
c	X^1A''	$\sigma_{mix} \rightarrow \pi_e^*$	9.11	135.98	0.0018	0.445 [1222222/100000] −0.358 [2212222/100000] −0.166 [2221000/100000]

[a]Seven higher energy-occupied molecular orbitals and six lower energy-unoccupied orbitals have been listed.
[b]Energy in Hartree.[c]Experimental value 212 nm (log $\varepsilon = 4.16$) from UV spectra [13].

correctness and appropriateness of the exchange–correlational (XC) functionals which can be broadly divided into local functionals, gradient-corrected functionals, and the hybrid functionals. At present TDDFT is the most prominent method for the calculation of excited states of medium- to large-sized molecules (such as biomolecules) for the valence-excited states which fall below the first ionization potential (IP). The results of TDDFT are comparable to the highly accurate methods such as MR-CI, equation of motion coupled-cluster singles and doubles (EOM-CCSD), and CASPT2, which are applicable for very small molecules with 15–20 atoms. However, in view of the approximate nature of the XC functionals, the method is suspect for the study of Rydberg states, doubly excited states, the charge-transfer states and systems with large π systems.

For a correct explanation and interpretation of electronic spectra of molecular systems it is necessary to have knowledge about the energetic position of electronically excited states relative to the ground state and also information about the geometric and electronic properties of the excited states. When a molecule is excited from the ground state to energetically higher lying excited state, the electronic many-body wavefunction changes. An analysis of the change in the wavefunction is necessary to gain insight into the nature of the corresponding electronic transition. For a direct analysis of this change, the techniques used for the analysis of the ground state wavefunction or electron density are applied to the excited state. A brief mention of the actual procedure used for TDDFT calculation has been given in Section 5.10.2. The calculation starts with DFT to calculate an initial set of orthonormal Kohn–Sham (KS) orbitals. These reproduce the exact density of the true ground state, which is the initial state for TDDFT calculations. The standard time-independent DFT exchange–correlational (XC) functionals for the ground state are then used in TDDFT calculations. Though this procedure is approximate and introduces errors in the calculations of the excitation energies and oscillator strengths yet the results are fairly accurate (within 0.5 eV of experimental values) for valence-excited states at a very low computational costs.

9.12.4 Illustrative Examples—Electronic Spectra of 2-Pyranone and Chloropyrimidine Derivatives by TDDFT

As illustrative examples, we shall consider the application of TDDFT for understanding the electronic spectra of 2-Pyranone derivatives [14] and chloropyrimidine [15] derivatives.

9.12.4.1 2-Pyranone Derivatives

In an attempt to understand the nature of electronic transitions in terms of their energies and oscillator strengths, Thul et al. [14] conducted TDDFT calculations on 6-phenyl-4-methylsulfanyl-2-oxo-2H-pyran (molecule **1**) and 6-phenyl-4-methylsulfanyl-2-oxo-2H-pyran-3-carbonitrile (molecule **2**) shown in Figure 9.5 and provided detailed assignment to the electronic transitions. Electronic transitions and oscillator strengths for molecules **1** and **2** were calculated by using 6-31+G** 5D basis set and B3LYP functionals after taking in to account configuration interaction between singly excited states.

Positions of experimental absorption peaks and calculated transition energies, optical strengths, main configurations, and mixing coefficients of the singlet ground and excited states and spectral assignments are given in Table 9.2. It follows from this table that the TDDFT calculations are able to provide a close agreement of less than 0.2 eV between the experimental and calculated values of transition energies in these molecules containing 25 atoms.

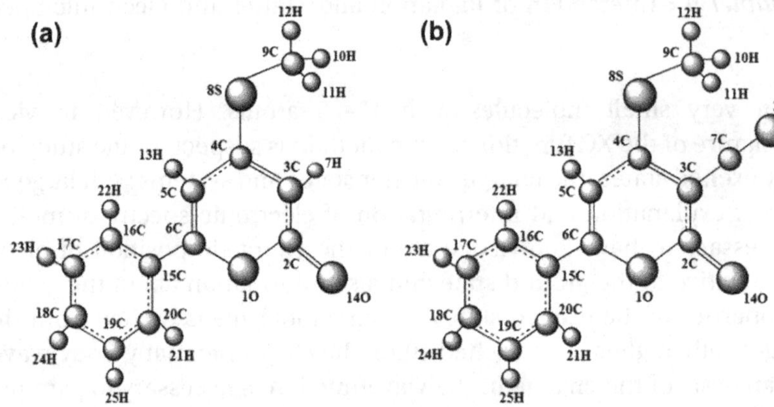

FIGURE 9.5 Structure of (a) 6-phenyl-4-methylsulfanyl-2-oxo-2*H*-pyran (molecule **1**) and (b) 6-phenyl- 4-methylsulfanyl-2-oxo-2*H*-pyran-3-carbonitrile (molecule **2**).

Table 9.2 Electronic transitions and assignments for 6-phenyl-4-methylsulfanyl-2-oxo-2*H*-pyran (molecule **1**) and 6-phenyl-4-methylsulfanyl-2-oxo-2*H*-pyran-3-carbonitrile (molecule **2**)

		Experimental		Calculated				
		Transition energy		Transition energy		Oscillator strength	Mixing coefficients and main	
S.No.	Compound	eV	nm	eV	nm	$f \times 10^3$	configurations	Assignments
Molecule 1								
Ground state $E = -1011.8209370$ au							1.000 [22222/0000]	
1		3.770	329	3.805	325.9	335.4	−0.643 [22221/1000]	π–π^*
2		4.189	295	4.250	291.7	219.2	0.652 [22212/1000]	n–π^*
3		–	–	4.515	274.6	12.8	0.512 [22122/1000]	π–π^*
							0.370 [2122/1000]	
4		–	–	4.547	272.7	9.4	0.566 [21222/1000]	σ–π^*
							0.316 [22122/1000]	
5		4.980	249	4.983	251.1	38.1	0.590 [22221/0100]	π–π^*
							0.239 [22122/1000]	
6		5.367	230	5.145	241.0	28.0	0.578 [12222/1000]	π–π^*
							0.271 [22221/0010]	
7		–	222	5.197	238.6	1.2	0.654 [22221/0001]	π–σ^*
							0.234 [22212/0001]	
Molecule 2								
Ground state $E = -1104.0685300$ au							1.000 [22222/0000]	
1		3.384	366	3.444	360.0	358.8	0.638 [22221/1000]	π–π^*
2		3.96	327	3.96	313.1	284.5	0.655 [22212/1000]	n–π^*
3		–	–	4.261	290.9	19.5	0.655 [22122/1000]	π–π^*
4		–	–	4.359	284.4	1.7	0.661 [21222/1000]	σ–π^*
5		–	–	4.829	256.7	47.7	0.507 [22221/0100]	π–π^*
							0.411 [12222/1000]	
6		4.907	252	4.889	253.6	74.3	0.479 [12222/1000]	π–π^*
							0.401 [22221/0100]	
7		5.21	238	5.109	242.7	14.6	0.627 [22221/0001]	π–σ^*

TDDFT calculations predict seven transitions in the near-ultraviolet region for molecule **1**. The strong transitions at 325.9 (0.3354), 251.1 (0.0381), and 241.0 (0.0280) nm have been experimentally observed at 329, 249, and 231 nm, respectively, in the UV spectra in methanol. Based on the molecular orbital coefficients and mixing coefficients, these bands may be assigned to a π–π^* transition. The numbers in parentheses represent oscillator strengths which too have a qualitative agreement with the experimental values of absorbance. The calculations also predict another strong band at 291.7 (0.2192) nm. This band corresponds to the experimental band at 295 nm and may be assigned to the n–π^* transition involving the lone pair of electrons of the sulfur atom. The intensity of an n–π^* band depends upon the local symmetry of the nonbonding orbital [16–18]. The large intensity of this band may, therefore, be due to the local symmetry of the n-orbital of the sulfur atom and the π^* orbital which makes it an allowed transition. The predicted weak transitions at 274.6 (0.0128), 272.7 (0.0094), and 238.6 (0.0012) nm could not be detected experimentally.

In molecule **2**, which differs from molecule **1** by a nitrile group substituent at position 3 in the pyran ring, the calculations predict transitions at 360.0 (0.3588), 313.1 (0.2845), 253.6 (0.0743), and 242.7 (0.0146) nm. These may be correlated to the experimental bands at 366, 327, 252, and 238 nm, respectively. The first two bands, which may be assigned to π–π^* and n–π^* transitions, experience a longer wavelength shift from their positions 329 and 249 nm in molecule **1**. On the basis of TDDFT calculations, this shift may be explained as arising out of a larger decrease in the energy of the lowest unoccupied molecular orbital (0.0278 eV) than those of the highest occupied molecular orbital (0.013 eV) and the nonbonding (0.0248 eV) orbitals relative to their positions in molecule **1**. The theoretically calculated transitions at 290.9 (0.0195), 284.4 (0.0017), and 256.7 (0.0477) nm are weak and were not observed experimentally.

9.12.4.2 Chloropyrimidine Derivatives

Gupta et al. [15] reported the experimental- and TDDFT-based calculated spectra of 4-chloro-2,6-dimethylsulfanyl pyrimidine-5-carbonitrile (molecule **1**) and 4-chloro-2-methylsulfanyl-6-(2-thienyl) pyrimidine-5-carbonitrile (molecule **2**) shown in Figure 9.6(a) and (b).

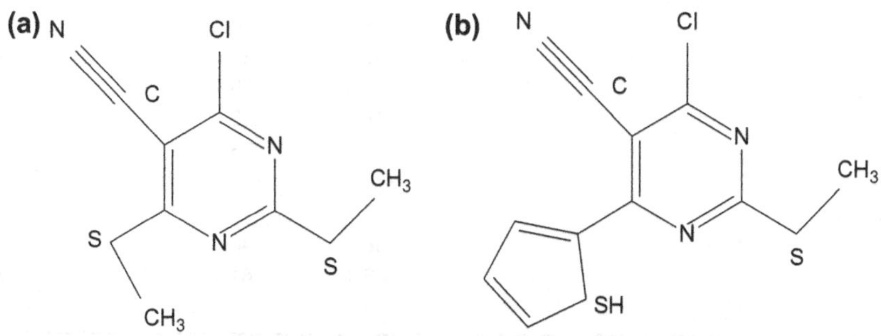

FIGURE 9.6 Structure of (a) 4-chloro-2,6-dimethylsulfanyl pyrimidine-5-carbonitrile (molecule **1**) and (b) 4-chloro-2-methylsulfanyl-6-(2-thienyl)pyrimidine-5-carbonitrile (molecule **2**).

Electronic transitions and oscillator strengths for the two molecules are calculated by TDDFT using 6-31+G* 5D basis set and B3LYP functionals after taking into account configuration interaction between singly excited states. Positions of experimental absorption peaks and the calculated transitions, optical oscillator strengths, and assignments based on the nature of involved molecular orbitals are given in Table 9.3.

As may be seen from this table a close agreement within 0.2 eV or about 8 nm has been obtained between the experimental and calculated values of transition energies. In the case of molecule **1**, the calculations predict seven electronic transitions in the near-UV region at 350.1 (0.0062), 331.5 (0.0281), 308.1 (0.3227), 280.1 (0.2488), 248.5 (0.1016), 235.7 (0.005), and 233.1 (0.0425) nm; the numbers in brackets represent oscillator strength (f). No experimental band has been observed corresponding to the 350.1 nm transition. The absorption bands corresponding to the other transitions appear at 324, 307, 285, 241, 236, and 232 nm, respectively. TDDFT calculations show that the absorption band at 324 nm may be assigned to n–π^* transition involving lone pair of electrons of the nitrogen atom and can be compared with similar transition at 322 nm in

Table 9.3 Electronic transitions and assignments for 4-chloro-2,6-dimethylsulfanyl pyrimidine-5-carbonitrile (molecule **1**) and 2-methylsulfanyl-6-(2-thienyl)-4-chloropyrimidine-5-carbonitrile (molecule **2**)

			Calculated			
			Energy		Oscillator strength	
S.No.	eV	Experimental λ (nm)	eV	λ (nm)	$f \times 10^3$	Assignment
Molecule 1						
1		–	3.5415	350.1	6.2	n–π^*
2	3.8266	324	3.7404	331.5	28.1	n–π^*
3	4.0385	307	4.0243	308.1	322.7	π–π^*
4	4.3503	285	4.4258	280.1	248.8	π–π^*
5	5.1445	241	4.9885	248.5	101.6	π–π^*
6	5.2535	236	5.2597	235.7	5.0	n–π^*
7	5.3441	232	5.3198	233.1	42.5	π–π^*
Molecule 2						
1		–	2.9519	420.0	20.6	n–π^*
2		–	3.4198	362.5	83.6	n–π^*
3	3.5123	353	3.5372	350.5	179.7	n–π^*
4	3.5833	346	3.6603	338.7	25.0	n–π^*
5	3.6900	336	3.7811	327.9	293.3	π–π^*
6	3.8032	326	3.8857	319.1	136.5	π–π^*
7	4.2315	293	4.3068	287.9	43.6	π–π^*
8	4.3050	288	4.8303	256.7	40.7	π–π^*
9	4.8431	256	4.9971	248.1	41.2	π–π^*
10	5.2535	236	5.0143	247.3	9.5	π–π^*

pyrimidine. The band at 236 nm also arises out of an n–π^* transition involving lone pair of electron of the sulfur atom. Absorption bands at 307, 285, 241, and 232 nm may be assigned to π–π^* transitions.

Several absorption bands, some of them with vibrational fine structure, are observed in the UV spectrum of molecule **2** in methanol in 220–360 nm range. These appear at 353, 346, 336, 326, 293, 288, 256, and 236 nm. The assignments of these bands are given in Table 9.3. While the bands at 353, 336, 293, 288, and 236 nm mainly arise out of electronic transitions in the pyrimidine ring, the bands at 346, 326, and 256 nm arise of transitions in the thiophene ring. Thiophene is known to have an absorption band at 231 nm in isooctane which experiences a bathochromic shift of 10–15 nm [19] on substitution by a methyl group. In addition, it may also have weak transitions, presumably n–π^*, in the 310–340 nm range [19]. Based on theoretical results and the experimental data, the absorption bands at 256 and 326 nm may be assigned to π–π^* transitions and the band at 346 nm to the n–π^* transition in the thiophene ring. It my however be noted that since the experimental spectrum has been recorded in a polar solvent methanol, the polarization effects cannot be ignored; close agreement of 1–3 nm may be accidental. One may however conclude that TDDFT calculations can provide quite accurate excitation energies and oscillator strengths for fairly large molecules.

9.13 Rydberg States

The various possible excited states in spectroscopy are usually distinguished according to their characteristics as either a valence shell or a Rydberg species. In a Rydberg state the excited electron has an energy very near the level of the continuum, i.e., it is almost detached. Thus, Rydberg states may be conveniently be thought of as an electron attached to a molecular cation. The molecular cation may be considered as an electron attractor in the same manner as the nucleus acts as a central attractor in an atom. Experimentally a Rydberg state can be easily identified by the fact that it is part of a series of states which converge toward a given ionization limit. The term value of a Rydberg state, i.e., the energy difference relative to the ionization potential can be described by a simple formula involving the ionization potential. The Rydberg states are by nature very much diffused than the valence states and have quite an expanded charge distribution in space. This is an important characteristic which distinguishes the Rydberg states from the valence states because the behavior of the diffuse charge distribution is a quite different from that of the contracted valence shell state. For this reason, in order to describe Rydberg states, atomic orbital basis sets that include diffuse functions on heavy atoms and preferably also on the hydrogen atoms are used. Looking to the nature of the Rydberg states, because of its atom-like charge distribution, in the molecular orbital theory, the upper orbital of the Rydberg state is generally denoted in terms of atomic orbitals. Another important characteristic of the Rydberg states is that the Rydberg transitions have very low intensity (f of the order of 0.08) and very small singlet–triplet splitting in the bands. This is the direct consequence of diffuse charge distribution in these states.

A review of the molecular orbital methods for calculations on Rydberg states has been given by Peyerimhoff and Buenker [20]. These authors have also reviewed the multi-reference double-excitation-CI (MRD-CI) procedures. It has been indicated that the *ab initio* treatments may use an extension of AO basis sets, instead of those used for ground state treatments, so as to include long range functions of proper symmetry. These may represent the extended charge density in the upper orbitals of the Rydberg states. While the configuration interaction treatments for the valence state and the Rydberg states are not formally different but they involve different reference configuration. The *ab initio* valence state CI-based calculations are also able to describe the inner shell excitations as well as the ionization from both the outer or inner shell valence states. This only requires different types of atomic orbitals.

A typical example for the occurrence of various types of valence states which are relatively separated from Rydberg states is the ozone molecule in which the transitions can be essentially divided into three groups: (1) transitions in the 1–5 eV arise mainly from excitations from the neighboring $1a_2, 4b_2$ and $6a_1$ MOs into the unoccupied $2b_1$ π-type species, (2) transitions in the energy range 5–8 eV arising out of double excitations involving the same MOs, and (c) transitions greater than 8 eV involving both the Rydberg upper states as well as excitations into antibonding σ^*-type orbitals. The MRD-CI calculations [20] show a very close agreement with the experimental values within 0.03 eV.

9.14 Studies of Core Electrons

While considering the electronic configuration of an atom, electrons can be divided into two categories: valence electrons and core electrons according to their position in an atom. The valence electrons occupy the outermost shell or highest energy levels of an atom whereas the core electrons occupy the lowest energy levels. The core electrons are so tightly bound to the nucleus that they are negligibly perturbed by the environment of the atom when it is in a molecular or solid state. As such, the binding properties are almost completely due to the valence electrons, the core electrons do not participate in bonding. Their main role is to screen the positive charge of the atomic nucleus, though they do influence chemical reactivity of an atom. In transition metals, however, the distinction between the core and valence electrons is very thin and the electrons in d-shells may be treated as valence rather than core electrons. The core electrons are involved in photoelectron emission and secondary events such as X-ray emission and Auger electron emission.

The electronic excitation from the core electrons can be of two types: a core electron can be removed from its core level upon absorption of electromagnetic radiation (X-rays) and emitted as a photoelectron or else can be excited to an empty outer shell of the molecules. Two possibilities may arise in the second case. The core to Rydberg transition may converge to a certain core ionization threshold (E_{Th}). Alternatively, since most molecules have unoccupied molecular orbitals of antibonding valence character such as σ^* and π^*, the electron from the inner shell may be excited to one of these orbitals. If the potential energy curve of the core to σ^* valence-excited state is repulsive for a given σ bond, then this state may either be lower in energy than the continuum and Rydberg region at longer bond distances or may be embedded in the continuum at shorter bond distances. Thus, two kinds of interactions may arise between the core electrons and the valence states for a given σ bond: Rydberg–valence interaction below E_{Th} and the continuum–valence interaction above E_{Th}.

The *ab initio* quantum chemical methods have been extensively used for the calculations and assignments of the vertical IPs for both core and valence electrons and the prediction and interpretation of the shifts in core electron binding energies (CEBEs) for a given atom with changing chemical environment. The terms bonding energies (BEs) and IPs are almost equivalent and are used interchangeably; while the former refers to low energies of a few electron volts to a few tens of electron volts, the latter refers to the high energy of a few tens to a few hundred electron volts IP. Due to large energy differences between the core orbitals of the different elements, the spectroscopy of core electrons has also been very attractive for atom-specific probe of electronic structure of molecules.

Theoretical quantum chemical studies of CEBEs have been extensively reviewed [21–26]. The core orbitals have a spatially localized nature. For a highly localized core MO which is essentially an AO, we expect a great deal of transferability from one molecular system to another. This has been confirmed by LCAO (linear combination of atomic orbital)-SCF molecular orbital calculations. X-ray photoelectron spectroscopy shows that ionizations in many molecules are characteristic of core electrons of a given atom and vary over a small range of few electron volts due to small changes in molecular environments. This feature makes photoelectron spectroscopy useful for chemical analysis (ESCA). Various techniques have been used for theoretical calculation of core electron excitations and CEBEs. These have been reviewed by Tolbatov and Chipman [26].

Post-HF methods can provide highly accurate CEBE. Thus, for example, observed shifts of the carbon 1*s* photoelectron energies in some organic compounds could be reproduced within 0.03 eV by Møller–Plesset and coupled-cluster approaches. EOM-CCSD [27] has been shown to provide very accurate core excitation energy. In this approach first a CCSD calculation on the core-ionized state is performed to describe the relaxation of the core electron and then, in the second step, an electron is added to the core-ionized state to get the core-excited states of the neutral molecule by the EOM-CCSD method. In the case of large molecules, the most commonly used methods are based on DFT with KS orbitals. The most efficient method for CEBE calculation in this category is based on Koopmann's theorem but it has limited accuracy [28]. Pettersson et al. [29] and Carroll et al. [30] obtained better results by using effective core potentials in this method. Another, simple approach is the Δ–Kohn-Sham method in which the core excitation energy is calculated as the difference in the expectation values of the neutral and core-excited KS Hamiltonians where the orbitals have been variationally optimized for the different states. This method is, however, not straightforward and suffers from many constraints [31]. In the case of first row atoms in many small gas phase molecules, it is found that the quality of results is very sensitive to the choice of the functional. The best agreement of the calculated CEBE by this method is about 0.16 eV. TDDFT is also widely used in CEBE calculations [32,33]. It is found that the best results by this method are obtained by using functionals having a long range correction [32,33].

References

[1] E.U. Condon, G.H. Shortley, The Theory of Atomic Spectra, Cambridge University Press, 1935.

[2] J.B. Foresman, H.B. Schlegel, in: R. Fausto (Ed.), Recent Experimental and Computational Advances in Molecular Spectroscopy, Kluwer, 1993.

[3] H.J. Werner, Adv. Chem. Phys. 69 (1987) 1.

[4] R. Shepard, Adv. Chem. Phys. 69 (1987) 63.

[5] B.O. Joos, Adv. Chem. Phys. 69 (1987) 399.

[6] J.B. Foresman, M. Head-Gordon, J.A. Pople, J. Phys. Chem. 96 (1992) 135.

[7] J.A. Pople, J.S. Binkley, R. Seeger, Int. J. Quantum Chem. Symp. 10 (1976) 1.

[8] Gaussian 09, Revision D.01, M.J. Frisch, G.W. Trucks, H.B. Schlegel, G.E. Scuseria, M.A. Robb, J.R. Cheeseman, G. Scalmani, V. Barone, B. Mennucci, G.A. Petersson, H. Nakatsuji, M. Caricato, X. Li, H.P. Hratchian, A.F. Izmaylov, J. Bloino, G. Zheng, J.L. Sonnenberg, M. Hada, M. Ehara, K. Toyota, R. Fukuda, J. Hasegawa, M. Ishida, T. Nakajima, Y. Honda, O. Kitao, H. Nakai, T. Vreven, J.A. Montgomery, Jr., J.E. Peralta, F. Ogliaro, M. Bearpark, J.J. Heyd, E. Brothers, K. N. Kudin, V.N. Staroverov, R. Kobayashi, J. Normand, K. Raghavachari, A. Rendell, J.C. Burant, S.S. Iyengar, J. Tomasi, M. Cossi, N. Rega, J.M. Millam, M. Klene, J.E. Knox, J.B. Cross, V. Bakken, C. Adamo, J. Jaramillo, R. Gomperts, R.E. Stratmann, O. Yazyev, A.J. Austin, R. Cammi, C. Pomelli, J.W. Ochterski, R.L. Martin, K. Morokuma, V.G. Zakrzewski, G.A. Voth, P. Salvador, J.J. Dannenberg, S. Dapprich, A. D. Daniels, Ö. Farkas, J.B. Foresman, J.V. Ortiz, J. Cioslowski, and D.J. Fox, Gaussian, Inc., Wallingford CT, 2009.

[9] M.W. Schmidt, K.K. Baldridge, J.A. Boatz, S.T. Elbert, M.S. Gordon, J.H. Jensen, S. Koseki, N. Matsunaga, K.A. Nguyen, S.J. Su, T.L. Windus, M. Dupuis, J.A. Montgomery, Computer program GAMESS, J. Comput. Chem. 14 (1993) 1347.

[10] R. Ahlrichs, M. Bar, M. Ḧaser, H. Horn, C. Kolmel, Chem. Phys. Lett. 162 (3) (1989) 165.

[11] R.D. Amos, I.L. Alberts, J.S. Andrews, S.M.C. Ell, N.C. Handy, D. Jayatilaka, P.J. Knowles, R. Kobayashi, G.J. Laming, A.M. Lee, P.E. Maslen, C.W. Murray, P. Palmieri, J.E. Rice, E.D. Simandiras, A.J. Stone, M.D. Su, D.J. Tozer, CADPAC 6.5, The Cambridge Analytic Derivatives Package, Cambridge, UK, 1998.

[12] A. Virdi, V.P. Gupta, A. Sharma, J. Mol. Struct. Theochem. 678 (2004) 239.

[13] R.F. Rekker, P.J. Brombacher, H. Hamann, W.T. Nauto, Rec. Trav. Chim. 73 (1954) 410.

[14] P. Thul, V.P. Gupta, V.J. Ram, P. Tandon, Spectrochim. Acta A75 (2010) 251.

[15] V.P. Gupta, A. Sharma, A. Virdi, V.J. Ram, Spectrochim. Acta A64 (2006) 57.

[16] J.N. Murrel, The Theory of Electronic Spectra of Organic Molecules, Chapman and Hall Ltd, London, 1971.

[17] J.R. Platt, J. Chem. Phys. 18 (1950) 1168.

[18] H.H. Orchin, M. Jaffe, Theory and Applications of Ultraviolet Spectroscopy, Wiley, New York, 1966.

[19] P. Ramart Lucas, Bull. Soc. Chem. (1954) 1017. France.

[20] S.D. Peyerimhoff, R.J. Buenker, Calculation of Electronically Excited States in Molecules, in: J. Bargon (Ed.), Computational Methods in Chemistry, Plenum Press, 1979.

[21] M. Schwartz, in: H.F. Schaefer (Ed.), Methods of Electronic Structure Theory and Application of Electron Structure Theory, Plenum, New York, 1977.

[22] D.A. Shirley, ESCA, in: I. Prigogine, S.A. Rice (Eds.), Advances in Chemical Physics, 23, Wiley, New York, 1973, p. 85.

[23] M.E. Schwartz, ESCA, in: C.A. Coulson, A.D. Buckingham (Eds.), MTP International Review of Sciences (Physical Chemistry Series 2: Theoretical Chemistry), 189, Butter-worths, London, 1975.

[24] M.E. Schwartz, J.D. Switalski, R.E. Stronski, Core-level binding energy shifts from molecular orbital theory, in: D.A. Shirley (Ed.), Electron Spectroscopy, North-Holland Publishing Co., Amsterdam, 1972, p. 605.

[25] H. Basch, J. Electron Spectros, Relat. Phenom. 5 (1974) 463.

[26] I. Talbatov, D.M. Chipman, Theor. Chem. Acc. 13 (2014) 1473 and references therein.

[27] M. Nooiin, R.J. Bartletl, J. Chem. Phys. 102 (1995) 6735.

[28] S. Shirai, S. Yamamoto, S. Hyodo, J. Chem. Phys. 121 (2004) 7586.

[29] L.G. Pettersson, U. Wahlgren, O. Gropen, Chem. Phys. 80 (1983) 7.

[30] T.X. Carroll, T.D. Thomas, L.J. Saethre, K.J. Borve, J. Phys. Chem. A113 (2009) 3481.

[31] A. Naves de Brito, N. Correia, S. Svensson, H.A. Ågren, J. Chem. Phys. 95 (1991) 2965.

[32] T. Yanai, D.P. Tew, N.C. Handy, Chem. Phys. Lett. 393 (2004) 51.

[33] N.A. Besley, M.J.G. Peach, D.J. Tozer, Phys. Chem. Chem. Phys. 11 (2009) 10350.

Further Reading

[1] P. Atkins, R. Friedman, Molecular Quantum Mechanics, fourth ed., Oxford University Press, 2005.

[2] H. Eyring, J. Walter, G.E. Kimball, Quantum Chemistry, John Wiley & Sons, 1944.

[3] W.S. Struve, Fundamentals of Molecular Spectroscopy, Wiley Interscience, 2010.

[4] M.E. Schwarz, in: H.F. Schaefer III (Ed.), Applications of Electronic Theory, 1977.

[5] M.A. Ratner, G.C. Schatz, Introduction to Quantum Mechanics in Chemistry, Prentice Hall, N. Jersey, 2002.

[6] E.G. Lewars, Computational Chemistry, Springer, 2011.

10

Energy and Force Concepts in Chemical Bonding

CHAPTER OUTLINE

10.1 Introduction

The concepts of energy and force are the two familiar concepts in chemistry that are used to explain almost every phenomenon or process in chemistry, such as chemical reactions, activation energy, mechanism of chemical bonding, etc. The concept of energy, however, lacks simplicity and elegancy. The force concept has its foundation in the observation that if the electron density distribution in a molecule is known from quantum mechanical methods, the actual force on the nuclei can be obtained in terms of classical electrostatics. These two concepts represented by virial theorem and the Hellmann–Feynman (H-F) theorem of electrostatic forces (ESFs), respectively, have been widely used to explain the various chemical phenomenon such as the nature of chemical bond, chemical reactions, shape of the molecule, and its change under external influences, etc. While the virial theorem is capable of providing the electronic kinetic and potential energies from experimental or calculated total energies and explain these contributions on the basis of structural features, the forces on the nuclei of molecules, calculated with the use of the H-F theorem provide great qualitative insight into the nature of the phenomena investigated. However, there are serious

Principles and Applications of Quantum Chemistry. http://dx.doi.org/10.1016/B978-0-12-803478-1.00010-8

limitations in quantitative applications of the H-F theorem with approximate wavefunctions, since the calculated forces are extremely sensitive to small inaccuracies in the wavefunctions, especially near the nuclei of interest. The H-F theorem is, however, a highly useful tool for developing qualitative chemical models based on firm quantum mechanical foundations and is also open to quantitative extension, at least in principle. In this chapter, we shall discuss the two approaches and consider their applications in molecular bond formation.

10.2 Virial Theorem

10.2.1 Virial Theorem for Atom and Bound Stationary States

In quantum mechanics, the equation of motion of a dynamical variable $F(q, p, t)$ which, for a system of f degrees of freedom, is a function of coordinates $(q_1 \ldots q_f)$, momenta $(p_1 \ldots p_f)$, and time t, is given as

$$\frac{dF}{dt} = \frac{\partial F}{\partial t} + \frac{1}{i\hbar}[F, H] \tag{10.2.1}$$

where H is the Hamiltonian of the system. This equation is equivalent to a similar equation in classical mechanics

$$\frac{dF}{dt} = \frac{\partial F}{\partial t} + \{F, H\} \tag{10.2.2}$$

where $\{F, H\}$ is the Poisson bracket of the two functions of coordinate and momentum. The first term on the right-hand side takes in to account the explicit time dependence of F.

A proof of the virial theorem of quantum mechanics can be given in analogy with the corresponding proof in classical mechanics. In classical mechanics the time average of the time derivative of the quantity $\mathbf{r}\cdot\mathbf{p}$ is zero for a periodic system. In quantum mechanics, the analogous quantity is the time derivative of the expectation value of $\mathbf{r}\cdot\mathbf{p}$ or the diagonal matrix element of the commutator of $\mathbf{r}\cdot\mathbf{p}$ and H in energy representation, which should also be zero. Hence, from Eq. (10.2.1)

$$\frac{d}{dt}\langle \mathbf{r}\cdot\mathbf{p}\rangle = \frac{1}{i\hbar}[\mathbf{r}\cdot\mathbf{p}, H] = 0 \tag{10.2.3}$$

For a system of 3 degrees of freedom

$$\begin{aligned}
[\mathbf{r}\cdot\mathbf{p}, H] &= \left[(xp_x + yp_y + zp_z), \left\{\frac{1}{2m}\left(p_x^2 + p_y^2 + p_z^2\right) + V(x, y, z)\right\}\right] \\
&= \frac{i\hbar}{m}\left(p_x^2 + p_y^2 + p_z^2\right) - i\hbar\left(x\frac{\partial V}{\partial x} + y\frac{\partial V}{\partial y} + z\frac{\partial V}{\partial z}\right) \\
&= 2i\hbar T - i\hbar(\mathbf{r}\cdot\nabla V)
\end{aligned} \tag{10.2.4}$$

where T is the kinetic energy, V is the potential energy, and $\mathbf{p} = -i\hbar\nabla$. Here, use has been made of the commutation relations

$$[x, p_x] = i\hbar, \quad [x, p_y] = [x, p_z] = 0$$

and, $[p_x, p_y] = [p_x, p_z] = 0$.

Thus, using Eq. (10.2.3), we get

$$2\langle T\rangle = \langle \mathbf{r}\cdot\nabla V\rangle \tag{10.2.5}$$

Equation (10.2.5) is the quantum mechanical virial theorem, which may also be written as

$$2\langle T\rangle = \left\langle \sum_j q_j \frac{\partial V}{\partial q_j}\right\rangle \tag{10.2.6}$$

where j represents the degree of freedom.

It may be noted that it is immaterial whether we start from $\mathbf{r}\cdot\mathbf{p}$ or $\mathbf{p}\cdot\mathbf{r}$ as the difference between them is a constant and hence commutes with H. The validity of the virial theorem is restricted to bound stationary states.

If V is spherically symmetric and proportional to r^n and the expectation values exist, then it can be shown using Eq. (10.2.6) that

$$2\langle T\rangle = n\langle V\rangle \tag{10.2.7}$$

For a stationary state, the total energy E is given as

$$E = \langle T\rangle + \langle V\rangle$$

Hence, $\langle T\rangle = \frac{nE}{n+2}$ and $\langle V\rangle = \frac{2E}{n+2}$.

Thus, for example, in the case of a hydrogen atom,

$$V = -\frac{Ze^2}{r}$$

and so, $n = -1$.

Hence, $\langle T\rangle = -E$ and $\langle V\rangle = 2E$, as expected.

Equation (10.2.7) is strictly valid for homogeneous functions of degree n, which follow Euler's theorem. According to this theorem, if $F(x_1, \ldots x_f)$ is a homogeneous function of degree n,

$$\text{then, } \sum_{j=1}^{f} q_j \frac{\partial F}{\partial q_j} = nF \tag{10.2.8}$$

10.2.2 Virial Theorem for Diatomic and Polyatomic Molecules

The virial theorem for a diatomic molecule was first obtained by Slater [1] who remarked that this theorem gives a means of finding kinetic and potential energy separately for all configurations of the nuclei, if the total energy is known from experiment or theory.

In the case of a molecule, the virial theorem is derived in the Born–Oppenheimer approximation which separates the electronic and nuclear motions.

The energy of a molecule is a sum of kinetic and potential energy contributions of the electrons and nuclei. Thus,

$$E = T_n + T_e + V_{nn} + V_{ne} + V_{ee}$$

where T_n and T_e are the kinetic energies of the nuclei and the electrons, respectively, V_{nn} is the nuclear–nuclear repulsion energy, V_{ne} is the electron–nuclear attraction energy, and V_{ee} is the electron–electron repulsion energy.

In the Born–Oppenheimer approximation, the molecular wavefunction is

$$\Psi = \Psi_e(q_i, q_\alpha)\Psi_n(q_\alpha) \tag{10.2.9}$$

where q_i and q_α symbolize the electronic and nuclear coordinates. The electronic wavefunction is obtained by solving the Schrödinger equation

$$\hat{H}_e\Psi_e(q_i, q_\alpha) = E_e(q_\alpha)\Psi_e(q_i, q_\alpha) \tag{10.2.10}$$

where E_e is the electronic energy and H_e is Hamiltonian for the electrons.

$$\widehat{H_e} = \widehat{T_e} + \widehat{V_e} \tag{10.2.11}$$

$$\text{where,}\quad \widehat{T_e} = -\frac{\hbar^2}{2}\sum_i \nabla_i^2 \tag{10.2.12}$$

$$\text{and,}\quad \widehat{V_e} = -\sum_\alpha\sum_i \frac{Z_\alpha}{r_{i\alpha}} + \sum_i\sum_{j>i}\frac{1}{r_{ij}} = V_{ne} + V_{ee} \tag{10.2.13}$$

where $r_{i\alpha}$ and r_{ij} are electron–nuclear and electron–electron distances. The first term on the right or Eq. (10.2.13) corresponding to electron–nuclear attraction and the second to electron–electron repulsion.

Since the electron is moving under the combined field of electrons and nuclei, the total potential energy is

$$V = V_{ee} + V_{ne} + V_{nn} \tag{10.2.14}$$

However, as usual, the nuclear–nuclear repulsion term V_{nn} is being ignored by considering the nuclei as fixed and is included later on.

If we view V_e as a function of both the electronic and nuclear coordinates then on applying Euler's theorem (Eq. (10.2.8))

$$\sum_i q_i\frac{\partial V_e}{\partial q_i} + \sum_\alpha q_\alpha\frac{\partial V_e}{\partial q_\alpha} = -V_e \tag{10.2.15}$$

A negative sign on the right-hand side appears because in this case V_e is inversely proportional to r and hence $n = -1$.

The virial Eq. (10.2.6) in this case becomes

$$2\langle T_e\rangle = \left\langle \sum_i q_i\frac{\partial V_e}{\partial q_i}\right\rangle \tag{10.2.16}$$

Using Eq. (10.2.15) in Eq. (10.2.16), we get

$$2\langle T_e\rangle = \left\langle -V_e - \sum_\alpha q_\alpha\frac{\partial V_e}{\partial q_\alpha}\right\rangle = -\langle V_e\rangle - \sum_\alpha\left\langle q_\alpha\frac{\partial V_e}{\partial q_\alpha}\right\rangle \tag{10.2.17}$$

The second term on the right-hand side may also written as

$$\sum_\alpha\int \Psi_e^* q_\alpha\frac{\partial V_e}{\partial q_\alpha}\Psi_e d\tau_e = \sum_\alpha\int \Psi_e^* q_\alpha\frac{\partial}{\partial q_\alpha}(E_e - T_e)\Psi_e d\tau_e = \sum_\alpha q_\alpha\int \Psi_e^*\frac{\partial E_e}{\partial q_\alpha}\Psi_e d\tau = \sum_\alpha q_\alpha\frac{\partial E_e}{\partial q_\alpha} \tag{10.2.18}$$

where $E_e = T_e + V_e$, and T_e is independent of q_α.

Hence, from Eqs (10.2.17) and (10.2.18)

$$2\langle T_e\rangle = -\langle V_e\rangle - \sum_{\alpha} q_\alpha \frac{\partial E_e}{\partial q_\alpha} \tag{10.2.19}$$

Either T_e or V_e can be eliminated from this equation by using

$$\langle T_e\rangle + \langle V_e\rangle = \langle E_e\rangle \tag{10.2.20}$$

If we now include the nuclear–nuclear repulsion term

$$V_{nn} = \sum_{\alpha}\sum_{\beta>\alpha} \frac{z_\alpha z_\beta}{r_{\alpha\beta}} \tag{10.2.21}$$

$$\text{then, the total potential, } V = V_e + V_{nn} \tag{10.2.22}$$

and the total electronic energy $E_e = T_e + V_e$, where V_e is given by Eq. (10.2.13).

Equation (10.2.19) can now be written as

$$2\langle T_e\rangle = -\langle V - V_{nn}\rangle - \sum_{\alpha} q_\alpha \frac{\partial E_e}{\partial q_\alpha} \tag{10.2.23}$$

From Euler's theorem,

$$\sum_{\alpha} q_\alpha \frac{\partial V_{nn}}{\partial q_\alpha} = -V_{nn} \tag{10.2.24}$$

Hence,

$$\begin{aligned} 2\langle T_e\rangle &= -\langle V\rangle - \sum_{\alpha} q_\alpha \frac{\partial V_{nn}}{\partial q_\alpha} - \sum_{\alpha} q_\alpha \frac{\partial E_e}{\partial q_\alpha} \\ &= -\langle V\rangle - \sum_{\alpha} q_\alpha \frac{\partial (V_{nn} + E_e)}{\partial q_\alpha} \end{aligned} \tag{10.2.25}$$

$$\text{Writing } U(q_\alpha) = E_e(q_\alpha) + V_{nn}$$

where $U(q_\alpha)$ is the potential energy function for nuclear motion, we get

$$2\langle T_e\rangle = -\langle V\rangle - \sum_{\alpha} q_\alpha \frac{\partial U}{\partial q_\alpha} \tag{10.2.26}$$

Equation (10.2.26) is the virial theorem for a diatomic molecule.

Taking the example of two nuclei having q_α represented by Cartesian coordinates (x_1, y_1, z_1) and (x_2, y_2, z_2) separated by a distance R, it can easily be shown that

$$\sum_{\alpha} q_\alpha \frac{\partial U}{\partial q_\alpha} = R\,\frac{dU}{dR}.$$

and so the virial theorem can be written in terms of internuclear separation R as

$$2\langle T_e\rangle = -\langle V\rangle - R\frac{dU}{dR} \tag{10.2.27}$$

Since, $U = T_e + V$, this equation may also be written as

$$\langle T_e \rangle = -U - R\frac{dU}{dR} \tag{10.2.28}$$

$$\text{and,}\ \ \langle V \rangle = 2U + R\frac{dU}{dR} \tag{10.2.29}$$

It should be noted that for $dU/d\mathrm{R} = 0$, as at equilibrium distance $R = R_e$ and at infinite separation of the nuclei $R = \infty$, Eq. (10.2.27) reduces to simpler form Eq. (10.2.7) for $n = -1$.

The importance of the virial theorem is that it provides a relationship between the changes of kinetic and potential energies during bond formation. If ΔE, ΔT, and ΔV are the differences between the total, kinetic, and potential energies of a diatomic molecule and of the separate atoms, then at any distance Eqs (10.2.27)–(10.2.29) may be written as

$$2\Delta T + \Delta V + R\frac{dE}{dR} = 0 \tag{10.2.30a}$$

$$\Delta T = -\Delta E - R\frac{dE}{dR} \tag{10.2.30b}$$

$$\text{and,}\ \ \Delta V = 2\Delta E + R\frac{dE}{dR}. \tag{10.2.30c}$$

At the equilibrium distance,

$$\Delta T_e = -\frac{1}{2}\Delta V_e = -\Delta E_e = D_e \tag{10.2.31}$$

where D_e is the dissociation energy of the molecule.

On the basis of Eq. (10.2.31) it is usually argued that the stability of a bond in a molecule is due solely to potential energy because it decreases during bond formation, whereas the kinetic energy increases by a factor of 2. This argument is, however, true only at the equilibrium position, $R = R_e$ or at $R = \infty$, when the molecule breaks into its constituent atoms, but for all other positions the contribution of the $\partial U/\partial R$ cannot be ignored as is clear from the above equations.

Equations (10.2.27)–(10.2.29) can be used to obtain expressions for the kinetic and potential energy in terms of the total energy and its first derivative. These equations are satisfied by true wavefunctions as well as by Hartree–Fock wavefunctions. Löwdin [2] has shown that by using a variational scaling procedure, an arbitrary approximate wavefunction can always be made to satisfy the virial theorem.

For polyatomic molecules, the second term in Eq. (10.2.26) is modified so that the summation is over all internuclear distances.

$$2\langle T_e \rangle = -\langle V \rangle - \sum_{\alpha}\sum_{\beta>\alpha} R_{\alpha\beta}\left(\frac{\partial U}{\partial R_{\alpha\beta}}\right) \tag{10.2.32}$$

Rodriguez et al. [3] have developed a virial theorem in the Kohn–Sham density functional theory formalism and used them to compute the atom-in-molecule (QTAIM) energies for a set of molecules.

10.2.3 Chemical Bonding in Diatomic Molecules

Virial theorem can help in understanding the process of bond formation in molecules and estimate the contribution of electronic potential and kinetic energies toward the total energy of the molecule. Let us first concentrate on the equilibrium position $R = R_e$ and at $R = \infty$, where the molecule dissociates into constituent atoms.

Since $dU/dR = 0$, both at $R = R_e$ and at $R = \infty$, we get from Eqs (10.2.27)–(10.2.29) the following relationships:

$$2\langle T_e\rangle|_{R_e} = -\langle V\rangle|_{R_e} \tag{10.2.33a}$$

$$\langle T_e\rangle|_{R_e} = -U(R_e) \tag{10.2.33b}$$

$$\langle V\rangle|_{R_e} = 2U(R_e) \tag{10.2.33c}$$

$$2\langle T_e\rangle|_{\infty} = -\langle V\rangle|_{\infty} \tag{10.2.33d}$$

$$\langle T_e\rangle|_{\infty} = -U(\infty) \tag{10.2.33e}$$

$$\langle V\rangle|_{\infty} = 2U(\infty) \tag{10.2.33f}$$

$$\text{Hence,}\quad \langle T_e\rangle|_{R_e} - \langle T_e\rangle|_{\infty} = U(\infty) - U(R_e) = D_e \tag{10.2.34}$$

$$\text{and,}\quad \langle V\rangle|_{R_e} - \langle V\rangle|_{\infty} = 2[U(R_e) - U(\infty)] = 2\left[\langle T_e\rangle|_{\infty} - \langle T_e\rangle|_{R_e}\right] \tag{10.2.35}$$

where D_e is the dissociation energy. Since $U(\infty)$ is the sum of energies of the two separated atoms, it should be greater than $U(R_e)$ for bond formation. It, therefore, follows from Eq. (10.2.35) that

$$\langle T_e\rangle|_{R_e} > \langle T_e\rangle|_{\infty} \tag{10.2.36}$$

$$\langle V\rangle|_{R_e} < \langle V\rangle|_{\infty} \tag{10.2.37}$$

So, at equilibrium internuclear distance (R_e), the average kinetic energy is greater than at infinity, whereas the reverse is the case with the average molecular potential energy. Also the decrease in potential energy is twice the increase in the kinetic energy. The decrease in potential energy may be attributed to the transfer of electron density from the region around the nuclear centers to the internuclear region so that the electron now gets attracted by both the nuclei. This information though important is still qualitative and does not reveal whether the increase or decrease of kinetic and potential energies be treated as an atomic or a molecular effect and what is the contribution of these energies toward the total energy. One would also like to know the changes that take place in these energies during the process of bond formation from the constituent atoms. In order to get these informations we need to get the energy profile for covalent bond formation. For this, virial Eqs (10.2.27)–(10.2.29) are solved to get the molecular kinetic and potential energies as a function of the internuclear distance R.

In order to get an understanding of the factors involved in bond formation, we shall here replace the *ab initio* $U(R)$ curve by Morse function which has a similar R dependence.

$$U(R) = D_e\left[1 - e^{-\beta(R-R_e)}\right]^2 - D_e \quad (10.2.38)$$

where D_e is the dissociation energy of the diatomic molecule. The parameters D_e and β can be determined experimentally from band spectra of molecules.

As an example, the energy profile for hydrogen molecule is given in Figure 10.1. This profile shows that as the internuclear distance decreases the potential energy first rises, then falls to a minimum, and then again rises. On the other hand, the kinetic energy first falls, reaches a minimum, and then rises at shorter internuclear distances. The total energy continuously decreases, takes a minimum value, and then again rises. This behavior of kinetic and potential energies can be interpreted in terms of two factors: (1) buildup of electron density in the internuclear region which results in charge delocalization (including polarization) and (2) contraction of atomic orbitals. Both these processes have atomic as well as molecular effects and provide vital information about molecular binding process. According to Feinberg and Ruedenberg [4] the contraction of the wavefunction, that is an increase in the orbital exponent ζ, is necessarily associated with a decrease in the potential energy (which occurs entirely in the atomic region) and an increase in the kinetic energy. The latter is due to the fact that an increase in ζ causes an increase in the average value of the wavefunction gradient. In other words, increasing localization of the wavefunction leads to an increase in the momentum in accordance with the uncertainty principle.

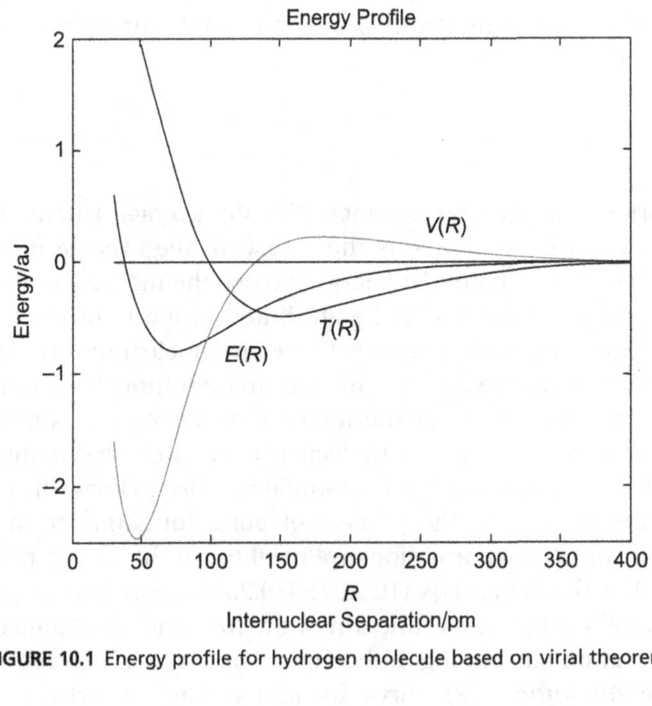

FIGURE 10.1 Energy profile for hydrogen molecule based on virial theorem.

10.3 Hellmann–Feynman Theorem

In Chapter 8, we had introduced the concept of H-F theorem in connection with geometry optimization by Hartree–Fock method. The H-F theorem has been a key ingredient of the quantum mechanical treatment of forces acting on nuclei in molecules and solids. The theorem states that if Ψ_λ is an exact wavefunction of a Hamiltonian H that depends on some parameter λ, and E is the eigenvalue of the Hamiltonian for the function Ψ_λ, then

$$\frac{dE}{d\lambda} = \langle \Psi_\lambda | H | \Psi_\lambda \rangle \tag{10.3.1}$$

The H-F theorem therefore relates the derivative of the total energy with respect to a parameter λ to the expectation value of the derivative of the Hamiltonian with respect to the same parameter. The parameter λ may be nuclear charge, nuclear coordinate, force constant for a harmonic oscillator or the electrical or magnetic fields, etc. The proof the H-F theorem is straightforward.

Consider a system with Hamiltonian $H(\lambda)$ that depends on some parameter λ. Let Ψ_λ be an eigenvector of $H(\lambda)$ with eigenvalue E_λ.

$$H(\lambda)|\Psi_\lambda\rangle = E_\lambda |\Psi_\lambda\rangle \tag{10.3.2}$$

We further assume that $|\Psi_\lambda\rangle$ is normalized, so that

$$\langle \Psi_\lambda | \Psi_\lambda \rangle = 1 \tag{10.3.3}$$

$$\text{and so,} \quad \frac{d}{d\lambda}\langle \Psi_\lambda | \Psi_\lambda \rangle = 0 \tag{10.3.4}$$

We multiply the two-side of Eq. (10.3.2) by $\langle \Psi_\lambda|$ and differentiate to get

$$\begin{aligned}
\frac{dE_\lambda}{d\lambda} &= \frac{d}{d\lambda}\langle \Psi_\lambda | H | \Psi_\lambda \rangle \\
&= \left\langle \frac{d\Psi_\lambda}{d\lambda} | H | \Psi_\lambda \right\rangle + \left\langle \Psi_\lambda \left| \frac{dH}{d\lambda} \right| \Psi_\lambda \right\rangle + \left\langle \Psi_\lambda | H | \frac{d\Psi_\lambda}{d\lambda} \right\rangle \\
&= E_\lambda \left\langle \frac{d\Psi_\lambda}{d\lambda} \middle| \Psi_\lambda \right\rangle + \left\langle \Psi_\lambda \left| \frac{dH}{d\lambda} \right| \Psi_\lambda \right\rangle + E_\lambda \left\langle \Psi_\lambda \middle| \frac{d\Psi_\lambda}{d\lambda} \right\rangle \\
&= E_\lambda \frac{d}{d\lambda}\langle \Psi_\lambda | \Psi_\lambda \rangle + \left\langle \Psi_\lambda \left| \frac{dH}{d\lambda} \right| \Psi_\lambda \right\rangle
\end{aligned} \tag{10.3.5}$$

Using Eq. (10.3.4), we get

$$\frac{dE_\lambda}{d\lambda} = \left\langle \Psi_\lambda \left| \frac{dH}{d\lambda} \right| \Psi_\lambda \right\rangle \tag{10.3.6}$$

If the quantity λ is a given degree of freedom of the system, then the quantity $-dE/d\lambda$ can be interpreted as generalized force associated with it and is known as the H-F force. Thus, if the nuclei are treated as classical particles and λ is the position vector of a nucleus, then $(-dE/d\lambda)$ is the classical force on that nucleus.

While Eq. (10.3.6) gives the H-F theorem for an exact eigenfunction, it is equally true for approximate wavefunctions (non-eigenfunctions) that are variationally determined. This can be shown as below.

Let $|\Psi_\lambda\rangle$ be the wavefunction associated with a Hamiltonian $(\widehat{H} + \lambda\widehat{H}')$, where $\widehat{H}'$ is the perturbation operator and $|\Psi_0\rangle$ be the unperturbed wavefunction at $\lambda = 0$. Then

$$\begin{aligned}\left.\frac{dE}{d\lambda}\right|_{\lambda=0} &= \frac{\partial}{\partial\lambda}\left.\frac{\left\langle\Psi_\lambda|\widehat{H} + \lambda\widehat{H}'|\Psi_\lambda\right\rangle}{\langle\Psi_\lambda|\Psi_\lambda\rangle}\right|_{\lambda=0} \\ &= 2Re\left\langle\left.\frac{\partial\Psi_\lambda}{\partial\lambda}\right|_{\lambda=0}|\widehat{H} - E_0|\Psi_0\right\rangle + \left\langle\Psi_0|\widehat{H}'|\Psi_0\right\rangle \\ &= \left\langle\Psi_0|\widehat{H}'|\Psi_0\right\rangle\end{aligned} \tag{10.3.7}$$

Thus, the first-order change in the energy due to a perturbation is equal to the expectation value of the perturbation operator $\widehat{H}'$.

For the H-F theorem to hold, the approximate wavefunctions should be optimized with respect to the changes induced by the perturbation.

$$|\Psi\rangle \rightarrow |\Psi\rangle + \lambda\left|\frac{\partial\Psi}{\partial\lambda}\right\rangle \tag{10.3.8}$$

It may also be noted that for wavefunctions that are not completely optimized with respect to all the parameters, as for example in configuration interaction, Möller–Plesset and coupled-cluster calculations, the H-F theorem does not hold, and first-order property calculated as expectation value of the perturbation from the perturbation theory shall not be identical to that obtained as an energy derivative from the H-F theorem. Since, H-F theorem holds for an exact wavefunction, the difference between the two values becomes smaller as the quality of the approximate wavefunction increases. However, for practical applications the difference is not negligible.

The practical importance of the H-F theory is that while energy derivatives are difficult to compute manually, especially in case of many-body Hamiltonian, the H-F force can be efficiently computed provided Hurley's condition given by Eq. (10.3.9) is satisfied. Hurley assumed dependence of the approximate wavefunction Ψ on λ through some parameters $\alpha_1, \alpha_2, \ldots\alpha_n$. Thus, for example, if the Hamiltonian contains nuclear charge or coordinates given by parameter λ explicitly, then the dependence of wavefunctions on these parameters is implicit. The wavefunction can be expressed as a linear combination of atomic orbitals (basis functions) whose coefficients now represent the parameters α_1, $\alpha_2, \ldots\alpha_n$. According to Hurley's conditions

$$\frac{\partial\alpha_i}{\partial\lambda} = 0 \tag{10.3.9}$$

that is, the basis functions coefficients do not depend on λ. When the parameters λ are nuclear position vectors, such basis sets are known as "floating sets"; for example, in solid state applications, the commonly used plane waves are such sets. The H-F theorem for many-body Hamiltonian is also valid if the derivatives of the basis functions with respect to λ are also part of the basis set.

H-F theorem is not valid for a general time-dependent wavefunction satisfying the time-dependent Schrödinger equation. However, in this case the following identity holds

$$\left\langle \Psi_\lambda(t) \left| \frac{\partial H(\lambda)}{\partial \lambda} \right| \Psi_\lambda(t) \right\rangle = i\hbar \frac{\partial}{\partial t} \left\langle \Psi_\lambda(t) \left| \frac{\partial \Psi_\lambda(t)}{\partial \lambda} \right. \right\rangle \tag{10.3.10}$$

$$\text{for } i\hbar \frac{\partial \Psi_\lambda(t)}{\partial t} = H(\lambda)\Psi_\lambda(t) \tag{10.3.11}$$

The proof of this identity is straightforward and relies on the assumption that partial derivatives with respect to λ and t can be interchanged.

10.4 Hellmann-Feynman Electrostatic Theorem

The most common application of the H-F theorem is for the calculation of intramolecular forces in molecules which allow the calculation of equilibrium geometries. At the equilibrium geometry the forces acting upon the nuclei due to the electrons and other nuclei vanish. Hellmann and Feynman independently applied the H-F theorem given by Eq. (10.3.6) to the case when the parameter λ corresponds to the nuclear coordinates. In a molecule containing N electrons and M nuclei, for a fixed nuclear configuration under the Born–Oppenheimer approximation, the Schrödinger equation can be written as,

$$\hat{H}\Psi_e = \left(\widehat{T_e} + \hat{V}\right)\Psi_e \tag{10.4.1}$$

where $\widehat{T_e}$ is the kinetic energy and $\hat{V}$ is the potential energy of the electrons. The potential energy consist of three parts,

$$V = V_{ee} + V_{nn} + V_{ne} \tag{10.4.2}$$

where $V_{ee} = \sum_i^N \sum_{j>i}^N \frac{1}{r_{ij}}$, is the electron–electron repulsion term and is independent of the nuclear coordinates.

$$V_{nn} = \sum_{\alpha=1}^{M} \sum_{\beta>\alpha}^{M} \frac{Z_\alpha Z_\beta}{|R_\alpha - R_\beta|}, \tag{10.4.3}$$

is the nuclear–nuclear repulsion term and

$$V_{ne} = -\sum_{i=1}^{N} \sum_{\alpha=1}^{M} \frac{Z_\alpha}{|r_i - R_\alpha|} \tag{10.4.4}$$

is the electron–nuclear attraction term.

Here, z_α and z_β are the charges of the nuclei located at points R_α and R_β with Cartesian coordinates $(x_\alpha, y_\alpha, z_\alpha)$ and $(x_\beta, y_\beta, z_\beta)$ while the position coordinate of the electron i is $r_i = (x_i, y_i, z_i)$. The Hamiltonian can therefore be written as

$$H = \widehat{T}_e + \widehat{V}_{ee} + \sum_{\alpha=1}^{M}\sum_{\beta>\alpha}^{M}\frac{Z_\alpha Z_\beta}{|R_\alpha - R_\beta|} - \sum_{i=1}^{N}\sum_{\alpha=1}^{M}\frac{Z_\alpha}{|r_i - R_\alpha|} \tag{10.4.5}$$

Since the x-component of the force acting on the a nucleus, say, δ is $F_{X_\delta} = -\frac{\partial E}{\partial X_\delta}$, using the H-F theorem (10.3.6) we get,

$$F_{X_\delta} = -\frac{\partial E}{\partial X_\delta} = -\left\langle \Psi_e \middle| \frac{\partial H}{\partial X_\delta} \middle| \Psi_e \right\rangle \tag{10.4.6}$$

A simple differentiation of Eq. (10.4.5) with respect to X_δ leads to,

$$\frac{\partial H}{\partial X_\delta} = Z_\delta \sum_{i=1}^{N} \frac{x_i - X_\delta}{|r_i - R_\delta|^3} - Z_\delta \sum_{\alpha\neq\delta}^{M} Z_\alpha \frac{X_\alpha - X_\delta}{|R_\alpha - R_\delta|^3} \tag{10.4.7}$$

Substituting the value of $\frac{\partial H}{\partial X_\delta}$ from Eq. (10.4.7) into Eq. (10.4.6), we get,

$$F_{X_\delta} = -\left\langle \Psi_e \middle| Z_\delta \sum_{i=1}^{N} \frac{x_i - X_\delta}{|r_i - R_\delta|^3} - Z_\delta \sum_{\alpha\neq\delta}^{M} Z_\alpha \frac{X_\alpha - X_\delta}{|R_\alpha - R_\delta|^3} \middle| \Psi_e \right\rangle \tag{10.4.8}$$

Since the second term in this equation does not involve electronic coordinates and the electronic wavefunction Ψ_e is normalized, i.e., $\langle \Psi_e | \Psi_e \rangle = 1$, we can write it as

$$F_{X_\delta} = -\left\langle \Psi_e \middle| Z_\delta \sum_{i=1}^{N} \frac{x_i - X_\delta}{|r_i - R_\delta|^3} \middle| \Psi_e \right\rangle + Z_\delta \sum_{\alpha\neq\delta}^{M} Z_\alpha \frac{X_\alpha - X_\delta}{|R_\alpha - R_\delta|^3} \tag{10.4.9}$$

In the integral form, this equation may be written as

$$F_{X_\delta} = -\sum_{i=1}^{N} Z_\delta \int |\Psi_e|^2 \frac{x_i - X_\delta}{|r_i - R_\delta|^3} d\tau_i + Z_\delta \sum_{\alpha\neq\delta}^{M} Z_\alpha \frac{X_\alpha - X_\delta}{|R_\alpha - R_\delta|^3} \tag{10.4.10}$$

The factor multiplying $|\Psi_e|^2$ in the integral term in this equation depends only on the coordinates of electron i, whereas the integral implies a summation over all spin variables as well as integration over all space coordinates. The first term in Eq. (10.4.10) can therefore be written as,

$$-\sum_{i=1}^{N} \int Z_\delta \left[\sum_{\text{all } m_s} \int' |\Psi_e|^2 d\tau'_e \right] \frac{x_i - X_\delta}{|r_i - R_\delta|^3} d\tau_i \tag{10.4.11}$$

Here, $\int' d\tau'_e$ means integration over the $3N-3$ spatial coordinates of all electrons except electron i, $d\tau_i$ refers to coordinates of the ith electron and the quantity in the square brackets is equal to $\frac{\rho}{N}$, where ρ is the electron probability density of an N-electron molecule. So, Eq. (10.4.10) can be written as

$$F_{X_\delta} = -\sum_{i=1}^{N} Z_\delta \frac{1}{N} \int \rho(x,y,z) \frac{x_i - X_\delta}{|r_i - R_\delta|^3} d\tau_i + Z_\delta \sum_{\alpha\neq\delta}^{M} Z_\alpha \frac{X_\alpha - X_\delta}{|R_\alpha - R_\delta|^3} \tag{10.4.12}$$

Due to the indistinguishability of the electrons, all the integration terms in the first summation will have the same value. Since i has N values, the total sum will be N times the value of a single term and hence Eq. (10.4.12) will reduce to

$$F_{X_\delta} = -Z_\delta \int \rho(x,y,z)\frac{x - X_\delta}{|r - R_\delta|^3}d\tau + Z_\delta \sum_{\alpha \neq \delta}^{M} Z_\alpha \frac{X_\alpha - X_\delta}{|R_\alpha - R_\delta|^3}$$

or,

$$F_{X_\delta} = -Z_\delta\left[\int \rho(\mathbf{r})\frac{x - X_\delta}{|r - R_\delta|^3}d\tau - \sum_{\alpha \neq \delta}^{M} Z_\alpha \frac{X_\alpha - X_\delta}{|R_\alpha - R_\delta|^3}\right] \tag{10.4.13}$$

where the superfluous subscript i has been dropped. $(r - R_\delta)$ is the distance between the electron at (x, y, z) and the nucleus δ and $R_\alpha - R_\delta$ is the distance between the two nuclei α and δ.

The net force on a nucleus, therefore, is simply a resultant of the attractive forces due to the surrounding electronic distribution (first term) and the repulsive force due to other nuclei (second term). This equation is sometimes called the electrostatic H-F theorem. If the electron charge density $\rho(\mathbf{r})$ can be found by solving the Schrödinger equation, one can calculate the effective force on a nucleus from simple electrostatics. We shall consider the application of (H-F) electrostatic theorem to the case of a diatomic molecule.

10.5 Forces in a Diatomic Molecule and Physical Picture of Chemical Bond

Consider a diatomic molecule AB, with internuclear axis as the x-axis (Figure 10.2). If an electron cloud of charge density $-e\rho(x, y, z)$ is located at the point i, then for symmetry reasons, the y and z components of the effective forces on the two nuclei is zero and so the force will be acting on the nuclei along the x-axis only. From H-F theorem, the force of nucleus A along the x-axis will be

$$F_{Ax} = -\left\langle \Psi \middle| \frac{\partial H}{\partial X_A} \middle| \Psi \right\rangle \tag{10.4.14}$$

Since, $\hat{H} = \hat{T} + \hat{V}$ and the kinetic energy operator is independent of the nuclear coordinates, we can write

$$F_{Ax} = -\left\langle \Psi \middle| \frac{\partial V}{\partial X_A} \middle| \Psi \right\rangle \tag{10.4.15}$$

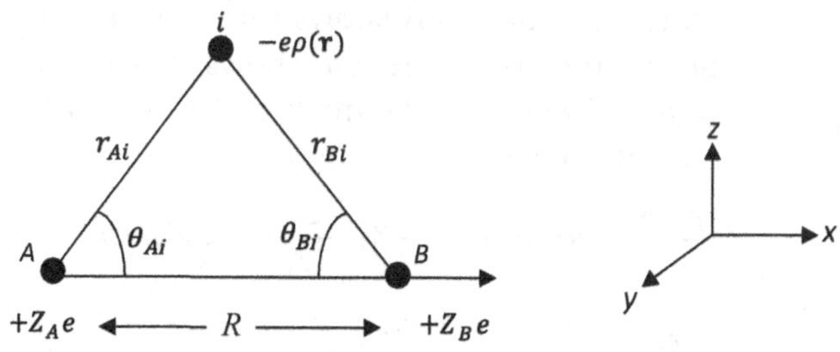

FIGURE 10.2 Electron and nuclear coordinates in diatomic molecule.

where $V = V_{ee} + V_{nn} + V_{en}$.

Since V_{ee} is independent of nuclear coordinates, we can write

$$F_{Ax} = -\left\langle \Psi \left| \frac{\partial}{\partial X_A}(V_{nn} + V_{en}) \right| \Psi \right\rangle \tag{10.4.16}$$

The force on nucleus A can then be calculated by using Eq. (10.4.13).

Let the coordinates of the point i where charge $-e\rho(\mathbf{r})$ is located be (x_i, y_i, z_i) and the coordinates of the nuclei A and B be $(X_A, 0, 0)$ and $(X_B, 0, 0)$. Then, from the Figure 10.2 the internuclear separation $R = X_B - X_A$.

Distance between the electron cloud at i and the nucleus A is $\mathbf{r}_{Ai} = \mathbf{r}_i - \mathbf{r}_A$, the distance between the electron cloud at i and the nucleus B, $\mathbf{r}_{Bi} = \mathbf{r}_i - \mathbf{r}_B$, and the internuclear distance $R = R_B - R_A$.

Equation (10.4.13) can therefore be written as

$$F_{Ax} = -Z_A\left[\int \rho \frac{x_i - X_A}{|r_i - r_A|^3} d\mathbf{r}_i - \frac{Z_B(X_B - X_A)}{|R_B - R_A|^3}\right] = -Z_A\left[\int \rho \frac{x_i - X_A}{r_{Ai}^3} d\mathbf{r}_i - \frac{Z_B}{R^2}\right] \tag{10.4.17}$$

But, from Figure 10.2, $\dfrac{x_i - X_A}{r_{Ai}} = \cos\theta_{Ai}$.

Hence,

$$F_{Ax} = -Z_A \int \frac{\rho \cos\theta_{Ai}}{r_{Ai}^2} d\mathbf{r}_i + \frac{Z_A Z_B}{R^2} \tag{10.4.18}$$

Similarly, we can find that

$$F_{Bx} = Z_B \int \frac{\rho \cos\theta_{Bi}}{r_{Bi}^2} d\mathbf{r}_i + \frac{Z_A Z_B}{R^2} \tag{10.4.19}$$

From Eq. (10.4.18) and Figure 10.2, the component of the effective force on nucleus A along the bond AB due to the electronic charge of density $-e\rho(x, y, z)$ in the region about the point (x_i, y_i, z_i) is

$$+Z_A \frac{\rho \cos\theta_{Ai}}{r_{Ai}^2} dx_i dy_i dz_i \tag{10.4.20}$$

and the component of the effective force on the nucleus B along the bond AB due to this charge is

$$-Z_B \frac{\rho \cos\theta_{Bi}}{r_{Bi}^2} dx_i dy_i dz_i \tag{10.4.21}$$

A positive value of the force on nucleus A shows that it is acting along $+x$ direction while the negative value of the force on nucleus B shows that it is acting in the $-x$ direction. The electronic charge at i will draw the nucleus A closer to B, when the force on A is greater than the force on B. Hence

$$\begin{aligned} &Z_A \frac{\rho \cos\theta_{Ai}}{r_{Ai}^2} dx_i dy_i dz_i > -Z_B \frac{\rho \cos\theta_{Bi}}{r_{Bi}^2} dx_i dy_i dz_i \\ &\text{or, } Z_A \frac{\cos\theta_{Ai}}{r_{Ai}^2} + Z_B \frac{\cos\theta_{Bi}}{r_{Bi}^2} > 0 \end{aligned} \tag{10.4.22}$$

The left side of this equation represents the net binding force ($f_{binding}$) on the nuclei. The region where Eq. (10.4.22) is satisfied is called the binding region. Although, to obtain net binding we need to sum Eq. (10.4.22) over the entire electron distribution, which can be obtained by solving Schrödinger wave equation for the molecule, yet some important insight into the nature of forces can be obtained from Eq. (10.4.22).

Thus,

1. If the point i, where the electronic charge is located, falls between the two nuclei, then $\theta_{Ai} < \frac{\pi}{2}$ and $\theta_{Bi} < \frac{\pi}{2}$. Both the terms in Eq. (10.4.22) are positive and hence the net binding force is greater than zero. The electrons located near the point i, therefore, tend to pull the two nuclei toward each other and the region between the two nuclei is a binding region.
2. If the point i lies beyond the nuclei on either side, say for example, behind nucleus B, Then $\theta_{Ai} < \frac{\pi}{2}$ but $\theta_{Bi} > \frac{\pi}{2}$. Also, r_{Ai} will be greater than r_{Bi} and so, if $Z_A = Z_B$, the binding force will be negative. The electronic charge at the point i thus pulls nucleus A toward B, but B away from A. Obviously, the electronic charge that lies behind the nuclei exerts a greater attraction on the nucleus that is closer to it than on the other nucleus and thus tends to pull the two nuclei apart leading to antibonding. Same will be true, if the electronic charge lies behind nucleus A. Such a region, where the net binding force is negative is called the antibinding region.
3. θ_{Ai} and θ_{Bi} may also take values such that binding force

$$f_{binding} = \frac{Z_A \cos\theta_i}{r_{Ai}^2} + \frac{Z \cos\theta_{Bi}}{r_{Bi}^2} = 0 \tag{10.4.23}$$

Surfaces that satisfy this equation are called the nonbinding surfaces. Such boundary surfaces divide the space into binding and antibinding regions. These nonbinding surfaces are hyperbolic contours and any charge residing on them will

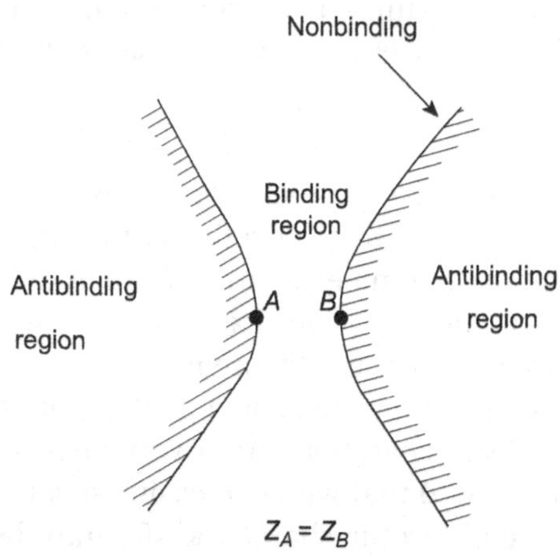

FIGURE 10.3 Berlin's binding and antibinding regions in a homonuclear diatomic molecule.

not lead to attraction or repulsion between the nuclei. The binding, antibinding regions, and the nonbinding surfaces for a homonuclear diatomic molecule are shown in the Figure 10.3.

The concept of binding and antibinding was introduced by Berlin [5] who also differentiated binding from bonding. According to Berlin the word binding relates to the forces acting on the nuclei in the molecule. Thus, binding can be distinguished from bonding which is usually related to the energy of the molecule. The bonding by a single electron is related to the energy of this electron in the molecule; in the same sense the binding by a single electron will be related to the forces exerted by this electron on the nuclei. In the molecular orbital theory, the bonding and antibonding electrons are distinguished on the basis of the criterion of lowering or rising, respectively, of the energy level of a given electron in transition from separated atoms to the molecule. Applying the consideration of the forces exerted by an electron on the nuclei in an analogous manner, we shall say that an electron moves predominantly in the binding or antibinding region when $\int f_{\text{binding}}\rho d\mathbf{r}$ is positive or negative, respectively. The quantity, the charge density of the single electron, is a function of the internuclear distance. The sign of $\int f_{\text{binding}}\rho d\mathbf{r}$ more clearly defines the binding region between the nuclei and the antibinding region beyond the nuclei.

10.6 Charge Density Maps

A described in Chapter 6 (Section 6.4.2), a visual picture of the shapes and sizes of the molecules is conveniently obtained from contour maps of total electron density $\rho(\mathbf{r})$. The contour lies in a plane containing the nuclei and the outermost contour includes over 95% of the electronic charge. Likewise, in order to get a detailed picture of the net charge redistribution resulting from the formation of a chemical bond, Bader et al. [6] plotted the density difference maps ($\Delta\rho$) for different molecules. These were obtained by subtracting the densities of the unperturbed constituent atoms, placed at appropriate positions, from the molecular density evaluated at the same distance

$$\Delta\rho(\mathbf{r}) = \rho_{\text{mol}}(\mathbf{r}) - \sum_{\text{atoms}} \rho_{\text{atom}}(\mathbf{r})$$

where the first term on the right side is molecular electron density at a point $\mathbf{r}$ in space and the second term is that obtained by the overlap of the unperturbed atomic densities. The overall charge distribution in a molecule may be analyzed in terms of charge found in different regions of space and the forces exerted by these charges on the nuclei in accordance with the Hellmann-Feynman theorem.

Thus, in the case of hydrogen molecule, Bader et al. [7] plotted the density difference maps for different internuclear separations from very large separation of 8 au to very small separation of 1 au and found that with decreasing separation, the atomic densities get so much distorted due to electronic charge transfer into the binding region that the individual character of the atomic densities is lost (Figure 10.4). At 2.1 au the electron

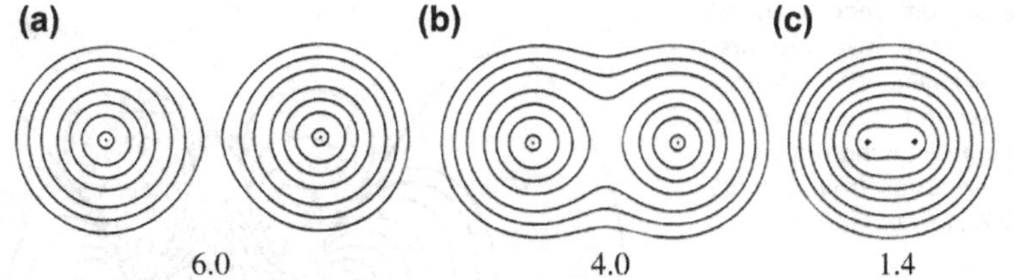

FIGURE 10.4 Electron density contour maps illustrating the changes in the electron charge distribution during the approach of two H atoms to form H_2. Internuclear distance (a) 6.00 au (b) 4.0 au (c) 1.4 au.

density transfer is maximum and is confined to the internuclear binding region. At this distance the attractive force on the nuclei is also maximum. At separations less than 2.0 au, an increasing amount of charge transfer takes place in the antibinding region behind each nucleus, leading to increasing nuclear repulsion and the corresponding decrease in the attractive force. At 1.4 au the attraction and repulsive forces on the nuclei balance each other and equilibrium is reached resulting in the formation of a chemical bond. At shorter distances, a greater buildup of charge in the antibinding region takes place resulting in increasing nuclear repulsion. This pattern is also seen in the potential energy curve for the hydrogen molecule (Figure 10.1) obtained using the virial theorem. However, it may be noted that while using H-F force concept the chemical bonding is explained solely on the basis of potential energy, the virial concept takes both the kinetic and potential energy into account. Thus, the H-F electrostatic theorem ignores the role of kinetic energy which, as we have seen earlier, plays an important role in bond formation.

Bader et al. [8] have plotted the contour map and density difference maps for N_2 shown in Figures 10.5 and 10.6. A buildup of the electronic charge up to 0.25*e* over that of

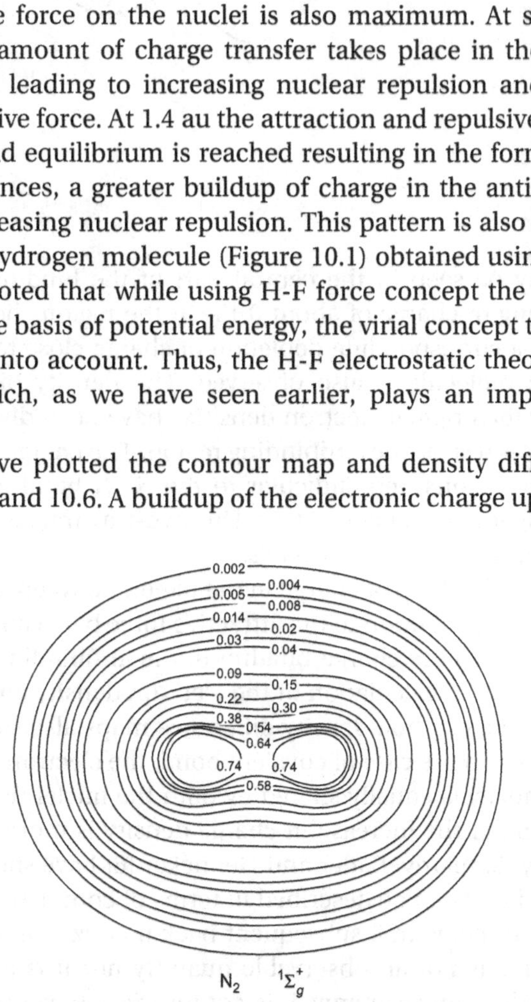

FIGURE 10.5 Total molecular charge density maps for the ground states of N_2. The innermost, circular contours centered on the nuclei have been omitted for clarity. *Reprinted with permission from Ref. [8]. Copyright @ 1967, AIP Publishing.*

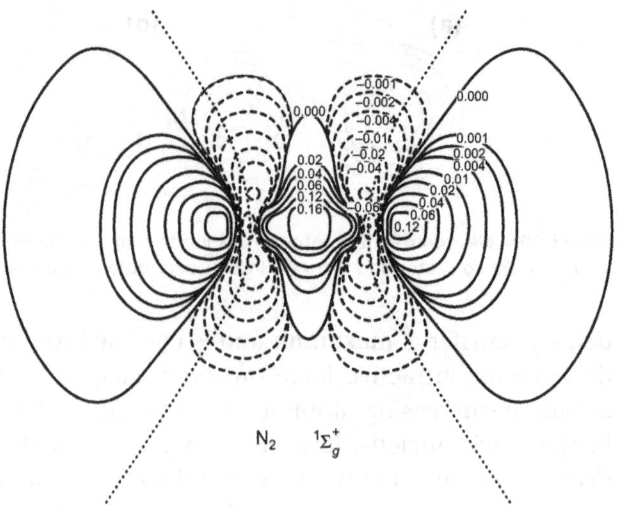

FIGURE 10.6 Density difference maps for N_2. The solid lines are positive contours, the dashed lines are negative contours. Dotted lines (shown in full) separate the binding from the antibinding regions. *Reprinted with permission from Ref. [8]. Copyright @ 1967, AIP Publishing.*

separated atoms may be seen in the central part of the binding region, as expected. Besides this, a building of charge of about 0.13e in the regions behind the nuclei (antibinding region) with a corresponding depletion of charge close to the nuclei and in the outer regions of the molecule is also observed. The density difference map ($\Delta\rho_{map}$) therefore shows that the original electron densities have been distorted to move charge in the binding region as well as the antibinding region. The charge density of the atoms is found to decrease in regions perpendicular to the N–N bond at the positions of the nuclei (shown by dashed line in Figure 10.6). This two-way transfer of charge is typical of increased participation of p_σ atomic orbitals.

In the case of N_2, the buildup of charge in the region between the two nuclei (binding region) is such that its contribution to the attractive force between the nuclei exceeds the H-F repulsion force due to the charge buildup in the antibinding region. The $\Delta\rho$ maps are sensitive to errors in the calculation of the electron density and hence on the nature of the wavefunction. The electron density difference maps also shed light on the nature of chemical bond. Thus, in the case of covalent bond, the electron density increase in the region between the nuclei is equally shared by both the nuclei (Figure 10.6), whereas in the case of an ionic bond the increase in charge density is more toward the atom with high electronegativity. Similarly, Bader and Henneker [6] have shown that in the case of LiF, the formation of bond can be described in terms of complete transfer of an electron from lithium onto fluorine, with a subsequent back polarization toward the lithium.

Unlike energy, force is not an observable quantity nor it is a quantum mechanical constant of motion. Also, unlike energy, it is not invariant to coordinate transformation. For these reasons the force method has received little attention in chemistry and its application has been confined basically to the ground state problems. The force concept based on the H-F electrostatic theorem, however, provides a perspective for the electronic origins of nuclear rearrangement processes on potential energy hypersurfaces. Honda and Nakatsuji [9,10,10a] have developed a force and density concept called electrostatic force (ESF) or H-F force theory for molecular geometries and chemical reactions which provides an intuitive and useful chemical concept for predicting the geometries of molecules in ground and excited states. This concept has also been extended to studies of the geometries of molecules in an external electric field.

Ventra and Pantelides [11] have, however, presented a rigorous derivation of a general H-F theorem that applies to all quantum mechanical systems with square integrable wavefunctions and reduces to well-known results for ground state problems. The generalized treatment of the H-F theorem is also applicable to problems related to molecules and solids in time-dependent external fields and to transport in nano structures and molecules.

References

[1] J.C. Slater, J. Chem. Phys. 1 (1933) 687.

[2] P.O. Löwdin, J. Mol. Spect. 3 (1959) 46.

[3] J.I. Rodriguez, P.W. Ayers, A.W. Götz, F.L. Castillo-Alvarado, J. Chem. Phys. 131 (2009) 021101.

[4] M.J. Feinberg, K. Ruedenberg, J. Chem. Phys. 54 (1971) 1495.

[5] T. Berlin, Chem. Phys. (1951) 208.

[6] R.F.W. Bader, W.H. Henneker, J. Am. Chem. Soc. 87 (1965) 3063.

[7] R.F. Bader, Lecture Notes. http://www.chemistry.mcmaster.ca/esam/credits.html.

[8] R.F. Bader, W.H. Henneker, P.E. Cade, J. Chem. Phys. 46 (1967) 3341.

[9] Y. Honda, H. Nakatsuji, Chem. Phys. Lett. 293 (1998) 230.

[10] H. Nakatsuji, J. Am. Chem. Soc. 95 (1973) 345;
[a] H. Nakatsuji, J. Am. Chem. Soc. 95 (1973) 354.

[11] M.D. Ventra, S.T. Pantelides, Phys. Rev. B61 (2000).

Further Reading

[1] B.M. Deb, The Force Concept in Chemistry, Van Nostrand-Reinhold, New York, 1981.

[2] I.N. Levine, Quantum Chemistry, fourth ed., Prentice Hall, Englewood, 1991.

[3] A.K. Chandra, Introductory Quantum Chemistry, McGraw Hill Publishing Co., New Delhi, 2003.

[4] R.F.W. Bader, Atoms in Molecules – A Quantum Theory, Oxford University Press, Oxford, UK, 1990.

11

Topological Analysis of Electron Density—Quantum Theory of Atoms in Molecules

CHAPTER OUTLINE

11.1 Introduction

Few ideas in chemistry have been as successful as the view that molecules consist of atoms held together by chemical bonds. A lot of work has therefore been devoted to the concept of the atoms in molecules (AIM). According to Parr and Yang [1] AIM is a noumenon, that is it is only an imaginary concept or an object of intellectual intuition

Principles and Applications of Quantum Chemistry. http://dx.doi.org/10.1016/B978-0-12-803478-1.00011-X

and so, despite its utility, neither one can directly observe an atom in a molecule by experiment nor can one measure enough properties of an AIM to define it unambiguously. An alternative physical observable that has been used to define an atom is the partial atomic charge or the electron density. In X-ray crystallography, the electron density is measured directly and may be defined experimentally by comparison to, say, spherically symmetric neutral atoms. Within the large set of different definitions of the AIM, two main groups of methods can be identified to define an atom. The first one describes an atomic density on the basis of the atom-centered basis functions. The most famous among these is the Mulliken technique. In Mulliken population analysis, the electron density is distributed equally among the atoms. This partitioning, as we noted earlier, is totally arbitrary. In another method, Bader and coworkers [2] have adopted an alternative way to partition the electrons between the atoms in a molecule. This theory, known as the quantum theory of atoms in molecules (QTAIM) is perhaps the most theoretically sound of all population analysis methods, mainly because it relies on the properties of the electron density alone and not on the basis set or integration grids. It treats the point charges determination as a mathematical problem and is able to answer some fundamental questions, namely: How to define the atom within a molecule? Is there a natural way to distribute the electrons in a molecule? Is it possible to identify atomic space within the molecular volume or draw lines separating atoms in a molecule?

The development of chemistry owes a lot to the observation that some properties attributed to atoms and functional groups are transferable from one molecule to another. These observations are based on thermodynamic and spectroscopic properties of molecules and provide a basis for group additivity schemes. The question, however, arises about the electronic basis of this empirical transferability. An answer to these questions has also been provided by QTAIM which relies on quantum observables such as electron density $\rho(\mathbf{r})$ and energy densities. An atom in QTAIM is defined as a proper open system that can share electron density and momentum with other atoms in a localized three-dimensional space. Quantum description of an open system is in agreement with the molecular structure hypothesis according to which a molecule is a collection of atoms each having a characteristic set of properties and linked by a network of bonds. In this chapter, we shall try to understand some basic concepts of QTAIM from the viewpoint of its applications to structural issues instead of developing its mathematical theory which is quite involved.

11.2 Topological Analysis of Electron Density

Matter is a distribution of charge in real space; point like nuclei are embedded in the diffuse density of electronic charge, $\rho(\mathbf{r})$. All properties of matter become apparent in the charge distribution, its topology that delineates atoms and the bonding between them. The electron density $\rho(x, y, z)$ is a function of three spatial coordinates and may be analyzed in terms of its topology (maxima, minima, and saddle points). Since nuclei of the atoms are the only source of positive charge, the electron density in a large majority of cases has its maxima at or near the nuclei. The nuclei act as attractor of electrons. In between the two attractors, i.e., atoms, the electron density is determined by the balance of forces which the electron experiences. The change in density is expressed in terms of a gradient vector. The gradient of a scalar function such as density $\rho(\mathbf{r})$ at a point in space is a vector pointing in the direction in which $\rho(\mathbf{r})$ undergoes the greatest rate of increase. The magnitude of the vector is equal to the rate of increase in that direction. So,

$$\nabla\rho = \mathbf{i}\frac{\partial\rho}{\partial x} + \mathbf{j}\frac{\partial\rho}{\partial y} + \mathbf{k}\frac{\partial\rho}{\partial z} \tag{11.2.1}$$

where **i**, **j**, **k** are unit vectors along the x, y, and z axis, respectively. If we follow a path outward from the nucleus, where the electron density is a maximum, then the direction that we are taking is opposite to the gradient of the density. Two possibilities can now arise. Either we shall proceed outward from the molecule indefinitely with density dropping exponentially but never becoming absolute zero or else, at some point between the atoms the gradient will reach a zero value because it passes from a negative value (constantly decreasing) to a positive value (increasing) as we move toward the other nucleus. Thus, two neighboring maxima are connected by a pass called the maximum electron density (MED) path. The electron density along this path is higher than in any other neighboring path. The minimum of an MED is especially important because at this point the density gradient (∇) has a zero value and it represents the minimum electron density point between the atoms, as shown in Figure 11.1(a). In some cases, however, the two peaks are not necessarily connected by an MED path, as in the case of the hydrogen atoms in oxirane Figure 11.2. The points at which the density gradient (∇) has a zero value are called "critical points" (CPs) and in QTAIM theory it is their existence that defines whether a bond between two atoms exists or not. Thus, in general

$$\begin{aligned}\nabla\rho(\mathbf{r}) &= 0 \text{ at critical points and at } \infty \\ &\neq 0 \text{ at all other points.}\end{aligned} \tag{11.2.2}$$

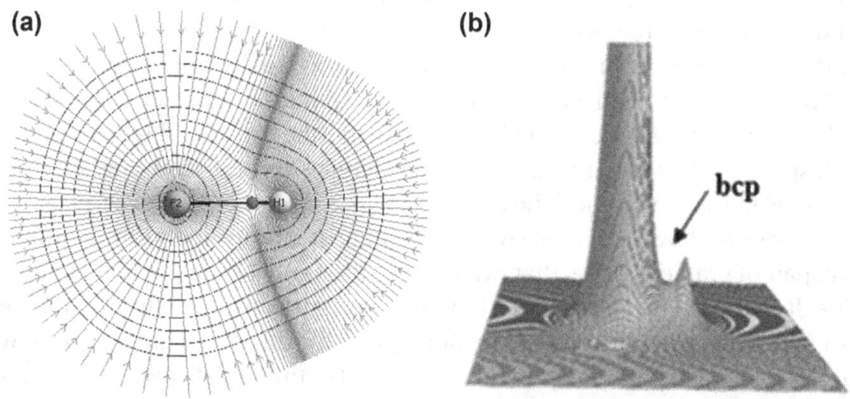

FIGURE 11.1 (a) A 2-D view of the topology of electron density of heteronuclear diatomic molecule HF, (b) Charge density display in the form of a relief map. The arrow indicates the bond critical point.

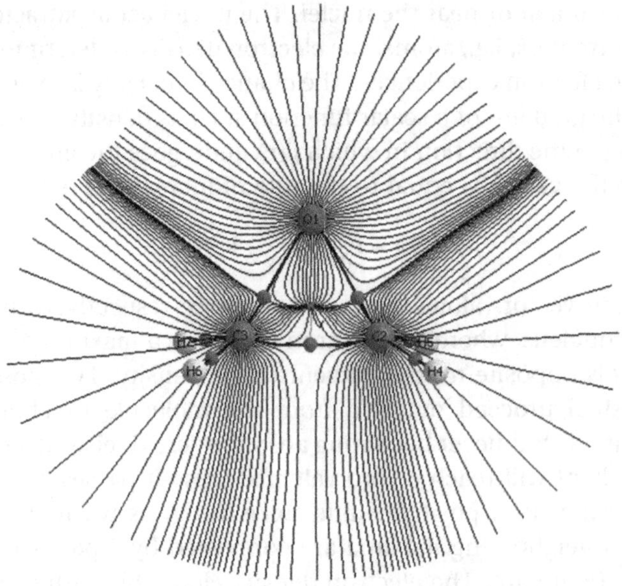

FIGURE 11.2 Gradient vector field map for oxirane showing trajectories which originate at infinity and terminate at the nuclei and the pairs of gradient paths which originate at each (3,−1) critical point and define the atomic interaction lines.

where the zero vector signifies that all components of the gradient, and not only their sum, are equal to zero.

A CP, as defined by Eq. (11.2.2), may correspond to a local maxima, local minima, or a saddle point. Thus, the maximum at the nucleus constitutes one type of CP called the nuclear critical point (NCP) and the CP between the atoms corresponds to a local minima called the bond critical point (BCP). In a formal mathematical sense NCP is not a true CP, because $\nabla\rho$ at the position of the nucleus is not defined, but topologically it does behave as a CP. The exact nature of the CP can, however, be determined by using the second derivative of the electron density ($\nabla^2\rho(\mathbf{r})$).

A two-dimensional view of the topology of electron density of heteronuclear diatomic molecule HF is shown in Figure 11.1(a). The bigger spheres are nuclei. It shows the trajectories of the gradient vector fields of electron density (the arrows), the bond path, and the BCP (small spheres). The contours of the charge density are superimposed on the gradient vector field of electron density and the basin in a plane containing the two nuclei. Two types of gradient fields are shown: (1) those which originate at infinity and terminate at the nuclei and (2) the pair of gradient paths that originate at BCP and define the line of atomic interaction. The dark line labeled S is a slice of the zero flux surface that defines spatial region for each atom. The MED path, also called the bond path, for this molecule is also shown in the Figure 11.1(a). The bond path provides an operational definition of chemical bonding which can be directly related to experimental observations. The charge density display in the form of relief map is shown in Figure 11.1(b), which also shows the position of the BCP.

Using the topology of electron density, QTAIM divides molecular space into atomic subspaces. Starting from a given point in space one may move in infinitesimal steps along the direction of the gradient until an attractor (nucleus) is encountered. The part of space from which all gradient paths end up at the same nucleus is called the basin of the given atom (Figure 11.1(a)). The molecular space is thus divided into separate mononuclear regions of volume Ω, the atomic basins. The border between two three-dimensional atomic basins identify atoms in the molecules. This border is a two dimensional zero flux surface. Thus, in the case of HF (Figure 11.1), we get two atomic basins associated with the nuclei of hydrogen and fluorine. Once the molecular volume is divided up, the electron density is integrated within each of the atomic basins and the atomic charges, dipoles, and multipoles can be determined.

In QTAIM, a molecule is uniquely divided into a set of atomic volumes. The zero flux surfaces in the gradient vector field of the electron density are not crossed by any of the gradient vectors $\nabla\rho(\mathbf{r})$ at any point. Mathematically, this may be written as

$$\nabla\rho(\mathbf{r}) \cdot n(\mathbf{r}) = 0 \tag{11.2.3}$$

where $\mathbf{r}$ is the position vector and $n(\mathbf{r})$ is a unit vector normal to the surface $S(\Omega)$. This relationship is true for all $\mathbf{r}$ belonging to the surface $S(\Omega)$. Such points clearly and unambiguously separate different parts of space, providing an appropriate definition for the atoms in a molecule. According to QTAIM, atoms are represented by centers of attraction of gradient paths and their neighboring basins.

The division of molecular volume into atomic basins follows from an analysis of the principle of stationary action [3]. This partitioning does take electronegativity into account. The shapes of the atomic basins, and the associated electron densities, in a given functional group are very similar in different molecules. This rationalizes the basis of organic molecules, i.e., functional groups react similarly in different molecules.

11.3 Hessian Matrix and Laplacian of Density

While the condition $\nabla\rho = 0$ (Eq. (11.2.2)) tells about the existence of a CP, it does not tell anything about its nature, i.e., whether it is a minimum, maximum, or a saddle point. In order to determine the nature of the CP, one has to take recourse to the second derivatives of the electron density. In three-dimensional space there can be nine second derivatives which can be arranged in a square matrix form, the so-called "Hessian matrix" or simply Hessian. For a CP located at r_c, this matrix may be written as

$$A(r_c) = \begin{pmatrix} \dfrac{\partial^2\rho}{\partial x^2} & \dfrac{\partial^2\rho}{\partial x\partial y} & \dfrac{\partial^2\rho}{\partial x\partial z} \\[2ex] \dfrac{\partial^2\rho}{\partial y\partial x} & \dfrac{\partial^2\rho}{\partial y^2} & \dfrac{\partial^2\rho}{\partial y\partial z} \\[2ex] \dfrac{\partial^2\rho}{\partial z\partial x} & \dfrac{\partial^2\rho}{\partial z\partial y} & \dfrac{\partial^2\rho}{\partial z^2} \end{pmatrix} \tag{11.3.1}$$

Since the Hessian is a real symmetric matrix, it can be diagonalized by carrying out a symmetry transformation of the type $\wedge = U^{-1}AU$, where U is a unitary matrix.

$$r(x', y', z') = U\, r(x, y, z) \tag{11.3.2}$$

The matrix U can rotate the coordinate system $r(x, y, z)$ to $r(x', y', z')$, where x', y', z' are the principal axes of curvature of the CP. These principal axes correspond to symmetry axes, if the CP lies on a symmetry element.

In the diagonalized form, Eq. (11.3.1) reduces to

$$\wedge = \begin{pmatrix} \frac{\partial^2 \rho}{\partial x'^2} & 0 & 0 \\ 0 & \frac{\partial^2 \rho}{\partial y'^2} & 0 \\ 0 & 0 & \frac{\partial^2 \rho}{\partial z'^2} \end{pmatrix}_{r'=r_c} = \begin{pmatrix} \lambda_1 & 0 & 0 \\ 0 & \lambda_2 & 0 \\ 0 & 0 & \lambda_3 \end{pmatrix} \tag{11.3.3}$$

where λ_1, λ_2, and λ_3 are the three eigenvalues of the Hessian matrix which represent curvature of the density with respect to the three principal axes x', y', z'.

A matrix has two important properties: (1) trace, which is the sum of the three diagonal terms and (2) the rank, which is the number of nonzero eigenvalues. In the case of Hessian matrix, the trace is invariant to rotation of the coordinate axes. In Eq. (11.3.3), the trace is,

$$\nabla^2 \rho(\mathbf{r}) = \frac{\partial^2 \rho(\mathbf{r})}{\partial x'^2} + \frac{\partial^2 \rho(\mathbf{r})}{\partial y'^2} + \frac{\partial^2 \rho(\mathbf{r})}{\partial z'^2}$$

Or, in general,

$$\nabla^2 \rho(\mathbf{r}) = \frac{\partial^2 \rho(\mathbf{r})}{\partial x^2} + \frac{\partial^2 \rho(\mathbf{r})}{\partial y^2} + \frac{\partial^2 \rho(\mathbf{r})}{\partial z^2}, \tag{11.3.4}$$

if we drop the primes on the principal axes, $\nabla^2 \rho(\mathbf{r})$ is the Laplacian of the density. If $\nabla^2 \rho(\mathbf{r}) \geq 0$, the CP is a minima and if $\nabla^2 \rho(\mathbf{r}) < 0$, it is a maxima. The Laplacian $\nabla^2 \rho(\mathbf{r})$ provides information on where the electron density is depleted or increased. Thus, a positive value of $\nabla^2 \rho(\mathbf{r})$ means local charge depletion—the electron density is depleted relative to the average distribution. A negative value of $\nabla^2 \rho(\mathbf{r})$ means local charge concentration—the region is tightly bound and compressed relative to its average distribution. A region of local charge depletion ($\nabla^2 \rho(\mathbf{r}) > 0$) also behaves as charge acceptor or as a Lewis acid (electron acceptor), whereas the region of local charge concentration ($\nabla^2 \rho(\mathbf{r}) < 0$) behaves as Lewis base (electron donor). This is shown in Figure 11.3 for benzene, where the (+)ve contours imply charge depletion (shown in blue (dark gray in print version)) and the (−)ve contours imply charge concentration (shown in red (gray in print version)). Often maps of $-\nabla^2 \rho(\mathbf{r})$, sometimes called as **L**, are drawn, where the (+) ve contours imply charge concentration and the (−)ve contours imply charge depletion.

At a BCP, the sign of the Laplacian has been used for characterizing nature of the bond. Thus, a negative value indicates a covalent bond, while a positive value indicates an ionic bond or a van der Waals interaction.

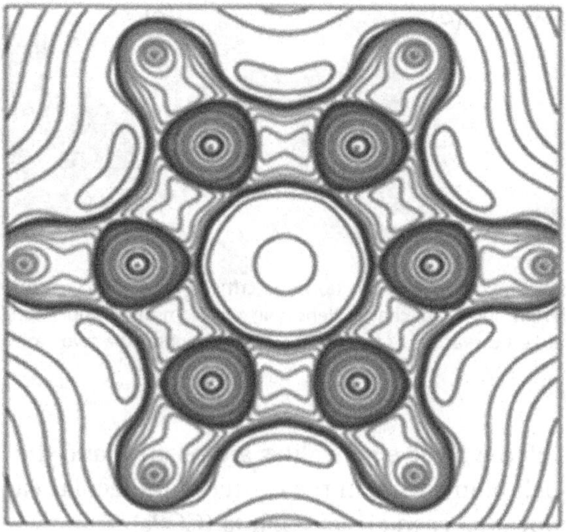

FIGURE 11.3 B3LYP/6-311G**-optimized structure of benzene showing regions of local charge concentration (red (gray in print versions)) and depletion (blue (dark gray in print versions)). *Courtesy Ref. [4].*

11.4 Critical Points

At a CP, the eigenvalues of the Hessian are all real and are generally nonzero. CPs are classified according to their rank (ω) and signature (σ) and are symbolized as (ω,σ). Rank is the number of nonzero curvatures (eigenvalues of the Hessian) and can have a maximum value of 3. Signature (σ) is the algebraic sum of the signs of the curvatures; each of the three curvatures (λ) contributes ± 1 depending on whether it is a positive curvature or a negative. A CP with $\omega < 3$ is mathematically unstable. The presence of such a CP indicates a change in the topology of the density and hence a change in the molecular structure. For a stable system, $\omega = 3$. For such systems, four kinds of CPs having (ω,σ) values between (3,−3) and (3,3) can be identified. These are the following:

(3,−3)—All the three eigenvalues are negative indicating that $\rho(\mathbf{r})$ is local maximum. This is the characteristic of the nuclei. The CP is therefore called NCP.

(3,−1)—One eigenvalue is positive and two are negative; the CP is a maximum from two perpendicular directions and a minimum from the third direction. This point corresponds to the minimum of the MED path. The electron density has a minimum value at this point in the direction of a chemical bond but maximum in other two directions. This point is called the BCP and is denoted by r_b (Figure11.4(a)).

(3,1)—Two eigenvalues are positive and one is negative. In this case the CP is a minimum in two perpendicular directions but a maximum in the third direction. Such a condition is found around the center of the ring in cyclic compounds. The CP is therefore called the ring critical point (RCP). The electron density at this point is a maximum from a direction perpendicular to the plane of the ring and is a minimum from all other directions (Figure 11.4(a)).

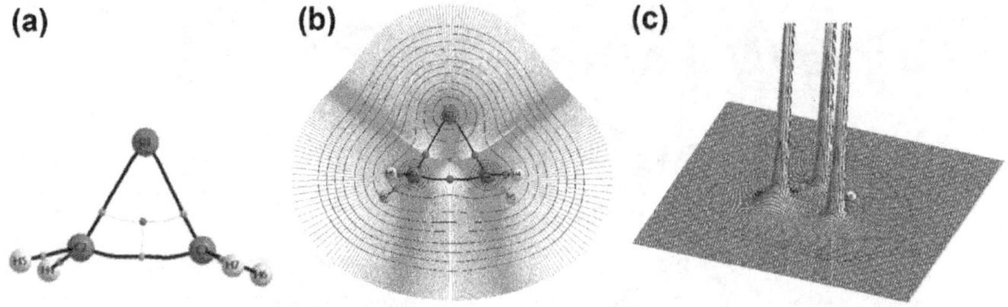

FIGURE 11.4 Molecular graph of oxirane showing (a) bond critical points (3,−1) and ring critical point (3,−2) and bond paths connecting different nuclei, (b) charge density in the form of contour map, and (c) charge density in the form relief map. The marked curvature of the bond path connecting the two carbon atoms is indicative of a ring strain in this molecule.

(3,3)—In this case all the three eigenvalues are positive and the CP is local minimum from all directions. Such points exist in the center of molecular cages, e.g., at the center of cubane, and are known as cage critical points (CCPs).

Results of QTAIM calculations on oxirane are graphically shown in Figure 11.4. The molecular graph (Figure 11.4(a)) shows the positions of the nuclear (3,3), bond (3,−1), and ring (3,1) CPs and the bond paths between the different atoms. The marked curvature of the bond path connecting the two carbon atoms is indicative of ring strain in the molecule. The charge density distribution as contour map and relief map are shown in Figure 11.4(b) and (c), respectively.

Each type of the above-described CPs is identified with an element of the chemical structure and the total number and types of CPs that can coexist in a molecule or crystal follows a strict topological relation.

If N_{NCP}, N_{BCP}, N_{RCP}, and N_{CCP} represent the number of nuclear, bond, ring, and cage CPs then, for isolated molecules

$$N_{NCP} - N_{BCP} + N_{RCP} - N_{CCP} = 1 \quad (11.4.1)$$

and for infinite crystals

$$N_{NCP} - N_{BCP} + N_{RCP} - N_{CCP} = 0 \quad (11.4.2)$$

The violation of these two equations implies that a CP has been missed and a search for the missing CP needs to be carried out. Thus, for example, in the case of oxirane C_2H_4O (Figure 11.4(a)), we have $N_{NCP} = 7$, $N_{BCP} = 7$, $N_{RCP} = 1$, and $N_{CCP} = 0$. Thus, Eq. (11.4.1) is satisfied indicating that all CPs have been identified.

11.5 Molecular Structure and Chemical Bond

The accepted concept of molecular structure is that it is a basic property of a system in which a network of bonds exists. These bonds exist even for large displacements till they are broken or formed to give a new structure. Molecular structure needs to be differentiated from molecular geometry. While the former is a generic property of the system which would remain unchanged till the same nuclei remain attached by the same network of bonds, the molecular geometry is a nongeneric property. It changes with every displacement of the nuclei. The essential point associated with the structure is that, like electrons, the nuclei can also not be localized in space and can be associated with a distribution function whose average value can be used to describe the system. It has been shown by Bader et al. [5] that the topological properties of charge distribution of a system enable the assignment of a molecular graph for a molecular structure. The collection of bond paths linking the nuclei of bonded atoms in an equilibrium geometry, with the associated CPs is called molecular graph. A molecular graph can be used to locate changes in structure along a reaction path, that is, when the system is undergoing a change of state.

11.5.1 Topological Characterization of Chemical Bond

It is important to emphasize that bond path, as described earlier, cannot be simply equated with a chemical bond. The QTAIM requires the replacement of the concepts of ball and stick model or its orbital equivalents of atomic and overlap contributions by a different concept, namely, two atoms are bonded if they share an interatomic surface and are consequently linked by a bond path. The entire network of bond paths defines the structure of the molecule. A bond path may or may not be coincident with the internuclear axis. If it is coincident, then both the bond length and bond path are equal. If it is not, the bond path is curved and the length of the bond path will exceed the bond length (Figure 11.4). Thus, in the case of oxirane (Figure 11.4), the bond path connecting the two carbon atoms is greater than the bond length. This is indicative of significant ring strain in the molecule.

The properties of chemical bonds are classified according to the properties of the electron and energy densities at the BCPs. There are several parameters which are used to characterize chemical bonds, such as electron density at BCP (ρ_b), bond order, bond radius, Laplacian $\nabla^2\rho_b$, bond path angle, bond ellipticity, and energy densities at BCP. These shall now be taken up in greater details.

11.5.2 Electron Density at BCP and Bond Order

The electron density at the BCP (ρ_b) is an indicator of the strength of a chemical bond. It is found to increase with the number of electron pair bonds. The ρ_b values for some bonds are given in Table 11.1. It is found that, in general, $\rho_b > 1.343$ in the case of shared interactions such as in covalent bonding and $\rho_b < 0.672$ in closed-shell interactions such as in van der Waals, hydrogen bonding, H–H bonding, ionic bonding, etc. An increase in ρ_b can also be correlated with the increase in strength of the hydrogen bond, as for example in $(H_2O)_2$ and $(HF)_2$, where the values of ρ_b are 0.134 and 0.177, respectively. ρ_b is also strongly correlated with the binding energy of several types of bonding interactions. For bonds between a given pair of atoms, one may define a bond order whose value is determined by charge density at the BCP (ρ_b). Bond order reflects

Table 11.1 Characterization of atomic interactions[a]

Molecule	Bond length R_e (C–C), Å	Electron density at BCP ρ_b, Å^{-3}	Hessian at BCP $\nabla^2\rho(r_b)$, Å^{-5}	Curvatures of density λ_1, Å^{-5}	λ_2, Å^{-5}	λ_3, Å^{-5}	Bond order n	Ellipticity ε
Ethane	1.527	1.702	−15.940	−11.500	−11.500	7.060	1.000	0.000
Benzene	1.386	2.205	−24.420	−17.040	−13.850	6.460	1.620	0.230
Ethylene	1.317	2.451	−28.730	−19.660	−13.600	4.540	2.050	0.450
Acetylene	1.185	2.820	−30.880	−14.160	−14.160	−2.560	2.920	0.000
CC bond in cyclopropane	1.498	1.680	−12.847	−11.789	−7.914	6.858	0.981	0.490
Oxirane	1.465	1.815	−16.122	−13.632	−10.514	8.027	1.11	0.2966
CH bond in CH_4	1.084	1.869	−23.578	−17.298	−17.298	11.016	1.176	0.000
OH bond in H_2O	0.943	2.638	−58.839	−49.783	−48.393	39.336	-	0.029
OH bond in proton donor of $(H_2O)_2$	0.947	2.581	−61.081	−51.058	−49.752	39.729	-	0.026
FH	0.895	2.728	−92.602	−69.602	−69.602	46.602	-	0.000
FH bond in proton donor of $(HF)_2$	0.906	2.595	−93.125	−67.963	−67.963	42.797	-	0.000
Closed-shell interactions								
Hydrogen bond in $(H_2O)_2$	2.039	0.134	1.501	−0.595	−0.578	2.675	0.223	0.029
Hydrogen bond in $(HF)_2$	1.778	0.177	2.887	−0.978	−0.868	4.805	0.233	0.128

[a]Data obtained at optimized geometries using 6-31G* basis set for hydrocarbons and 6-31G** basis set for monomer and dimer of H_2O and HF.

the strength of a chemical bond. Wiberg et al. [6] have given an empirical relationship for bond order (BO),

$$\text{BO} = \exp[A(\rho_b - B)] \tag{11.5.1}$$

where A and B are empirical constants which depend upon the nature of the bonded atoms. Bader et al. [7] took the values of these constants A and B to be 0.957 and 1.70, respectively, for the hydrocarbons when densities are obtained using 6-31G* basis set. They fitted the ρ_b values of C–C bonds (Table 11.1) in hydrocarbons such as ethane, benzene, ethylene, and acetylene and obtained bond orders of 1.0, 1.6, 2.0, and 3.0, respectively. The sequence agrees with the experimentally determined heats of reaction and bond lengths.

11.5.3 Bond Radius of an Atom (r_b) and Bond Path Length

The bond path, also called the MED path, is the line connecting the two nuclei and the bond existence of BCP lying on this path, is the essential condition of forming a bond. The distance of BCP from the nucleus of an atom (say A) determines its bonded radius denoted as $r_b(A)$. In cases where the bond path coincides with internuclear axis, the sum of the two associated bond radii (called the bond path length), equals the bond length. In curved bond paths, the bond path length R_b exceeds the corresponding internuclear separation or bond length R_e. Unless dictated so by symmetry the bond path does not coincide with the internuclear axis and so in all such as $R_b > R_e$. Curved bond paths are present in those systems where some kind of strain is present in those systems where some kind of strain is present such as in small ring compounds—hydrocarbons and others like oxirane (Figure 11.2(a)). The presence of bent bond paths is quite frequent and the degree of bending is a useful parameter in understanding structural effects in molecules. However, in order to understand the effect of curvature of the bond paths, a more frequently used concept is of bond path angle.

11.5.4 Bond Path Angles

In contrast to bond angles (α_e) which is the geometrical construct of the angle subtended by two bonds at a nucleus, bond path angle (α_b) is the limiting value of the angle subtended by two bond paths at the nucleus. The difference $\Delta\alpha = \alpha_b - \alpha_e$ provides a measure of the degree of relaxation of the charge density away from the geometrical constraints imposed by the nuclear framework. Thus, $\Delta\alpha > 0$ suggest that the bonds in a molecule are less strained than what the bond angles α_e would suggest. Thus, for example, in cyclobutane (Table 11.2), the CCC bond path angle is 95.73° which exceeds

Table 11.2 Geometry and bond path angles

	Molecule	Angle	Geometric angle (α_e)	Bond path angle (α_b)	$\Delta\alpha$ $\alpha_b - \alpha_e$	Strain energy (kcal)
Cyclopropane		1	60.0	78.84	18.84	27.5
Cyclobutane	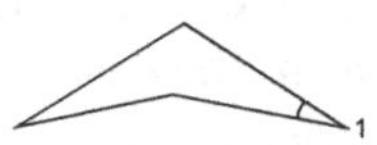	1	89.01	95.73	6.72	26.5
Cyclopentane	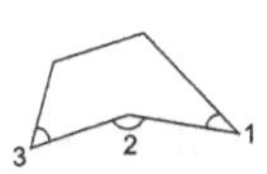	1	104.55	104.02	−0.53	6.2
		2	105.36	104.43	−0.93	
Cyclohexane	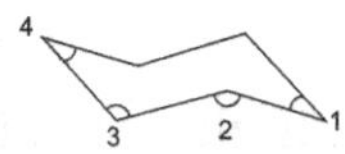	3	106.50	105.03	−1.54	
		1	111.41	110.08	−1.34	0.0
Oxirane		1	58.80	72.39	13.59	-

by 6.72° the CCC bond angle having value 89.01°. The difference between the conventional bond angles and bond path angles ($\Delta\alpha$) may be due to a combination of steric and electronic interactions. A positive value of $\Delta\alpha$ may indicate an attractive interaction between the outer atoms among those forming the bond angle whereas a negative value of $\Delta\alpha$ could indicate a repulsive interaction between these atoms. The positive $\Delta\alpha$ are generally found with large bond path angles (α_b), where repulsion between the end atoms would be small, and the negative $\Delta\alpha$ are found with small bond path angles that would tend to bring the end atoms closer together. A direct connection between bond path angles and electronegativity of atoms has also been suggested by Wiberg and Laidig [8] and Wiberg and Breneman [9]. Thus, in a variety of methyl derivatives CH_3X, large negative values of $\Delta\alpha$ are found for the HCX angle in those cases where X is more electronegative and large positive value when X is less electronegative. Likewise, in case of carbonyl derivatives such as, 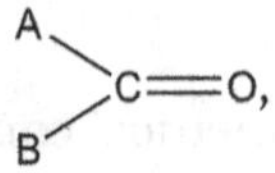

where A and B are combinations of H, F, Cl, OH, NH_2, and CH_3 groups. The angle opposite the more electronegative atom has larger bond angle.

11.5.5 Bond Ellipticity

Bond ellipticity ε is a measure of the extent to which charge is preferentially accumulated in a given plane and also of the deviation of the charge distribution of a bond path from axial symmetry. It is defined as

$$\varepsilon = \frac{\lambda_1}{\lambda_2} - 1 \qquad (11.5.2)$$

where λ_1 and $\lambda_2(|\lambda_1| > |\lambda_2|)$ are the two negative curvatures of electron density ρ at BCP (r_b). While the magnitude of electron density ρ_b at BCP gives the strength of the bond, the anisotropy of the electron density distribution determined by ε could point to the presence of different bond types, e.g., a σ or a π-bond. If $\lambda_1 = \lambda_2$, as in the case of a CC bond in ethane where the axes associated with λ_1 and λ_2 are symmetrically equivalent, then

$$\varepsilon = \frac{\lambda_1}{\lambda_2} - 1 = 0, \qquad (11.5.3)$$

and a display of ρ in the plane perpendicular to the bond axis will field circular contours centered at BCP (r_b). The bond in this case has cylindrical symmetry. Single and triple bonds are found to have cylindrical symmetry though in the latter case ρ_b is much higher.

If $|\lambda_1| > |\lambda_2|$, $\varepsilon \neq 0$, as in the case of CC bond in ethylene. In this case, the contour of electron density in the plane perpendicular to the bond axis is elliptical. The eigenvectors associated with λ_1 and λ_2 define a unique pair of orthogonal axes perpendicular to the bond path. The minor axis of the ellipse lies along the axis associated with λ_1 and the major axis along that associated with λ_2 (Figure 11.5).

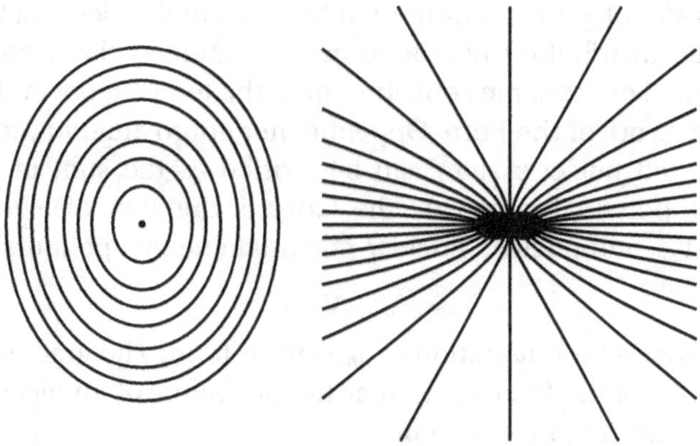

FIGURE 11.5 Contour map of charge density and gradient vector field for the plane containing the C–C interatomic surface in ethene. This plane bisects and is perpendicular to the C–C bond path. Elliptical nature of the contours with major axis perpendicular to the plane containing the nuclei may be noted. *Reprinted with permission from Ref. [3]. Copyright 1991 American Chemical Society.*

Ellipticity is taken to be a measure of π character of bonding up to a double bond and takes maximum value for a double bond (bond order 2). It decreases thereafter and at bond order 3 it is again 0. The ellipticity of some bonds is given in Table 11.1. While it is zero for C–C bond in ethane, it is 0.23 in benzene, and 0.45 in ethylene. The ε value in a double bond is around 0.4–0.6.

11.6 Energy of Atom in Molecule

In Section 11.2 it was mentioned that QTAIM uniquely divides a molecular space into a set of mononuclear regions or atomic volumes Ω, identified as atom in the molecule. Each of these atomic volumes is bounded by a surface of zero flux. Thus the molecule is divided by a series of surfaces through which the gradient vector field of the electron density has no flux and satisfies the condition given by Eq. (10.2.3).

The atomic properties such as atomic charge, dipole moment, and energies can be calculated by integrating the corresponding operator over the atomic volume Ω. An important consequence of the definition of the atomic property is that the average value of an observable molecular property can be written as a sum of corresponding atomic contributions, $A(\Omega_i)$

$$\langle \hat{A} \rangle_{\text{molecule}} = \sum_{i}^{\text{all atoms}} A(\Omega_i) \tag{11.6.1}$$

This equation is true for both one-particle and two-particle operators.

Thus the total energy can be written as a sum of atomic energies. This equation states that each atom makes an additive contribution to the value of every property of the system and represents one of the most important landmarks of chemistry which explains the transferability of properties of atoms and functional groups in molecules. According to QTAIM, the transfer of atomic information between molecules is at the level of charge density. Also, when distribution of charge over an atom is the same in two different molecules, then it makes the same contribution to the total energy in the two molecules.

Within the framework of the Born–Oppenheimer approximation, the Hamiltonian of a many-electron fixed nuclei system can be written as the sum of electronic kinetic energy T_e and the potential energy V; the latter is the sum of electron–nuclear V_{ne}, electron–electron V_{ee}, and nuclear–nuclear potential energy operators.

$$\widehat{V} = \widehat{V}_{ne} + \widehat{V}_{ee} + \widehat{V}_{nn} \tag{11.6.2}$$

where for a fixed nuclear configuration, $\widehat{V}_{nn}$ is fixed term. The gradient of the potential energy operator $-\nabla\widehat{V}$ is the force exerted at the position r of an electron (say, 1) by all other electrons and nuclei in the system

$$\mathbf{F}(\mathbf{r}) = N\int \Psi^*\left(-\nabla\widehat{V}\right)\Psi d\tau' \tag{11.6.3}$$

Integration over $d\tau'$ averages this force overall positions of the remaining particles and so gives the force on the electron due to the average distribution of the remaining particles of the system. This force is called the Ehrenfest force and its integration over the atomic basin gives average electronic force exerted on the atom in the system. The direct evaluation of the force operator $-\nabla\widehat{V}$ requires information contained in the one-electron density matrix. The virial of the Ehrenfest force $\mathbb{V}$ given as

$$\mathbf{r}\cdot\mathbf{F}(r) = \mathrm{N}\int \Psi^*\left(-\mathbf{r}\cdot\nabla\widehat{V}\right)\Psi d\tau' \tag{11.6.4}$$

determines the potential energy of the electrons. Thus the potential energy of the electrons in the atomic basin is

$$\mathbb{V}_b(\Omega) = N\int_{\Omega} d\mathbf{r}\int \Psi^*\left(-\mathbf{r}\cdot\nabla\widehat{V}\right)\Psi d\tau' \tag{11.6.5}$$

The total potential energy of the electrons in the atom is the sum of basin ($\mathbb{V}_b$) and surface ($\mathbb{V}_s$) terms; the latter is a virial of the Ehrenfest force exerted on the surface of the atom.

$$\mathbb{V}(\Omega) = \mathbb{V}_b(\Omega) + \mathbb{V}_s(\Omega) \tag{11.6.6}$$

Partitioning of the virial $\mathbb{V}$ into the basin and surface contributions depend upon the choice of the region. An origin can always be found such that $\mathbb{V}_s = 0$. The average effective potential field experienced by an electron at a point $\mathbf{r}$ in a many-particle system is also known as virial field or potential energy density, $\mathbb{V}(\mathbf{r})$.

From the mathematical formalism based on the Lagrangian approach and the extension of the variation of action to an open system, Srebrenik and Bader [10] identified two forms of kinetic energy. These are the following:

1. Schrödinger kinetic energy

$$K(\mathbf{r}) = -\frac{\hbar^2}{4m}N\int \left|\Psi^*\nabla^2\Psi + \Psi\nabla^2\Psi^*\right|d\tau' \tag{11.6.7}$$

2. Gradient kinetic energy

$$G(\mathbf{r}) = \frac{\hbar^2}{2m} N \int \nabla\Psi^* \cdot \nabla\Psi d\tau' \tag{11.6.8}$$

The two kinetic energy densities differ by a term $L(\mathbf{r})$

$$K(\mathbf{r}) = G(\mathbf{r}) + L(\mathbf{r}) \tag{11.6.9}$$

where,

$$L(\mathbf{r}) = -\frac{\hbar^2}{4m} N \int \nabla^2(\Psi^*\Psi) d\tau' = -\frac{\hbar^2}{4m} N \int \nabla^2 \rho(\mathbf{r}) d\tau' \tag{11.6.10}$$

Integration of these equations over a region of space Ω gives

$$K(\Omega) = G(\Omega) + L(\Omega) \tag{11.6.11}$$

where,

$$K(\Omega) = -\frac{\hbar^2}{4m} N \int_\Omega d\mathbf{r} \int \left|\Psi^* \nabla^2 \Psi + \Psi \nabla^2 \Psi^*\right| d\tau' \tag{11.6.12}$$

and,

$$G(\Omega) = \frac{\hbar^2}{2m} N \int_\Omega d\mathbf{r} \int (\nabla_i \Psi^* \cdot \nabla_i \Psi) d\tau' \tag{11.6.13}$$

When the integration is carried over an atomic basin then $L(\Omega)$ vanishes because of the zero flux condition (Eq. (11.2.3)) resulting in $K(\Omega) = G(\Omega)$.

Both these kinetic energy densities reduce to the average kinetic energy when integrated over all space. For a proper open quantum system,

$$K(\Omega) = G(\Omega) = T(\Omega) \tag{11.6.14}$$

where $T(\Omega)$ is the electronic kinetic energy of an atom, which is a well-defined quantity.

In model calculations, any small difference in the values of $G(\Omega)$ and $K(\Omega)$ points toward inaccuracy in atomic integrations.

Using Eqs (11.6.6) and (11.6.14), the total electronic energy of an atom in a molecule, $E_e(\Omega)$ is defined as;

$$E_e(\Omega) = T(\Omega) + \mathbb{V}(\Omega) \tag{11.6.15}$$

According to atomic virial theorem,

$$-2T(\Omega) = \mathbb{V}_b(\Omega) + \mathbb{V}_s(\Omega) + L(\Omega) \tag{11.6.16}$$

where,

$$L(\Omega) = -\frac{\hbar^2}{4m} \oint \nabla\rho(\mathbf{r}) \cdot n(\mathbf{r}) ds\,(\Omega) \tag{11.6.17}$$

By virtue of the zero flux condition, Eq. (11.2.3), $L(\Omega) = 0$, and we get

$$-2T(\Omega) = \mathbb{V}_b(\Omega) + \mathbb{V}_s(\Omega) \text{ or } -2T(\Omega) = \mathbb{V}(\Omega) \tag{11.6.18}$$

The vanishing of $L(\Omega)$ is necessary for the kinetic energy $T(\Omega)$ to be well defined. So the negative of twice the average kinetic energy of electrons equals the virial of forces exerted on them, which is the same as their potential energy.

From Eqs (11.6.15) and (11.6.18) it follows that for the atomic volume Ω, the electronic energy of an atom in a molecule satisfies the relation

$$E_e(\Omega) = -T(\Omega) = \frac{1}{2}\mathbb{V}(\Omega) \qquad (11.6.19)$$

that is, the total virial $\mathbb{V}$ is twice the kinetic energy. This relationship, known as atomic virial theorem, is identical to the virial theorem for the total system.

The local form of the virial theorem for a stationary state is written as

$$\frac{\hbar^2}{4m}\nabla^2\rho(\mathbf{r}) = 2G(r) + \mathbb{V}(r) \qquad (11.6.20)$$

This relationship is important because it relates a property of charge density to the local contributions to the energy and is useful in understanding the energy contributions in the formation of chemical bonds.

From the above discussion, it is clear that like other atomic properties the sum of the energies of atoms in a system is equal to the total electronic energy

$$E = \sum_i E_e(\Omega_i) \qquad (11.6.21)$$

where E would also be the total molecular energy if the system is in equilibrium and no force is acting on any of the nuclei.

11.6.1 Energy Densities at BCPs and Chemical Binding

When the atoms in a molecule are bonded to one another, the bond paths, that is, paths connecting points of MED, meet all physical requirements set by Ehrenfest, Hellmann–Feynman and virial theorems. Ehrenfest force exerted on the electron density makes the two atoms experience an attractive force that draws their atomic basins together. The virial of this force determines the virial field $\mathbb{V}(\mathbf{r})$ which is a local description of the systems potential energy. The Hellmann–Feynman force is exerted on the nuclei. In an equilibrium state, due to the accumulation of the electron density in the binding region, no Hellmann–Feynman force, either attractive or repulsive, acts on the nuclei as they balance each other. The electron density accumulation in the binding region results in a reduction of the electron–nuclear potential energy (V_{ne}) and increase in the electron–electron (V_{ee}) and nuclear–nuclear (V_{nn}) repulsion energies. However, decrease in V_{ne} exceeds increase in V_{ee} and V_{nn}, with the result that in accordance with the virial theorem, the decrease in potential energy equals twice the decrease in the total energy. The bond path connecting two atoms is therefore indicative of the accumulation of electron density between the nuclei. The presence of electron charge between the nuclei is both necessary and sufficient condition for two atoms to be bonded together.

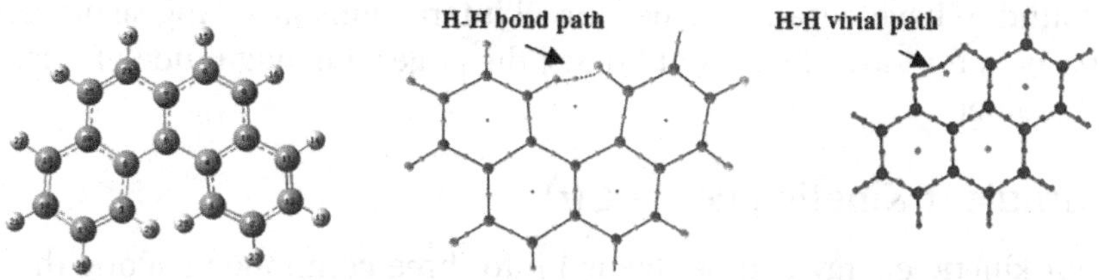

FIGURE 11.6 Molecular structure, molecular graph, and virial graph of benzo[c]phenanthrene.

The line linking two nuclei along which the virial field is maximally negative with respect to any other neighboring line is called the virial Path. Since the potential energy density $\mathbb{V}(\mathbf{r})$ along this path has a minimum value (maximally negative), the path is most stabilized. A one-to-one relationship exists between the *bond path* and the *virial path* and so the electron density and virial field are said to be homeomorphic.

The chemically bonded nuclei of the atoms are linked by a single bond path and a corresponding virial path. In other words every bond path is mirrored by a virial path. In analogy with molecular graph, the collection of virial paths and the associated CPs form a virial graph. Both these graphs describe molecular structure equivalently. Molecular structure of benzo[c]phenanthrene and its molecular and virial graphs are shown in Figure 11.6.

The formation of bond in a molecule results from two competing processes: (1) perpendicular compression of electron density toward the bond path and (2) its expansion along the bond path. The stresses in the density in these processes can be determined by the Laplacian $\nabla^2\rho$ at the BCP. Since $\nabla^2\rho$ is associated with the curvature of the density along the bond path and is related to the virial field $\mathbb{V}(\mathbf{r})$ by virtue of the virial theorem Eq. (11.6.20), a relationship exists between the bond properties and the virial field.

The virial field $\mathbb{V}(\mathbf{r})$, whose integral over all space yields the total potential energy of the molecule is always negative whereas the gradient kinetic energy $G(\mathbf{r})$ is always positive. The sign of $\nabla^2\rho_b$ at the BCP therefore determines which of these contributions to the total energy is in excess of the 2:1 virial ratio. Thus, the closed interactions in the region of interacting surfaces where potential energy is lowered are characterized by $\nabla^2\rho_b < 0$, whereas those where the kinetic energy dominates are characterized by $\nabla^2\rho_b > 0$.

When $\nabla^2\rho_b < 0$ and is large in magnitude, there is large concentration of electronic charge in the internuclear region— as in the case of covalent or polar interactions. Such interactions are also called shared interactions. When $\nabla^2\rho_b > 0$, but ρ_b has a relatively low value, the interactions are characterized by contraction of charge away from the interatomic surface toward each of the nuclei. Such a situation is found in closed-shell systems such as ionic bonds or van der Waals interactions.

In order to compare the kinetic and potential energies at equal footing, in place of the virial theorem where they are in the 2:1 ratio, Cremer and Kraka [11] proposed evaluation of total electronic energy density $[E(\mathbf{r}) = G(\mathbf{r}) + \mathbb{V}(\mathbf{r})]$ at the BCP,

$$E_{BCP} = G_{BCP} + V_{BCP} \tag{11.6.22}$$

$E(\mathbf{r})$ is found to have negative values for all interactions involving significant sharing of electrons as in the case of covalent bonds; the larger the magnitude of $E(\mathbf{r})$, larger the covalent character.

11.6.2 Gradient Kinetic Energy *G*(*r*)

The gradient kinetic energy can be divided into three components along three orthogonal axes—a component parallel to the internuclear axis ($G_{\parallel}$) and two components perpendicular to it ($G_{\perp}$). Bader and Preston [12] made some observations regarding these parallel and perpendicular components and their relationship with the gradients and curvatures of the electron density ρ in the molecular system. It is found that in the case of shared interactions, $G_{\perp} > G_{\parallel}$ at the BCP whereas for closed-shell interactions $G_{\perp} < G_{\parallel}$. Also, the kinetic energy per electronic charge density at the BCP, $G_{\mathrm{BCP}}/\rho_{\mathrm{BCP}}$ is less than unity for shared interactions and greater than unity for closed-shell interactions. The relationship between the local behavior of kinetic energy density with the gradient ($\nabla\rho$) and curvature ($\nabla^2\rho$) of electron density ρ is to some extent accounted for by the virial Eq. (11.6.20). It is found that in regions of space where $\nabla^2\rho(\mathbf{r}) > 0$ and $G_{\parallel}$ is large, the kinetic energy dominates the contribution to the virial and to the energy of the system. On the other hand, in regions where the perpendicular components $G_{\perp}$ are the largest, the potential energy makes the major contribution to the virial and to the energy of the system.

11.7 Applications

In the field of organic chemistry, the atoms and molecules model has found applications in molecular structural analysis, chemical reaction studies, and the formation of hydrogen bonds and van der Waals intermolecular bonds. The atomic properties defined within QTAIM are useful to describe and predict phenomena in molecules and materials in various fields ranging from solid state physics and X-ray crystallography to drug design and biochemistry. Though the list of applications is very large, we shall consider here only three applications related to the study of bond formations in chemical reactions and the formation of weak hydrogen bonds namely, (1) formation of 2-imino-malononitrile in nitrile-rich atmospheres, (2) formation of hydrogen bonds in EDNPAPC (ethyl 3,5-dimethyl-4[3-(2-nitro-phenyl)-acryloyl]-1H-pyrrole-2-carboxylate), and (3) formation of intramolecular H···H bonds in the Fjord region of helicenes.

11.7.1 Formation of 2-Imino-Malononitrile in Nitrile-Rich Atmospheres

It has been suggested that in nitrile-rich environment, such as found in the Saturnian moon Titan, the formation of molecules like diaminomaleonitrile (DAMN) and diaminofumaronitrile (DAFN) results from the addition of the hydrogen and cyanide (CN•) radicals. This process, shown in Figure 11.7, may ultimately lead to the formation of the nucleic base adenine which is one of the building blocks of life. Gupta et al. [13] analyzed

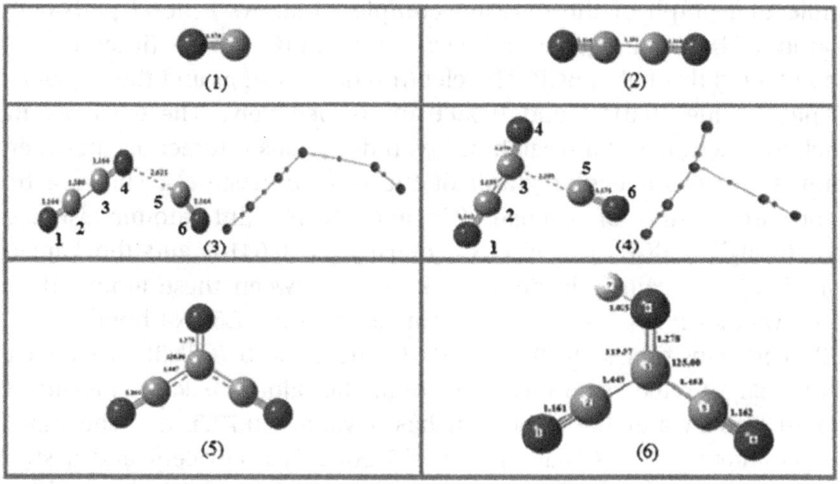

FIGURE 11.7 Formation of 2-imino-malononitrile by addition reactions of hydrogen and cyanide radicals. Molecular graphs of 3 and 4 are shown side by side with their molecular structure. Small-sized spheres show BCPs.

this process by using the QTAIM. This technique has additionally been used to locate the changes in the structure along the reaction path that leads to the formation of a neutral intermediate 2-imino-malononitrile which, being a neutral species, has greater stability than the corresponding ions and radicals. The addition of two cyanide radicals **1** (Figure 11.7) may result in the formation of cyanogen (preferably in the presence of a condensate), a molecule which is found in abundance in the Titan atmosphere. A CN radical is characterized by a triple bond between the carbon and nitrogen atoms with the radical electron residing in a *sp* orbital on the carbon atom which sticks out along the axis of the CN bond, away from the triple-bond electron. As such, two CN radicals easily combine to form cyanogen (NCCN) **2**.

The addition reaction of cyanogen leading to the formation of molecule **5** proceeds in two steps, involving a reaction complex **3** (RC) and the transition state **4** (TS). The possible bonding feature in reaction complex and the transition state following it are given by QTAIM-based molecular graphs shown in Figure 11.7. The graph shows bond paths linking the nuclei of bonded atoms in equilibrium geometry and the associated CPs. The structural and topological parameters at the BCPs are given in Table 11.3.

Table 11.3 Structural and topological parameters at the bond critical points (BCPs) (in au) in the reaction complex (RC) and the transition state (TS) during the formation of 2-imino-malononitrile, based on QTAIM calculations

Compounds	Bonds (length in Å)	ρ_b	$\nabla^2\rho_b$	λ_1	λ_2	λ_3
3 (RC)	C5–N4 (2.631)	0.0183	0.0586	−0.0171	−0.0167	0.0924
4 (TS)	C3–N5 (2.289)	0.0418	0.0795	−0.0456	−0.0428	0.1678
5 C3N3	C3–C5 (1.460)	0.2753	−0.7278	−0.5452	−0.5140	0.3315

Abbreviations: ρ_b, electron density; $\nabla^2\rho_b$, Laplacian of electron density; λ_1, λ_2, λ_3, eigenvalues of the Hessian matrix at BCP.

The molecular graph of the reaction complex **3** shows a bond path between the nitrogen atom (N4) of cyanogen and the carbon atom (C5) of CN (interatomic distance 2.631 Å) connected through a BCP. The electron density at ρ_b and the Laplacian $\nabla^2\rho_b$ at the BCP have values 0.0183 and 0.0586 au, respectively. These values indicate a closed-shell interaction and a weak bond (van der Waals interaction) between the two atoms. Similarly, the molecular graph of the transition state **4** shows a bond path between the carbon atoms of cyanogen (C3) and CN (N5) (interatomic distance 2.289 Å) connected through a BCP. The electron density ρ_b (0.0418) and the Laplacian $\nabla^2\rho_b$ (0.0795) at this BCP again indicate a weak bond between these atoms; the ρ_b value, however, shows that this bond is much stronger than the C5–N4 bond in the reaction complex **3**. The transition state then leads to molecule **5** (C3N3) in which the bond C3–C5 (bond length 1.460) is now a stable bond. The value of electron density at the BCP in C3N3 is 0.2753 au and the Laplacian has a value −0.7278 au. The magnitude of these two parameters shows that the C3–N5 bond is a covalent and a stable bond. The protonation of molecule **5** results in the formation of 2-imino-malononitrile **6** (Figure 11.7).

11.7.2 Hydrogen Bonding and Weak Interactions in EDNPAPC

Hydrogen bonds are of versatile importance in the fields of chemistry and biochemistry, which govern chemical reactions, supramolecular structures, molecular assemblies, and life processes. Intra- and intermolecular hydrogen bonds are classified into two categories: (1) classical or conventional hydrogen bonds and (2) nonconventional or improper hydrogen bonds, depending upon the nature of atoms involved in the hydrogen bridges. The nonconventional hydrogen bonds are further classified into three categories: (1) those in which the nature of hydrogen bond donor is nonconventional, as a C–H group, (2) the nature of hydrogen bond acceptor is nonconventional, as a C atom or a π system, (3) both the donor and acceptor are nonconventional groups.

The presence of hydrogen bonds as well as weak interactions can be detected and analyzed by using geometrical as well as topological parameters. The geometrical criterion for the existence of H-bond of the type D–H···A, where the three letters denote donor, proton, and acceptor, respectively, are the following:

1. The distance between the proton and the acceptor A is less than the sum of their van der Waals radii;
2. The D–H···A angle is greater than 90°; and
3. An elongation of the donor D-proton H bond length takes place.

As the above criteria are frequently considered insufficient, the existence of a hydrogen bond could be supported further by using the Koch and Popelier criteria [14] which is based on the QTAIM. The highlights of these criteria are the following:

1. The existence of BCP for the "proton (H) and acceptor (A)" contact is a confirmation of the existence of hydrogen bonding interaction
2. The value of electron density ($\rho_{H\cdots A}$) at the BCP should be within the range 0.002–0.040 au

According to Rozas et al. [15], it is possible to correlate the charge density parameters at the BCP to the local energy density E_{BCP} of the electrons by evaluating the local kinetic energy density G_{BCP} and the local potential energy density V_{BCP} by using the relations

$$G_{BCP} = \left(\frac{3}{10}\right)\left(3\pi^2\right)^{2/3}\rho_{BCP}^{5/3} + \left(\frac{1}{6}\right)\nabla^2\rho_{BCP},$$

$$V_{BCP} = \left(\frac{1}{4}\right)\nabla^2\rho_{BCP} - 2G_{BCP}$$

and,

$$E_{BCP} = G_{BCP} + V_{BCP}$$

It has been suggested that $E_{BCP} < 0$ indicates a medium to strong hydrogen bond while $E_{BCP} > 0$ indicates a weak H-bond or van der Waals interactions.

3. The nature of the H-bond can be ascertained by knowing the value of the Laplacian of the electron density at the BCP $\nabla^2\rho_{BCP}$. Thus strong H-bonds are characterized by $\nabla^2\rho_{BCP} < 0$ and the medium and weak H-bonds are characterized by $\nabla^2\rho_{BCP} > 0$. In general, the following conditions are satisfied:
   Strong H-bond $\nabla^2\rho_{BCP} < 0$, $E_{BCP} < 0$
   Medium H-bond $\nabla^2\rho_{BCP} > 0$, $E_{BCP} < 0$
   Weak H-bond $\nabla^2\rho_{BCP} > 0$, $E_{BCP} > 0$
   and weak interactions (van der Waals interactions) $\nabla^2\rho_{BCP} > 0$, $E_{BCP} > 0$.
   While the strong H-bonds are covalent, the medium bonds are partially covalent in nature. The weak H-bonds are mainly electrostatic.
4. Another criterion for H-bonding deals with the mutual penetration of the hydrogen and acceptor atoms. This criterion, considered as a necessary and sufficient condition, compares the nonbonded radii of the donor H atom (r_0^D) and the acceptor atom (r_0^A) with their corresponding bonding radii. The nonbonding radius (r) is taken to be equivalent to the gas phase van der Waals radius of the participating atoms. The bonding radius is the distance between the nucleus and the BCP. If $\Delta r_D = (r_0^D - r_D)$ and $\Delta r_A = (r_0^A - r_A)$, then in a typical H-bond, $\Delta r_D > \Delta r_A$ and $\Delta r_D + \Delta r_A > 0$ indicate positive interpenetration. If either or both of these conditions are violated, the interactions are essentially van der Waals in nature.

As an illustration of the weak molecular interactions we consider the structures of the stable conformers of EDNPAPC [16], given in Figure 11.8, and its dimer.

Molecular geometries of both the monomer and the dimer were optimized by molecular orbital calculations at the B3LYP/6-31G(d,p) level and the molecular graph of the dimer (Figure 11.9) was plotted using AIMALL program. Geometrical as well as topological parameters for bonds of interacting atoms in dimer are given in Table 11.4. On the basis of these parameters and using the Koch and Popelier criterion, it can be seen that O7···H69 and O50···H26 are weak hydrogen bonds whereas O68···C58, O25···C15, O8···H36, O51···H79, O17···C11, O60···C54, C12···H38, C55···H81, O7···H76, O50···H33

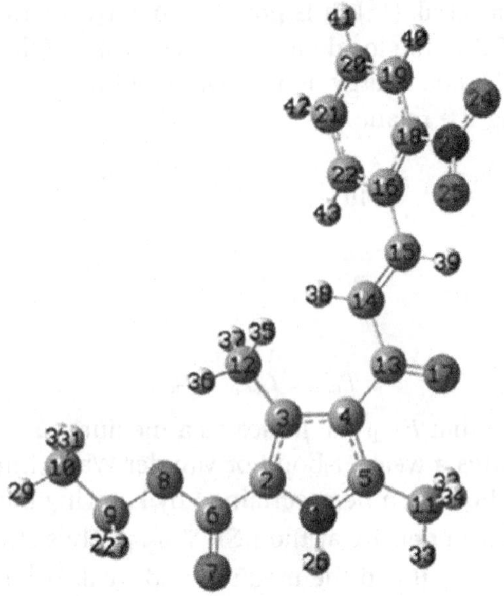

FIGURE 11.8 Optimized Geometry of EDNPAPC with atomic numbering.

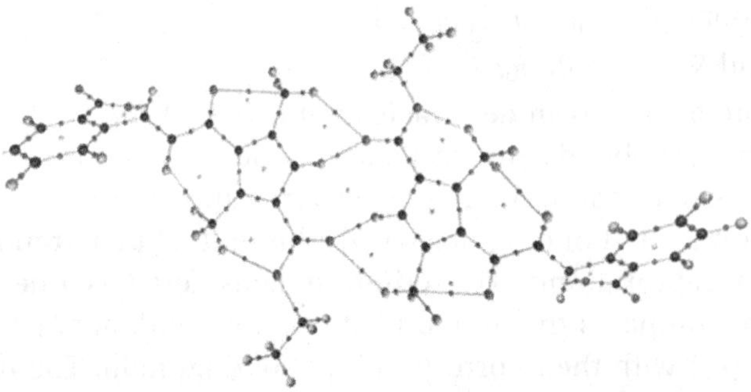

FIGURE 11.9 Molecular graph of the dimer of EDNPAPC.

are weak interactions. The various type of interactions visualized in molecular graph are classified on the basis of geometrical, topological, and energetic parameters.

Espinosa et al. [17] proposed proportionality between hydrogen bond energy (E) and potential energy density (V_{BCP}) at H–O contact: $E = 1/2(V_{BCP})$. According to AIM calculations, the binding energy of dimer is sum of the energies of all intermolecular interactions and this is calculated to be 15.22 kcal/mol. The intermolecular hydrogen bond energy of dimer of both heteronuclear intermolecular hydrogen bonds (N–H···O) is calculated to be 12.11 kcal/mol.

The ellipticity (ε) at BCP is a sensitive index to monitor the π-character of a bond. The analysis of bond ellipticity of the dimer was performed to investigate the effect of π-electron

Table 11.4 Geometrical parameters—contact distance, angle; topological parameters—electron density (ρ_{BCP}), Laplacian of electron density ($\nabla^2\rho(r_{BCP})$); energetic parameters—electron kinetic energy density (G_{BCP}), electron potential energy (V_{BCP}), total local energy density (G_{BCP}); interaction energy (E_{int}) at bond critical point (BCP); ellipticity ε for dimer of EDNPAPC

Interactions	Bond length	$\rho_{(BCP)}$	$\nabla^2\rho_{(BCP)}$	G_{BCP}	V_{BCP}	E_{BCP}	Ellipticity (ε)	E_{int}
N1 H26···O50	1.9085	0.02535	0.08172	0.01987	−0.0193	0.00056	0.02186	−6.0579
N44 H69···O7	1.9087	0.02534	0.0817	0.01986	−0.0193	0.00056	0.02186	−6.0563
H82 C58···O68	2.7165	0.01697	0.06804	0.01463	−0.0123	0.00238	1.20767	−3.8437
H35 C15···O25	2.7167	0.01697	0.06805	0.01463	−0.0123	0.00238	1.20767	−3.8437
C12 H36···O8	2.3072	0.01466	0.05067	0.0117	−0.0107	0.00097	0.13234	−3.3683
C55 H7···O5	2.3072	0.01466	0.05067	0.0117	−0.0107	0.00097	0.13234	−3.3678
O17···C11	2.8812	0.01174	0.04709	0.01029	−0.0088	0.00148	0.27250	−2.7656
O60···C54	2.8812	0.01174	0.04709	0.01029	−0.0088	0.00148	0.27250	−2.7655
C14 H38···C12	2.5519	0.01023	0.04262	0.00825	−0.0058	0.00241	1.40142	−1.8338
C57 H81···C55	2.5519	0.01022	0.04262	0.00825	−0.0058	0.00241	1.40142	−1.8337
C11 H33···O50	2.5861	0.0076	0.02658	0.0058	−0.005	0.00085	0.16687	−1.554
C54 H76···O7	2.5862	0.0076	0.02657	0.0058	−0.005	0.00085	0.16687	−1.5537

delocalization in bonds associated with N and O atoms of the two (N–H···O=C) intermolecular hydrogen bonds. These hydrogen bonds form 10-membered ring. The values of ε for bond O7–C6, C6–C2, C2–C3, C3–C4, C4–C5, C5–N1, O50–C49, C49–C45, C45–C46, C46–C47, C47–C48, and C48–N44 in this ring are in the range of 0.1007–0.2875 (Tables 11.4 and 11.5). These values of ε correspond to those suggested for the aromatic bonds in the literature. This confirms that these intermolecular hydrogen bonds are resonance-assisted hydrogen bonds, except the N–H bonds, because both N and O atoms are interconnected by a system of π-conjugated double bonds. Supplementary data associated with this work can be found at http://dx.doi.org/10.1016/j.mol-struc.2013.01.013.

Table 11.5 Ellipticity values for bonds involved in 16-membered pseudo-ring of dimer of EDNPAPC using AIM calculations

Bond	Ellipticity	Bond	Ellipticity
O7–C6	0.1007	O50–C49	0.1007
C6–C2	0.2240	C49–C45	0.2240
C2–C3	0.2875	C45–C46	0.2875
C3–C4	0.2093	C46–C47	0.2093
C4–C5	0.2485	C47–C48	0.2485
C5–N1	0.1471	C48–N44	0.1471
N1–H26	0.0303	N44–H69	0.0303
H26···O50	0.0219	H69···O7	0.0219

11.7.3 Intramolecular H···H Bonds in the Fjord Region of Helicenes

In a DFT study on BCPh and its derivatives using 6-31G** basis set, Gupta et al. [18] reported that a close H···H contact occurs between the terminal hydrogen atoms in the crowded Fjord region. In the case of BCPh, the distance between hydrogen atoms H_{28} and H29 is found to be 1.956 Å. The intramolecular H···H distance is therefore significantly shorter than twice the van der Waals radius of the hydrogen atom (2.40 Å). The molecular graph of BCPh obtained from QTAIM calculations (Figure 11.10) shows the bond paths and BCPs between the atoms. It also shows the formation of a seven-membered ring in the Fjord region with a RCP. It is seen that H28 and H29 are connected by a bond path of length 1.979 Å through a (3,−1) BCP. The bond path length is larger than the bond length between the two hydrogen atoms by about 0.02 Å, indicating ring strain in the Fjord region. The calculated electron density at the BCP is 0.0156 au, the Laplacian $\nabla^2\rho(b)$ is 0.0673, the total energy density at the BCP, E(BCP), is 0.0033, and the ellipticity ε is 0.5766. All these parameters at the BCP indicate the presence of a stable but weak bond between the two hydrogen atoms (H28···H29), which is a typical H···H bond. Likewise, in the case of some other substituted BCPhs, such as 6-(piperidine-1-yl)-1,2,3,4,7,8-hexahydrobenzo[c]phenanthrene-5-carbonitrile (**1**), 6-(piperidine-1-yl)-1,2,7,8-tetrahydrobenzo[c]phenanthrene-5-carbonitrile (**2**), 2-oxo-6-(piperidine-1-yl)-1,2,3,4,7,8-hexahydrobenzo[c]phenanthrene-5-carbonitrile (**3**) shown in Figure 11.11, the interatomic distance between the terminal hydrogen atoms is also

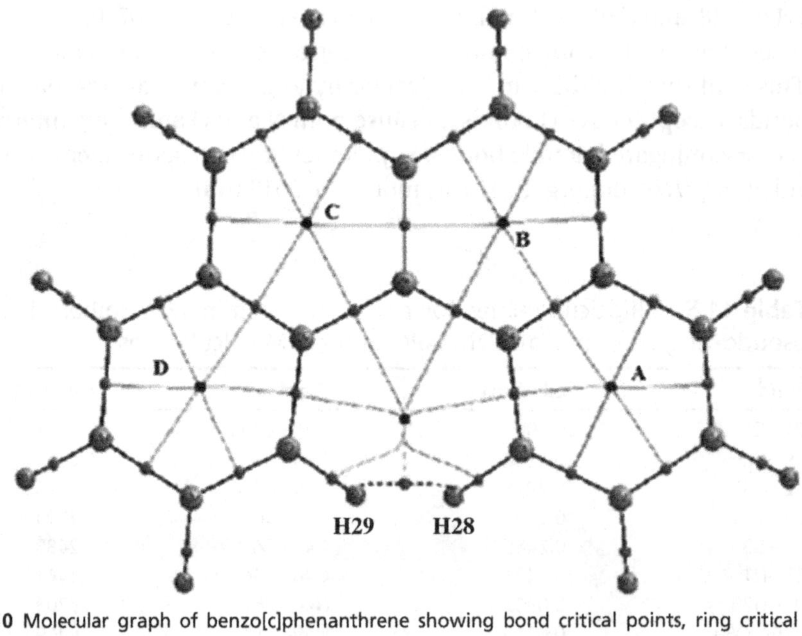

FIGURE 11.10 Molecular graph of benzo[c]phenanthrene showing bond critical points, ring critical points, bond paths, and H···H bonding between atoms H28 and H29.

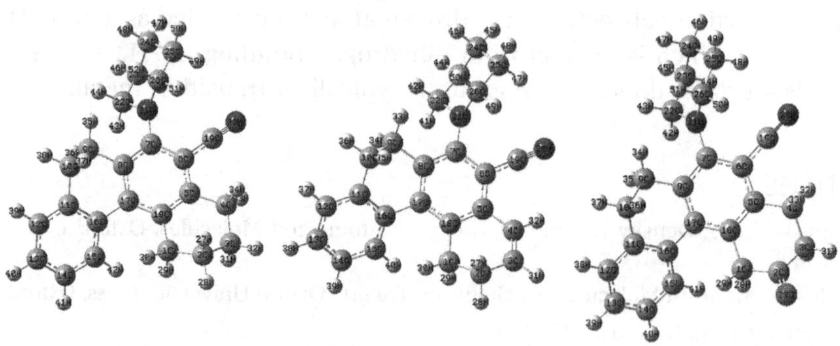

FIGURE 11.11 Molecular geometries of 6-(piperidine-1-yl)-1,2,3,4,7,8-hexahydrobenzo[c]phenanthrene-5-carbonitrile (**1**), 6-(piperidine-1-yl)-1,2,7,8-tetrahydrobenzo[c]phenanthrene-5-carbonitrile (**2**), 2-oxo-6-(piperidine-1-yl)-1,2,3,4,7,8-hexahydrobenzo[c]phenanthrene-5-carbonitrile (**3**).

found to be less than the sum of the van der Waals radii of the hydrogen atoms (2.40 Å). Thus, while in BCPh, the distance between the terminal hydrogen atoms, as may be seen from Table 11.6, is 1.955 Å, in **1** it is 2.070 Å, in **2** it is 2.252 Å, and in **3** it is 1.989 Å. This shows the possibility of existence of H···H bond in all these cases.

Wolstenholme et al. [19] have experimentally and theoretically determined intramolecular bond paths between the terminal hydrogen atoms in 4-methyl-[4]helicene. The intramolecular H···H distance in this case is 1.985 Å. They have shown that this closed-shell bond path exists in equilibrium geometry and is shared between two similar

Table 11.6 Approximate mean angle (°) between rings (A,C) and (A,D) and distances (Å) between terminal hydrogen atoms in the Fjord region in benzo[c]phenanthrene

Mean angle between rings	BCPh		1		2	3
	Calculated	Experimental[a]	Calculated	Experimental[b]	Calculated	Calculated
A,C	15.8	18.1[a]	35.0	32.6[b]	34.8	36.0
A,D	27.3	26.7	37.6	33.2[b]	46.0	34.6
Distance between terminal hydrogen atoms						
H(28)···H(29)	1.955	–	–	–	–	–
H(29)···H(42)	–	–	2.070	–	–	–
H(30)···H(42)	–	–	2.471	–	–	1.989
H(28)···H(41)	–	–	–	–	–	1.989
H(29)···H(41)	–	–	–	–	–	2.688
H(29)···H(40)	–	–	–	–	2.252	–
H(30)···H(40)	–	–	–	–	2.301	–

(1) 6-(piperidine-1-yl)-1,2,3,4,7,8-hexahydrobenzo[c]phenanthrene-5-carbonitrile.

(2) 6-(piperidine-1-yl)-1,2,7,8-tetra hydrobenzo[c]phenanthrene-5-carbonitrile.

(3) 2-oxo-6-(piperidine-1-yl)-1,2,3,4,7,8-hexahydrobenzo[c]phenanthrene-5-carbonitrile.

[a]Reference [20].

[b]Reference [21].

atoms. The interaction between such hydrogen atoms is classified as a weak H-bonding ($C-H^{\delta+}\cdots{}^{\delta+}H-C$) which is distinct from dihydrogen bonding ($X-H^{\delta+}\cdots{}^{\delta-}H-E$), where X–H group is a proton donor and *E* group is typically a transition metal.

References

[1] R.G. Parr, W. Yang, Density-Functional Theory of Atoms and Molecules, Oxford University Press, Oxford, 1989.

[2] R.F.W. Bader, Atoms in Molecules – A Quantum Theory, Oxford University Press, Oxford, UK, 1990.

[3] R.W.F. Bader, Chem. Rev. 91 (1991) 893.

[4] L.J. Farrugia, Topological Analysis of Electron Density, Jyväskylä Summer School on Charge Density, University of Glasgow, August 2007.

[5] R.F.W. Bader, T.T. Nguyen-Den, Y. Tal, Rep. Prog. Phys. 44 (1981) 893.

[6] K.B. Wiberg, R.F.W. Bader, C.D.H. Lau, J. Am. Chem. Soc. 109 (1987) 985.

[7] R.F.W. Bader, T.S. Slee, D. Cremer, E. Kraka, J. Am. Chem. Soc. 105 (1983) 5061.

[8] K.B. Wiberg, K.E. Laidig, J. Am. Chem. Soc. 109 (1987) 5935.

[9] K.B. Wiberg, C.J. Breneman, J. Am. Chem. Soc. 112 (1990) 8765.

[10] S. Srebrenik, R.F.W. Bader, J. Chem. Phys. 63 (1975) 3945.

[11] D. Cremer, E. Kraka, Angew. Chem. Int. Ed. Engl. 23 (1984) 627.

[12] R.F.W. Bader, H.J.T. Preston, Int. J. Quant. Chem. 3 (1969) 327.

[13] V.P. Gupta, P. Rawat, R.N. Singh, Poonam Tandon, Comput. Theor. Chem. 983 (2012) 7.

[14] U. Koch, P. Popelier, J. Phys. Chem. A 99 (1995) 9747.

[15] I. Rozas, I. Alkorta, J. Elguero, J. Am. Chem. Soc. 122 (2000) 11154.

[16] R.N. Singh, V. Baboo, P. Rawat, V.P. Gupta, J. Mol. Struct. 1037 (2013) 338. Supplementary data available at: http://dx.doi.org/10.1016/j.mol-struc.2013.01.013.

[17] E. Espinosa, E. Molins, C. Lecomte, Chem. Phys. Lett. 285 (1998) 170.

[18] V.P. Gupta, P. Thul, S. Misra, R. Pratap, V.J. Ram, Spectrochim. Acta A 72 (1) (2009) 82.

[19] D.J. Wolstenholme, C.F. Matta, T.S. Cameron, J. Phys. Chem. A111 (2007) 8803.

[20] S.A. Bae, H. Mah, S. Chaturvedi, J.M. Jeknic, W.M. Baird, A.K. Katz, H.L. Carrell, J.P. Glusker, T. Okazaki, K.K. Laali, B. Zalc, M.K. Lakshman, J. Org. Chem. 72 (2007) 7625.

[21] D.J. Wolstenholme, C.F. Matta, T.S. Cameron, J. Phys. Chem. A111 (2007) 8803.

Further Reading

[1] R.F.W. Bader, Atoms in Molecules – A Quantum Theory, Oxford University Press, Oxford, UK, 1990.

[2] C.F. Matta, R.J. Boyd (Eds.), The Quantum Theory of Atoms in Molecules: From Solid State to DNA and Drug Design, WILEY-VCH Verlag GmbH & Co, Weinheim, 2007.

[3] T. Veszpremi, M. Feher, Quantum Chemistry – Fundamentals to Applications, Springer, USA, 1999.

12

Characterization of Chemical Reactions

CHAPTER OUTLINE

Principles and Applications of Quantum Chemistry. http://dx.doi.org/10.1016/B978-0-12-803478-1.00012-1

12.1 Introduction

Chemical process can be characterized through reaction profiles that illustrate the way in which the properties of the reacting system change as a function of reaction coordinate (RC). Characterization of a chemical reaction involves product analysis, kinetics, and investigation of reactive intermediates. Quantum mechanics is an important tool to understand at the theoretical level the electronic structure of chemical compounds and the mechanism, thermodynamics, and kinetics of chemical reactions. It also provides reactivity parameters to understand a reaction process. Reactivity, selectivity, and site activation are classical concepts in chemistry which are amenable to quantitative representation in terms of static global and local density response functions. Some of these aspects shall be discussed in this chapter from a theoretical point of view.

12.2 Types of Chemical Reaction Mechanisms

In order to understand the mechanism of a chemical reaction one needs to have several important information like which bonds break and in what order, how many steps are involved, the relative rate of each step, etc. One also needs to know the positions of all atoms and the energy of the system at every point in the reaction process. In most reactions of organic compounds one or more covalent bonds are broken. Depending upon how the bonds break, the organic reaction mechanisms can be divided in to three basic types.

1. ***Heterolytic mechanism***: In a chemical reaction, one of the reactant is called the attacking agent and the other the substrate. If a bond breaks in such a way that both the electrons remain with one fragment, the reaction mechanism is called heterolytic. Such reactions do not necessarily involve ionic intermediates, though they usually do. A reagent that brings an electron pair is called a nucleophile and the reaction is nucleophilic. On the other hand, a reagent that takes an electron pair is called electrophile and the reaction is called electrophilic. Perhaps the two most important heterolytic bond dissociation reactions are those used to define "absolute" acidity and basicity.

$$A-H \rightarrow A:^{-} + H^{+}$$

$$B-H \rightarrow B: + H^{+}$$

Both reactions involve dissociation of a polar covalent bond to hydrogen and both lead to a "free" proton. While absolute acidities and basicities are rarely if ever (directly) measured experimentally, they provide a good opportunity to assess the performance of different models with regard to the energetics of heterolytic bond dissociation.

2. *Homolytic or free radical mechanism*: In this case a bond breaks up in a way that each of the fragments gets one electron leading to the formation of free radicals. Example:

$$CH_3{-}CH_3 \rightarrow CH_3^{\cdot} + CH_3^{\cdot}$$

3. *Pericyclic mechanism*: In this mechanism the electrons (usually six but sometimes it may be a different number) move in a closed ring. There are no intermediates, ions, or free radicals and it is not possible to say whether the electrons are paired or unpaired.

Understanding of chemical reactivity allows one to predict the reaction mechanism and to determine how the activation energy depends on specific properties of the reactants and products. However, for a chemical reaction to take place, it must also meet some thermodynamic and kinetic requirements. It may be noted that the study of kinetics is concerned with the details of how one molecule is transformed into another and the timescale of this transformation. This is in stark contrast to thermodynamics which is solely concerned with the initial and final states of a system, i.e., the reactant and product for chemical reactions. The mechanism involved in the transformation is not considered in thermodynamics and hence, time is not a factor. Kinetics and thermodynamics are, however, highly interrelated.

12.3 Thermodynamic Requirements for Reactions

Chemical reactions are determined by the laws of thermodynamics. In order for a reaction to take place spontaneously, the free energy of the products must be less than the free energy of the reactants. The associated free energy (ΔG) of the reaction is composed of two different thermodynamic quantities, enthalpy and entropy. Enthalpy (H) is a measure of the total energy or heat content of a thermodynamic system. It includes the internal energy, which is the energy required to create a system, and the amount of energy required to make room for it by displacing its environment and establishing its volume and pressure. Enthalpy change accounts for energy transferred to the environment at constant pressure through expansion or heating. The enthalpy change in a physical or chemical transformation is written ΔH, defined as $H_{\text{products}} - H_{\text{reactants}}$. This means that the change in enthalpy under such conditions is the heat absorbed (or released) by the material through a chemical reaction or by external heat transfer. Entropy, on the other hand, is a measure of the number of specific ways in which a system may be arranged. It is often taken to be a measure of disorder or randomness of the system; the higher the entropy, the higher the disorder. The entropy change in a physical or chemical transformation is written ΔS and defined as $S_{\text{products}} - S_{\text{reactants}}$. This is the same notation used to represent an enthalpy change, and the sign of ΔS

indicates a similar directional change. Thus, a positive ΔS denotes an increase in entropy on going from reactants to products, while a negative sign is associated with a decrease in entropy. It should be noted that a given sign for ΔS carries a different interpretation than it does for ΔH. Accordingly, a negative ΔH is associated with an exothermic and energetically favorable transformation; but a negative ΔS indicates an increase in system order (a less random system), and this is entropically unfavorable. The preferred conditions in nature are low enthalpy and high entropy.

The free energy (or Gibbs free energy) of a reaction is written as:

$$\Delta G = \Delta H - T\Delta S \tag{12.3.1}$$

Gibb's free energy ($\Delta G = \Delta H - T\Delta S$) determines the direction and extent of chemical change. It serves as the single master variable that determines whether a given chemical change is thermodynamically possible. Thus, if the free energy of the reactants is greater than that of the products, the entropy will increase when the reaction takes place as written, and so the reaction will tend to take place spontaneously. Conversely, if the free energy of the products exceeds that of the reactants, then the reaction will not take place in the direction written, but it will tend to proceed in the reverse direction. If the free energy of the reactants is greater than that of the products, the entropy will increase when the reaction takes place as written, and so the reaction will tend to take place spontaneously. Thus,

$\Delta G < 0$ reaction can spontaneously proceed to the right: $A \rightarrow B$

$\Delta G > 0$ reaction can spontaneously proceed to the left: $A \leftarrow B$

$\Delta G = 0$ reaction is at equilibrium; the quantities of A and B will not change

In order to make use of Gibbs energies to predict chemical changes, we need to know the free energies of the individual components of the reaction. For this purpose we can combine the standard enthalpy of formation and the standard entropy of a substance to get its standard free energy of formation

$$\Delta G_f^0 = \Delta H_f^0 - T\Delta S_f^0 \tag{12.3.2}$$

and then determine the standard Gibbs energy of the reaction according to

$$\Delta G^0 = \sum \Delta G_f^0\,(\text{products}) - \sum \Delta G_f^0\,(\text{reactants}) \tag{12.3.3}$$

The symbol ^{0} refers to the standard state of a substance measured under the conditions of 1 atm pressure or an effective concentration of 1 mol l^{-1} and a temperature of 298 K.

12.4 Kinetic Requirements for Reaction

A negative value of ΔG is a necessary but not a sufficient condition for a reaction to take place spontaneously. Thus, for example, reaction between H_2 and O_2 to give water has large negative ΔG, but mixtures of these two gases can be kept at room temperature for long periods without reacting to a significant extent. To understand why some reactions occur readily (almost spontaneously), whereas other reactions are slow, even to the point of being unobservable, we need to consider the intermediate stages through which reacting molecules pass on the way to products. Every reaction in which bonds are broken will necessarily have a higher energy *transition state* on the reaction path that must be traversed before products can form. This is true for both exothermic and endothermic reactions. In order for the reactants to reach this transition state, energy must be supplied from the surroundings and reactant molecules must orient themselves in a suitable fashion. The heat energy needed to raise the reactants to the transition state energy level is called the **activation enthalpy**, $\Delta H^{\#}$. As expected, the rate at which chemical reactions proceed is, in large part, inversely proportional to their activation enthalpies and is dependent on the concentrations of the reactants. For a simple transformation that interconverts A and B,

$$\ln K_{eq} = -\frac{\Delta G^{\#}}{KT} \quad \text{or,} \quad K_{eq} = e^{-\frac{\Delta G^{\#}}{KT}} \tag{12.4.1}$$

and,

$$K_{eq} = \frac{[B]}{[A]}, \tag{12.4.2}$$

the square brackets show concentrations.

In order for a reaction to take place, free energy of activation $\Delta G^{\#}$ must be added. This is illustrated in the figure which shows the energy profile for a reaction having no intermediate product.

In Figure 12.1, the horizontal axis is the reaction coordinate which is an abstract coordinate used to measure the progress of a chemical reaction. As a reaction progresses, reactants become products. To measure this progress, a reaction coordinate is chosen from a list of measurable coordinates such as bond length, bond angle, a combination of bond length and bond angle, or bond order. The vertical axis shows the free energy. $\Delta G_f^{\#}$ is the free energy of activation for the forward reaction. If the reaction is reversible, $\Delta G_r^{\#}$ for the reverse path must be greater than $\Delta G_f^{\#}$ for the forward path.

Like ΔG, $\Delta G^{\#}$ is made up of enthalpy and entropy components

$$\Delta G^{\#} = \Delta H^{\#} - T\Delta S^{\#} \tag{12.4.3}$$

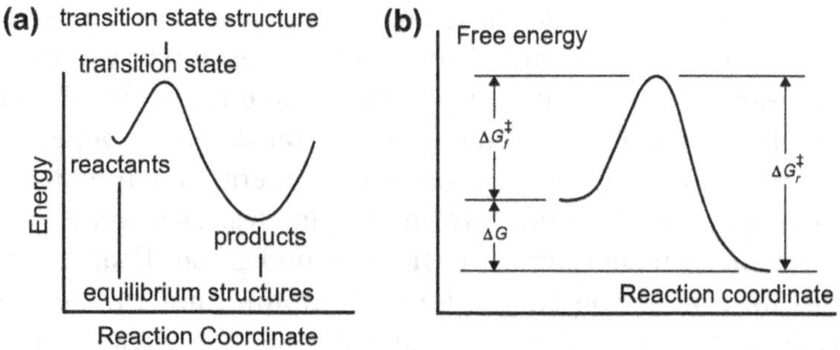

FIGURE 12.1 Energy profile for a reaction having no intermediate product (a) and free energy of activation (b).

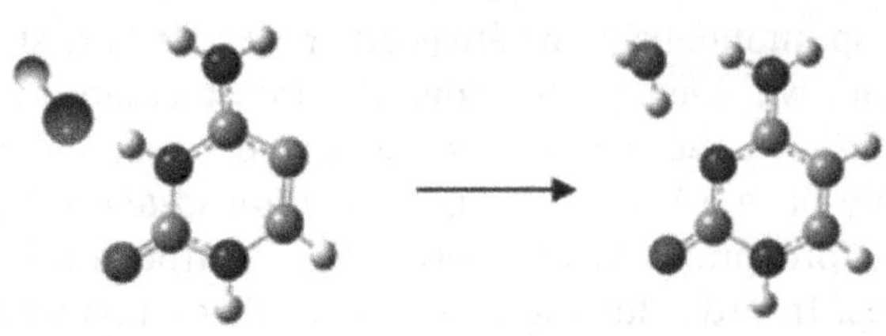

FIGURE 12.2 Removal of water in the formation of cytosine in interstellar space (representations—Gray dot C; Black dot N; Red (dark gray in print versions) dot O, and White dot H).

where $\Delta H^{\#}$, the enthalpy, is the difference in bond energies, including strain, resonance, and solvation energies, between the starting compounds and the transition state. $\Delta S^{\#}$, the entropy of activation is the difference in entropy between the starting compounds and the transition state.

In order to understand why reactions have activation energy, we must understand the mechanism of the reaction process. First and foremost, two molecules have to collide with sufficient kinetic energy, thereby organizing the system. Not only do they have to be brought together, they have to be held in exactly the right orientation relative to each other to ensure that reaction can occur. Both of these factors raise the free energy of the system by lowering the entropy. For such reactions to occur, the molecules need to surrender the freedom that they normally have to assume many possible arrangements in space and adopt a particular orientation that leads to reaction. This involves a considerable loss of entropy with the result that $\Delta S^{\#}$ is negative.

As an example, we may consider the formation of cytosine in the interstellar space [1] where an $OH^{\cdot}$ radical must come closer to a hydrogen atom attached to the nitrogen atom in the heterocyclic ring to remove a water molecule (Figure 12.2).

12.5 Potential Energy Surfaces and Related Concept

During a reaction process, the molecules undergo structural changes that change their energies. The way the energy of a molecule changes with small changes in its structure is specified by its potential energy surface (PES). In the case of a diatomic molecule *A–B*, the change in energy with the change in bond length between atoms *A* and *B* can be represented by a two-dimensional curve, known as the potential energy curve, used earlier. For larger systems, the surface has as many dimensions as the degrees of freedom within the molecule. Thus, if there are *N* nuclei, the dimensionality of the PES is 3*N*, i.e., there are 3*N* nuclear coordinates that define the geometry. In the absence of fields, a molecule's potential energy does not change if it is translated or rotated in space. Thus, the potential energy only depends on a molecule's internal coordinates which may be represented by simple stretch, bend, torsion coordinates, or symmetry-adapted linear combinations or redundant coordinates, or normal modes coordinates, etc. There are 3*N* total coordinates for a molecule (*x*, *y*, *z* for each atom), minus three translations and three rotations with respect to three axis; in the case of a linear molecule, only two coordinates are necessary to describe the rotation, and so only $3N - 6$ coordinates (or $3N - 5$ for a linear molecule) are needed to describe the internal motion of the nuclei. Such a PES is a hypersurface defined by the potential energy of a collection of atoms over all possible atomic arrangements. For a 3-atom molecule there are only three degrees of freedom which rises very fast as the number of atoms increases—six for four atoms, nine for five atoms, etc. The construction of such surfaces for all but the smallest molecules is therefore impossible.

Topologically, a PES is like a mountain range that has multiple hills and valleys along with mountain passes that lead from one valley to another (Figure 12.3(a)).

Particularly interesting points in this description are:

1. The valleys, that correspond to the minima of the PES, have molecules with relatively low energies. In a chemical reaction, this is the place where we expect to find the stable or quasi-stable species, i.e., reactants, products, and intermediates.
2. The saddle points, which resemble the center of a horseback riding saddle. In our topological description, if we look from the peak of the mountain pass, it is the place where on the forward and backward sides the trail goes downhill but on the left and right sides the mountains rise up. Thus, it is a point which is maximum in one direction but a minimum in the other direction (Figure 12.3(a) and (b)). In the PES, the saddle points are the points characterized by having no slope in any direction, downward curvature for a single coordinate, and upward curvature for all the other molecular coordinates. The saddle points which are minima in all directions but one (a maximum in that direction) are known as first-order saddle points or transition states. Besides, there are also the higher-order saddle points. An n-order saddle point, where $n > 1$, shall be maximum in n-directions and minimum in all others. Thus, a second-order saddle point would be a maximum in two directions and a minimum in all others.
3. The peaks of the mountains and the passes. These correspond to the maxima on the PES. A two-dimensional slice of a three-dimensional PES is given in Figure 12.3(c). The very tops of the mountains represent extremely high energy chemical structures that are seldom achieved, and the tops of the mountain passes (saddle points) represent the highest energy points that must be traversed for the transformation of one molecule into another.

As a chemical reaction/structural transformation occurs, the internal energy of the molecules rise to a peak at the saddle point and then drops off when the molecules fall into the neighboring valley. The molecular structure at the peak of the PES is called an activated complex or a *transition state* (strictly speaking the two terms are different). The transition state possesses a definite geometry and charge distribution but has no definite existence; the system passes through it. The height of the peak of the PES from its minimum is called *activation barrier* to the reaction and the amount of energy required to achieve the transition state is called the *activation energy*.

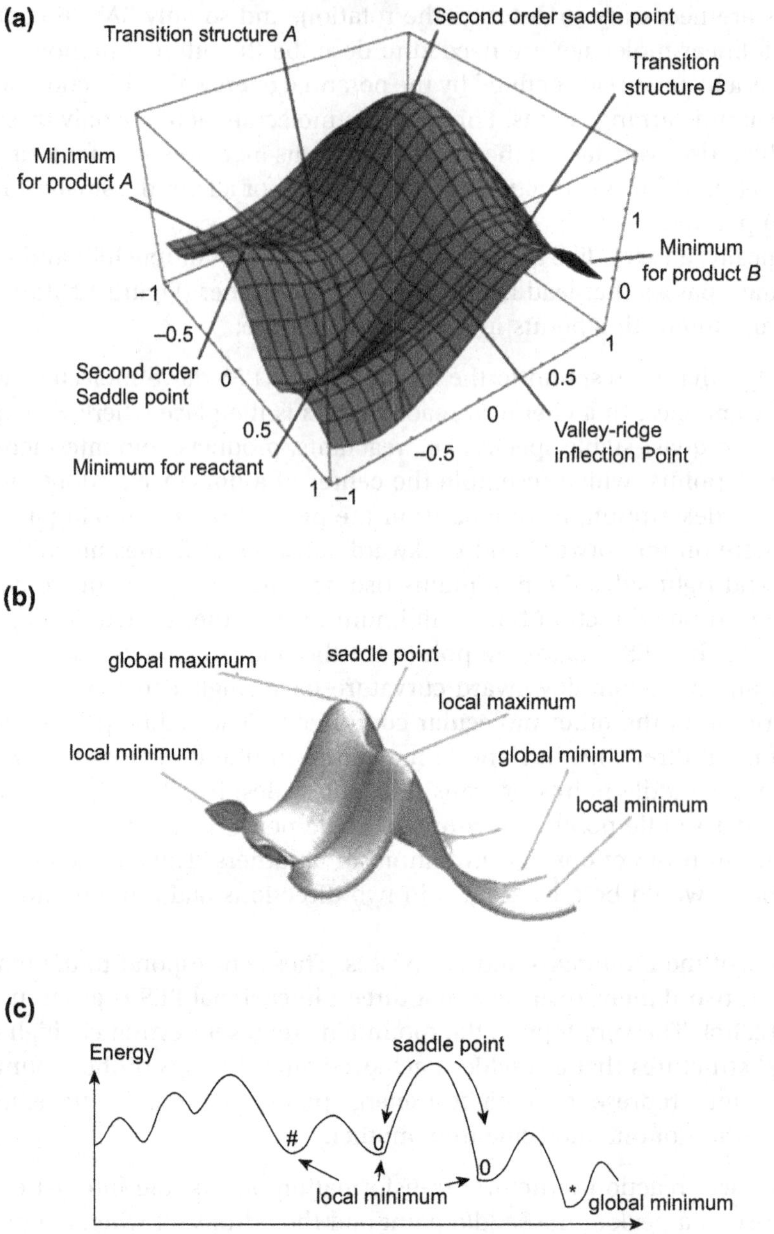

FIGURE 12.3 Energy surfaces showing (a) multiple valleys, (b) passes, and (c) saddle points. *Figure 12.3(a) reprinted with permission from H.B. Schlegel, Modern Electronic Structure Theory, Part I, Ed. D. R. Yarkony, Copyright @ 1995, World Scientific Publishing, Singapore.*

12.6 Stationary Points and Their Characteristics

Mathematically, a PES can be expressed in a functional form *V(R)* as Taylor expansion about the equilibrium configuration (eq)

$$V(R) = V(R^{\text{eq}}) + \sum_{n=1}^{3N-6}(R_n - R_n^{\text{eq}})\left(\frac{\partial V}{\partial R_n}\right)^{\text{eq}} + \frac{1}{2!}\sum_{m,n=1}^{3N-6}(R_m - R_m^{\text{eq}})(R_n - R_n^{\text{eq}})\left(\frac{\partial^2 V}{\partial R_m \partial R_n}\right)^{\text{eq}}$$
$$+ \frac{1}{3!}\sum_{m,n,l=1}^{3N-6}(R_l - R_l^{\text{eq}})(R_m - R_m^{\text{eq}})(R_n - R_n^{\text{eq}})\left(\frac{\partial^3 V}{\partial R_l \partial R_m \partial R_n}\right)^{\text{eq}} + \ldots \tag{12.6.1}$$

Since $V(R)^{\text{eq}}$ is a constant quantity, it may be included in *V(R)* to set the origin for the potential energy. The second term is the energy gradient at the equilibrium which vanishes and so the equation somewhat simplifies to

$$V(R - R^{\text{eq}}) = \frac{1}{2!}\sum_{m,n=1}^{3N-6}(R_m - R_m^{\text{eq}})(R_n - R_n^{\text{eq}})\left(\frac{\partial^2 V}{\partial R_m \partial R_n}\right)^{\text{eq}} + \frac{1}{3!}\sum_{m,n,l=1}^{3N-6}(R_l - R_l^{\text{eq}})(R_m - R_m^{\text{eq}})$$
$$\times (R_n - R_n^{\text{eq}})\left(\frac{\partial^3 V}{\partial R_l \partial R_m \partial R_n}\right)^{\text{eq}} \ldots$$

or,

$$V(q) = \frac{1}{2}\sum_{i,j=1}^{3N-6} k_{ij} q_{ij}^{\ 2} + \frac{1}{3!}\sum_{i,j,l=1}^{3N-6} k_{ijl} q_{ijl}^3, \tag{12.6.2}$$

where R_n is the internal coordinate that may take $3N - 6$ values and $q = R - R^{\text{eq}}$. k_{ij}, $k_{ijl}\ldots$ are the second, third, etc., order derivatives of the potential energy also known as quadratic force constant, cubic force constant, etc.

In the case of finite displacement, as in the case of the harmonic vibrational motion, where the geometries are not very different from the equilibrium geometries, one can ignore terms higher than the second order. However, during a chemical reaction, several interatomic distances change by large amounts (several angstroms) while others may change by few tenths of an angstrom. In such cases, one may have to retain the higher terms.

A stationary point on the PES is one that has no forces on it. The force, which is a vector quantity, corresponds to $\left(\frac{\partial V}{\partial r}\right)^{\text{eq}}$ and so the first derivatives with respect to all the internal coordinates are zero at the stationary point. Also, from analytical geometry, a point on the PES, for which $\left(\frac{\partial V}{\partial r}\right)^{\text{eq}} = 0$ is either a maximum or a minimum. If $\frac{\partial^2 V}{\partial R^2}$ is positive, it is a minimum and if negative, it is a maximum. The quantity $\left(\frac{\partial^2 V}{\partial r^2}\right)^{\text{eq}}$ corresponds to the force constant and is a scalar quantity. Thus, for a point on the PES if,

$$\left(\frac{\partial V}{\partial r}\right) = 0 \quad \text{and} \quad \frac{\partial^2 V}{\partial r^2} \geq 0, \quad \text{it is a potential energy minimum and,}$$

$$\left(\frac{\partial V}{\partial r}\right) = 0 \quad \text{and} \quad \frac{\partial^2 V}{\partial r^2} < 0, \quad \text{it is a potential energy maximum.}$$

The same criteria apply for the transition state, which is a minimum in all directions except along a reaction coordinate, where it is a maximum. The above conditions are helpful in characterizing potential energy minima and transition state structures by the classical (molecular mechanics) as well as by quantum mechanical computational methods.

As discussed in Chapter 7, the second derivatives of energy with respect to the internal coordinates $\left(\frac{\partial^2 V}{\partial R_m \partial R_n}\right)$ (also known as Hessian) as well as the vibrational frequencies are used to characterize the stationary points. The numbers of negative eigenvalues of the Hessian matrix are known as Hessian index. The Hessian index is 0 for minima, 1 for transition states, and more than 1 for higher-order saddle points. The Hessian index also corresponds to the number of imaginary vibrational frequencies. Thus, if there are no imaginary frequencies, the molecular structure is a minimum; if it has one imaginary frequency, it is a transition state connecting two minima; and if has more than one imaginary frequency, it is a higher-order saddle point and not a transition structure.

Several theoretical methods exist for the determination of stationary points. Many optimization methods determine the nearest stable point, but a multidimensional function may contain many different stationary points of the same kind. The minimum having the lowest value is called the global minimum, while the others are called local minima (Figure 12.3(b) and (c)). Similarly, a large number of different sophisticated methods exist to locate the transition states which are very important in the study of conformational changes, reaction mechanisms, reaction kinetics and dynamics. Some of the methods require knowledge of the two minima that may be connected to the transition states while the others allow specification of a particular coordinate with respect to which the energy is maximized while minimizing it with respect to all other coordinates. A good initial guess for the structures and the availability of good Hessian matrix are the necessary conditions for locating a good transition state.

12.7 Determination of Potential Energy Surfaces

The nature of the PESs can be probed either experimentally or by using theoretical methods. These may be classified as follows depending upon the methods used for their determination.

1. *Classical (molecular mechanics) surfaces*: The first PESs (or potential energy curves) were constructed by molecular spectroscopists for heterodiatomic molecules as they have only one degree of freedom. These were used to understand the energy separation between the vibrational states and to workout anharmonicity effects. Molecular mechanics provides this energy as a function of stretches, bends, torsions, etc. This is an approximate model that breaks down in some cases such as breaking of bonds and works only when parameters are available.

2. *Ab initio* PESs: *Ab initio* quantum mechanics provides an energy function which can be exact in principle and works for any molecule but it is calculated within the framework of the Born–Oppenheimer approximation namely, the electrons are much lighter than nuclei, and so they move much faster and adjust adiabatically to any change in nuclear configuration. This means that a separate PES is defined for each possible electronic state. Generally, the dynamics are studied on the ground electronic state surface.

PESs may be determined by *ab initio* electronic structure calculations. A large number of electronic structure calculations are performed and the results are fitted using a least squares procedure. The reliability of the PES depends on the basis set completeness and how well electron correlation is accounted. This method is computationally too expensive as the size and the number of calculations necessary to characterize the entire surface for large systems is too large. It is difficult to obtain a good fit when using general least squares fitting of *ab initio* results unless physically motivated functional forms are used.

3. *Analytic PESs*: There have been considerable advances in recent years in developing analytical functions for the PESs of small molecules Analytical methods use available experimental and *ab initio* information to calibrate functional forms based on simple valence theory or bond functions. A standard formulation is used as a starting point for designing a polyatomic PES. It is followed by using additional terms or by providing more flexibility to the adjustable parameters to remedy deficiencies or to improve selected areas of the surface.

Information about the PES has considerably improved over the years. However, it is still a major task to gather available information to construct functional representations which can be used for dynamical calculations. In general, different experiments provide information about different parts of the surface and highly accurate calculations are expensive to perform. Therefore, it is only by using chemical judgment that one can combine information from different sources to produce a satisfactory function for the whole surface. Moreover, there is often a conflict between the accuracy of a function to represent a surface and the simplicity of the function. The use of a more mathematically elaborate function increases the complexity of the dynamical calculations.

12.8 Potential Energy Surfaces in Molecular Mechanics

If we consider four atoms connected in sequence, *ABCD*, the Figure 12.4 shows that a convenient means to describe the location of atom *D* is by means of a *CD* bond length, a

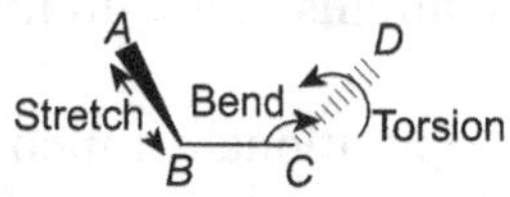

FIGURE 12.4 Stretch, bend, and torsion modes.

BCD valence angle, and the torsional angle (or dihedral angle) associated with the *ABCD* linkage. As depicted in the figure, the torsional angle is defined as the angle between bonds *AB* and *CD* when they are projected into the plane bisecting the *BC* bond.

The total energy of such a system can be written as

$$V = V_{\text{str}} + V_{\text{bend}} + V_{\text{tors}} + V_{\text{vdw}} + V_{\text{es}} + V_{\text{cross}}, \tag{12.8.1}$$

where V_{str} is the energy function for stretching a bond between two atoms, V_{bend} represents the energy required for bending an angle, and V_{tors} is the torsional energy for rotation around a bond. Additionally, there may be contributions from the nonbonded atom–atom interactions such as van der Waals V_{vdw} and the electrostatic energy V_{es} arising out of internal redistribution of the electrons, creating positive and negative parts of the molecule, and finally V_{cross} describes coupling between the first three terms.

12.8.1 Potential Energy Surface for Bond Stretching

The effect of bond stretching on potential energy has been discussed earlier in Chapter 3. In a diatomic molecule when the atomic displacement during stretch is small it is reasonable to consider terminating Eq. (12.6.2) for the potential energy after the quadratic term. The equation then reduces to

$$V(AB) = \frac{1}{2}K_{AB}q_{AB}^2, \tag{12.8.2}$$

which represents a parabola (Figure 3.2 Chapter 3). It shows that as the bond *A–B* is stretched longer and longer, the energy goes on increasing and becomes infinitely positive which is certainly chemically unrealistic.

The practical solution to such inaccuracy is to include cubic and quartic terms in the Taylor expansion leading to the equation

$$V(r_{AB}) = \frac{1}{2}\left[k_{AB}q_{AB}^2 + k_{AB}^{(3)}q_{AB}^3 + k_{AB}^{(4)}q_{AB}^4\right] \tag{12.8.3}$$

Diatomic molecules differ from harmonic oscillators mainly in that they may dissociate on large stretching of the bond. The correct limiting behavior for a bond stretched to infinity is that the energy should converge toward the dissociation energy (Figure 3.2 Chapter 3). A simple function that satisfies this criterion is the Morse potential function with a parameter $\alpha > 0$.

$$V(q) = D[1 - e^{-\alpha q}]^2 \tag{12.8.4}$$

This is plotted in Figure 12.5 for two different values of D and α. D represents the well depth and is the dissociation energy while the parameter α is the Morse parameter that decides its width and can be determined from spectroscopic measurements. When the displacement $q = 0$, the function attains the minimum $V = -D$, and when $q \to \infty$, $V \to 0$.

Morse potential is, however, computationally much less efficient to evaluate because of the exponential. If we approximate the exponential in Eq. (12.8.4) as its infinite series

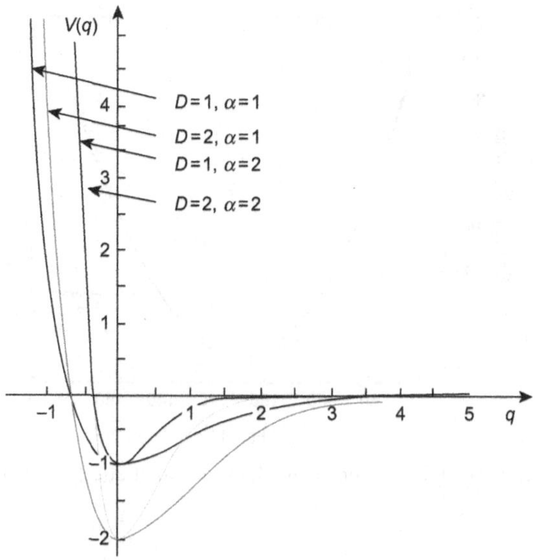

FIGURE 12.5 Morse potential energy curves for $D = 1, 2$ and $\alpha = 1, 2$.

expansion truncated at the cubic term, and compare it with Eq. (12.8.3), we get the relationship between the various force constants and the parameters D and α of the Morse potential, particularly the relation

$$\alpha = \sqrt{\frac{k}{2D}} \tag{12.8.5}$$

The simplest parameters to determine from experiment are k and D. With these two parameters available, α can be determined from Eq. (12.8.5), and thus the cubic and quartic force constants can also be determined from Eqs (12.8.2) and (12.8.3). Direct measurement of cubic and quartic force constants requires more spectral data than are available for many kinds of bonds, so this derivation facilitates parameterization.

12.8.2 Potential Energy Surface for Bending of an Angle

Equation (12.8.3) can be used to obtain potential energy curve for the bending of an angle, say θ_{ABC}, between bonds AB and BC by taking the internal coordinate as $q = \theta_{ABC} - \theta_{ABC}^{eq}$.

While the simple harmonic expansion is adequate for most applications, there may be cases where higher accuracy is required. The next improvement is to include a third-order term, analogous to stretching. Cramer [2] has shown that in CH_4 this can give a very good description over a large range of angles. Thus the quadratic approximation is seen to be accurate to about ±30° from the equilibrium geometry while the cubic approximation up to ±70° (Figure 12.6). Higher order terms are often included in order also to reproduce vibrational frequencies.

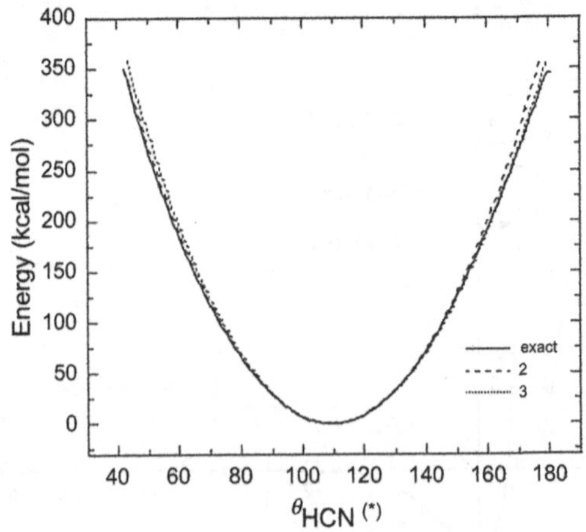

FIGURE 12.6 Potential energy curve for bending in methane in quadratic (---) and cubic (···) approximations. Compared with exact (—) curve.

While these equations are valid for the case of bending of angles, certain difficulties arise in their implementation. No power expansion having the form of these equations will show the appropriate chemical behavior as the bond angle becomes linear, i.e., at $\theta = \pi$. Also, particularly in inorganic systems, it is possible to have multiple equilibrium values; for instance, in the trigonal bipyramidal system PCl_5 there are stable Cl–P–Cl angles of $\pi/2$, $\pi/3$, and π for axial/equatorial, equatorial/equatorial, and axial/axial combinations of chlorine atoms, respectively. Complications also arise in the case of out-of-plane bending modes.

12.8.3 Potential Energy Surface for Torsion and Internal Rotation

The torsional energy is fundamentally different from the stretch and bend energies in three aspects, namely,

1. A rotational barrier has contributions from both the nonbonded (van der Waals and electrostatic) terms, as well as the torsional energy. The torsional parameters are therefore intimately coupled to the nonbonded parameters.
2. Since the torsion itself is periodic, so shall also be the torsional potential energy. If the bond is rotated through 360° the energy should return to the same value. The energy needed for distorting a molecule by rotation around a bond is often very low, i.e., large deviations from the minimum energy structure may occur, and as such Taylor expansion in torsional angle α is not very appropriate. As such, to encompass the periodicity, V_{tors} is written as a Fourier series

$$V_{tors} = \sum_{i=1}^{N} [a_i \cos(i\alpha) + b_i \sin(i\alpha)] \tag{12.8.6}$$

where N represents the number of potential maxima or minima generated in a 360° rotation.

For a symmetric barrier where the potential energy is an even function of α,

$$V_{\text{tors}} = \sum_{i=1}^{N} V_i \cos{(i\alpha)} \tag{12.8.7}$$

where N represents the number of potential minima generated in a 360° rotation. Thus, $N = 1$ term describes a rotation that is periodic by 360°, the $N = 2$ term is periodic by 180°, the $N = 3$ term is periodic by 120°, and so on. The V_i constants determine the size of the barrier for rotation around the B–C bond (Figure 12.4). Depending on the situation, some of these V_n constants may be zero. It is customary to shift the zero point of the potential by adding a factor of one to each term. A factor 1/2 is included such that the V_i parameters directly give the height of the barrier if only one term is present. The V_i parameters may also be negative which corresponds to changing the minima on the rotational energy profile to a maxima or vice versa.

The torsional energy is often given by the expression

$$V_{\text{tors}}(\alpha) = \frac{1}{2}\sum_{i=1}^{6} V_i\left(1 + (-1)^{i+1}\cos{i\alpha}\right) \tag{12.8.8a}$$

The term $(-1)^{i+1}$ is included so that the function in the brackets within the sum is zero for $\alpha = \pi$. So,

$$V_{\text{tors}}(\alpha) = \frac{1}{2}V_1(1 + \cos{\alpha}) + \frac{1}{2}V_2(1 - \cos{2\alpha}) + \frac{1}{2}V_3(1 + \cos{3\alpha}) + \ldots \tag{12.8.8b}$$

It follows from this equation that the choice of the + and − signs is made such that the onefold rotational term has a minimum for an angle of 180°, the twofold rotational term has minima for angles of 0° and 180°, and the threefold rotational term has minima for angles of 60°, 180°, and 300° (−60°).

While the torsional potential energy function is in general an infinite Fourier series, in practice it is limited up to $i = 6$, particularly in case of internal rotation about a bond connecting sp^2- and sp^3-, or sp^2- and sp^2-hybridized atoms such as in acetaldehyde and acrolein. The threefold term 3α is important for the sp^3-hybridized system, the 2α term for the sp^2-hybridized system and the α term for large systems having rotational periodicity of 360°, as in fluoromethanol. Thus for a molecule having threefold internal rotation barrier such as CH_3Cl, ethane, etc. V_{tors} may be represented by the equation

$$V_{\text{tors}}(\alpha) = \frac{1}{2}V_3(1 + \cos{3\alpha}) \tag{12.8.9}$$

where V_3 is the maximum value of the potential barrier (Figure 12.7).

The torsional potential energy curves help in providing invaluable information about the geometrical, electronic, and thermodynamic parameters of the isomeric forms of molecules. Thus, for example, from the potential energy curves, it is possible to determine the relative stability, enthalpy difference, and rotational barrier between the conformers of a molecule and their torsional frequencies. The fundamental torsional frequencies and anharmonicity constants can be calculated by using the following relations [3–5]:

$$V_{\text{tors}} = FV^* \tag{12.8.10}$$

where $V^* = \sum_{m=1}^{6} m^2 V_m$ and $F = \frac{h}{8\pi^2 Ic}$ cm^{-1}.

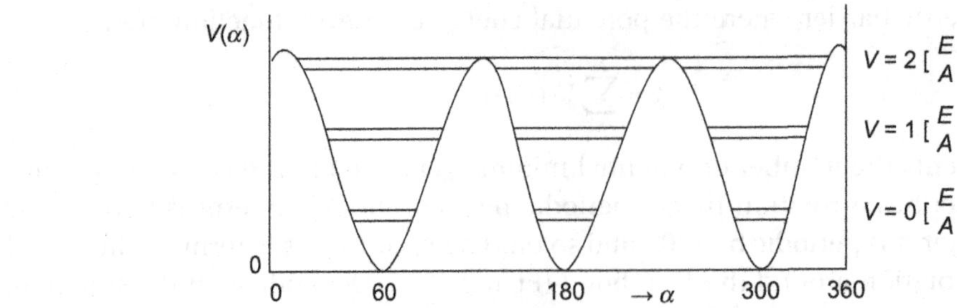

FIGURE 12.7 A cosine potential energy curve and torsional energy levels for internal rotation with a threefold barrier.

I is the moment of inertia which, for a diatomic molecule, is

$$I = m_1 r_1^2 + m_2 r_2^2 \tag{12.8.11}$$

Here, m_1 and m_2 are the atomic masses and r_1 and r_2 are the distances of the nuclei relative to the center of mass. In the case of a polyatomic molecule, Eq. (12.8.11) is replaced by

$$I = \begin{bmatrix} \sum_i m_i(y_i^2 + z_i^2) & -\sum_i m_i x_i y_i & -\sum_i m_i x_i z_i \\ -\sum_i m_i x_i y_i & \sum_i m_i(x_i^2 + z_i^2) & -\sum_i m_i y_i z_i \\ -\sum_i m_i x_i z_i & -\sum_i m_i y_i z_i & \sum_i m_i(x_i^2 + y_i^2) \end{bmatrix} \tag{12.8.12}$$

The atomic coordinates (x_i, y_i, z_i) are again relative to the center of mass. By choosing a suitable coordinate transformation, this matrix may be diagonalized with the eigenvalues being the *moments of inertia* and the eigenvectors called *principal axes of inertia.*

For infinitely high barriers, the internal rotation is confined to small torsional oscillations about the potential minima. In the case of a threefold rotation the torsional frequency for harmonic oscillations is given by the relation

$$\nu_{\text{tors}} = \frac{3}{2\pi}\sqrt{\frac{V_3}{2I}} \tag{12.8.13}$$

where I is the reduced moment of inertia for relative motion of the two groups. Using this relation the internal rotation barrier can be calculated from a knowledge of the torsional frequency which are obtained from vibrational spectroscopy studies.

12.8.4 Illustration

As an illustration, we can take the case of methyl vinyl ketone studied by Gupta and Thakur [6] by the semiempirical method AM1 (Austin Model-1).

Methyl vinyl ketone molecule (Figure 12.8) may be considered as a two rotor system in which internal rotation may take place about the $C_1{-}C_3$ single bond connecting its

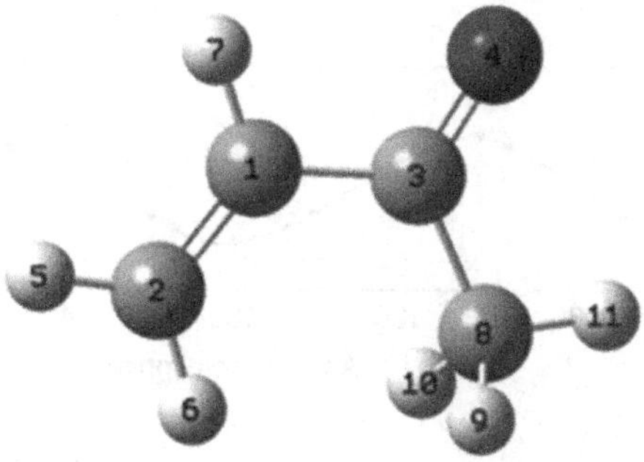

FIGURE 12.8 Methyl vinyl ketone molecule.

two completely asymmetric moieties $CH{=}CH_2$ and $O{=}CCH_3$ and also about the $C{-}CH_3$ bond. Thus, a number of conformations can theoretically arise for the molecule due to rotation about the two C–C bonds. In order to obtain the potential energy curve for asymmetric torsion, the dihedral angle ϕ ($C_2{-}C_1{-}C_3{-}O_4$) was changed from 0° to 360° in intervals of 20° (intervals of 10° at the turning points). Threefold potential function for the internal rotation of the methyl group about $C_3{-}C_8$ bond was obtained by changing the dihedral angle ϕ ($H_{11}{-}C_8{-}C_3{-}C_1$) in intervals of 20°.

A plot of energy versus angle of rotation ϕ shows that in the electronic ground state (S_0), the total energy has minima at 0° and 180° suggesting that the molecule exists in two stable forms—*trans* and *cis*; the *trans* conformer is more stable than *cis* by 0.95 kcal/mol. This value is close to 0.80 kcal/mol from experimental infrared studies by Durig and Little [7]. The two conformers are separated by a rotational barrier of 2.76 kcal/mol (experimental value 2.36 kcal/mol).

The potential energy curves of methyl vinyl ketone (Figure 12.9(a)) can be fitted with potential function of the form given in Eq. (12.8.8b) extended to include six terms.

The values of the potential constants V_1–V_6 are found to be 0.7937, 2.1237, −0.0871, −0.0106, 0.2397, and 0.1673, respectively. Using potential constants in the relationship Eq. (12.8.10) suggested by Margolin et al. [4], the torsional frequency and the anharmonicity constants were calculated for the more abundant *trans* conformer. The calculated value of the torsional frequency is 114.9 cm^{-1} which is close to the experimental value 101 ± 2 cm^{-1}. The anharmonicity constant for the torsional mode as calculated from Eq. (12.8.11) is found to be 4.5 cm^{-1}. These results show that no more than six Fourier terms are necessary in the potential function (Eq. (12.8.8)) for reproducing experimental results [8].

Likewise, potential constants and torsional frequencies of the methyl group in the *trans* and *cis* conformers of methyl vinyl ketone can be calculated from the potential energy curves given in Figure 12.9(b). Threefold potential function given by Eq. (12.8.9) can been fitted in this case to obtain V_3 and the torsional frequency calculated using Eq. (12.8.13). The reduced moment of inertia for the methyl group was obtained from Eq. (12.8.12). The calculated barrier to its internal rotation in the *trans* conformer is 0.96 kcal/mol against

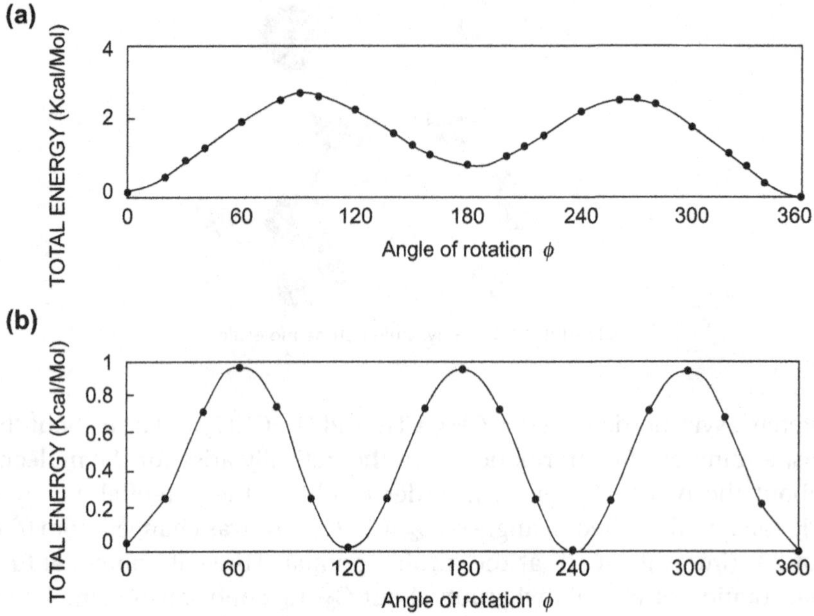

FIGURE 12.9 (a) Potential energy curves of methyl vinyl ketone in ground electronic state based on MO calculations. In (a) abscissa is the angle of rotation (ϕ) about the C_1–C_3 bond relative to the trans conformation for which the dihedral angle (C_2=C_1–C_3=O_4) is 180°, (b) Potential energy curve for rotation of methyl group. The abscissa is the angle of rotation ϕ about C_3–C_8 bond relative to the dihedral angle (H_{11}–C_8–C_3–C_1) = 180°.

the experimental value of 1.25 kcal/mol from microwave measurements [9]. The calculated torsional frequency of the methyl group in the trans conformer is 157 cm^{-1} against the experimental value 175 cm^{-1} [7].

12.9 Prediction of Activation Barrier

PESs can be used for predicting the activation energies of chemical reactions. The energies of the reactants, products, and transition states are calculated at their optimized geometries. Using the intrinsic reaction coordinate (IRC) calculations, it is also verified that the transition structure does connect the reactants and products for the reaction. For better results, frequency calculations at the optimized geometries are also carried out and the zero-point vibrational energies (ZPVE) obtained. Electron correlation methods such as MP2, MP4, and the density functional theory (DFT) methods are superior to those which ignore the correlation effects. As an example, the reaction involving transformation of diaminomaleonitrile (DAMN) to diaminofumaronitrile (DAFN) (Figure 12.10) may be considered [10].

It follows from Table 12.1 that the barriers in both directions of the reaction are essentially equal. Good barriers to internal rotation can usually be obtained by *ab initio*

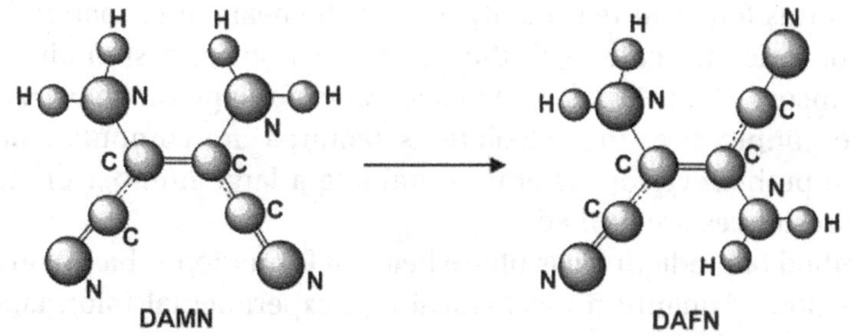

FIGURE 12.10 Molecular geometries of diaminomaleonitrile (DAMN) and diaminofumaronitrile (DAFN).

Table 12.1 Forward and backward activation energies for DAMN to DAFN transformation using B3LYP/6-31G(d,p)

Molecule	Electronic energy (au) (*a*)	ZPVE (au) (*b*)	Total energy (au) C = (*a*) + (*b*)	Forward activation energy (TS − DAMN) (kcal/mol)	Reverse activation energy (TS − DAFN) (kcal/mol)
DAMN	−373.79113	0.08439	−373.70674	47.02	46.85
DAFN	−373.79044	0.08406	−373.70638		
TS	−373.71309	0.08140	−373.63169		

SCF MO calculations, because of a near cancellation of correlation energies between different molecular conformations.

12.10 Heats and Free Energies of Formation and Reaction

Heat of formation is the heat evolved or absorbed during the formation of 1 mol of a substance from its component elements. Following this definition, the experimental heat of formation for a molecule is the (molar) enthalpy change associated with removing each of the atoms in the molecule from its elemental standard state and assembling them into the molecule. Direct computation of a molecular heat or free energy of formation is not feasible, as it would imply computing the difference in H or G for some molecule compared to the reference elemental standard states. Such a calculation might readily be imagined for a diatomic molecule like HCl, but not so for polyatomic molecules. The three quantities that are most routinely employed in practice are the 0 K enthalpy of formation, the 298 K enthalpy of formation, and the 298 K free energy of formation. Enthalpies of formation are also commonly known as heats of formation.

There are a variety of methods to predict gas phase heats of formation from quantum mechanical information. One method uses the known heats of formation of isolated atoms and calculated atomization energies (D_0) to predict gas phase heats of formation of molecules.

$$\Delta_f H^0(A_xB_y) = x\Delta_f H^0(A) + y\Delta_f H^0(B) - \sum D_0 \tag{12.10.1}$$

This method was found to reasonably predict the heats of formation for a variety of organic and inorganic molecules with the best predictions corresponding to those using the G2 level of theory. While the degree of accuracy of the predictions using this level of theory has been impressive, the calculations required are computationally expensive which might be prohibitive for systems containing a large number of atoms or where computational resources are limited.

Another method of predicting gas phase heats of formation is based on Hess' law and uses a combination of quantum mechanical and experimental information. Hess' law states that the standard reaction enthalpy is expressed as:

$$\Delta H^0_{\text{Reaction}} = \Delta H^0_f(\text{Products}) - \Delta H^0_f(\text{Reactants}) \qquad (12.10.2)$$

Using this method, the standard heat of formation of a single component of a reaction (either product or reactant) can be determined using the reaction enthalpy (which can be obtained from quantum mechanical calculations) and reliable values of heats of formation of the remaining products and reactants. The accuracy of the prediction using Eq. (12.10.2) is increased if the reaction is *isodesmic*. B3LYP/6-31G* predictions of heats of formation of hydrocarbons had a maximum difference of 6.2 kcal/mol with experiment which improved to 2.34 kcal/mol in 6-311+G** basis set [11].

12.10.1 Isodesmic Reactions

An isodesmic reaction is one in which the number of and types of bonds in the reactants and products is the same. It may even be an hypothetical reaction which may never occur. Isodesmic reactions represent a subclass of isogyric reactions, that is, reactions in which the reactants and products have the same number of electron pairs. Some of the examples of isodesmic reactions are:

Formation of ethane and propene from ethene and propane

$$C_2H_4 + C_3H_8 \rightarrow C_2H_6 + C_3H_6$$

Deprotonation of methyl halide by a methyl ion

$$CH_3^- + CH_3X \rightarrow CH_4 + CH_2X^-, \ X = F, Cl, Br, I$$

and substitution reaction

$$R(NH_2)C{=}C{=}O + CH_3CH{=}CH_2 \rightarrow CH_3(NH_2)C{=}C{=}O + RCH{=}CH_2;$$

$$R = Cl, Br$$

As may be seen, in the last reaction involving a substituted aminoketene, both the reactant and product have 11 single and 3 double bonds.

However, a reaction like $H_3C{-}CH_3 + H_2 \rightarrow 2CH_4$ is, strictly speaking, not isodesmic because although the number of single bonds on the two sides of the reaction is the same, the type of bonds are different as there is no H–H bond on the right-hand side.

The quantum mechanical determination of the heat of formation by isodesmic reaction method is superior to other methods due to cancellation of errors on the two sides of the reaction. Also, bond-by-bond errors in correlation energy largely cancel in the computed heat of reaction.

If the isodesmic reaction may be written in the form

$$mA + nB \rightarrow rC + sD \quad (12.10.3)$$

where *A*, *B*, *C*, and *D* are molecules and *m*, *n*, *r*, and *s* indicate the number of moles of each in the balanced equation, then following Eq. (12.10.2), the heat of reaction for the above chemical transformation can be calculated.

$$\Delta H^0_{\text{reaction}} = \left[r\Delta H^0_f(C) + s\Delta H^0_f(D)\right] - \left[m\Delta H^0_f(A) + n\Delta H^0_f(B)\right] \quad (12.10.4)$$

From a knowledge of the calculated value of the heat of reaction from Eq. (12.10.4) and the experimental heats of formation for all but one of the species in Eq. (12.10.3), it is then possible to determine the heat of formation of one of the species (say *B*), by using the equation:

$$\Delta H^0_f(B) = \frac{1}{n}\left[-\Delta H^0_{\text{reaction}} - \left\{r\Delta H^0_f(C) + s\Delta H^0_f(D) - m\Delta H^0_f(A)\right\}\right] \quad (12.10.5)$$

As an illustration, we may calculate the heat of formation of formaldehyde (H_2CO) in the reaction

$$CO_2 + CH_4 \rightarrow 2H_2CO$$

The following procedure is involved:

1. Geometries of all the species, both reactants and products are optimized at B3LYP/6-311G(d,p) level.
2. The frequency calculations conducted for all the species at the same level to obtain the ZPVE.

Energies corrected for the ZPVE for the reactants and the product and the experimental values of the heats of formation of CO_2 and methane are given in Table 12.2.

From Eqs (12.10.4) and (12.10.5) and Table 12.2, we get the calculated value of $\Delta H^0_{\text{reaction}}$ and the heat of formation of formaldehyde if the experimental values of this parameter for carbon dioxide and methane are known. Thus,

$$\Delta H^0_{\text{reaction}} = 2E^0(H_2CO) - E^0(CO_2) - E^0(CH_4) \quad \text{and,}$$

Table 12.2 Energies for the reactants and the product and the experimental values of the heats of formation of CO_2 and methane

	E (au)	ZPVE (au)	E^0 (au)	ΔH_f^{exp} (kcal/mol)
CO_2	−188.641139	0.01172	−188.629418	−93.96
Formaldehyde (H_2CO)	−114.536341	0.026475	−114.509866	?
Methane (CH_4)	−40.5337483	0.044604	−40.489145	−16.0

$$\Delta H_f^0(\mathrm{H_2CO}) = \frac{1}{2}\left[\Delta H^0_{\text{reaction}} + \left\{\Delta H_f^{\text{exp}}(\mathrm{CH_4}) + \Delta H_f^{\text{exp}}(\mathrm{CO_2})\right\}\right]$$

This gives calculated values

$$\Delta H^0_{\text{reaction}} = 62.019 \text{ kcal/mol} \quad \text{and,} \quad \Delta H_f^0(\mathrm{H_2CO}) = -23.97 \text{ kcal/mol}.$$

The calculated value of heat of formation of formaldehyde is close to the experimental value −25.0 kcal/mol.

12.10.2 Substitution Effect on Stabilization Energy of Chemical Reaction by Isodesmic Reaction Methods

Isodesmic reactions can be used to compare the effect of different substituents on the stabilization energies of molecules and correlate the group electronegativities and stabilization energies. Gupta et al. [12] have used isodesmic reactions to compare the effect of substituents on a group of 24 mono- and disubstituted ketenes. This approach offers the advantage that the effect of the substituent on the property in question is isolated and systematic errors are minimized. As an illustration, we consider this study in greater details.

In order to compare the effect of substituents on aminoketene relative to those on alkenes, with CH_3 as a reference substituent the following isodesmic reactions can be utilized

$$\mathrm{R(NH_2)C{=}C{=}O + CH_3CH{=}CH_2 \underset{\Delta E}{\overset{SE}{\longrightarrow}} CH_3(NH_2)C{=}C{=}O + RCH{=}CH_2} \tag{12.10.6}$$

A more general expression may be written as

$$\mathrm{M_2(M_1)C{=}C{=}O + CH_3CH{=}CH_2 \underset{\Delta E}{\overset{SE}{\longrightarrow}} CH_3(M_1)C{=}C{=}O + M_2CH{=}CH_2} \tag{12.10.7}$$

where $M_1 = NH_2$ and $M_2 = H, CH_3, NH_2, OH,$ and OCH_3 (electron donor groups), CHO, NO, and NO_2 (π-electron acceptor groups), and $HC{=}CH_2$, $C{\equiv}CH$, and $C{\equiv}N$ (conjugated substituents).

As a first step, the geometrical parameters of all the reactants and products are fully optimized and used to compute electronic energies, harmonic vibrational frequencies, and ZPVE. The results based on MP2/6-31G** calculations are given in Table 12.3. This table also contains stabilization energies calculated by using Eq. (12.10.2) for substituted aminoketenes and alkenes.

It follows from Table 12.3 that with electropositive groups like H, CH_3, NH_2, and OH the stabilization energy (SE) has negative values ranging from −7.26 to −2.29 kcal/mol. These groups, therefore, have destabilizing effect on ketenes. A number of conjugating substituents like $C{\equiv}CH$, $C{\equiv}N$, and $HC{=}CH_2$ have the ability to act both as electron donors and electron acceptors. While the SE energies for the first two substituents are found to be −0.66 and −0.31 kcal/mol, respectively, showing that they too have a very mild destabilizing effect, the SE energy for $HC{=}CH_2$ is 12.99 kcal/mol and points toward a strong stabilizing influence of this group. In contrast, electron acceptor groups like NO, NO_2, CHO have positive values of SE ranging from 4.25 to 18.29 kcal/mol and hence show strong stabilizing influence. The stabilization energies are found to depend both

Table 12.3 Total energies, zero-point vibrational energies (ZPVE) and stabilization energies (SE) for isodesmic reactions for substituted aminoketenes ($\phi_{NH_2} \sim 120°$) and alkenes, and group electronegativities (χ)

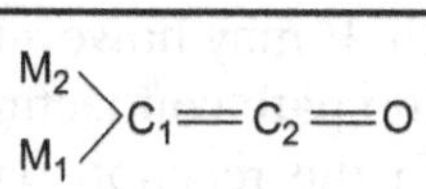

S.No	M_1	M_2	Energy (au)— $E(M_2(M_1)C{=}C{=}O)$	ZPVE (au)	Energy (au)— $E(M_2CH{=}CH_2)$[a]	ZPVE (au)	SE (kcal/mol)	Relative SE (kcal/mol)	χ
1	NH_2	H	−207.3222549	0.051137	78.2983	0.0495	−7.26	0.0	3.10
2	NH_2	CH_3	−246.4923871	0.080177	117.4697	0.0775	0.0	7.26	2.56
3	NH_2	C≡CH	−283.2438222	0.059559	154.2249	0.0596	−0.66	6.6	2.66
4	NH_2	$HC{=}CH_2$	−284.4619602	0.084949	155.4226	0.0864	12.99	5.73	2.58
5	NH_2	NO(t)	−336.3116671	0.047958	207.2655	0.0494	17.30	10.04	3.12
6	NH_2	NO(c)	−336.3162770	0.048160					3.12
7	NH_2	CN	−299.3359892	0.049872	170.3161	0.0495	−0.31	6.95	2.69
8	NH_2	NH_2	−262.5097603	0.068902	133.4913	0.0668	−2.29	4.97	3.12
9	NH_2	CHO(t)	−320.3595112	0.061004	191.3116	0.0623	6.15	13.41	2.60
10	NH_2	CHO(c)	−320.3571222	0.060975	191.3093	0.0623	18.29	11.03	2.60
11	NH_2	NO_2	−411.3382322	0.054343	282.3113	0.0542	4.25	11.51	3.22
12	NH_2	OCH_3(c)	−321.4990160	0.084627	–	–	–	–	3.53
13	NH_2	OCH_3(g)	−321.5015142	0.084824	–	–	–	–	3.53
14	NH_2	OH	−282.3455589	0.055511	153.3322	0.0545	−4.81	2.45	3.64

[a]R.D. Brown, P.D. Godfrey, M. Woodruff, Aust. J. Chem. 32 (1979) 2103.

upon the conformation of the amino group ($\phi_{NH_2} \sim 120°$ or $60°$) as well as the *cis* and *trans* orientations of the substituents.

It is found that in the case of most stable conformers of substituted aminoketenes with $\phi_{NH_2} \sim 120°$, a correlation exists between the stabilization energies and group electronegativities χ of the substituents. In the case of electron-donating groups this relationship is linear.

12.11 Reaction Pathways and Intrinsic Reaction Coordinates

It may be said at the outset that there is not one single valid definition of reaction path. Depending upon how one wishes to perform dynamical calculations, various definitions of the reaction path might be appropriate. As discussed above, adiabatic potential of a chemically reacting system can be theoretically obtained as a function of the variables which can determine the atomic arrangement. Such a potential function, *V*(*R*), possesses several equilibrium points where all of their derivatives vanish. Among these equilibrium points, those of interest are the initial stable point, the transition-state point, and the final stable point of a reaction.

Any individual step in a chemical process normally involves some movement of atoms and changes in the structural parameters of the participants. In one definition, the paths of minimum potential energy linking the transition state to the reactants and products can be called reaction path. It may however be noted that two minima on the PES may have more than one reaction path connecting them, corresponding to different transition structures through which the reaction passes. Thus, to confirm a reaction mechanism, it may be necessary to prove that the particular transition structure found in the optimization connects the desired reactants and products. This can be done by following the path of steepest descent downhill from the transition structure toward the reactants and toward the products. Following a reaction path can also show whether the mechanism involves any intermediates between reactants and products. Although the path of steepest descent depends on the coordinate system, a change in the coordinate system does not change the nature of the stationary points and does not alter the fact that the energy decreases monotonically along the reaction path from the transition structure toward reactants and products. Thus any coordinate system can be used to explore the mechanism of a reaction. One system, mass-weighted Cartesian coordinates, has special significance for reaction dynamics, and the path of steepest descent in this coordinate system is called the intrinsic reaction coordinate (IRC). It provides an effective and convenient basis for analyzing the course of the reaction step. However, reaction paths have also been analyzed in terms of Cartesian coordinates and internal coordinates comprised of the stretching, bending, torsion of bonds and adiabatic local modes.

12.11.1 Path of Steepest Decent and Intrinsic Reaction Coordinate

The classical equations of motion can be used to establish the paths of minimum potential energy linking the transition state to the reactants and products. A fuller treatment of these equations is given by Collins [13].

If $\mathbf{x}(i)$ denotes the Cartesian coordinate vector (x_i, y_i, z_i) of the ith atom of mass m_i in a molecule of N atoms, we define the mass-weighted Cartesian coordinate $\xi(i)$, as

$$\xi(i) = \sqrt{m_i}\,\mathbf{x}(i) \tag{12.11.1}$$

The Lagrangian is then

$$L = \sum_{i=1}^{N} \frac{1}{2} m_i \left|\frac{dx(i)}{dt}\right|^2 - V[x(1), x(2), \ldots x(N)]$$

$$= \sum_{i=1}^{N} \frac{1}{2} \left|\frac{d\xi(i)}{dt}\right|^2 - V[\xi(1), \xi(2), \ldots \xi(N)] \tag{12.11.2}$$

On solving the Lagrangian equation

$$\frac{d}{dt}\left(\frac{\partial L}{\partial \dot{\xi}}\right) - \frac{\partial L}{\partial \xi} = 0,$$

We get,

$$\frac{d^2\xi(i)}{dt^2} = -\frac{\partial V}{\partial \xi(i)}, \quad i = 1, 2, \ldots N \tag{12.11.3}$$

If ζ represents the 3N-dimensional vector $[\xi(1), \xi(2), \ldots \xi(N)]$, this equation may be written as

$$\frac{d^2\zeta}{dt^2} = -\frac{\partial V}{\partial \zeta} \tag{12.11.4}$$

If the motion starts at the saddle point $\zeta(t = 0)$ and the initial velocities contain a component either toward the reactant or the product, the solution of the above equation will lead to a vibratory motion. $\zeta(t)$ will trace out a trajectory such that it shall accelerate down the valley while oscillating from end to end (Figures 12.11 and 12.12).

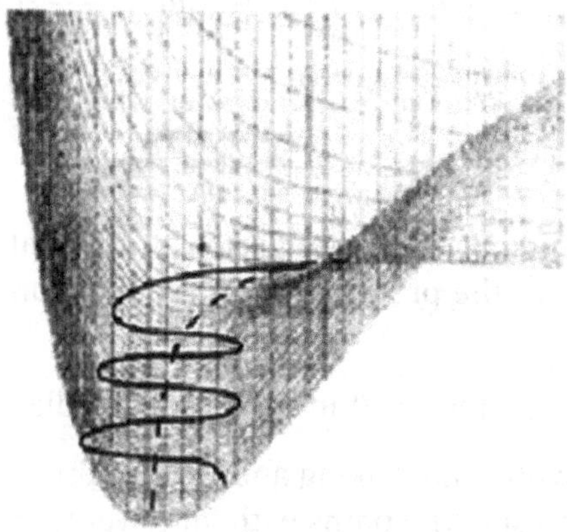

FIGURE 12.11 PES sketching the projection of classical trajectory onto the PES (solid line) and a similar projection of the path of steepest descent (dashed line). *Reprinted with permission from Ref. [13], Copyright @ 1996, John Wiley and Sons Publishing.*

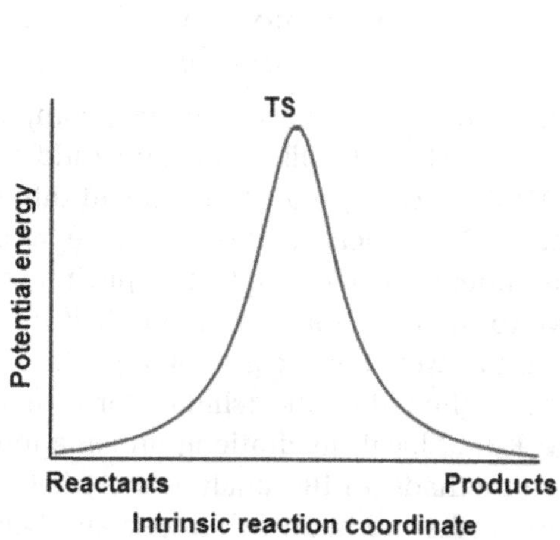

FIGURE 12.12 Intrinsic reaction path connecting reactants and products with the transition state.

Changing the initial atomic velocities may generate similar but different oscillatory trajectories or paths. None will pass directly down the middle of the valley. If the velocities of atoms become very small (almost zero), say under heavy friction, the above equation leads to

$$\frac{d\zeta}{dt} = -\frac{\partial V}{\partial \zeta} \tag{12.11.5}$$

This is an equation of the *path of steepest decent.* Obviously, there are two such paths from the saddle point—one toward the reactants and the other toward the products. The combined path from reactants to products through the transition state is called *the intrinsic reaction path* (IRP).

Equations (12.11.4) can also be expressed in terms of distance ds along the IRP. If $\zeta(t)$ and $\zeta(t+dt)$ are neighboring points on the IRP separated by a distance ds, then

$$ds^2 = |\zeta(t+dt) - \zeta(t)|^2 = \left|\frac{d\zeta}{dt}\right|^2 dt^2 = \left|\frac{\partial V}{\partial \zeta(t)}\right|^2 dt^2 \tag{12.11.6}$$

Using Eqs (12.11.5) and (12.11.6), we get,

$$\frac{d\zeta}{ds} = -\frac{\partial V}{\partial \zeta} \Big/ \left|\frac{\partial V}{\partial \zeta}\right| \tag{12.11.7}$$

The distance s has units of (length mass$^{\frac{1}{2}}$) and is called the *intrinsic reaction coordinate.* It is a measure of the progress along the reaction path. By convention,

$s = 0$ at the saddle point,
$s < 0$ in the reactant valley and $s > 0$ in the product valley.

Since reactions often involve the making and/or breaking of a small number of bonds, it should be possible to represent the paths with far fewer than $3N - 6$ coordinates. If s' is a new reaction coordinate (say, the distance along the path of steepest decent in interatomic distance coordinate R), then Eq. (12.11.7) may be written as

$$\frac{dR}{ds'} = -\left(\frac{\partial V}{\partial R}\right) \Big/ \left|\frac{\partial V}{\partial R}\right| \tag{12.11.8}$$

The steepest decent path on a PES in a well-defined, complete set of coordinates is also the minimum energy path (MEP). It will connect the saddle point to the appropriate energy minima and is a MEP. Since Eq. (12.11.7) is a stiff differential equation, care is taken during the integration. Numerical methods for integrating ordinary differential equations are classified as either explicit or implicit. Explicit methods use information at the current point to move to the next point, while implicit methods require derivative information at the end point as well. For integrating Eq. (12.11.7), the common explicit algorithms used are Euler's method, the IMK (Ishida–Morokuma–Komornicki) stabilized Euler method, the Runge–Kutta, local quadratic approximation (LQA) method and its modifications. The implicit methods for IRC analysis include the Müller–Brown method (implicit Euler), the second-order method of Gonzalez and Schlegel (GS2), and higher order methods by the same authors. A review of different methods available for calculating MEPs has been given by Collins [13].

12.11.2 Steps for IRC Calculations

The IRC calculations start from the transition state and move in both directions—toward the reactants and the products. Following steps are involved in these calculations:

1. The first step for the IRC calculations is to optimize the starting transition state which is performed in the usual manner of geometry optimization except that now the potential energy maximum instead of the minimum is being searched.
2. A frequency calculation is then performed on this optimized structure so as to generate force constants required for IRC calculations. The frequency calculation also helps to verify that the structure selected is indeed a true transition state having a single negative frequency. It also provides the zero-point energy of the transition state.
3. Starting from the transition state the path of steepest decent is then calculated using an appropriate algorithm. In the Gaussian suite of programs, the IRC keyword is used for this purpose. By default it performs calculations in both the directions for a number of points which can be defined to come closer to the potential minima.
4. The IRC calculations may not necessarily lead to the original structures of the reactants and products and are more of suggestive nature.
5. In order to accurately predict the barrier for reaction, additional calculations using higher level basis sets are needed. This may also require optimizations of the reactants, products, and transition states at a higher level of theory and frequency calculations to get the zero-point energies.

Reaction coordinate diagrams show clear distinctions between transition states, stable states, and transient intermediates. A relatively stable structure is given by a low energy depression in the curve, reactive intermediates are high-energy shallow wells, and transition states are peaks. Any chemical structure that lasts longer than the time for a typical bond vibration (10^{-13}–10^{-14} s) can be considered an intermediate. When a reaction involves more than one chemical step, one or more intermediates are formed. This means that there is more than one energy barrier that must be traversed during the reaction and there is more than one transition state. Almost all computer softwares meant for electronic structure calculations have algorithm for performing IRC calculations.

12.11.3 Illustration

These steps are illustrated by the following example from the work of Gupta et al. [1] suggesting the formation of cytosine in interstellar space and nitrile rich atmospheres through interaction between propylidine (CCCH) and isocyanic acid (HNCO) in the presence of NH and NH_2. The example depicts formation of the ring structure in molecule **2** from the open structure in molecule **1** (Figure 12.13(c)). Starting from the optimized transition state, the IRC calculations were conducted by DFT at B3LYP/6-31G** level using step size of 0.1 Bohr (0.053 Å). The reaction path is followed by integrating the intrinsic reaction coordinate using the mass-weighted Cartesian coordinates. Hessian-based predictor–corrector integrator, a very accurate algorithm that uses the Hessian-based

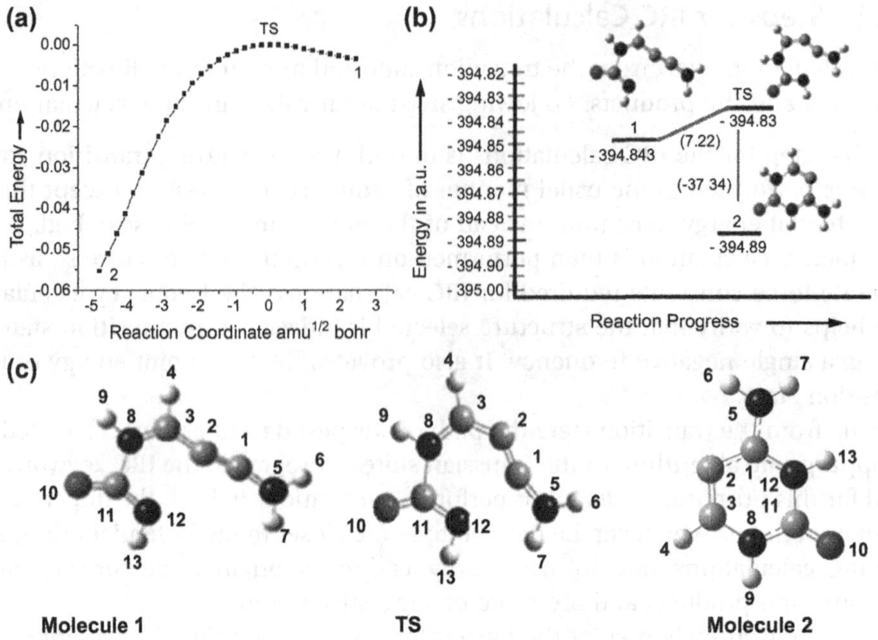

FIGURE 12.13 (a) Reaction path for molecule 1→2 relative to transition state energy of −394.380 au (b) Energy level diagram and (c) Geometrics of molecules 1,2 and transition state (TS).

LQA as the predictor component and a modified Bulrisch–Stoer integrator for the corrector portion, is used in the calculations. The results are given in Figure 12.13(a–c). A rearrangement of atoms of molecule **1** in the formation of the pyrimidine ring **2** passes through a transition state (TS) which offers a potential barrier of 7.22 kcal/mol. The reaction path as calculated from IRC calculations connects the transition state to the reactant **1** and the product **2** (Figure 12.13(a)).

12.12 Photodissociation of Molecules and Bond Dissociation Energies

Photodissociation, photolysis, or photodecomposition is a chemical reaction in which a chemical bond is broken down by photons. Any photon with sufficient energy can affect the chemical bonds of a chemical compound. Since a photon's energy is inversely proportional to its wavelength, electromagnetic waves with the energy of visible light or higher, such as ultraviolet light, X-rays, and gamma rays are usually involved in such reactions. In astrophysical and atmospheric chemistry, photodissociation is one of the major processes through which molecules are broken down (but new molecules are being formed). Photodissociation rates are important in the study of the composition of interstellar clouds.

Thus, for example, study of dissociation mechanism in molecules of the type *ABC* and AB_2, where *A*, *B*, and *C* may represent molecular or atomic species, has received considerable attention recently. A few studies of this type have been reported for acetone $CO(CH_3)_2$, azomethane $N_2(CH_3)_2$, halomethanes CF_2I_2, and carbonyl cyanide $CO(CN)_2$. It is possible to study bond rupture processes and photodissociation and rearrangement mechanisms in short-lived molecules by quantum chemical methods. It is also possible to correlate the photodecomposition process with the electronic spectral and structural characteristics of the constituent groups.

A three-body photodecomposition event in a molecule may proceed either via sequential reaction mechanism such as

$$ABC + h\nu \rightarrow AB + C$$

$$AB + C \rightarrow A + B + C$$

or by a concerted mechanism such as

$$ABC + h\nu \rightarrow A + B + C$$

Thus, in molecules of the type AB_2 such as $CO(CN)_2$ and $CO(CCH)_2$ ($A = CO$, $B = CN$, CCH) the routes can be

$$AB_2 \rightarrow AB + B \rightarrow A + B + B \quad \text{(sequential)}$$

or

$$\text{or } AB_2 \rightarrow A + B + B \quad \text{(concerted)}$$

A detailed theoretical study on AB_2 systems has been reported by Gupta et al. [14] using restricted and unrestricted DFT and MP2 theories. Since the character of the wavefunction may change while going from a closed-shell electronic state to an open-shell state along the reaction course, the quantum chemical calculations are conducted using both the restricted and unrestricted methods. As a first step, geometrical parameters are fully optimized using split-valence basis sets involving diffuse and polarization functions such as 6-311++G(2df,2p), 6-311+G*, and 6-31G*, which are also used to compute electronic energies, harmonic vibrational frequencies, and zero point energies (ZPE). A search for the transition state structure is carried out by the STQN (QST2) procedure suggested by Peng et al. [15]. The transition state is characterized by harmonic frequency analysis. Intrinsic reaction coordinate (IRC) studies are conducted to analyze the reaction pathways from the transition states. PES scans (Figure 12.14) are performed to get an estimate of the dissociation energy and understand bond rupture processes.

From studies on molecules and radicals like carbonyl cyanide ($CO(CN)_2$), diethynyl ketone, acetyl cyanide (CH_3COCN), formyl cyanide (HOCN), and CH_3CO radical, it is found that stepwise decomposition process is preferred over the concerted reaction process. The dissociation process in carbonyl cyanide is given in Figure 12.15. Based on potential energy curves for bond dissociation and the transition state and IRC studies, it is found that besides the direct dissociation of carbonyl cyanide, a

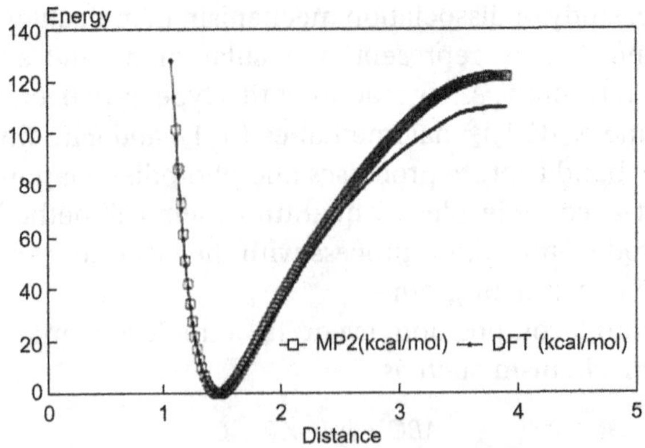

FIGURE 12.14 Potential energy curves for C–CN bond dissociation in carbonyl cyanide. C–CN bond length in Å and energies relative to minima in DFT (−187,657.7 kcal/mol) and MP2 (−187,159.9 kcal/mol) are given along the abscissa and ordinate.

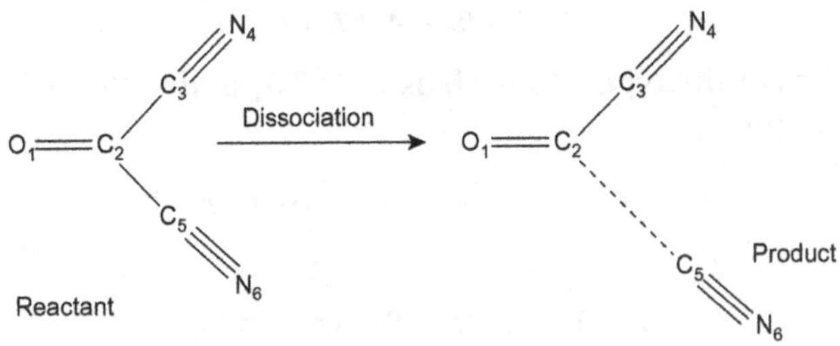

FIGURE 12.15 Dissociation of carbonyl cyanide.

photoisomerization process through a nonplanar transition state may also occur resulting in the formation of a stable and planar isomer CNC(O)CN.

12.13 Chemical Reactivity and Its Indicators

The behavior of an atom or a molecule is often characterized by some parameters that chemists have extracted from their experience and used for the prediction of chemical reactivity. After the introduction of the concept of sharing of electrons by G.N. Lewis in 1916, the electron theory of chemical reactivity has been developed by a number of chemists. Ingold [16] developed a theory in which the electron density distribution and its change in a chemical compound played an important role in determining the chemical reactivity of organic compounds. Various chemical concepts were understood in terms of the electronic charge. The electronic theory was able to explain a majority of complicated experimental results in a unified manner. The static and dynamic behaviors of molecules could be explained by the electronic effects which are based on nothing but the distribution of electrons in a molecule. The mode of charge distribution in a molecule can be sketched to some extent by the use of the *electronegativity* concept of atoms through organic chemical experience. This leads to the formulation of group electronegativity concept. At the same time, it is given foundation and made quantitative by theoretical calculations based on quantum theory. Thus, for example, the position of attack in a chemical reaction was discussed by electrostatic principle. However, the problem arose in the case of substitution reaction of the nonsubstituted aromatic hydrocarbons such as naphthalene which reacts with electrophiles (electron-deficient reagents) and nucleophiles (electron-rich reagents) at the same position **1**. This leads to the formulation of another concept of reactivity by Coulson and Longuet-Higgins [17,18] which involved change of electron density under the influence of the attacking reagent.

Realizing the principal role of valence electrons in the formation of a molecule from constituent atoms, Fukui correlated chemical activity with the density distribution in the highest occupied molecular orbital (HOMO) and lowest unoccupied molecular orbital (LUMO). These orbitals are known as "frontier orbitals." Fukui obtained an agreement between the position of attack and the site of largest density of the HOMO for the attack of an electrophilic reagent. In the case of a nucleophilic attack, he calculated the distribution of the LUMO and found a correlation with reactivity. In the radical substitution, the distribution of (HOMO + LUMO) possessed a correlation with reactivity [19,20]. Based on electron delocalization between the frontier orbitals of the substrate and the reagent, this approach was soon generalized to all types of molecules and reactions [21,22]. The electron delocalization between HOMO and LUMO was found to be the principal factor in all cases. So, the easiness of a chemical reaction could be decided on the basis of the HOMO–LUMO interaction. It was also used to decide about the stereoselective path [23–29] and the intramolecular as well as intermolecular processes. Some reactivity indices were derived from this delocalization picture.

Based on the DFT a large number of chemical reactivity indices have been proposed by Pople and coworkers and others. Prominent among these are group electronegativities, global hardness and softness [30], local hardness [31,32], local softness [32], Fukui functions [19] and the corresponding condensed forms [20], relative electrophilicity and relative nucleophilicity [21], global electrophilicity [33], and more recently the philicity index [34]. Although, the global reactivity descriptions are supposed to provide intermolecular reactivity trends, the local counterpart generates the intramolecular reactivity sequence or site selectivity of an individual chemical system. These concepts have played a key role in the organization and rationalization of chemical facts and observations [35–37].

12.14 Electronegativity and Group Electronegativity

12.14.1 Electronegativity

Electronegativity is a chemical property that describes the tendency of an atom or a functional group to attract electrons toward itself. The electronegativity of an atom is affected by both its atomic number and the distance that its valence electrons reside from the charged nuclei. The concept of electronegativity was first proposed by Pauli in 1932 as an explanation of the fact that the covalent bond between two different atoms (*A–B*) is stronger than would be expected by taking the average of the strengths of the *A–A* and *B–B* bonds. This additional stabilization of the heteronuclear bond has been explained by the valence bond theory due to the contribution of the ionic canonical forms to the bonding. Electronegativity is a relative term and only the differences in the electronegativity are defined. Thus, Pauli proposed

$$\chi_A - \chi_B = \left[D(AB) - \frac{D(AA) + D(BB)}{2}\right]^{1/2} (\text{eV})^{-1/2}, \tag{12.14.1}$$

where χ_A and χ_B are electronegativities of atoms *A* and *B* and *D*(*AB*), *D*(*AA*), and *D*(*BB*) are the dissociation energies of bonds *A–B*, *A–A*, and *B–B*, respectively, expressed in units of electron volts (eV). The inclusion of $(\text{eV})^{-1/2}$ expresses $\chi_A - \chi_B$ as a dimensionless quantity. In order to construct an electronegativity scale, Pauli chose hydrogen as reference as it forms covalent bonds with a large variety of elements and fixed its electronegativity at 2.1, which was later revised to 2.20. The atomic electronegativity scale so constructed is known as Pauli's electronegativity scale. Electronegativity was originally defined as an invariant property of atoms. Recently several workers have suggested that electronegativity of an atom depends upon its environment in the molecule. Thus, it depends on hybridization and oxidation state of the element.

Mulliken [38] introduced the concept of absolute electronegativity and proposed that the arithmetic mean of the first ionization energy (*I*) and the electron affinity (*A*) of an element should be the measure of the tendency of an atom to attract electrons.

$$\chi = \frac{(I + A)}{2} \tag{12.14.2}$$

This expression based on the definitions of *I* and *A*, follows from simple considerations. Thus, if *X* and *Y* are two systems of interest, then the condition that they have equal electronegativity is that the energy changes for the reactions

$$X + Y \rightarrow X^+ + Y^-$$

and

$$X + Y \rightarrow X^- + Y^+$$

are equal.

$$\text{or, } [E(X^+) - E(X)] + [E(Y^-) - E(Y)] = [E(X^-) - E(X)] + [E(Y^+) - E(Y)] \tag{12.14.3}$$

By definition, $E(X^+) - E(X) = I(X)$ is the ionization energy of *X* and $E(X) - E(X^-) = A(X)$ is its electron affinity.

Hence from equation (12.14.3), $I(X) - A(Y) = I(Y) - A(X)$

$$\text{or, } [I(X) + A(X)] \text{ for system } X = [I(Y) + A(Y)] \text{ for system } Y \tag{12.14.4}$$

It, therefore, makes sense to define electronegativity as $(I + A)$. The factor 1/2 in Eq. (12.14.2) was introduced by Mulliken as he considered that χ as an arithmetic mean of *I* and *A* is an easily grasped concept.

12.14.2 Group Electronegativity

Most of the attempts to develop electronegativity scale considered electronegativity as an atomic property although from Pauli's original definition, electronegativity is the power or tendency of a group of atoms in a molecule to attract electrons to themselves. In organic chemistry, electronegativity is associated more with different functional groups rather than with individual atoms. The electronegativity associated with functional groups is therefore called group electronegativity or substituent electronegativity. Despite computational advances and availability of large number of experimental data [39,40], not many theoretical studies have been reported for group electronegativity. Relatively few methods have been proposed for the evaluation of group electronegativity and most of them have been used to evaluate only small subsets of the chemically interesting groups of atoms. It is common to distinguish between the inductive effect and the resonance effect which are described as σ- and π-electronegativities, respectively. The quantization of the induction and resonance effects is most often done by using Hammett's equations. Gupta et al. [12] studied isodesmic substituent stabilization energies in disubstituted ketenes relative to alkenes and correlated them with group electronegativities. They also analyzed the role of induction effect (F) and resonance effect (R) parameters of the substituent groups on charge distribution. A similar study has been reported by McAllister and Tidwell [41]. Boyd and Boyd [42,43] introduced the bond critical point model for the determination of group electronegativities and reported values for over 100 groups from *ab initio* calculations.

The electronegativity of a group A was calculated from the properties associated with the A–H bond critical point in AH. For example, the electronegativity χ of the methoxy group CH_3O– was obtained by determining the position of the O–H bond critical point in CH_3OH and substituting the appropriate values in the following equations:

$$F_A = r_H / [N_A \rho(r_c) r_{AH}] \tag{12.14.5a}$$

$$\chi_A = 1.938 F_A^{-0.2502}, \tag{12.14.5b}$$

where F_A is the electronegativity factor of the atom A, r_H is the distance from the bond critical point to hydrogen nucleus, N_A is the number of valence electrons of atom A (atom O in case of CH_3OH), $\rho(r_c)$ is the density at the bond critical point, and r_{AH} is the internuclear distance. The bond critical point and the electron density at this point are determined by using the quantum theory of atoms-in-molecules. The group electronegativities of some of the important functional groups determined by Boyd and Boyd [43] using Eq. (12.14.5b) at the HF/6-31G*//HF/6-31G* level are given in Table 12.4. In this table, the values in the parenthesis separated by a comma were obtained using HF/6-31G*//experimental geometries and extended basis set// experimental geometry.

Table 12.4 Group electronegativities of some important functional groups

Group	χ	Group	χ
−CH$_3$	2.55(2.55,2.56)	−CCl$_2$F	2.71(2.67, –)
−CH$_2$CH$_3$	2.55(2.55,2.56)	−CCl$_3$	2.70(2.66, –)
−CH$_2$CH$_2$CH$_3$	2.55(2.55, –)	−NH$_2$	3.12(3.10,3.10)
−CH$_2$CH(CH$_3$)$_2$	2.54(2.53, –)	−NHCH$_3$	3.13(3.11, –)
−CH$_2$C≡N	2.58(2.56, –)	−NHCOH	3.18(3.17, –)
−CH$_2$COH	2.58(2.56, –)	−N(CH$_3$)$_2$	3.13(3.08, –)
−CH$_2$COO$^-$	2.52(–, –)	−N=C=O	3.18(3.20,3.22)
−CH$_2$COOH	2.58(–, –)	−N≡C	3.26(3.26,3.30)
−CH$_2$COF	2.59(2.58, –)	−NHNH$_2$	3.13(3.12, –)
−CH$_2$CSH	2.58(2.57, –)	−N=N=N	3.15(3.23, –)
−CH$_2$COCl	2.59(2.57, –)	−N=O	3.12(3.05,3.06)
−CH(CH$_3$)$_2$	2.55(2.54, –)	−NHF	3.19(–, –)
−CH(CH$_2$)$_2$	2.57(2.56, –)	−NF$_2$	3.25(–, –)
−CH=CH$_2$	2.58(2.57,2.61)	−O$^{\cdot}$	3.36(–, –)
−CH=C=CH$_2$	2.58(2.57, –)	−OH	3.55(3.52,3.64)
−CH=C=C=O	2.61(–, –)	−OH$_2^+$	3.57(–, –)
−CH=C=O	2.58(2.57, –)	−OCH$_3$	3.53(3.51,3.70)
−C(CH$_3$)$_3$	2.55(2.53, –)	−OC≡N	3.57(–, 3.73)
−C$_6$H$_5$	2.58(2.58, –)	−OCOH	3.56(3.50,3.65)
−C≡CH	2.66(2.65,2.66)	−OCOCH$_3$	3.57(–, –)
−C≡CF	2.66(2.66, –)	−ONH$_2$	3.58(3.54, –)
−C≡CCl	2.66(2.66, –)	−ON=O	3.55(3.49, –)
−CH$_2$NH$_2$	2.55(2.54, –)	−OO$^{\cdot}$	3.58(3.51, –)
−CH$_2$NHCH$_3$	2.57(2.57, –)	−Of	3.60(3.56, –)
−CH$_2$NO$_2$	2.62(2.61, –)	−OPH$_2$=O	3.54(–, –)
−CH=NH	2.59(2.56, –)	−MgH	1.30(1.31,1.33)
−CH=N=N	2.60(2.60, –)	−AlH$_2$	1.60(1.61,1.62)
−C≡N	2.69(2.68,2.69)	−SiH	1.87(1.87, –)
−C≡NO	2.69(2.73, –)	−SiH$_3$	1.90(1.89,1.91)
−CH$_2$OH	2.59(2.58,2.59)	−SiF	1.88(–, –)
−CH$_2$OOCH$_3$	2.61(2.60, –)	−SiF$_2$H	1.93(–, –)
−COH	2.60(2.58,2.60)	−SiCl	1.89(1.85, –)
−COCH$_3$	2.59(2.59, –)	−SiClH$_2$	1.91(1.90, –)
−CONH$_2$	2.61(2.60, –)	−SiCl$_2$H	1.93(1.92, –)
−CO$_2^-$	2.49(–, –)	−SiCl$_3$	1.95(1.94, –)
−COOH	2.63(2.62,2.66)	−PH$_2$	2.17(2.16,2.17)
−COOCH$_3$	2.64(2.62, –)	−PH$_2$=O	2.21(–, –)
−C≡O	2.57(2.54,2.57)	−PH(OH)=O	2.22(–, –)
−COF	2.67(2.65, –)	−P(OH)$_2$=O	2.25(–, –)
−COCl	2.66(2.64, –)	−S$^{\cdot}$	2.52(–, –)
−CFH$_2$	2.60(2.61,2.61)	−SH	2.65(2.66,2.63)
−CF$_2$H	2.65(2.64, –)	−SCH$_3$	2.65(2.65, –)
−CF$_3$	2.71(2.68, –)	−SCN	2.70(–, –)
−CClH$_2$	2.61(2.60, –)	−SSH	2.68(2.67, –)
−CClFH	2.66(2.63, –)		

Adapted with permission from [43]. Copyright @ 1988, American Chemical Society.

12.15 Chemical Reactivity Indices and Their Mathematical Formulation

In order to understand reaction between two molecules, it is important to know their individual properties as isolated molecules. During the interaction process these properties change either due to change in the number of electrons or due to change in the external potential resulting in the change of chemical potential. A number of chemical reactivity indices have therefore been identified and their rigorous formal foundation has been laid on the basis of the DFT. Here we shall briefly review their definition and give their mathematical foundation.

12.15.1 Electronegativity

The concept of electronegativity and its evolution has been described in the last section. The popularity of the concept is due to its simplicity and to the availability of numerical values for most of the elements. Numerous correlations between atomic electronegativities and a variety of chemical and physical properties have played a key part in the organization and interpretation of chemical facts and observations. It is found that Eq. (12.14.2) given by Mulliken [38] to define electronegativity χ as the average of ionization potential I and electron affinity A, holds for any chemical system—atom, ion, molecule, or radical. It is a useful measure of the tendency of a species to attract electrons.

The concept of electronegativity can be quantitatively analyzed using the wave-function theorem as well as the DFT; in fact, much of the formulation in this regard has been from the later by Parr and Yang [44]. This can easily be seen if we consider the effect of adding electrons to the energy of a molecule, atom, or ion. Thus, for example, we start with bromine cation Br^+ whose energy is −2571.1517 au. Its energy decreases to −2571.6569 au in Br on the addition of one electron and to −2571.7613 au in the bromine anion Br^- on the addition of the second electron. The numbers of electron N that have been added to Br^+ are integral numbers. Thus, we can take $N=0$ for Br^+, $N=1$ for Br, and $N=2$ for Br^-. In general, we can add electronic charge continuously with the result that we shall have a continuous curve $E=f(N)$ as in Figure 12.16.

We can then examine the derivative of E with respect to N at a constant nuclear charge $\left(\frac{\partial E}{\partial N}\right)_Z$. By analogy to the term

$$\mu = \left(\frac{\partial E}{\partial N}\right)_{T,P} \tag{12.15.1}$$

in thermodynamics, which is the variation of energy at constant temperature and pressure caused by small change in the composition (number of molecules N) of the system and known as chemical potential, the quantity

$$\mu = \left(\frac{\partial E}{\partial N}\right)_Z \tag{12.15.2}$$

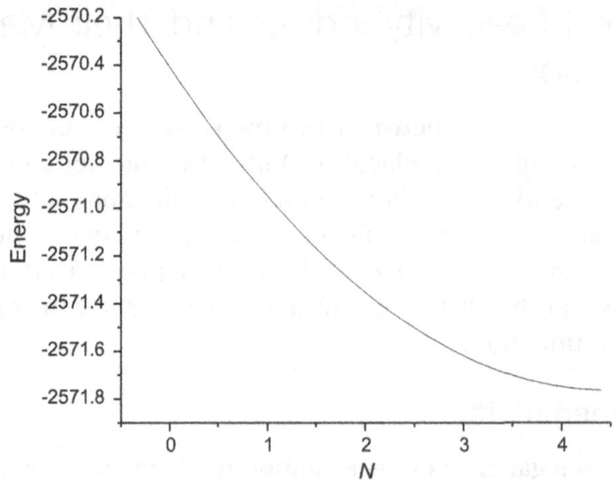

FIGURE 12.16 Effect of adding electrons to the energy of a Bromine atom. $N=0$ for Br^{+}, $N=1$ for Br, and $N=2$ for Br^{-}.

is termed as electronic chemical potential of an atom. For a molecule, the differentiation in Eq. (12.15.2) is at a constant nuclear framework; the charges and their potential being constant. Thus, for a molecule, atom, or ion at a constant external potential $V(r)$, this expression may be written as

$$\mu = \left(\frac{\partial E}{\partial N}\right)_{V(r)} \tag{12.15.3}$$

Intuitively one feels that the more electronegative a species is, the more should its energy drop when it acquires electrons. Thus, one can suspect a link between the electronic chemical potential (μ) and the electronegativity. Since $\left(\frac{\partial E}{\partial N}\right)_{V(r)}$ has a negative value due to decrease in E with increasing N (Figure 12.16), we may define the electronegativity as the negative of the electronic chemical potential so as to tag it with a positive sign. Thus,

$$\chi = -\mu = -\left(\frac{\partial E}{\partial N}\right)_{V(r)} \tag{12.15.4}$$

From this equation, we may infer that the electronegativity of a species is the drop in energy when an infinitesimal amount of electronic charge enters it.

It can easily be shown as below that the Mulliken formula

$$\chi = \frac{1}{2}(I + A)$$

is a finite difference approximation of Eq. (12.15.4).

Thus, as mentioned above, on going from X^+ to X^-, N changes from $N=0$ to $N=2$. Hence, from Eq. (12.15.4)

$$\left(\frac{\partial E}{\partial N}\right)_{V(r)} = \frac{[E(X^-) - E(X^+)]}{[2-0]} = \frac{-[E(X^+) - E(X)] - [E(X) - E(X^-)]}{2} = -\frac{(I+A)}{2}$$

$$\text{or,}\ \chi = -\left(\frac{\partial E}{\partial N}\right)_{V(r)} = \frac{(I+A)}{2} \qquad (12.15.5)$$

A theoretical and practically more useful basis for the electronegativity χ is found by expressing it in the molecular orbital (MO) framework.

Within the validity of the Koopmann's theorem [45] the frontier orbital energies are given as:

$$-E(\text{HOMO}) = I \text{ and } -E(\text{LUMO}) = A$$

where HOMO is the highest occupied and LUMO is the lowest unoccupied molecular orbital. Thus,

$$\chi = -\frac{E_{\text{HOMO}} + E_{\text{LUMO}}}{2} \qquad (12.15.6)$$

Equation (12.15.6) refers to a system where the HOMO is filled. It may be noted that the use of negative electron affinities (A) is essential for most of the molecules. These can be measured in many cases [46]. Following Pearson [47], if we consider a molecule for which $I = 10$ eV and $A = -2$ eV (Figure 12.17), then $-E_{\text{HOMO}} = I = 10$ eV and $-E_{\text{LUMO}} = A = -2$ eV. Hence, $\chi = 4$ eV.

Also, in this case,

$$\eta = \frac{I-A}{2} = 6 \text{ eV.}$$

This quantity η, called the chemical hardness, shall be discussed in the next subsection. The hardness for the molecule is shown in Figure 12.17. It follows that the energy gap $(E_{\text{HOMO}} - E_{\text{LUMO}}) = 2\eta$.

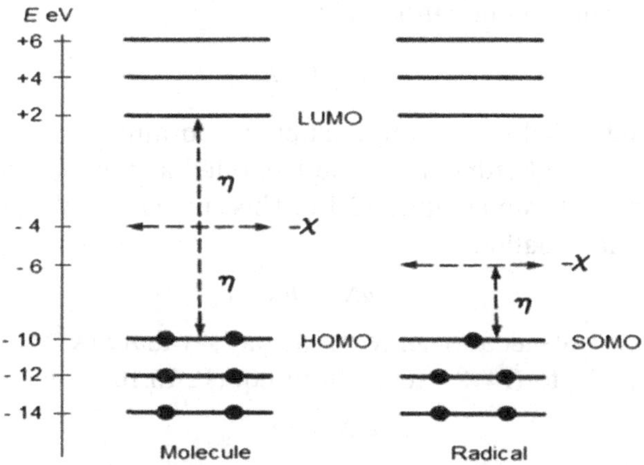

FIGURE 12.17 Energy level diagram for molecule, radical showing χ and η.

The situation in the case of radicals, where the frontier orbital is a singly occupied molecular orbital (SOMO) and is half-filled, is somewhat different. Thus, for example, in the case of a radical, if $I = 10$ eV and $A = +2$ eV, then the energy of the singly unoccupied molecular orbital E (SOMO) $= -10$ eV and from equation (12.15.5), $\chi = 6$ eV. In this case, the unknown energy of LUMO plays no role. Here, the quantity $(I - A) = 2\eta$ is just the mean repulsion energy of two electrons. The chemical hardness for the radical is also shown in Figure 12.17.

The identification of a molecule as a Lewis acid or base can be made by its χ value. Large χ values characterize acids, while the small χ values characterize bases. For any two molecules, electrons will be transferred from the one having small χ to the other having large χ. Thus, the electrons flow from high chemical potential to low chemical potential.

The determination of electronegativity from Eq. (12.15.6) requires a knowledge of HOMO and LUMO which can be obtained from a single-step quantum chemical calculation for the energy of the system.

12.15.2 Chemical Hardness and Softness

The concept of hardness (η) was introduced by Parr [30] and Pearson [47] for the rationalization of acid–base reactions. Softness (S), also called as global softness, is just the reverse of hardness $\left(S = \frac{1}{\eta}\right)$.

Parr and Pearson defined global hardness or absolute hardness as

$$\eta = \frac{1}{2}\left(\frac{\partial^2 E}{\partial N^2}\right)_{V(r)} \quad (12.15.7)$$

which is related to the electronegativity χ and electronic potential μ through the identity

$$2\eta = \left(\frac{\partial \mu}{\partial N}\right)_{V(r)} = -\left(\frac{\partial \chi}{\partial N}\right)_{V(r)} \quad (12.15.8)$$

The nonchemical meaning of the word "hardness" is resistance to deformation or change. Equation (12.15.8) shows that chemical hardness is resistance of the chemical potential to change in the number of electrons.

The operational definition of hardness is

$$\eta = \frac{1}{2}(I - A) \quad (12.15.9)$$

which can be derived as below as a finite difference formula.

From Eq. (12.15.7), the hardness may be identified as the curvature (or second derivative) of the E versus N graph (Figure 12.16). This curve $E = f(N)$ may be approximated by a general quadratic equation

$$E = aN^2 + bN + c \quad (12.15.10)$$

As earlier in the case of electronegativity, we may say that $E(X^+)$ corresponds to $N = 0$, $E(X)$ to $N = 1$, and $E(X^-)$ to $N = 2$. Hence from Eq. (12.15.10)

$$E(X^+) = c$$

$$E(X) = a + b + c \quad (12.15.11)$$

$$\text{and, } E(X^-) = 4a + 2b + c$$

which gives,

$$2a = E(X^+) + E(X^-) - 2E(X) = [E(X^+) - E(X)] - [E(X) - E(X^-)] = I - A \tag{12.15.11}$$

Also, from Eq. (12.15.10),

$$\frac{\partial^2 E}{\partial N^2} = 2a \tag{12.15.12}$$

Comparing Eqs (12.15.7), (12.15.11), and (12.15.12), we get

$$\eta = \frac{1}{2}\left(\frac{\partial^2 E}{\partial N^2}\right)_{V(r)} = \frac{1}{2}(I - A) \tag{12.15.13}$$

A relationship may also be established between η, μ, and χ as follows:

$$\eta = \frac{1}{2}\left(\frac{\partial^2 E}{\partial N^2}\right)_{V(r)} = \frac{1}{2}\left(\frac{\partial \mu}{\partial N}\right)_{V(r)} = -\frac{1}{2}\left(\frac{\partial \chi}{\partial N}\right)_{V(r)} \tag{12.15.14}$$

Since, within the Koopmann's approximation

$$-E(\text{HOMO}) = I \text{ and } -E(\text{LUMO}) = A$$

we may also write

$$\eta = \frac{1}{2}(E_{\text{LUMO}} - E_{\text{HOMO}}) \tag{12.15.15}$$

From Eq. (12.15.15) it follows that the hard molecules have a large energy gap and the soft molecules have a small energy gap between HOMO and LUMO.

Pearson [47] explained high polarizability of soft acids and bases on the basis of small energy gap between HOMO and LUMO, because a small value of this gap means a smaller excitation energy and hence a larger mixing coefficient between the two states. The soft molecules will, therefore, undergo unimolecular reactions more readily than hard molecules. Since the processes like dissociation and isomerization proceed by mixing of the ground and excited state wavefunctions, a smaller energy gap is more favorable for easy reaction. The hard molecules, on the other hand, resist changes in their electron number and distribution. The idea of absolute hardness is commonly used as a criterion of chemical reactivity and stability and has been discussed in details by Profit and Geerlings [48].

Pearson [49] used the concept of hardness to predict the direction of acid–base reactions. "Hard acids" are acceptor species of small size, low polarizability, having no unpaired electrons in the valence shell. "Soft acids" are acceptor acids of large size, high polarizability, and often have unshared pair of electrons in the valence shell. On the other hand, "hard bases" are donor species with small size and low polarizability, have tightly bound outer electrons, and are hard to oxidize. "Soft bases" are donor species with large size, high polarizability, an easily deformable electron cloud, and are easy to oxidize. Hard–hard reactions are charge controlled and the soft–soft reactions are of covalent type. The concept of hardness is the basis of the "maximum hardness principle (MHP)" of Pearson according to which "as a rule of nature, molecules arrange themselves to be as hard as possible." Pearson's hard soft acid base (HSAB) principle is a corollary of the MHP principle.

In the case of aromatic systems, it is found that high aromaticity and hardness are measures of high stability and low reactivity [50]. A direct relationship has been shown to exist between the resonance stabilization energies (RE) and the HOMO–LUMO gap in azulenes [51,52].

$$\mathrm{RE} = \frac{\pi^2 \rho_{rs}^2 (E_{\mathrm{LUMO}} - E_{\mathrm{HOMO}})}{24} = \frac{\pi^2 \rho_{rs}^2 \eta}{12} \tag{12.15.16}$$

where ρ_{rs} is the bond order of the *r–s* bond. A correlation has also been shown to exist between the thermodynamic stability and kinetic stability (reactivity) of aromatic compounds. While thermodynamic stability is a unique property of the ground state, the kinetic stability measures how fast a reaction goes.

12.15.3 Fukui Functions

Encouraged by the success of electron density redistribution concept under the influence of a reacting agent put forth by Coulson and Longuet-Higgins [17,18] and its relationship with chemical reactivity, Fukui [53] used the perturbation theory–Frontier orbital approach to calculate the density distribution of the HOMO for the attack of an electrophilic reagent and found an agreement between the position of attack and the site of the large density of this particular orbital. Similarly, the electron density distribution of a LUMO was calculated for nucleophilic attack and correlation with reactivity was established. In a radical substitution, the distribution of (HOMO + LUMO) was found to correlate with reactivity [19]. Parr and Yang [32,54–56] and Yang and Mortier [57] rationalized the frontier orbital theory of reactivity of Fukui by using DFT and introduced Fukui functions (f) for a molecule. Thus, the reactivity of a molecule at a site **r** is reflected in the Fukui function

$$f(\mathbf{r}) = \left(\frac{\partial \rho(\mathbf{r})}{\partial N}\right)_{V(\mathbf{r})} = \left[\frac{\partial \mu}{\partial V(\mathbf{r})}\right]_N \tag{12.15.17}$$

From this relationship it follows that the Fukui function $f(\mathbf{r})$ is the change in electron density at a given position caused by the change in the number of electrons. This is also the functional derivative of the electronic chemical potential μ with respect to the change in the external potential. One important property of the Fukui function is that it obeys the normalization condition

$$\int f(\mathbf{r}) d\mathbf{r} = 1 \tag{12.15.18}$$

As against electronegativity (χ) and hardness (η), which have global character, the Fukui functions are local in nature and reflect the properties of different sites within the molecule. Since $\rho(\mathbf{r})$ changes from point to point in a molecule, so does the Fukui function.

Parr and Yang have given three numerical definitions for the Fukui functions:

$$f^+(r) = (\partial \rho / \partial N)^+_{V(r)} \tag{12.15.19a}$$

$$f^-(r) = (\partial \rho / \partial N)^-_{V(r)} \tag{12.15.19b}$$

$$f^0(r) = (\partial \rho / \partial N)^0_{V(r)} \tag{12.15.19c}$$

Equations (12.15.19a), (12.15.19b), and (12.15.19c) govern nucleophilic, electrophilic, and radical attacks, respectively. A molecule readily accepts electrons in regions where $f^+(r)$ is large and likewise, donates electrons in regions where $f^-(r)$ is large.

Since Eq. (12.15.19) is discontinuous for a molecular or ionic system, it is difficult to evaluate it. Hence, as in the case of electronegativity (χ) and hardness (η), it is expressed in the finite difference approximation as follows:

$$f_k^+ = \rho_k(N+1) - \rho_k(N) \tag{12.15.20a}$$

$$f_k^- = \rho_k(N) - \rho_K(N-1) \tag{12.15.20b}$$

$$f_k^0 = \frac{1}{2}(f_k^+ + f_k^-) = \frac{1}{2}[\rho_k(N+1) - \rho_k(N-1)] \tag{12.15.20c}$$

The electron density $\rho(\mathbf{r})$ is such that

$$\int \rho(\mathbf{r})d\mathbf{r} = N,$$

where N being the total number of electrons in the system.

These equations can also be written as

$$f_k^+ = q_k(N) - q_k(N+1), \quad \text{for nucleophilic attack} \tag{12.15.21a}$$

$$f_k^- = q_k(N-1) - q_k(N), \quad \text{for electrophilic attack} \tag{12.15.21b}$$

$$f_k^0 = \frac{1}{2}(f_k^+ + f_k^-) = \frac{1}{2}[q_k(N-1) - q_k(N+1)], \quad \text{for radical attack} \tag{12.15.21c}$$

Here, $q_k(N)$ is the charge on atom k having N electrons. q_k is related to ρ_k by the relation $q_k = Z - \rho_k$, Z being the atomic number.

In order to calculate the Fukui functions, a set of three calculations have to be performed: on the given molecule/radical, its anion and on its cation, to obtain electron population charges on the atoms. These are then used in Eq. (12.15.21) to obtain the Fukui functions f_k^+, f_k^-, and f_k^0. The method of calculation of the different chemical indices shall be explained through an illustrative example in Section 12.15.6.

12.15.4 Global and Local Electrophilicity Indicators

Prompted by a qualitative suggestion by Maynard et al. [58], a new reactivity indicator named as electrophilicity index (ω) was proposed by Parr et al. [59]. This is defined as

$$\omega = \frac{\mu^2}{2\eta} = \frac{(I+A)^2}{4(I-A)} \tag{12.15.22}$$

where μ is the chemical potential, η is hardness, and I and A are the vertical ionization potential and electron affinity, respectively. By analogy with classical electrostatics (Power $= V^2/R$, where V is the electrostatic potential and R is the resistance), ω may be considered as a measure of electrophilic power or the total ability of a chemical species to attract electrons. The electrophilicity index and electron affinity are the two measures of the power to attract electrons. These indices are not equal but are correlated to each other. Since ω depends both on I and A, it is expected that A (electron affinity) can provide similar qualitative trends as ω wherever variation in I is not very significant. This is commonly observed for the elements belonging to the same group in the periodic table and the functional groups made from them. Electrophilicity is a global descriptor.

The local reactivities of a chemical species are represented by local reactivity descriptors. One such descriptor Fukui function has been discussed above. Another descriptor, local electrophilicity (ω_k) has been introduced to analyze the electrophilic–nucleophilic reactions better. The global trend of electrophilicity or nucleophilicity (ω) of a molecule may be explained in terms of the local behavior of those atomic sites of the molecule that are prone to electrophilic or nucleophilic attack. The local electrophilicity index (ω_k) is one such descriptor that varies from point to point in an atom, molecule, ion, or solid and has been used to define the global electrophilicity index ω, in Eq. (12.15.23).

We start with the Fukui function which is a normalized function satisfying the relation

$$\int f(\mathbf{r})d\mathbf{r} = 1$$

Multiplying the two sides by ω, the global philicity index

$$\omega = \omega\int f(\mathbf{r})d\mathbf{r} = \int \omega f(\mathbf{r})d\mathbf{r} = \int \omega(\mathbf{r})d\mathbf{r} \tag{12.15.23a}$$

where,

$$\omega(\mathbf{r}) = \omega f(\mathbf{r}) \tag{12.15.23b}$$

In order to take care of all types of reactions, three different forms of $\omega(\mathbf{r})$ are defined,

$$\omega^{\alpha}(\mathbf{r}) = \omega f^{\alpha}(\mathbf{r}) \tag{12.15.24a}$$

where $\alpha = +, -,$ or 0 for attacks by a nucleophile, electrophile, or a radical, respectively.

The corresponding condensed to atom forms of the local philicity index (ω_k) for atom k may be written as

$$\omega_k^{\alpha}(\mathbf{r}) = \omega_k f_k^{\alpha}(\mathbf{r}); \quad \alpha = +, -, \text{or } 0 \tag{12.15.24b}$$

In order to understand hard–soft interactions, another parameter *local softness* (s) has been introduced [60]. This is defined as

$$s^{\alpha}(\mathbf{r}) = \mathbf{S} f^{\alpha}(\mathbf{r}) \tag{12.15.25a}$$

where S is the global softness and $\alpha = +, -,$ or 0 refer to nucleophilic, electrophilic, and radical reactions. In the condensed to atom form, the local softness may be written as

$$s_k^{\alpha}(\mathbf{r}) = S f_k^{\alpha}(\mathbf{r}); \quad \alpha = +, -, \text{or } 0 \tag{12.15.25b}$$

Using relations (12.15.20) or (12.15.21) for the Fukui functions, the local softness $s_k^{\alpha}(\mathbf{r})$ may be written as:

$$s_k^+(\mathbf{r}) = S f_k^+ = (\rho_k(N+1) - \rho_k(N))S \tag{12.15.26a}$$

$$s_k^-(\mathbf{r}) = S f_k^- = (\rho_k(N) - \rho_k(N-1))\, S \tag{12.15.26b}$$

$$s_k^0(\mathbf{r}) = S f_k^0 = \frac{1}{2}[\rho_k(N+1) - \rho_k(N-1)]S \tag{12.15.26c}$$

where $\rho_k(N)$, $\rho_k(N+1)$, and $\rho_k(N-1)$ represent the condensed electronic populations of atom k for the neutral, anionic, and cationic systems, respectively.

Thus, s_k^+, s_k^- and s_k^0 represent the condensed local softness values indicating whether atom k is more susceptible to attack by a nucleophile, electrophile, or a radical.

12.15.5 Local Parameters and Site Selectivity

As seen above, the local softness and Fukui functions are closely related to each other; the local softness $s(\mathbf{r})$ combines the site reactivity index $f(\mathbf{r})$ with global softness S. The indices S and ω remain unchanged in a molecule except when it is undergoing an intramolecular process such as vibration, internal rotation, rearrangement, or interaction with a solvent or an external field. Hence local reactivity in a molecule can be equivalently understood in terms of the local parameters $f_k^{\alpha}, s_k^{\alpha}$ or ω_k^{α}. However, Geerlings [36] has suggested that local softness should be used as an intermolecular reactivity index whereas Fukui function as an intramolecular index. It is generally accepted that the most natural density functional concept for characterizing a site should be local softness. Electrophilicity and local softness essentially give the same information about intramolecular reactivity trends as is provided by the Fukui functions. The exceptions are cases where the above mentioned intramolecular processes take place. For intermolecular activity, the use of the Fukui functions f_k^{α} is not enough and both s_k^{α} and ω_k^{α} should be used to compare the hard–soft or electrophilic–nucleophilic behavior of the given atomic sites of the two molecules. It has also been suggested that the best information about the site reactivity is obtained through a combined use of atom-in-molecule (QTAIM) calculations and the Fukui functions [61].

12.15.6 Application—Illustrative Example

As an illustration of the use of the above given formula for the reactivity parameters, we shall take the examples of cyanate anion (OCN^-) and cyanoacetylene (HCCCN), both of which are interstellar molecules and play important role in chemical reactions in the outer space and in the formation of cytosine, a component of DNA (diribonucleic acid) in the interstellar space [1,62]. The results of calculation are given in Table 12.5 which contains chemical reactivity indices in two parts—the energy-based indices and the charge density- or electron population-based indices. Global reactivity indices such as hardness (η), electronic chemical potential (μ), electrophilicity (ω), and softness (s) belong to the first category while the indices such as Fukui functions (f^+, f^-, f^0), local philicity indices (ω^+, ω^-, ω^0), and local softness (s^+, s^-, s^0) belong to the second category. The energy-based indices need a knowledge of the energies of the HOMO and LUMO, whereas the charge density-based indices need the electronic population/atomic charge on the neutral atom, its anion and cation. Thus, the calculations are performed on $N-1$, N, and $N+1$ electron systems to obtain charges $q_k(N-1)$, $q_k(N)$, and $q_k(N+1)$, which are then used in Eq. 12.15.21 (a, b, c) to obtain f_k^+, f_k^-, and f_k^0. Instead of atomic charge (q_k) on each atom, electron population (ρ_k) can also be used.

Usually, the single point calculations on the anion and cation are conducted by using the optimized geometry of the neutral molecule.

It may be noted from Figure 12.18 and Table 12.5 that the atomic charges and the electron populations depend on the method/basis set and the way they are calculated (e.g., Mulliken, APT, AIM, or ESP). Since most of the reactivity parameters depend upon the atomic charges, it may be concluded that the reactivity indices too may vary identically.

The energy-based global indices like hardness (η), softness (S), and electrophilicity (ω) are obtained from frontier orbitals HOMO and LUMO, which need a single point calculation on the optimized geometry of the molecule. A higher value of HOMO indicates that the molecule is more reactive in reactions with electrophiles while a lower value of LUMO is essential for reactions with nucleophiles. A harder molecule, that is, the one having higher hardness index η, has a larger HUMO–LUMO gap than a softer molecule. As B3LYP/6-311++G** calculations show, the (η) values for OCN^- and HCCCN are 0.6623 eV and 3.4458 eV, respectively (Table 12.5); the cyanoacetylene molecule is therefore more stable than OCN^-.

Hardness measures the resistance to change in electron distribution in a molecule. Vektariene et al. [63] have found a direct correlation between hardness and aromaticity. It follows from Table 12.5 that the values of Fukui function f^+ for the C, O, and N atoms of OCN^- using the 6-31G** basis set are 0.5678, 0.1632, and 0.2690 and the corresponding value of f^- are 0.3102, 0.3288, and 0.3610. The nucleophilicity of atoms C, O, and N, therefore, follows the sequence N > O > C and their electrophilicity follows the sequence C>N>O. Thus, while C atom has the highest electrophilicity ($f^+ = 0.5678$) the N atom has the highest nucleophilicity ($f^- = 0.3610$). The nitrogen atom has therefore, greater chance of attacking the electrophilic site of other molecules. This has been shown by Gupta et al. [1] to happen in a reaction between OCN^- and cyanoacetylene. It may be noted from Table 12.5 that f^- values for the five atoms 1C, 2C, 3C, 4N, and 5H follow the sequence $1C>3C>4N>2C>1H$ with the $1C$ atom having the highest nucleophilicity. Thus the N atom of OCN^- is more likely to attack cyanoacetylene (HCCCN) at the $1C$ position rather than the $3C$ position.

Table 12.5 Chemical reactivity indices for cyanate anion OCN^- and cyanoacetylene (CA)

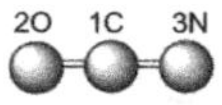

OCN^-

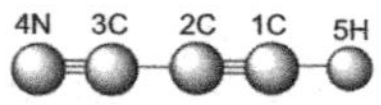

CA

	HOMO	LUMO	$\eta = (L-H)/2$	η(ev)	$\mu = (L+H)/2$	μ^2	$\omega = \mu^2/(2\eta)$	ω (ev)	S (ev)
Energy-based indices									
Basis 6-31G**									
OCN^-	0.0137	0.3697	0.1780	4.8416	0.1917	0.0367	0.1032	2.8072	0.2065
CA	−0.3089	−0.0552	0.1269	3.4510	−0.1820	0.0331	0.1306	3.5516	0.2898
Basis 6-311++G**									
OCN^-	−0.0318	0.0169	0.0243	0.6623	−0.0075	0.0001	0.0011	0.0311	1.5100
CA	−0.3242	−0.0709	0.1267	3.4458	−0.1975	0.0390	0.1540	4.1889	0.2902

Charge density-based indices

	Charge density			f^+	f^-	f^0	ω	$\omega^+ = \omega f^+$	$\omega^- = \omega f^-$	$\omega^0 = \omega f^0$	S	$s^+ = Sf^+$	$s^- = Sf^-$	$s^0 = Sf^0$
Basis 6-31G**														
Atom	**OCN^-**	**Anion**	**Cation**				2.8072				0.2065			
1C	0.3248	−0.2430	0.6350	0.5678	0.3102	0.4390		1.5939	0.8708	1.2324		0.1173	0.0641	0.0907
2 O	−0.6428	−0.8060	−0.3140	0.1632	0.3288	0.2460		0.4581	0.9230	0.6906		0.0337	0.0679	0.0508
3 N	−0.6820	−0.9510	−0.3210	0.2690	0.3610	0.3150		0.7551	1.0134	0.8843		0.0556	0.0745	0.0650
Atom	**CA**	**Anion**	**Cation**				3.5516				0.2898			
1C	−0.4813	−0.7042	−0.1462	0.2229	0.3351	0.2790		0.7917	1.1902	0.9910		0.0646	0.0971	0.0809
2C	0.6530	0.4972	0.8084	0.1558	0.1554	0.1556		0.5533	0.5520	0.5526		0.0451	0.0450	0.0451
3C	0.0905	−0.1699	0.2921	0.2603	0.2016	0.2310		0.9246	0.7159	0.8203		0.0754	0.0584	0.0669
4 N	−0.4717	−0.6665	−0.2975	0.1949	0.1742	0.1845		0.6921	0.6186	0.6553		0.0565	0.0505	0.0535
5H	0.2095	0.0434	0.3432	0.1661	0.1337	0.1499		0.5898	0.4748	0.5323		0.0481	0.0387	0.0434
Basis 6-311++G**														
Atom	**OCN^-**	**Anion**	**Cation**				0.0311				1.5100			
1C	0.1973	−3.0697	0.3636	3.2670	0.1663	1.7166		0.1016	0.5906	6.0968		4.9331	0.2511	2.5921
2 O	−0.5410	0.1439	−0.1558	−0.6849	0.3852	−0.1498		−0.0213	1.3681	−0.5322		−1.0342	0.5817	−0.2263
3 N	−0.6563	0.9258	−0.2078	−1.5821	0.4485	−0.5668		−0.0492	1.5929	−2.0130		−2.3890	0.6773	−0.8559
Atom	**CA**	**Anion**	**Cation**				4.1889				0.2902			
1C	−0.8664	−1.4127	−0.5379	0.5463	0.3285	0.4374		2.2883	1.3760	1.8321		0.1585	0.0953	0.1269
2C	2.7345	2.7286	2.9822	0.0060	0.2477	0.1268		0.0250	1.0375	0.5313		0.0017	0.0719	0.0368
3C	−1.8352	−1.9464	−1.8321	0.1112	0.0031	0.0572		0.4659	0.0130	0.2395		0.0323	0.0009	0.0166
4 N	−0.1836	−0.4494	0.1165	0.2658	0.3001	0.2829		1.1134	1.2569	1.1852		0.0771	0.0871	0.0821
5H	0.1507	0.0799	0.2713	0.0707	0.1207	0.0957		0.2963	0.5054	0.4009		0.0205	0.0350	0.0278

(a)

	Cyanate anion			Cyanoacetylene				
Atomic charge (*q*)	−0.541	0.197	−0.656	−0.184	−1.835	2.734	0.866	0.151
	O =	C =	N^-	N4 ≡	C5 —	C2 ≡	C1 —	H5
Electron population (*p*)	8.541	5.803	7.656	7.184	7.835	3.266	6.866	0.849

Basis set 6–311 ++ G**

(b)

Atomic charge (*q*)	−0.643	0.325	−0.682	−0.472	0.091	0.653	−0.481	0.209
	O =	C =	N^-	N4 ≡	C5 —	C2 ≡	C1 —	H5
Electron population (*p*)	8.643	5.675	7.682	7.472	5.909	5.347	6.481	0.791

Basis set 6–31 G**

FIGURE 12.18 The electron population (ρ) and atomic charge (q) on the OCN^- and HCCCN molecules based on (a) B3LYP/6–311++G** and (b) B3LYP/6-31G** level calculations.

As mentioned, the electronic chemical potential (μ) and the global electrophilicity index (ω) measure the propensity of a species to accept electrons. While higher values of μ and ω indicate a higher electrophilicity, the lower values indicate greater nucleophilicity. In the case of OCN^- and HCCCN, the values of μ^2 and ω are 0.0001 and 0.0311 eV and 0.0390 and 4.1889 eV, respectively. Thus HCCN is more electrophilic in nature than OCN^-.

Similar conclusions can be drawn on the basis of the other local reactivity indices like $\omega^{0,\pm}$ and $s_k^{0,\pm}$ given in Table 12.5. Thus OCN^- has two heteroatoms oxygen and nitrogen for which the local philicity ω_k^- has values 0.5906 and 1.5929 (6-311++G**). Against this, the carbon atoms 1*C*, 2*C*, and 3*C* of cyanoacetylene have ω^+ values 2.2883, 0.0250, and 0.4659, respectively. It is therefore expected that the N atom of the OCN^- shall have a greater chance of attacking the 1*C* atom of cyanoacetylene. Similar conclusion can be drawn on the basis of large s_k^- value of nitrogen of OCN^- ($s^- = 0.6773$) and large s_k^+ value of 1*C* in HCCCN ($s^+ = 0.1585$).

Sometimes, during calculations, negative values of Fukui functions are obtained. These values are usually discarded. However, it has been suggested that these negative values can be useful in understanding reactions in which oxidation of an entire molecule leads to the reduction of a part of it, for example, removing electrons from alkynes can result in an increase in the electron density in the CC bond [64]. Padmanabhan et al. [65] have extended the Fukui function concept from local philicity and dual philicity descriptor to be that of a multiphilicity descriptor which may simultaneously reflect the nucleophilicity and electrophilicity of a given site in a molecule.

References

[1] V.P. Gupta, P. Tandon, P. Mishra, Adv. Space Res. 51 (2013) 797.

[2] C.J. Cramer, Essentials of Computational Chemistry Theories and Models, second ed., John Wiley & Sons, Ltd, 2004.

[3] S. Thakur, PhD thesis, University of Jammu, Jammu, India, (1998) 67.

[4] L.N. Margolin, A. Yu, Pentin, Opt. Spectros. 40 (1976) 461.

[5] W.G. Fatley, R.K. Harris, F.A. Miller, R.E. Witkowski, Spectrochim. Acta 21A (1965) 231.

[6] V.P. Gupta, S. Thakur, Indian J. Phys. 73 (1999) 651.

[7] J.R. Durig, T.S. Little, J. Chem. Phys. 75 (1981) 3660.

[8] Yu.A. Pentin, J. Mol. Struct. 46 (1970) 149.

[9] P.D. Foster, V.V.M. Rao, R.F. Curl Jr., J. Chem. Phys. 43 (1965) 1064.

[10] V.P. Gupta, P. Rawat, R.N. Singh, P. Tandon, Comput. Theoret. Chem. 983 (2012) 7.

[11] D. Habibollahzadeh, M.E. Grice, M.C. Concha, J.S. Murray, P. Politzer, Comput. Chem. 16 (1995) 654.

[12] V.P. Gupta, A. Sharma, S.G. Agrawal, Bull. Korean Chem. Soc. 27 (2006) 1297.

[13] M.A. Collins, The interface between electronic structure theory and reaction dynamics by reaction path methods in new methods of computational quantum mechanics, in: I. Prigogine, S.A. Rice (Eds.), Advances in Chemical Physics, vol. 93, John Wiley & Sons, Inc, 1996.

[14] V.P. Gupta, A. Sharma, A. Virdi, Pramana J Phys. 67 (2006) 487.

[15] C. Peng, P.Y. Ayala, H.B. Schlegel, M.J. Frisch, J. Comput. Chem. 17 (1996) 49.

[16] C.K. Ingold, Structure and Mechanism in Organic Chemistry, Cornell University Press, Ithaca, NY, 1953.

[17] C.A. Coulson, H.C. Longuet-Higgins, Proc. Roy. Soc. (London) A191 (1947) 39.

[18] C.A. Coulson, H.C. Longuet-Higgins, Proc. Roy. Soc. (London) A192 (1947) 16.

[19] K. Fukui, T. Yonezawa, H. Shingu, J. Chem. Phys. 20 (1952) 722.

[20] K. Fukui, T. Yonezawa, C. Nagata, H. Shingu, J. Chem. Phys. 22 (1954) 1433.

[21] K. Fukui, H. Kato, T. Yonezawa, Bull. Chem. Soc. Jpn. 34 (1961) 1112.

[22] K. Fukui, Theory of Orientation and Stereoselection, Springer-Verlag, Berlin, 1970.

[23] L. Salem, Chem. Brit. 5 (1969) 449.

[24] G. Klopman, Chemical Reactivity and Reaction Paths, John Wiley, New York, NY, 1974.

[25] R.G. Pearson, Symmetry Rules for Chemical Reactions, John Wiley, New York, NY, 1976.

[26] W.C. Herndon, Chem. Rev. 72 (1972) 157.

[27] R.F. Hudson, Angew. Chem. Mt. Ed. Eng. 12 (1973) 36.

[28] K.N. Houk, Acc. Chem. Res. 8 (1975) 361.

[29] I. Fleming, Frontier Orbitals and Organic Chemical Reactions, John Wiley, New York, NY, 1976.

[30] R.G. Parr, R.G. Pearson, J. Am. Chem. Soc. 105 (1983) 7512.

[31] S.K. Ghosh, M.J. Berkowitz, Chem. Phys. 83 (1985) 2976.

[32] W. Yang, R.G. Parr, Proc. Natl. Acad. Sci. USA 82 (1985) 6723.

[33] N. Inamoto, S. Masuda, Tetrahedron Lett. (1977) 3287.
[34] R.P. Iczkowski, J.L. Margrave, J. Am. Chem. Soc. 83 (1961) 3547.
[35] F. DeProft, W. Langenaeker, P. Geerlings, J. Phys. Chem. 97 (1993) 1826.
[36] P. Geerlings, F. De Proft, W. Langenaeker, Chem. Rev. 103 (2003) 1793.
[37] K.D. Sen, C.K. Jorgensen (Eds.), Structure and Bonding, vol. 66, Springer Verlag, Berlin, 1987, p. 99.
[38] R.S. Mulliken, J. Chem. Phys. 2 (1934) 782.
[39] B. Wrackmeyer, Annu. Rep. NMR Spectrosc. 16 (1985) 73.
[40] B. Wrackmeyer, Annu. Rep. NMR Spectrosc. 38 (1999) 204.
[41] M.A. McAllister, T.T. Tidwell, J. Org. Chem. 59 (1994) 4506.
[42] R.J. Boyd, S.L. Boyd, J. Am. Chem. Soc. 114 (1992) 1652.
[43] R.J. Boyd, K.E. Edgecombe, J. Am. Chem. Soc. 110 (1988) 4182.
[44] R.G. Parr, W. Yang, Density-functional Theory of Atoms and Molecules, Oxford, New York, 1989. Chapters 4 and 5.
[45] T. Koopmans, Physics 1 (1934) 104.
[46] J.C. Gordon, J.H. Moore, J.A. Tossel, W. Kaim, J. Am. Chem. Soc. 107 (1985) 5600.
[47] R.G. Pearson, Proc.Natl. Acad.Sci. USA 83 (1986) 8440.
[48] F. de Profit, P. Geerlings, Phys. Chem. Chem. Phys. 6 (2004) 2426.
[49] R.G. Pearson, J. Am. Chem. Soc. 85 (1963) 3533.
[50] R.K. Roy, F. de Profit, P. Geerlings, J. Phys. Chem. 12 (1999) 503.
[51] R.C. Haddon, T. Fuguhata, Tetrahedron Lett. 21 (1980) 1191.
[52] R.C. Haddon, J. Am. Chem. Soc. 101 (1979) 1722.
[53] K. Fukui, Pure Appl. Chem. 34 (1982) 1825.
[54] R.G. Parr, W. Yang, J. Am. Chem. Soc. 106 (1984) 4049.
[55] W. Yang, R.G. Parr, R. Pucci, J. Chem. Phys. 81 (1984) 2862.
[56] R.G. Parr, in: R.M. Dreizler, J. Da Providencia (Eds.), Density Functional Theory in Chemistry. Density Functional Methods in Physics, Plenum Press, New York, 1985, p. 141.
[57] W. Yang, W.J. Mortier, J. Am. Chem. Soc. 108 (1986) 5708.
[58] A.T. Maynard, M. Huang, W.G. Rice, D.G. Covel, Proc. Natl. Acad. Sci. USA 95 (1998) 11578.
[59] R.G. Parr, S. Liu, L. von Szentpaly, J. Am. Chem. Soc. 121 (1999) 1992.
[60] P.K. Chattaraj, Chem. Rev. 106 (2006) 2065.
[61] F.A. Bulat, E. Chamorro, P. Fuentealba, A. Torro - Labbe, J. Phys. Chem. A 108 (2004) 342.
[62] R.A. Sanchez, J.P. Ferris, L.E. Orgel, Science 154 (1966) 784.
[63] A. Vektariene, G. Vektaris, J. Svoboda, Arkivoc (2009) 311.
[64] J. Melin, P.W. Ayers, J.V. Oritz, J. Phys. Chem. A 111 (2007) 10017.
[65] J. Padmanabhan, R. Parthasarthi, M. Elango, V. Subramanian, B.S. Krisnamoorthy, S. Gutierrez-Oliva, A. Torro-Labbe, D.R. Roy, P.K. Chattaraj, J. Phys. Chem. A108 (2007) 342.

Further Reading

[1] K.D. Sen (Ed.) Reviews of Modern Quantum Chemistry: A Celebration of the Contributions of Robert G. Parr, vol. II, Ed., World Scientific, Singapore, 2002

[2] I. Prigogine, S.A. Rice, New method in computational quantum mechanics, in: Adv. In Chem. Phys., vol. XC111, John Wiley & Sons, Inc, 1996.

[3] K. Fukui, in: P.O. Löwdin, B. Pullman (Eds.), Molecular Orbitals in Chemistry, Physics and Biology, Academic Press, New York, NY, 1964, 513.

[4] K. Fukui, Theory of Orientation and Stereo-Selection, Springer-Verlag, Berlin, 1970.

[5] E.G. Lewars, Computational Chemistry: Introduction to the Theory and Applications of Molecular and Quantum Mechanics, Kluwer Academic Publishers, 2011.

Index

Note: Page numbers followed by "f" and "t" indicate figures and tables, respectively.

R

S

Printed in the United States
By Bookmasters